New Approaches in Water Purification

New Approaches in Water Purification

Editor:

Quan Cui, Ph.D.

Delve Publishing

www.delvepublishing.com

New Approaches in Water Purification

Editor: Quan Cui, Ph.D.

708 3rd Avenue, 6th Floor

New York, NY 10017

United States of America

www.delvepublishing.com

ISBN: 978-1-68095-835-5

Library of Congress Control Number: 2016951577

About the Editor

Quan Cui, Ph.D.

Quan Cui obtained her PhD from Beijing Normal University in 2015. She worked as environmental impact assessment engineer for years. Aiming at linking science to policy, her study interest spans from ecosystem modelling to water resources management. She published peer-reviewed articles about using machine learning models in inflow forecasting, CO2 management and watershed ecosystem health assessment.

Contents

List of Contributors ..... *xi*

Preface ..... *xvii*

Introduction ..... *xix*

Chapter 1 **Optimum Design of Cogeneration for Power and Desalination to Satisfy the Demand of Water And Power** **1**

1. Introduction 1
2. The Schematic of the Cogeneration System 2
3. Mathematical Model 2
4. Modified Genetic Algorithm 3
5. Case Study 4

Acknowledgments 8

References 8

Chapter 2 **Fresh Groundwater for Wajir—Ex-Ante Assessment of Uncertain Benefits for Multiple Stakeholders in a Water Supply Project in Northern Kenya** **9**

1. Introduction 9
2. Decision Modeling 10
3. The Habaswein-Wajir Water Supply Project 11
4. Methods 13
5. Results 15
6. Discussion 20

Conclusions 23

Conflict of Interest Statement 23

Acknowledgments 23

Supplementary Material 23

References 23

Chapter 3 **Graphene Oxide: A Promising Nanomaterial for Energy and Environmental Applications** **25**

1. Introduction 25
2. Applications of Go in Energy Conversion and Storage 26
3. Applications of Go in Environmental Protection 40
4. Summary and Perspective 46

Acknowledgements 47

References 47

Chapter 4 **Graphene-Based Microbots for Toxic Heavy Metal Removal and Recovery From Water** **55**

Acknowledgment 60

References 60

Chapter 5 **Brackish Groundwater Membrane System Design for Sustainable Irrigation** **63**

1. Introduction 63
2. Materials and Method 64
3. Case Studies 67

Conclusions 69

Conflict of Interest Statement 69

Acknowledgment 69

References 69

Chapter 6 **Nano-Enabled Membranes Technology** **73**
1. Introduction: Sustainability of Membrane Desalination 73
2. Innovative Nanomaterials: Revolutionary Solutions for Sustainable Membrane Desalination? 74
3. Nano-Enabled Membrane Desalination: Transforming Challenges Into Opportunities 75
4. Future Outlook and Concluding Remarks 76
References 76

Chapter 7 **Coagulation-Microfiltration Water Treatment** **79**
1. Introduction 79
2. Materials and Methods 80
3. Results and Discussion 82
Conclusions 87
Acknowledgment 87
Appendix A. Supplementary Data 87

Chapter 8 **Portable Led Fluorescence Instrumentation for the Rapid Assessment of Potable Water Quality** **89**
1. Introduction 89
2. Background 90
3. Methodology 91
4. Results 93
5. Discussion 97
Conclusions 97
Acknowledgements 98
References 98

Chapter 9 **The Use of Simulated Whole Effluents in Toxicity Assessments: A Review of Case Studies From Reverse Osmosis Desalination Plants** **101**
1. Introduction 101
2. Case Studies 102
3. Discussion 107
Conclusions 108
Acknowledgements 109
References 109

Chapter 10 **Ultraviolet Radiation as a Ballast Water Treatment Strategy** **111**
1. Introduction 111
2. Material and Methods 112
3. Results 113
4. Discussion 115
Acknowledgments 117
References 117

Chapter 11 **Significance of Coagulation Kinetics on Maintaining Membrane Permeability in an MBR Following Reactive Coagulant Dosing** **121**
1. Introduction 121
2. Material and Methods 122
3. Results 123
4. Discussion 130
Conclusions 131
Acknowledgements 131
References 131

Chapter 12 **Controlling Molecular Weight Cut-Off of Peek Nanofiltration Membranes Using a Drying Method ... 133**

1. Introduction ... 133
2. Methods ... 134
3. Results and Discussion ... 136
4. Modelling the Post-Phase Inversion Drying Process of Peek Nanofiltration Membranes ... 145

Conclusions ... 147

Author Contributions ... 148

Acknowledgements ... 148

Appendix ... 148

References ... 149

Chapter 13 **High-Precision Measurements of Seawater PB Isotope Compositions by Double Spike Thermal Ionization Mass Spectrometry ... 151**

1. Introduction ... 151
2. Experimental ... 152
3. Results aAnd Discussion ... 156

Conclusions ... 162

Acknowledgements ... 162

References ... 162

Chapter 14 **Degrading Organic Micropollutants: The Next Challenge in the Evolution of Biological Wastewater Treatment Processes ... 165**

1. Wastewater, A Wasted Resource ... 165
2. Current Wastewater Treatment Approaches ... 166
3. Enhancing Biological Wastewater Treatment ... 166
4. Implementation ... 167

Conflict Of Interest Statement ... 167

References ... 168

Chapter 15 **Reclaimed Water as a Reservoir of Antibiotic Resistance Genes: Distribution System and Irrigation Implications ... 171**

1. Introduction ... 171
2. Materials and Methods ... 172
3. Results ... 173
4. Discussion ... 176

Conclusion ... 177

Conflict of Interest Statement ... 177

Acknowledgments ... 177

References ... 178

Chapter 16 **Efficient Purification and Concentration of Viruses from a Large Body of High Turbidity Seawater ... 181**

1. Method Details ... 181
2. Summary ... 185
3. Additional Information ... 185

Acknowledgments ... 186

References ... 186

Chapter 17 **High Salinity Tolerance of the Red Sea Coral Fungia Granulosa Under Desalination Concentrate Discharge Conditions: An in Situ Photophysiology Experiment ... 189**

1. Introduction ... 189
2. Materials and Methods ... 190
3. Results ... 191

4. Discussion 194
Conflict of Interest Statement 195
Acknowledgments 195
References 195

Chapter 18 Drivers And Economic Aspects for the Implementation of Advanced Wastewater Treatment and Water Reuse in a PVC Plant 197
1. Introduction 197
2. Materials and Methods 198
3. Results and Discussion 200
Conclusions 202
Acknowledgements 202
References 202

Chapter 19 Solar Uv-Treatment of Water Samples for Stripping-Voltammetric Determination of Trace Heavy Metals 203
1. Introduction 203
2. Experimental 204
3. Results and Discussion 206
Conclusions 209
Declarations 209
References 209

Index 211

List of Contributors

Wu Lianying
Key Laboratory of Marine Chemistry Theory and Technology, Ministry of Education, Ocean University of China, China

Hu Yangdong
Key Laboratory of Marine Chemistry Theory and Technology, Ministry of Education, Ocean University of China, China

Gao Congjie
Key Laboratory of Marine Chemistry Theory and Technology, Ministry of Education, Ocean University of China, China

Eike Luedeling
Land Health Decisions, World Agroforestry Centre (ICRAF), Nairobi, Kenya
Centre for Development Research, University of Bonn, Bonn, Germany

Arjen L. Oord
Acacia Water, Gouda, Netherlands

Boniface Kiteme
Centre for Training and Integrated Research in ASAL Development, Nanyuki, Kenya

Sarah Ogalleh
Centre for Training and Integrated Research in ASAL Development, Nanyuki, Kenya

Maimbo Malesu
Land Health Decisions, World Agroforestry Centre (ICRAF), Nairobi, Kenya

Keith D. Shepherd
Land Health Decisions, World Agroforestry Centre (ICRAF), Nairobi, Kenya

Jan De Leeuw
Land Health Decisions, World Agroforestry Centre (ICRAF), Nairobi, Kenya

Fen Li
Key Laboratory of Materials Modification by Laser, Ion and Electron Beams (Dalian University of Technology), Ministry of Education, Dalian 116024, China

Xue Jiang
Key Laboratory of Materials Modification by Laser, Ion and Electron Beams (Dalian University of Technology), Ministry of Education, Dalian 116024, China

Jijun Zhao
Key Laboratory of Materials Modification by Laser, Ion and Electron Beams (Dalian University of Technology), Ministry of Education, Dalian 116024, China
Beijing Computational Science Research Center, Beijing 100089, China

Shengbai Zhang
Department of Physics, Applied Physics, and Astronomy, Rensselaer Polytechnic Institute, Troy, NY 12180, USA

Diana Vilela
Max-Planck Institute for Intelligent Systems, Heisenbergstr. 3, 70569 Stuttgart, Germany

Jemish Parmar
Max-Planck Institute for Intelligent Systems, Heisenbergstr. 3, 70569 Stuttgart, Germany
Institute for Bioengineering of Catalonia (IBEC), Baldiri I Reixac 10-12, 08028 Barcelona, Spain

Yongfei Zeng
Division of Chemistry and Biological Chemistry, School of Physical and Mathematical Sciences, Nanyang Technological University, 21 Nanyang Link, 637371 Singapore, Singapore

Yanli Zhao
Division of Chemistry and Biological Chemistry, School of Physical and Mathematical Sciences, Nanyang Technological University, 21 Nanyang Link, 637371 Singapore, Singapore

Samuel Sánchez
Max-Planck Institute for Intelligent Systems, Heisenbergstr. 3, 70569 Stuttgart, Germany
Institute for Bioengineering of Catalonia (IBEC), Baldiri I Reixac 10-12, 08028 Barcelona, Spain
Institució Catalana de Recerca i Estudis Avançats (ICREA), Psg. Lluís Companys, 23, 08010 Barcelona, Spain

Beni Lew
Department of Civil Engineering, Ariel University, Ariel, Israel
Agriculture Research Organization Volcani Center, Institute of Agricultural Engineering, Bet Dagan, Israel

Lolita Trachtengertz
The Department of Chemical Engineering, Shenkar College of Engineering and Design, Ramat-Gan, Israel

Shany Ratsin
The Department of Chemical Engineering, Shenkar College of Engineering and Design, Ramat-Gan, Israel

Gideon Oron
Department of Environment Water Resources, The Jacob Blaustein Institutes for Desert Research, Ben-Gurion University of the Negev, Midrashet Ben-Gurion, Israel
The Department of Industrial Engineering and Management, and the Environmental Engineering Program, Ben-Gurion University of the Negev, Beer Sheva, Israel

Amos Bick
Bick & Associates, Ganey-Tikva, Israel

P.S. Goh
Advanced Membrane Technology Research Centre, Universiti Teknologi Malaysia, 81310 Johor, Malaysia

A.F. Ismail
Advanced Membrane Technology Research Centre, Universiti Teknologi Malaysia, 81310 Johor, Malaysia

N. Hilal
Centre for Water Advanced Technologies and Environmental Research (CWATER), College of Engineering, Swansea University, Singleton Park, Swansea SA2 8PP, UK
Qatar Energy and Environment Research Institute (QEERI), Doha, Qatar

Q. Ding
Faculty of Science and Engineering, Chuo University, 1-13-27 Kasuga, Bunkyo-ku, Tokyo 112-8551, Japan

H. Yamamura
Faculty of Science and Engineering, Chuo University, 1-13-27 Kasuga, Bunkyo-ku, Tokyo 112-8551, Japan

N. Murata
R&D Center, METAWATER Co., Ltd, 1 Maegata-cho, Handa, Aichi 475-0825, Japan

N. Aoki
Business Strategy Division, METAWATER Co., Ltd, JR Kanda Manseibashi Bldg, 1-25, Kanda-sudacho, Chiyoda-ku, Tokyo 101-0041, Japan

H. Yonekawa
Drinking Water Treatment Engineering Department, METAWATER Co., Ltd, 2-56 Suda-cho, Mizuho, Nagoya 467-8530, Japan

A. Hafuka
Faculty of Science and Engineering, Chuo University, 1-13-27 Kasuga, Bunkyo-ku, Tokyo 112-8551, Japan

Y. Watanabe
Research and Development Initiatives, Chuo University, 1-13-27 Kasuga, Bunkyo-ku, Tokyo 112-8551, Japan

J. Bridgeman
School of Civil Engineering, University of Birmingham, Birmingham, United Kingdom

A. Baker
Connected Waters Initiative Research Centre, University of NSW, Sydney, NSW 2052, Australia

D. Brown
School of Civil Engineering, University of Birmingham, Birmingham, United Kingdom
Department of Civil and Structural Engineering, Sir Frederick Mappin Building, Mappin Street, Sheffield S1 3JD, United Kingdom

J.B. Boxall
Department of Civil and Structural Engineering, Sir Frederick Mappin Building, Mappin Street, Sheffield S1 3JD, United Kingdom

Laura J. Falkenberg
School of Energy and Resources, UCL Australia, University College London, Torrens Building, 220 Victoria Square, Adelaide, SA 5000, Australia

Craig A. Styan
School of Energy and Resources, UCL Australia, University College London, Torrens Building, 220 Victoria Square, Adelaide, SA 5000, Australia

Ranveig Ottoey Olsen
Stord/Haugesund University College, Klingenbergvegen 8, 5414 Stord, Norway

Friederike Hoffmann
University of Bergen, P.O. Box 7800, 5020 Bergen, Norway
Uni Research Environment, Thormoehlensgt. 49b, 5006 Bergen, Norway

Ole-Kristian Hess-Erga
Norwegian Institute for Water Research, Thormoehlensgt. 53 D, 5006 Bergen, Norway

Aud Larsen
Uni Research Environment, Thormoehlensgt. 49b, 5006 Bergen, Norway

Gunnar Thuestad
Stord/Haugesund University College, Klingenbergvegen 8, 5414 Stord, Norway

Ingunn Alne Hoell
Stord/Haugesund University College, Klingenbergvegen 8, 5414 Stord, Norway

O. Autin
Cranfield Water Science Institute, Vincent Building, Cranfield University, Bedfordshire, MK43 0AL, UK

F. Hai
Strategic Water Infrastructure Laboratory, School of Civil, Mining and Environmental Engineering, University of Wollongong, Wollongong, NSW 2522, Australia

S. Judd
Cranfield Water Science Institute, Vincent Building, Cranfield University, Bedfordshire, MK43 0AL, UK Cranfield Water Science Institute, Vincent Building, Cranfield University, Bedfordshire, MK43 0AL, UK

E.J. McAdam
Cranfield Water Science Institute, Vincent Building, Cranfield University, Bedfordshire, MK43 0AL, UK

João da Silva Burgal
Department of Chemical Engineering and Chemical Technology, Imperial College, Exhibition Road, London SW7 2AZ, UK

Ludmila Peeva
Department of Chemical Engineering and Chemical Technology, Imperial College, Exhibition Road, London SW7 2AZ, UK

Patrizia Marchetti
Department of Chemical Engineering and Chemical Technology, Imperial College, Exhibition Road, London SW7 2AZ, UK

Andrew Livingston
Department of Chemical Engineering and Chemical Technology, Imperial College, Exhibition Road, London SW7 2AZ, UK

Maxence Paul
Department of Earth Science & Engineering, Imperial College London, London SW7 2AZ, UK

Luke Bridgestock
Department of Earth Science & Engineering, Imperial College London, London SW7 2AZ, UK

Mark Rehkämper
Department of Earth Science & Engineering, Imperial College London, London SW7 2AZ, UK

Tina van DeFlierdt
Department of Earth Science & Engineering, Imperial College London, London SW7 2AZ, UK

Dominik Weiss
Department of Earth Science & Engineering, Imperial College London, London SW7 2AZ, UK

Naresh Singhal
Department of Civil and Environmental Engineering, the University of Auckland, Auckland, New Zealand

Octavio Perez-Garcia
Department of Civil and Environmental Engineering, the University of Auckland, Auckland, New Zealand

Nicole Fahrenfeld
Department of Civil and Environmental Engineering, Virginia Tech, Blacksburg, VA, USA

Yanjun Ma
Department of Civil and Environmental Engineering, Virginia Tech, Blacksburg, VA, USA

Maureen O'Brien
Department of Civil and Environmental Engineering, Colorado School of Mines, Golden, CO, USA

Amy Pruden
Department of Civil and Environmental Engineering, Virginia Tech, Blacksburg, VA, USA

Guowei Sun
Laboratory of Quality and Safety Risk Assessment for Aquatic Products on Storage & Preservation (Shanghai), Ministry of Agriculture, China
College of Food Science and Technology, Shanghai Ocean University, Shanghai, China

Jinzhou Xiao
Laboratory of Quality and Safety Risk Assessment for Aquatic Products on Storage & Preservation (Shanghai), Ministry of Agriculture, China
College of Food Science and Technology, Shanghai Ocean University, Shanghai, China

Hongming Wang
Laboratory of Quality and Safety Risk Assessment for Aquatic Products on Storage & Preservation (Shanghai), Ministry of Agriculture, China
College of Food Science and Technology, Shanghai Ocean University, Shanghai, China

Chaowen Gong
Laboratory of Quality and Safety Risk Assessment for Aquatic Products on Storage & Preservation (Shanghai), Ministry of Agriculture, China
College of Food Science and Technology, Shanghai Ocean University, Shanghai, China

Yingjie Pan
Laboratory of Quality and Safety Risk Assessment for Aquatic Products on Storage & Preservation (Shanghai), Ministry of Agriculture, China
College of Food Science and Technology, Shanghai Ocean University, Shanghai, China

Shuling Yan
Institute of Biochemistry and Molecular Cell Biology, University of Göttingen, Germany

Yongjie Wang
Laboratory of Quality and Safety Risk Assessment for Aquatic Products on Storage & Preservation (Shanghai), Ministry of Agriculture, China
College of Food Science and Technology, Shanghai Ocean University, Shanghai, China

Riaan van der Merwe
Water Desalination and Reuse Center, Biological and Environmental Sciences and Engineering Division, King Abdullah University of Science and Technology, Thuwal, Saudi Arabia

Till Röthig
Red Sea Research Center, Biological and Environmental Sciences and Engineering Division, King Abdullah University of Science and Technology, Thuwal, Saudi Arabia

Christian R. Voolstra
Red Sea Research Center, Biological and Environmental Sciences and Engineering Division, King Abdullah University of Science and Technology, Thuwal, Saudi Arabia

Michael A. Ochsenkühn
Biological and Organometallic Catalysis Laboratories, Physical Sciences and Engineering Division, King Abdullah University of Science and Technology, Thuwal, Saudi Arabia

Sabine Lattemann
Water Desalination and Reuse Center, Biological and Environmental Sciences and Engineering Division, King Abdullah University of Science and Technology, Thuwal, Saudi Arabia

Gary L. Amy
Water Desalination and Reuse Center, Biological and Environmental Sciences and Engineering Division, King Abdullah University of Science and Technology, Thuwal, Saudi Arabia

David Prieto
INOVYN Spain, Marie Curie 1, 3, 5, 08760 Martorell, Spain

Nathalie Swinnen
Solvay, Rue de Ransbeek 310, 1120 Bruxelles, Belgium

Laura Blanco
Departamento de Ingeniería Química, Facultad de Ciencias Químicas, Universidad Complutense de Madrid, Ciudad Universitaria s/n, 28040 Madrid, Spain

Daphne Hermosilla
Departamento de Ingeniería Química, Facultad de Ciencias Químicas, Universidad Complutense de Madrid, Ciudad Universitaria s/n, 28040 Madrid, Spain

Peter Cauwenberg
VITO nv, Boeretang 200, 2400 Mol, Belgium

Ángeles **Blanco**
Departamento de Ingeniería Química, Facultad de Ciencias Químicas, Universidad Complutense de Madrid, Ciudad Universitaria s/n, 28040 Madrid, Spain

Carlos Negro
Departamento de Ingeniería Química, Facultad de Ciencias Químicas, Universidad Complutense de Madrid, Ciudad Universitaria s/n, 28040 Madrid, Spain

Gelaneh Woldemichael
Adama Science and Technology University, Adama, Ethiopia

Taffa Tulu
Addis Ababa University, Addis Ababa, Ethiopia

Gerd-Uwe Flechsig
University at Albany, SUNY, Department of Chemistry, 1400 Washington Avenue, Albany, NY 12222, United States
University of Rostock, Department of Chemistry, A.-Einstein-Str. 3a, D-18059 Rostock, Germany

Preface

In the past decade, practical applications of a variety of new wastewater-treatment technologies, such as membrane filtration systems and advanced oxidation, have led to new ways of managing urban water systems and water resources. These new treatment regimes, especially the integration of urban-water and waste-management systems, promise to dramatically improve the sustainability of our water resources. These new systems are creating new needs, which are driving further technology development. This book describes some of the approaches in water purification that are already being implemented; and describe technological advances that are under development. The major driver of change is, quite simply, population growth coupled with a rising global standard of living, a combination that has resulted in resource consumption (including water use) that exceeds the current resources of planet Earth. By about 2050, when the population is expected to reach about 9 billion, and if standards of living continue to rise, the amount consumed will be the resources of about three planet Earths. Obviously, this scenario is not sustainable. When the population of Earth was much smaller (e.g., fewer than 2 billion) and when per capita use of resources was much smaller, our traditional "take, make, waste" pattern of resource consumption was sustainable. Now, however, we need to recycle and reuse all types of resources (including water), and we must increase our use of renewable resources. Water purification means to remove undesirable chemicals, contaminants, suspended particles, parasites, bacteria, algae, viruses, fungi, dissolved material and gases from source water to gain water qualified for a variety of different purpose. The most desired use of purified water is drinking water. Besides this, purposes of purified water demands also include medical use, industrial use, chemical use, et al. The sources of purified water come from both inland water, including surface water and groundwater, and oceans. Topics in this book covered the developments and applications of water purification systems and water purifications technologies, including coagulation, filtration, disinfection, ultraviolet radiation, osmosis, distillation and nanotechnology.

Editor
Quan Cui, Ph.D.

INTRODUCTION

Water purification is the process of removing contaminants from a raw water source. The goal is to produce water for a specific purpose with a treatment profile designed to limit the inclusion of specific materials; most water is purified for human consumption (drinking water). Water purification may also be designed for a variety of other purposes, including to meet the requirements of medical, pharmacology, chemical and industrial applications. It may be carried out by a variety of methods, including treatment with ultraviolet light, filtration, water softening, reverse osmosis, ultrafiltration, molecular stripping, deionization, and carbon treatment.

Water purification may remove such contaminants as particulate sand, suspended particles of organic material, parasites (such as Giardia and Cryptosporidium), bacteria, algae, viruses, fungi, and toxic metals (such as lead, copper, and chromium). Minerals such as calcium, silica, and magnesium may also be removed. Some purification may be elective in the purification process, including the removal of smell (hydrogen sulfide remediation), taste (mineral extraction), and color (from iron incapsulation).

Governments usually dictate the standards for drinking water quality. These standards require minimum and maximum set points of contaminants and the inclusion of control elements that produce drinking water. Quality standards in many countries require specific amounts of disinfectant (such as chlorine or ozone) in the water after it leaves the water treatment plant (WTP), to reduce the risk of re-contamination while the water is in the distribution system.

Ground water (usually supplied as well water) is typically a more economical choice than surface water (from rivers, lakes and streams) as a source for drinking, as it is inherently pre-filtered by the aquifer from which it is extracted. Over large areas of the world, aquifers are recharged as part of the hydrologic cycle. In more arid regions, water from an aquifer will have a limited output and can take thousands of years to recharge. Surface water is locally more abundant where subsurface formations do not function as aquifers; however, ground water is far more abundant than the more-visible surface water. Surface water is a typical raw water source used to make drinking water where it is abundant and where ground water is unavailable or of poor quality. However, it is much more exposed to human activity and its byproducts. As a water source, it is carefully monitored for the presence of a variety of contaminants by the WTP operators.

It is not possible to tell whether water is safe to drink just by looking at it. Simple procedures such as boiling or the use of a household activated carbon filter are not sufficient for treating all the possible contaminants that may be present in water from an unknown source. Even natural spring water—considered safe for all practical purposes in the 1800s—must now be tested before determining what kind of treatment, if any, is needed. Chemical analysis, while expensive, is the only way to obtain the information necessary for deciding on method of purification.

Sources of Drinking Water

The water emerging from some deep ground water may have fallen as rain many decades, hundreds, thousands or in some cases millions of years ago. Soil and rock layers naturally filter the ground water to a high degree of clarity before it is pumped to the treatment plant. Such water may emerge as springs, artesian springs, or may be extracted from boreholes or wells. Deep ground water is generally of very high bacteriological quality (i.e., pathogenic bacteria such as Campylobacter or the pathogenic protozoa Cryptosporidium and Giardia are typically absent), but the water typically is rich in dissolved solids, especially carbonates and sulfates of calcium and magnesium. Depending on the strata through which the water has flowed, other ions may also be present including chloride, and bicarbonate. There may be a requirement to reduce the iron or manganese content of this water to make it pleasant for drinking, cooking, and laundry use. Disinfection may also be required. Where groundwater recharge is practiced, it is equivalent to lowland surface waters for treatment purposes.

- Shallow groundwaters: Water emerging from shallow groundwaters is usually abstracted from wells or boreholes. The bacteriological quality can be variable depending on the nature of the catchment. A variety of soluble materials may be present including (rarely) potentially toxic metals such as zinc, copper or arsenic. Arsenic contamination of groundwater is a serious problem in some areas, notably from shallow wells in Bangladesh and West Bengal in the Ganges Delta.
- Upland lakes and reservoirs: Typically located in the headwaters of river systems, upland reservoirs are usually sited above any human habitation and may be surrounded by a protective zone to restrict the opportunities for contamination. Bacteria and pathogen levels are usually low, but some bacteria, protozoa or algae will be present. Where uplands are forested or peaty, humic acids can color the water. Many upland sources have low pH which require adjustment.
- Rivers, canals and low land reservoirs: Low land surface waters will have a significant bacterial load and may also contain algae, suspended solids and a variety of dissolved constituents.
- Atmospheric water generation is a new technology that can provide high quality drinking water by extracting water from the air by cooling the air and thus condensing water vapor.
- Rainwater harvesting or fog collection which collects water from the atmosphere can be used especially in areas with significant dry seasons and in areas which experience fog even when there is little rain.

Water Treatment

Pre-treatment

- Pumping and containment: The majority of water must be pumped from its source or directed into pipes or holding tanks. To avoid adding contaminants to the water, this physical infrastructure must be made from appropriate materials and constructed so that accidental contamination does not occur.
- Screening (see also screen filter): The first step in purifying surface water is to remove large debris such as sticks, leaves, trash and other large particles which may interfere with subsequent purification steps. Most deep groundwater does not need screening before other purification steps.
- Storage: Water from rivers may also be stored in bankside reservoirs for periods between a few days and many months to allow natural biological purification to take place. This is especially important if treatment is by slow sand filters.

Storage reservoirs also provide a buffer against short periods of drought or to allow water supply to be maintained during transitory pollution incidents in the source river.

- Pre-conditioning: Many waters rich in hardness salts are treated with soda-ash (Sodium carbonate) to precipitate calcium carbonate out utilizing the common ion effect.
- Pre-chlorination: In many plants the incoming water was chlorinated to minimize the growth of fouling organisms on the pipe-work and tanks. Because of the potential adverse quality effects (see chlorine below), this has largely been discontinued.

Widely varied techniques are available to remove the fine solids, micro-organisms and some dissolved inorganic and organic materials. The choice of method will depend on the quality of the water being treated, the cost of the treatment process and the quality standards expected of the processed water.

pH adjustment

Distilled water has an average pH of 7 (neither alkaline nor acidic) and sea water has an average pH of 8.3 (slightly alkaline). If the water is acidic (lower than 7), lime or soda ash is added to raise the pH. Lime is the more common of the two additives because it is cheap, but it also adds to the resulting water hardness. Making the water slightly alkaline ensures that coagulation and flocculation processes work effectively and also helps to minimize the risk of lead being dissolved from lead pipes and lead solder in pipe fittings.

Flocculation is a process which starts with clarifying the water. Clarifying means removing any turbidity or color so that the water is clear and colorless. Clarification is done by causing a precipitate to form in the water. Initially the precipitate forms as very small particles but as the water is gently stirred, these particles stick together to form bigger particles—this process is sometimes called flocculation. Many of the small particles that were originally present in the raw water absorb onto the surface of these small precipitate particles and so get incorporated into the larger particles that coagulation produces. In this way the coagulated precipitate takes most of the suspended matter out of the water and is then filtered off, generally by passing the mixture through a coarse sand filter or sometimes through a mixture of sand and granulated anthracite (high quality coal). Anthracite, with its high carbon content, is able to absorb much of the organic matter present in solution and this can remove odor and taste from the water. A precipitate that is widely used to clarify water is iron (III) hydroxide. This is formed first by adjusting (if necessary) the pH of the incoming water to above 7 (by adding lime or sodium hydroxide), then by adding a solution of an iron (III) compound such as iron (III) chloride. Iron (III) hydroxide is extremely insoluble and forms even at a pH as low as 7. Aluminium hydroxide is also widely used as the flocculating precipitate.

Sedimentation

Water exiting the flocculation basin may enter the sedimentation basin, also called a clarifier or settling basin. It is a large tank with slow flow, allowing flock to settle to the bottom. The sedimentation basin is best located close to the flocculation basin so the transit between does not permit settlement or flock break up. Sedimentation basins can be in the shape of a rectangle, where water flows from end to end, or circular where flow is from the center outward. Sedimentation basin outflow is typically over a weir so only a thin top layer-furthest from the sediment-exits. The amount of flock that settles out of the water is dependent on the time the water spends in the basin and the depth of the basin. The retention time of the water must therefore be balanced against the cost of a larger basin. The minimum clarifier retention time is normally 4 hours. A deep basin will allow more flock to settle out than a shallow basin. This is because large particles settle faster than smaller ones, so large particles bump into and integrate smaller particles as they settle. In effect, large particles sweep vertically though the basin and clean out smaller particles on their way to the bottom.

As particles settle to the bottom of the basin a layer of sludge is formed on the floor of the tank. This layer of sludge must be removed and treated. The amount of sludge that is generated is significant, often 3 percent to 5 percent of the total volume of water that is treated. The cost of treating and disposing of the sludge can be a significant part of the operating cost of a water treatment plant. The tank may be equipped with mechanical cleaning devices that continually clean the bottom of the tank or the tank can be taken out of service when the bottom needs to be cleaned.

Filtration

After separating most flock, the water is filtered as the final step to remove remaining suspended particles and unsettled flock. The most common type of filter is a rapid sand filter. Water moves vertically through sand which often has a layer of activated carbon or anthracite coal above the sand. The top layer removes organic compounds, which contribute to taste and odor. The space between sand particles is larger than the smallest suspended particles, so simple filtration is not enough. Most particles pass through surface layers but are trapped in pore spaces or adhere to sand particles. Effective filtration extends into the depth of the filter. This property of the filter is key to its operation: if the top layer of sand were to block all the particles, the filter would quickly clog.

To clean the filter, water is passed quickly upward through the filter, opposite the normal direction (called back flushing or backwashing) to remove embedded particles. Prior to this, compressed air may be blown up through the bottom of the filter to break up the compacted filter media to aid the backwashing process; this is known as air scouring. This contaminated water can be disposed of, along with the sludge from the sedimentation basin, or it can be recycled by mixing with the raw water entering the plant.

Some water treatment plants employ pressure filters. These work on the same principle as rapid gravity filters, differing in that the filter medium is enclosed in a steel vessel and the water is forced through it under pressure.

Advantages:

- Filters out much smaller particles than paper and sand filters can.
- Filters out virtually all particles larger than their specified pore sizes.
- They are quite thin and so liquids flow through them fairly rapidly.
- They are reasonably strong and so can withstand pressure differences across them of typically 2-5 atmospheres.
- They can be cleaned (back flushed) and reused.

Membrane filters are widely used for filtering both drinking water and sewage (for reuse). For drinking water, membrane filters can remove virtually all particles larger than 0.2 um—including Giardia and cryptosporidium. Membrane filters are an effective form of tertiary treatment when it is desired to reuse the water for industry, for limited domestic purposes, or before discharging the water into a river that is used by towns further downstream. They are widely used in industry, particularly for beverage preparation (including bottled water). However no filtration can remove substances that are actually dissolved in the water such as phosphorus, nitrates and heavy metal ions.

Slow Sand Filters

Slow sand filters may be used where there is sufficient land and space. These rely on biological treatment processes for their action rather than physical filtration. Slow sand filters are carefully constructed using graded layers of sand with the coarsest at the top and finest at the base. Drains at the base convey treated water away for disinfection. Filtration depends on the development of a thin biological layer on the surface of the filter. An effective slow sand filter may remain in service for many weeks or even months if the pre-treatment is well designed and produces an excellent quality of water which physical methods of treatment rarely achieve.

Ultrafiltration

Ultrafiltration membranes are a relatively new development; they use polymer film with chemically formed microscopic pores that can be used in place of granular media to filter water effectively without coagulants. The type of membrane media determines how much pressure is needed to drive the water through and what sizes of micro-organisms can be filtered out.

Disinfection

Disinfection is normally the last step in purifying drinking water. Water is disinfected to kill any pathogens which pass through the filters. Possible pathogens include viruses, bacteria, including Escherichia coli, Campylobacter and Shigella, and protozoans, including G. lamblia and other Cryptosporidia. In most developed countries, public water supplies are required to maintain a residual disinfecting agent throughout the distribution system, in which water may remain for days before reaching the consumer. Following the introduction of any chemical disinfecting agent, the water is usually held in temporary storage—often called a contact tank or clear well to allow the disinfecting action to complete.

- Chlorination- The most common disinfection method is some form of chlorine or its compounds such as chloramine or chlorine dioxide. Chlorine is a strong oxidant that kills many micro-organisms.

- Because chlorine is a toxic gas, there is a danger of a release associated with its use. This problem is avoided by the use of sodium hypochlorite, which is either a relatively inexpensive solid that releases free chlorine when dissolved in water or a liquid (bleach) that is typically generated on site using common salt and high voltage DC. Handling the solid, however, requires greater routine human contact through opening bags and pouring than the use of gas cylinders which are more easily automated. The generation of liquid sodium hypochlorite is both inexpensive and safer than the use of gas or solid chlorine. Both disinfectants are widely used despite their respective drawbacks. One drawback to using chlorine gas or sodium hypochlorite is that they react with organic compounds in the water to form potentially harmful chemical by-products trihalomethanes (THMs) and haloacetic acids (HAAs), both of which are carcinogenic in large quantities and regulated by the United States Environmental Protection Agency (EPA). The formation of THMs and haloacetic acids is minimized by effective removal of as many organics from the water as possible prior to chlorine addition. Although chlorine is effective in killing bacteria, it has limited effectiveness against protozoans that form cysts in water. (Giardia lamblia and Cryptosporidium, both of which are pathogenic.)
- Chlorine dioxide is another fast-acting disinfectant. It is, however, rarely used, because it may create excessive amounts of chlorate and chlorite, both of which are regulated to low allowable levels. Chlorine dioxide also poses extreme risks in handling: not only is the gas toxic, but it may spontaneously detonate upon release to the atmosphere in an accident.
- Chloramines are another chlorine-based disinfectant. Although chloramines are not as strong of an oxidant or provide a reliable residual, as compared to chlorine gas or sodium hypochlorite, they are less prone to form THMs or haloacetic acids. It is possible to convert chlorine to chloramine by adding ammonia to the water along with the chlorine: The chlorine and ammonia react to form chloramine. Water distribution systems disinfected with chloramines may experience nitrification, wherein ammonia is used a nitrogen source for bacterial growth, with nitrates being generated as a byproduct.
- Ozone (O3) is a relatively unstable molecule "free radical" of oxygen which readily gives up one atom of oxygen providing a powerful oxidizing agent which is toxic to most water borne organisms. It is a very strong, broad spectrum disinfectant that is widely used in Europe. It is an effective method to inactivate harmful protozoans that form cysts. It also works well against almost all other pathogens. Ozone is made by passing oxygen through ultraviolet light or a "cold" electrical discharge. To use ozone as a disinfectant, it must be created on site and added to the water by bubble contact. Some of the advantages of ozone include the production of relatively fewer dangerous by-products (in comparison to chlorination) and the lack of taste and odor produced by ozonation. Although fewer by-products are formed by ozonation, it has been discovered that the use of ozone produces a small amount of the suspected carcinogen Bromate, although little Bromine should be present in treated water. Another one of the main disadvantages of ozone is that it leaves no disinfectant residual in the water. Ozone has been used in drinking water plants since 1906 where the first industrial ozonation plant was built in Nice, France. The U.S. Food and Drug Administration has accepted ozone as being safe; and it is applied as an anti-microbiological agent for the treatment, storage, and processing of foods.
- UV radiation (light) is very effective at inactivating cysts, as long as the water has a low level of color so the UV can pass through without being absorbed. The main disadvantage to the use of UV radiation is that, like ozone treatment, it leaves no residual disinfectant in the water.

  Because neither ozone nor UV radiation leaves a residual disinfectant in the water, it is sometimes necessary to add a residual disinfectant after they are used. This is often done through the addition of chloramines, discussed above as a primary disinfectant. When used in this manner, chloramines provide an effective residual disinfectant with very little of the negative aspects of chlorination.

Additional Treatment Options

- Fluoridation: in many areas fluoride is added to water for the purpose of preventing tooth decay. This process is referred to as water fluoridation. Fluoride is usually added after the disinfection process. In the United States, fluoridation is usually accomplished by the addition of hexafluorosilicic acid, which decomposes in water, yielding fluoride ions.
- Water conditioning: This is a method of reducing the effects of hard water. Hardness salts are deposited in water systems subject to heating because the decomposition of bicarbonate ions creates carbonate ions which crystallise out of the saturated solution of calcium or magnesium carbonate. Water with high concentrations of hardness salts can be treated with soda ash (sodium carbonate) which precipitates out the excess salts, through the Common-ion effect, producing calcium carbonate of very high purity. The precipitated calcium carbonate is traditionally sold to the manufacturers of toothpaste. Several other methods of industrial and residential water treatment are claimed (without general scientific acceptance) to include the use of magnetic or/and electrical fields reducing the effects of hard water
- Plumbosolvency reduction: In areas with naturally acidic waters of low conductivity (i.e surface rainfall in upland mountains of igneous rocks), the water may be capable of dissolving lead from any lead pipes that it is carried in. The addition of small quantities of phosphate ion and increasing the pH slightly both assist in greatly reducing plumbo-solvency by creating insoluble lead salts on the inner surfaces of the pipes.

- Radium Removal: Some groundwater sources contain radium, a radioactive chemical element. Typical sources include many groundwater sources north of the Illinois River in Illinois. Radium can be removed by ion exchange, or by water conditioning. The back flush or sludge that is produced is, however, a low-level radioactive waste.
- Fluoride Removal: Although fluoride is added to water in many areas, some areas of the world have excessive levels of natural fluoride in the source water. Excessive levels can be toxic or cause undesirable cosmetic effects such as staining of teeth. One method of reducing fluoride levels is through treatment with activated alumina.

Additional Water Purification Techniques

Other popular methods for purifying water, especially for local private supplies are listed below. In some countries some of these methods are also used for large scale municipal supplies. Particularly important are distillation (de-salination of seawater) and reverse osmosis.

- Boiling: Water is boiled for a sufficient duration to inactivate or kill microorganisms that normally live in water at room temperature. Near sea level, a vigorous rolling boil for at least one minute is sufficient. At high altitudes (higher than two kilometers or 5,000 feet), three minutes is recommended. In areas where the water is "hard" (that is, containing significant concentrations of dissolved calcium salts), boiling decomposes the bicarbonate ions, resulting in partial precipitation as calcium carbonate. This is the "fur" that builds up on kettle elements in hard water areas. With the exception of calcium, boiling does not remove solutes of higher boiling point than water and in fact increases their concentration (due to some water being lost as vapor). Boiling does not leave a residual disinfectant in the water. Therefore, water that has been boiled and then stored for any length of time may have acquired new pathogens.
- Carbon filtering: Charcoal, a form of carbon with a high surface area, absorbs many compounds including some toxic compounds. Water passing through activated charcoal is common in household water filters and fish tanks. Household filters for drinking water sometimes contain silver to release silver ions which have an anti-bacterial effect.
- Distillation involves boiling the water to produce water vapor. The vapor contacts a cool surface where it condenses as a liquid. Because the solutes are not normally vaporized, they remain in the boiling solution. Even distillation does not completely purify water, because of contaminants with similar boiling points and droplets of unvaporised liquid carried with the steam. However, 99.9 percent pure water can be obtained by distillation. Distillation does not confer any residual disinfectant and the distillation apparatus may be the ideal place to harbor Legionnaires' disease.
- Reverse osmosis: Mechanical pressure is applied to an impure solution to force pure water through a semi-permeable membrane. Reverse osmosis is theoretically the most thorough method of large scale water purification available, although perfect semi-permeable membranes are difficult to create. Unless membranes are well-maintained, algae and other life forms can colonize the membranes.
- Ion exchange: Most common ion exchange systems use a zeolite resin bed to replace unwanted $Ca2+$ and $Mg2+$ ions with benign (soap friendly) $Na+$ or $K+$ ions. This is the common water softener.
- Electrodeionization: Water is passed between a positive electrode and a negative electrode. Ion selective membranes allow the positive ions to separate from the water toward the negative electrode and the negative ions toward the positive electrode. High purity deionized water results. The water is usually passed through a reverse osmosis unit first to remove non-ionic organic contaminants.
- The use of iron in removing arsenic from water.

Portable techniques

Portable techniques for purifying water are used for hiking or camping, or for use in rural areas or emergency situations. Common techniques include boiling, disinfection with tablets or ultra-filtration using a small hand pump.

Water Purification for Hydrogen Production

For the small scale production of hydrogen, water purifiers are installed to prevent formation of minerals on the surface of the electrodes and to remove organics and chlorine from utility water. First, the water passes through a 20-micrometer interference (mesh or screen filter) filter to remove sand and dust particles, then a charcoal filter using activated carbon to remove organics and chlorine and, finally, a de-ionizing filter to remove metallic ions. Testing can be done before and after the filter to verify the proper removal of barium, calcium, potassium, magnesium, sodium, and silicon.

CHAPTER 1

OPTIMUM DESIGN OF COGENERATION FOR POWER AND DESALINATION TO SATISFY THE DEMAND OF WATER AND POWER

Wu Lianying, Hu Yangdong, Gao Congjie

Keywords: Cogeneration for power and desalination, Desalination, Genetic algorithm, Optimum design

INTRODUCTION

Cogeneration is defined as the combined production of two or more useful forms of energy from the same primary energy, thus allowing financial return and less impact with respect to the atmospheric emissions. The thermal power plants not only produce the electric energy, but also supply large amount low pressure steam. At the same time, the thermal power plants require a mass of freshwater to generate the high pressure and high temperature steam. The MSF and the RO are respectively driven by thermal energy and electricity to produce freshwater. So the integrating thermal and membrane desalination process with power generation in the same site are currently considered as a viable alternative. The advantages of the triple hybrid power–MSF–RO over the dual power–MSF and single purpose MSF or RO plants were reported [1], [2] and [3].

Many researchers have investigated in the cogeneration systems. A state-of-the-art review for simple and fully integrated hybrid desalination systems is presented in [4]. An overview of research endeavors carried out by hybrid desalination systems is also presented. A small-scale cogeneration system based on reciprocating engine is coupled in [5]. An exergoeconomic method is proposed in [6] for a combined gas/steam cycle associated with a MSF-RO desalination system. A thermodynamic model for integrated multi-effect evaporation thermal vapor compression (METVC) and humidified gas turbine cycle is presented in [7], [8] and [9]. Nevertheless, there is no economic analysis and optimization approach in their researches. The performance of a cogeneration plant (combined power plant and desalination) is analyzed by R. Chacartegui [10] with a stationary lumped volume model and the design and optimization guidelines are presented. A dual purpose plant for supplying given amount of power and fresh water is modeled and optimized from thermodynamic and economic point of view [11]. The thermodynamic and economic aspect of METVC is studied without performing a complete economical analysis [12].

Some optimization models of combined desalination and cogeneration systems associated to the superstructure concept are presented in the literature [13], [14], [15],[16], [17], [18] and [19]. An evolution of the modeling processes can be observed in the technical literature in which mixed integer non-linear programming technique is proposed for associating desalination systems and combined gas/steam cycles and culminates by presenting a disjunctive programming model for optimizing such systems.

Despite that many contributions dealing with the design of dual purpose desalination plants have been published, only few of them focus on the simultaneous optimization of the configuration and operating conditions.

Different arrangements are possible in order to satisfy electricity and freshwater demands and the selection of the optimal system is a difficult task because it depends strongly on many factors such as the power to water ratio, cost of fuel,

Citation: Lianying Wu, Yangdong Hu, Congjie Gao, Optimum design of cogeneration for power and desalination to satisfy the demand of water and power, Desalination, Volume 324, 2 September 2013, Pages 111-117, ISSN 0011-9164, http://dx.doi.org/10.1016/j.desal.2013.06.006.

capital cost and local requirements. Alternative configuration must be considered in order to select the most suitable cogeneration desalting plants. Therefore, the formulation of models for the synthesis and analysis of different design alternatives are very important and useful.

In this paper, the superstructure model of a triple hybrid power–MSF–RO system is developed and described as a MINLP problem, which contains all the potential interaction of material streams and energy streams involved in the combined systems. The optimization is performed based on economic points of view and a heuristic optimization algorithm, namely, modified genetic algorithm is applied. Afterward, the optimal configuration and operation condition is obtained to satisfy electricity and fresh water demand simultaneously.

THE SCHEMATIC OF THE COGENERATION SYSTEM

Fig. 1 illustrates the schematic of the combined power–desalination system for simultaneous generation of the electric power and fresh water. The power generation cycle includes boiler, steam turbine that produces the electric energy and the thermal energy. The desalination plant includes the MSF and RO. All parts of the cogeneration system were modeled and simulated to evaluate performance of the combined system.

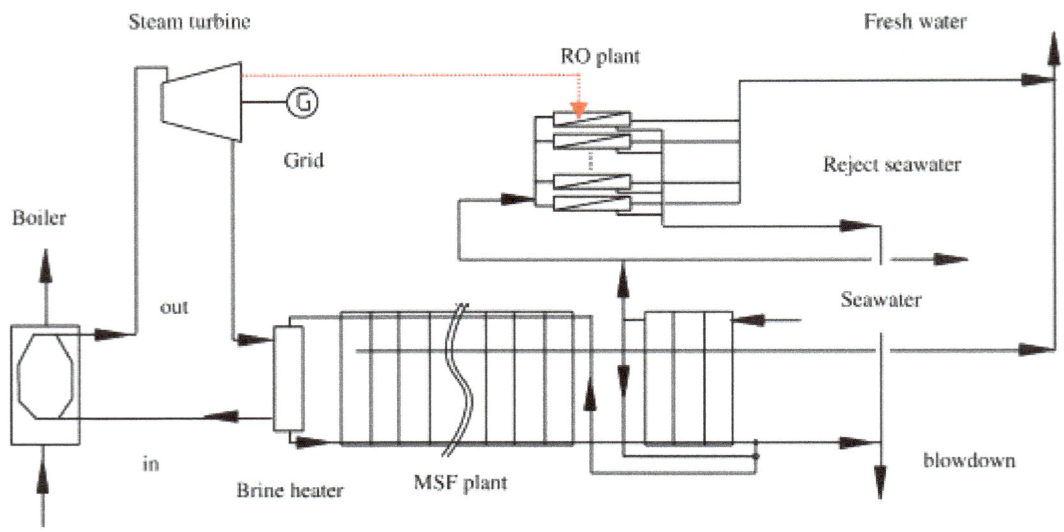

Figure 1: Schematic of cogeneration of power and desalination.

The fuel is combusted in the boiler to produce the superheated steam. The steam is sent to the steam turbine to generate the electric energy, which is used for MSF, RO and the grid. The whole or part steam extracted from turbine is feed to the brine heater of MSF and condensed to recycle the boiler. The seawater is sent to the MSF and heated as they flow through the pre-heater tubes of the heat rejection section. Subsequently, it is divided into two parts. One part is used as the makeup water to the heat recovery section and the other part is sent to RO as the feed water and return to the sea. The fresh water produced by MSF and RO are blended as the total water production. The blow-down brine of MSF and RO is feed to the common processing system.

MATHEMATICAL MODEL

To simplify the cogeneration system, this model is based on the following assumptions:

(1) Boiler, steam turbine and the MSF unit are adiabatic. The heat losses are neglected.

(2) The isentropic efficiency of the steam turbine is calculated. The working mode of steam turbine could be selected back pressure or extracted steam based on the demand of desalination.

(3) The cogeneration system for power and desalination, including steam turbine, MSF and RO, works in steady condition.

(4) Physical properties of all streams, such as enthalpy, specific heat capacity and so on, are calculated in mean inlet and outlet temperatures.

(5) Water is assumed as working fluid on the boiler and steam turbine.

(6) The distillate product of MSF is assumed to be salt free.

(7) The distillate of the MSF plant section can be blended with the RO permeate to obtain suitable water quality. So the single stage RO process can be used.

The plant consists of two desalination sections (MSF and RO). The energy demand is supplied by power plant where electric and thermal energy are co-produced. The detailed equations of the model are as the follows.

Co-generation plant

The co-generation plant consists of a boiler, a steam turbine, and an alternator. The superheated steam produced by boiler is sent to the steam turbine, which is connected with an alternator in order to produce electric energy. The outlet steam or extracted steam from steam turbine is feed to the brine heater of MSF. The electric energy produced is used for MSF, RO and grid.

Boiler

The water is heated to superheated steam in the boiler. The heat supplied by boiler is obtained from the following equation.

$Qbo = M_{s,}bo(hbo_{,}out - hbo_{,}in)$ (1)

The necessary fuel mass to be admitted to the boiler is determined as:

$$M_c = \frac{Q_{bo}}{Hc \cdot \eta_{bo}}. \quad (2)$$

Steam turbine

The steam turbine comprises back pressure steam turbine and the condensing/extraction steam turbine. It should be realized that only one of the turbines would be selected to be used in the desalination. The mass and energy balance could be written as follows.

If the back pressure steam turbine is selected.

$WeBP = M_{s,}BP(hbo_{,}out - hBP_{,}out) \cdot \eta BP$ (3)

$M_{s,}bo = M_{s,}BP$ (4)

Otherwise, the condensing/extraction steam turbine is selected.

$M_{s,}bo = M_{s,}co + M_{s,}ex$ (5)

$Weco = (M_{s,}bohbo_{,}out - M_{s,}exhex - M_{s,}cohco).\eta co$ (6)

Total electric energy produced by steam turbine can be defined as follows.

$We = \gamma WeBP + \beta Weco$ (7)

The steam flow rate from turbine to MSF can be described as Eq. (8).

$M_{s,}bh = \gamma \cdot M_{s,}bo + \beta \cdot M_{s,}ex$ (8)

$\gamma + \beta = 1$ (9)

Desalination Plant

The desalination plant considers the hybrid system of MSF and RO. The required energy is supplied by power plant with co-production of electric and thermal energy. The desalination plant model includes three parts: MSF unit, RO unit and the integration system of RO and MSF. The detailed equations of mass balance, energy balance and momentum balance for MSF and RO unit are described in paper [20]. Here, only the additional balance equations of hybrid desalination system are given as follows:

The total intake flow rate of seawater is equal to the sum of make-up flow rate, feed rate of RO and the blow-down brine.

$M_F = M_m + M_{F,}RO + Mbd$ (10)

The total freshwater product and the blow-down flow rate are obtained from Eqs. (11) and(12) respectively.

$M_d = M_{d,}MSF + M_{d,}RO$ (11)

$Mbd = Mbd_{,}MSF + Mbd_{,}RO$ (12)

The energy balance of MSF brine heater is given by Eq. (13).

$M_{s,}bth_s = M_R CPR(TBT - TMSF_{,}out)$ (13)

The concentration of freshwater and blow-down brine is calculated by Eqs. (14) and (15).

$M_d C_d = M_{d,}MSFC_{d,}MSF + M_{d,}ROC_{d,}RO$ (14)

$MbdCbd = Mbd_{,}MSFCbd_{,}MSF + Mbd_{,}ROCbd_{,}RO$ (15)

The distillate product of MSF is assumed to be salt free. $Cd_{,}MSF = 0$, so Eq. (14) could be simplified as follows:

$M_d C_d = M_{d,}ROC_{d,}RO$ (16)

The product ratio of MSF and RO is defined as follows.

$$R_{MSF,RO} = \frac{M_{d,MSF}}{M_{d,RO}} \quad (17)$$

The power/water ratio (PWR) indicates the ratio between the water produced and the electricity energy consumed, expressed in kWh/m³.

$$PWR = \frac{W_e}{M_d} \quad (18)$$

The economic model

The cost is calculated as the sum of the specific cost of the three sections presented in the cogeneration plant. The cost of each section includes the investment cost and operation cost, which is described as Eqs. (19), (20) and (21), respectively.

$Cpower = Ccap_{,}po + Cop_{,}po + C_c$ (19)

$CMSF = Ccap_{,}MSF + Cop_{,}MSF$ (20)

$CRO = Ccap_{,}RO + Cop_{,}RO$ (21)

The optimization objective of the cogeneration system is to minimize the TAC while meeting the demand of electric energy and fresh water.

Objective function: $\text{Min } TAC = Cpower + C_{MSF} + C_{RO}$ (22)

Where, TAC is the total annual cost of the cogeneration system. Cpower, CMSF, and CROis the cost of power plant, MSF plant and RO plant, respectively. The detailed description of the operation cost and investment cost can be found in [20].

The following limitations, which include the equipment and operation condition except equations of the material balance and energy balance, are considered for decision variable in the optimization study.

The production of power and fresh water for the cogeneration plant is described by Eqs. (23) and (24), respectively.

$We_G \geq Wedem$ (23)

$M_d \geq M_{d,}dem$ (24)

The restrictions defined in Eqs. (25) and (26) are included for the concentration of blow-down brine and fresh water.

$C_{bd} \leq 70,000\,\text{mg/L}$ (25)

$C_d \leq 500\,\text{mg/L}$ (26)

The difference temperature between intake seawater and blow-down brine of MSF is described by Eq. (27).

$Tbd_{,}MSF - T0 \leq 10°C$ (27)

MODIFIED GENETIC ALGORITHM

The coding design for a special problem is the key to use genetic algorithm effectively. For the above problem, there exist three kinds of variables in the model: bool variable, integer variable and real variable, which are inconvenient to be coded using binary coding simultaneously. To overcome this problem, here a modified genetic algorithm featuring mixed coding, which uses different crossover and mutation operators for the above three kinds of variables, is shown as follows.

Coding

In the cogeneration system, the chromosome based on mixing coding can be described as:

$\{(a_1,a_2),(z_1,z_2\ldots\ldots),(b_1,b_2,b_3,\ldots\ldots)\}$.

Where the bool coding (a_1,a_2), the first part of chromosome, represents the equipment type of steam turbine, and the integer coding $(z_1,z_2\ldots\ldots)$, the second part of chromosome, represents stage number of MSF, module number of RO, etc. The real coding $(b_1,b_2,b_3,\ldots\ldots)$, the third part of chromosome, represents the flow rate, operation pressure, production ratio of MSF and RO, and so on.

Initializing the population

For all independent variables, the initial values are generated within the feasible region randomly by the following five steps.

Step 1 Initialize the population size L, population Q = {Φ} and let index j = 1.

Step 2 Randomly generate the equipment type $A_j = (a_1,a_2)$, the integer string. $Z_j = (z_1,z_2\ldots\ldots)$ and the real string $B_j = (b_1,b_2\ldots\ldots)$, thus the individual $Q_j = \{A_j, Z_j, B_j\}$.

Step 3 If Q_j can meet the above constraints, then go to step 4, else go to step 2.

Step 4 If Q_j is new individual different from all the previous individuals, then $Q = Q + \{Q_j\}$, and $j = j + 1$; else go to step 2.

Step 5 If j > L, then $Q = \{Q_1,Q_2,\ldots Q_L\}$and stop; else go to step 2.

The genetic algorithm performance is influenced heavily by the population size. It is possible to run the risk of serious under-covering the solution space and result in a local optimum when the population size is comparatively small, whereas it would increase the computation load when the population size is comparatively large. Consequently, the population size should be chosen according to the problem scale. In this paper, the population size is set to 50.

Crossover

The direct exchange is adopted for the bool and integer variable. The arithmetic crossover is used to real variable in solving process. Here the arithmetic crossover is only described.

The two individuals V_1, and V_2 (two parents) are selected randomly and the offspring could be found as follows.

$$\text{offspring 1: } V_1' = V_1 + (1-\lambda)V_2 \quad (28)$$

$$\text{offspring 2: } V_2' = V_2 + (1-\lambda)V_1 \quad (29)$$

The λ is a random number between 0 and 1.

The arithmetic crossover could hold the diversity of offspring and avoid prematurely convergence and local optimum. The offspring generated by crossover replaced the parent regardless their fitness to maintain enough diversity in the population.

Mutation

The selection of mutation operator depended on the coding type. In this section, the mutation strategy for the mixed coding is adopted by the following step: if the selected gene is an integer, a new integer would be created to replace the old integer. If the selected gene is bool, the gene would be mutated 0 or 1. If the selected gene is real, the heteropic mutation is selected to solving process.

For example, first a random integer "temp" (0 or 1) would be created, then

If temp = 0

$$a = a + (a\_max - a)\cdot temp\cdot(1 - t/TT)^2. \quad (30)$$

If temp = 1

$$a = a - a\cdot temp\cdot(1 - t/TT)^2. \quad (31)$$

Where a represents the mutation gene, a_max is the maximum of a; the t is the iterative number, and TT is the total iterative number.

The structure of the MGA can be described as Fig. 2.

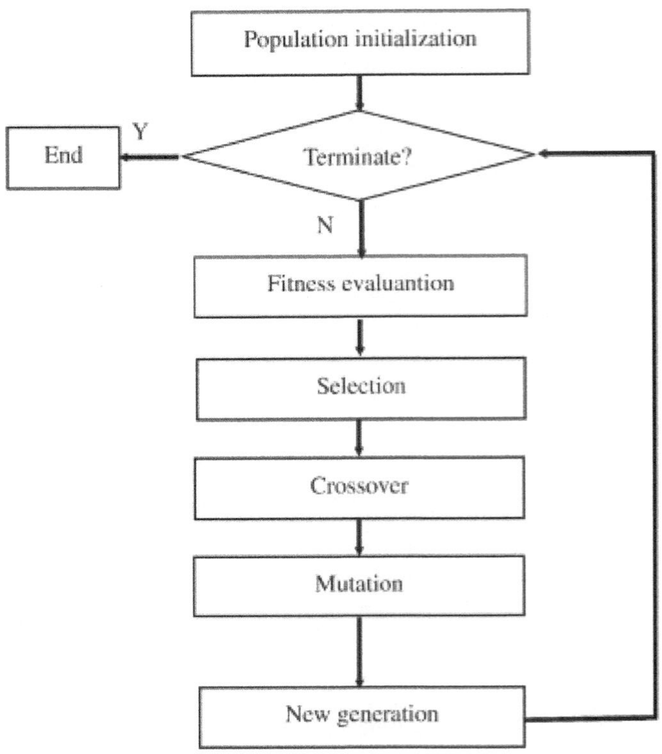

Figure. 2: Scheme of modified genetic algorithm.

CASE STUDY

A plant consisting of power plant coupled with hybrid MSF/RO desalination has been considered. The water production is given 12,000 m³/h while the total demand electric energy of 250 MW must be satisfied. The blended water concentration of RO and MSF should be lower than 500 mg/L and the blow-down brine concentration is less than 70,000 mg/L. The main optimization variables are: the heat consumption of and the total transfer area of MSF, module number of RO, and water production ratio of MSF and RO. The objective is to determine the optimal configuration of the cogeneration system and to select the optimal operation conditions of the cogeneration system in order to satisfy the water production and electric power demand while the total annual cost is minimal.

The main problem parameters are given in Table 1, and the optimal result is shown in Table 2.

Table 1: Input design parameter.

Parameter	Value	Parameter	Value
Power plant		Temperature of brine blow-down, °C	35
Outlet steam temperature of boiler, °C	566	Steam temperature of brine heater, °C	120
Outlet steam pressure of boiler, Mpa	24.2	RO plant	
Lower heat value of fuel, $kJ \cdot kg^{-1}$	25,200	Pure water permeability $m \cdot s^{-1} \cdot Pa^{-1}$	8.33×10^{-13}
Efficiency of boiler	0.92	Solute transport parameter, $m \cdot s^{-1}$	3.51×10^{-8}
Efficiency of steam turbine	0.97	Active area of membrane, m^2	34.5
MSF plant		Membrane lifetime, y	5
Stage number of MSF	21	Efficiency of HP pump	0.64
TBT, °C	110	Efficiency of energy recovery device	0.9
Feed concentration of seawater, $mg \cdot L^{-1}$	42,000	Price of membrane, $	500
Feed temperature of seawater, °C	25	Total product of fresh water, m^3/h	12,000

Table 2: Result of optimal design.

Parameter	Value	Parameter	Value
Power plant		RO plant	
Fuel consume rate, 10^3 kg/h	125.8	Operation pressure, MPa	0.7
Steam flow rate of boiler, 10^3 kg/h	787.7	Feed rate of RO, m^3/h	12,980
MSF plant		W_{d_RO}, m^3/h	3946
Stage number of MSF	21	Concentration of permeate, mg/L	788
Makeup seawater rate, m^3/h	21,000	Module number	13,684
Concentration of blow-down, mg/L	68,093	Reject brine concentration, mg/L	60,000
W_{d_MSF}, m^3/h	8054	Concentration of final fresh water, mg/L	259
Heat steam rate, 10^3 kg/h	787.7	Total feed rate of seawater, m^3/h	33,980
		Total recovery rate, %	35

As shown in Table 2 the optimal configuration is the combination of power plant, MSF and RO plant. The water production ratio of MSF and RO is 2.04. Namely, the water production of MSF is greater than that of RO. The steam flow rate from the boiler is equal to that feeding to the brine heater. Therefore the back pressure steam turbine is the optimal selection for the power plant. So MSF plant would be run on full load and the lacking amount of the total required water is made up by the RO plant.

In addition, the different demand of fresh water is considered. Here, four cases, 3000 m^3/h, 5000 m^3/h, 8000 m^3/h, and 12,000 m^3/h, are considered to observe the variation of the cogeneration system configuration. And the specific cost of the cogeneration system is compared with independent working MSF plant or RO plant. The result is showed in Table 3, and Fig. 3 and Fig. 1.

Table 3: The optimal result of cogeneration system at different water products.

	Water production, m^3/h			
	3000	5000	8000	12,000
W_{d_MSF}, m^3/h	3000	5000	7290	8054
W_{d_RO}, m^3/h	0	0	710	3946
W_{d_MSF}/W_d	1	1	0.91	0.67
Cost of water, $\$/m^3$	0.88	0.87	0.82	0.81
TAC, $10^8 \$ \cdot y^{-1}$	1.35	1.49	1.70	1.99
W_c, 10^3 kg/h	95.4	103.1	113.9	125.8
Steam flow rate of boiler, 10^3 kg/h	597.3	645.5	713.2	787.7

W_s, 10^3 kg/h	291.5	486.6	713.2	787.7
W_e, MW	254.8	258.0	267.4	295.4
Operating mode	MSF	MSF	MSF + RO	MSF + RO

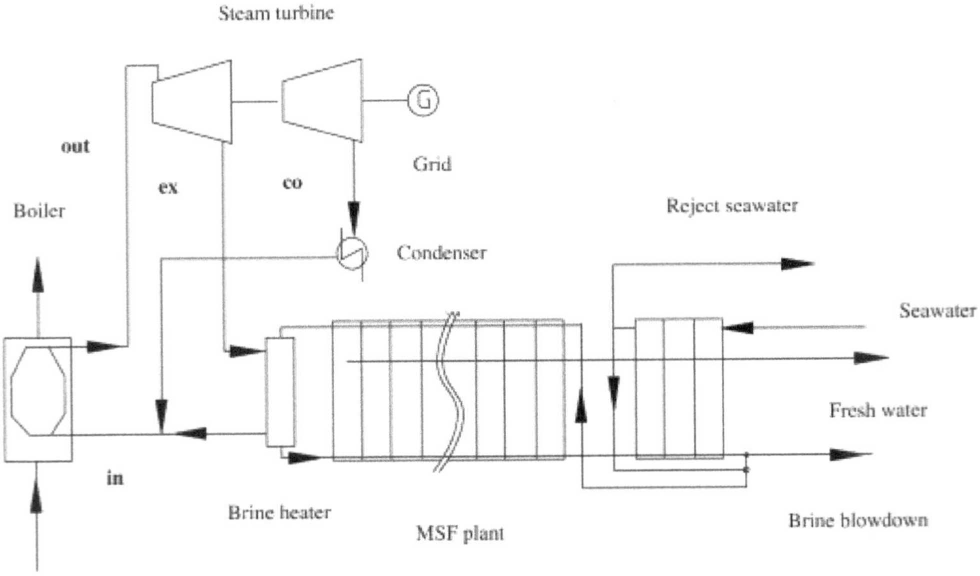

Figure 3: Schematic of cogeneration of extra-condensed steam turbine with MSF.

Fig. 1 and Fig. 3 show the optimal solutions for the case of the high demand of fresh water (more than 8000 m^3/h) and the low demand of fresh water (lower than 8000 m^3/h), respectively. When the demand of fresh water is low, the product of MSF plant is enough to satisfy the water demand and the RO plant is not a requisite (see Fig. 3). On the contrary, when the demand of fresh water is high, the MSF plant and RO plant run at the same to supply fresh water to consumer (see Fig. 1).

As shown in Table 3,the operation pattern of the cogeneration system converts from combination of power plant and MSF (Fig. 3) to the combination of power plant, MSF and RO (Fig. 1) when the water production increases from 3000 m^3/h to 12,000 m^3/h. At the same time, the water production ratio of MSF and RO is reduced and the cost of water production decreases gradually. On the one hand, it is caused by the scale effect of the desalination plant. On the other hand, the cost of RO is lower than that of MSF commonly, the increasing of RO permeate flow rate results in the combined freshwater cost decreased. Although the required electric energy of the grid is constant, the more quantity of the required freshwater, the more the total electric energy produced by power plant and the fuel consume rate. So the TAC would be increased.

Any more, the production ratio between MSF and RO goes down gradually while the required freshwater increases. Therefore properly enhanced permeate rate of RO is advantageous for the cogeneration system when the demand of freshwater is increased.

In addition, when the required freshwater is comparatively small (3000 m^3/h and 5000 m^3/h), the demand steam of MSF is less than the steam flow rate of boiler. Consequently, the condensing-extraction steam turbine could be selected. As the required freshwater rate exceeds to 8000 m^3/h, the demand steam of MSF is equal to the steam flow rate of boiler and the back pressure steam turbine is the optimal selection.

A comparison of fresh water cost for cogeneration plant, MSF plant and RO plant is shown in Fig 4. The fresh water cost of the three independent systems reduces generally along with the increasing of fresh water production. The cost of MSF exceeds that of RO and cogeneration system at the same yield of fresh water. When the fresh water production is greater than 5000 m^3/h, the cost of cogeneration system is superior to MSF and RO system. In other words, the optimal operation pattern is independent RO system or combination of power and MSF when the quantity demand of fresh water is lower. However, the combination of power and MSF not only satisfied quantity demand of the fresh water, but also supplied the electricity to consumers. So the combination of power and MSF is superior to alone RO system. Conversely, when the quantity demand of fresh water is high, the optimal operation pattern is the combination of power, MSF and RO.

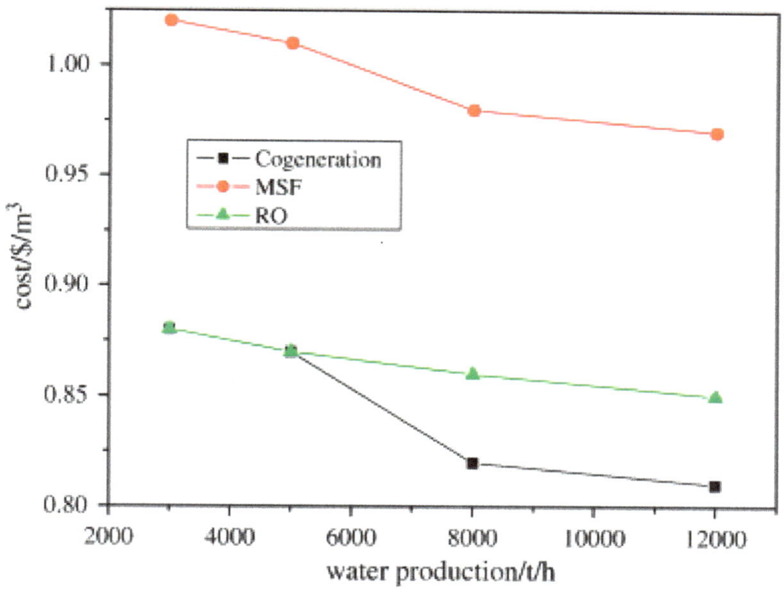

Figure 4: Water cost of cogeneration system VS. MSF and RO.

CONCLUSIONS

In this paper, a cogeneration plant consisting of power, MSF and RO units was considered, which provides the required electricity and fresh water simultaneously. The superstructure of tri-function power–MSF–RO plant was developed as a MINLP model. The objective function is to minimize the TAC of cogeneration system. The proposed model was solved by the modified genetic algorithm featuring mixed coding, where bool variables were used to select the equipment type, integer variables were used to represent stage number of MSF, module number of RO, etc. and the real variables were used to represent the flow rate, operation pressure, production ratio of MSF and RO, and so on. An example, which is supposed to supply 250 MW power and 12,000 m/h for Huangdao district of Qingdao City, has been successfully solved by applying the modified genetic algorithm. The optimization result showed: the optimal pattern of the cogeneration system for power and desalination could be varied in terms of the water demand. When the water demand is lower than 8000 m^3/h, the combination of power plant associated with MSF is adopted, in which the condensing-extraction steam turbine is used for the power plant. But when the water demand of water is higher than 8000 m^3/h, the tri-combination of power plant, MSF and RO is the optimal choice, in which back pressure steam turbine is used for the power plant.

NOMENCLATURE

C concentration, mg/L

Cp specific heat capacity,kJ·kg^{-1}·K^{-1}

Cpower power plant cost, \$·y^{-1}

C_{MSF} MSF plant cost, \$·y^{-1}

C_{RO} RO plant cost, \$·y^{-1}

Hc lower heat value of coal, kJ·kg^{-1}

h enthalpy, kJ·kg^{-1}

M flow rate, kg·h^{-1}

Q heat duty, kW

T temperature, K

T0 seawater temperature, K

TAC total annual cost, \$·y^{-1}

TBT the top brine temperature, K

We the electricity generating capacity, MW

R water product ratio

PWR the power/water ration, kWh/m^3

Greek

η efficiency

γbool variable, 0或1

β bool variable, 0或1

Subscripts

MSF multi stage flash

RO reverse osmosis

out outlet

in inlet

F seawater

s steam

G grid

dem the demand

c coal

ex extraction steam turbine

bh brine heater

cap capital

bd blow-down

m makeup

d desalination

po power plant

BP back pressure turbine

R recycle brine

Co zcondense steam turbine

bo the boiler

op operation

ACKNOWLEDGMENTS

This work was financially supported by the National Nature Science Fund Program of China (21076202). This is MCTL Contribution No. 19.

REFERENCES

1. B.M. Misra, P.K. Tewari, B. Bhattacharjee, Futuristic trends in hybrid system for desalination, IDA Conference, San Diego, 1999, pp. 311–320.
2. L. Awerbuch, Power–desalination and the importance of hybrid ideas, IDA World Congress, Madrid, 1997.
3. M.A.K. Al-Sofi, A.M. Hassan, E.F. El-Sayed, Integrated and non-integrated power/MSF/RO plants, Desalin. Water Reuse 3 (1992) 10–16.
4. O.A. Hamed, Overview of hybrid desalination systems—current status and future prospects, Desalination 186 (2005) 207–214.
5. E. Cardona, A. Piacentino, F. Marchese, Performance evaluation of CHP hybrid seawater desalination plants, Desalination 205 (2007) 1–14.
6. E. Cardona, A. Piacentino, Optimal design of cogeneration plants for seawater desalination, Desalination 166 (2004) 411–426.
7. Y. Wang, N. Lior, Performance analysis of combined humidified gas turbine power generation and multi-effect thermal vapor compression desalination systems. Part 1: the desalination unit and its combination with a steam-injected gas turbine power system, Desalination 196 (2006) 84–104.
8. Y. Wang, N. Lior, Fuel allocation in a combined steam-injected gas turbine and thermal seawater desalination system, Desalination 214 (2007) 306–326.
9. Y. Wang, N. Lior, Performance analysis of combined humidified gas turbine power generation and multi-effect thermal vapor compression desalination systems. Part 2: the evaporative gas turbine based system and some discussions, Desalination 207 (2007) 243–256.
10. R. Chacartegui, D. Sanchez, N. di Gregorio, F.J. Jiménez-Espadafor, A. Munoz, T. Sanchez, Feasibility analysis of a MED desalination plant in a combined cycle based cogeneration facility, Appl. Therm. Eng. 29 (2009) 412–417.
11. S.E. Shakib, S.R. Hosseini, M. Amidpour, C. Aghanajafi, Multi-objective optimization of a cogeneration plant for supplying given amount of power and fresh water, Desalination 286 (2012) 225–234.
12. S.E. Shakib, M. Amidpour, C. Aghanajafi, A new approach for process optimization of a METVC desalination system, Desalin. Water Treat. 37 (2012) 1–13.
13. S.F. Mussati, P.A. Aguirre, N.J. Scenna, Dual-purpose desalination plants. Part I. Optimal design, Desalination 153 (2003) 179–184.
14. S.F. Mussati, P.A. Aguirre, N.J. Scenna, Dual-purpose desalination plants. Part II. Optimal configuration, Desalination 153 (2003) 185–189.
15. S.F. Mussati, P.A. Aguirre, N.J. Scenna, A rigorous, mixed-integer, nonlinear programming model (MINLP) for synthesis and optimal operation of cogeneration seawater desalination plants, Desalination 166 (2004) 339–345.
16. S.F. Mussati, P.A. Aguirre, N.J. Scenna, Optimization of alternative structures of integrated power and desalination plants, Desalination 182 (2005) 123–129.
17. S.F. Mussati, M. Barttfeld, P.A. Aguirre, N.J. Scenna, A disjunctive programming model for superstructure optimization of power and desalting plants, Desalination 222 (2008) 457–465.
18. Eduardo Manfredini Ferreira, José Antônio Perrella Balestieri, Maurício Araujo Zanardi, Optimization analysis of dual-purpose systems, Desalination 250 (2010) 936–944.
19. J. Uche, L. Serra, A. Valero, Thermoeconomic optimization of a dual-purpose power and desalination plant, Desalination 36 (2001) 147–158.
20. Lianying Wu, Optimization and Design of Integrated Desalination System Based on the Genetic Algorithm. Ph.D Ocean University of China, 2012.

CHAPTER 2

FRESH GROUNDWATER FOR WAJIR—EX-ANTE ASSESSMENT OF UNCERTAIN BENEFITS FOR MULTIPLE STAKEHOLDERS IN A WATER SUPPLY PROJECT IN NORTHERN KENYA

Eike Luedeling, Arjen L. Oord, Boniface Kiteme, Sarah Ogalleh, Maimbo Malesu, Keith D. Shepherd and Jan De Leeuw

Keywords: Pipeline, risk, uncertainties, stakeholder, decision-making, dryland, Merti aquifer

INTRODUCTION

Historically, decisions about which development interventions to pursue have been based mainly on their perceived technical and economic feasibility. It is now generally acknowledged, however, that *ex-ante* evaluation of development decisions should also consider impacts on the wider natural and social environment. Such an inclusive approach is critical for ensuring that the consequences of the intervention are environmentally sustainable, economically viable, and socially just. Consequently, many countries have developed legislative instruments requiring *ex-ante*environmental and social impact assessments for major interventions (Jay et al., 2007).

Despite the recognition that holistic impact assessment is highly desirable in intervention planning, a number of limitations hinder its application in practice. First, available data is often insufficient for arriving at unambiguous conclusions about the impacts of an intervention (Peterman and Anderson, 1999; Regan et al., 2005). This is a particular problem in data-scarce environments such as Sub-Saharan Africa, but data on some aspects, such as social and off-site environmental impacts, tend to be limiting everywhere. Second, even where some information is available, decision-makers often lack appropriate tools for integrating different types of information into a context-specific forecast of decision outcomes (Peterman and Anderson, 1999). Third, decision-makers lack appropriate tools for adequately addressing uncertainties of individual variables and the effects of their accumulation in the decision making process. In cost-benefit analyses, uncertain variables describing costs and benefits are often substituted by precise best-bet estimates, which only represent one possible value out of a larger population of plausible values. Fourth, as a result of this, intervention planning often fails to assess risks that could jeopardize success of an intervention or compromise its performance.

In day-to-day practice decisions are frequently made without attempt to address these limitations. Consequently, when risks remain unconsidered, there is a high chance that decision outcomes will not meet stakeholder expectations.

Aware of the above limitations, decision-makers often commission measurements or detailed scientific studies to close perceived knowledge gaps. However, decisions on which knowledge gaps should be closed or narrowed are rarely guided by a structured analysis of information value, or which uncertainties most affect project outcomes (Clapp et al., 2013; Hubbard, 2014). In the absence of such prioritization, commissioned research will tend to follow intuitive preference for information that is readily available or easy to generate, leading to what Hubbard (2014) describes as

Citation: Luedeling E, Oord AL, Kiteme B, Ogalleh S, Malesu M, Shepherd KD and De Leeuw J (2015) Fresh groundwater for Wajir—ex-ante assessment of uncertain benefits for multiple stakeholders in a water supply project in Northern Kenya. Front. Environ. Sci.3:16. doi: 10.3389/fenvs.2015.00016.

the "measurement inversion." Sometimes, the knowledge gaps that most strongly constrain the capacity to make informed decisions include variables that are difficult to quantify or not easily measured, such as the future market price of a commodity or cultural acceptability of an innovation. Such variables are easily overlooked in project feasibility studies. Interestingly, reducing uncertainty on such variables can be quite cheap and easy, given that little work has often been done on them. For example, if cultural acceptability of a new crop is an uncertainty in project planning, a small number of interviews may suffice for establishing whether this crop has a chance of success. In comprehensive decision evaluation, this evidence can easily produce a higher gain in certainty about the project's prospective outcomes than a large-scale soil survey or detailed modeling of crop performance. Since time and resources for decision-relevant research are normally limited, structured prioritization of knowledge gaps can make a critical contribution to the effectiveness of decision-support research.

In this study, we propose a holistic decision analysis approach that allows comprehensive integration of all relevant variables, even those with uncertain and missing information, and that considers risk factors that may jeopardize project success or affect project performance. The proposed approach also allows users to identify and prioritize those knowledge gaps that should most urgently be narrowed in order to improve certainty about decision outcomes.

DECISION MODELING

The approach taken in this study is inspired by the principles of Applied Information Economics outlined by Hubbard (2014). We start by thoroughly framing the decision to be modeled, followed by identification of all relevant stakeholders (Figure 1). We then identify a small group of experts with a good understanding of the decision in question as well as the local environmental and socioeconomic context of the target area and population. It is desirable that this expert group also includes the decision-makers or their close advisors and other major stakeholders, because their personal involvement in the modeling team may help them refine their understanding of the implications of their decision and ensure that study results are duly considered in the decision-making process.

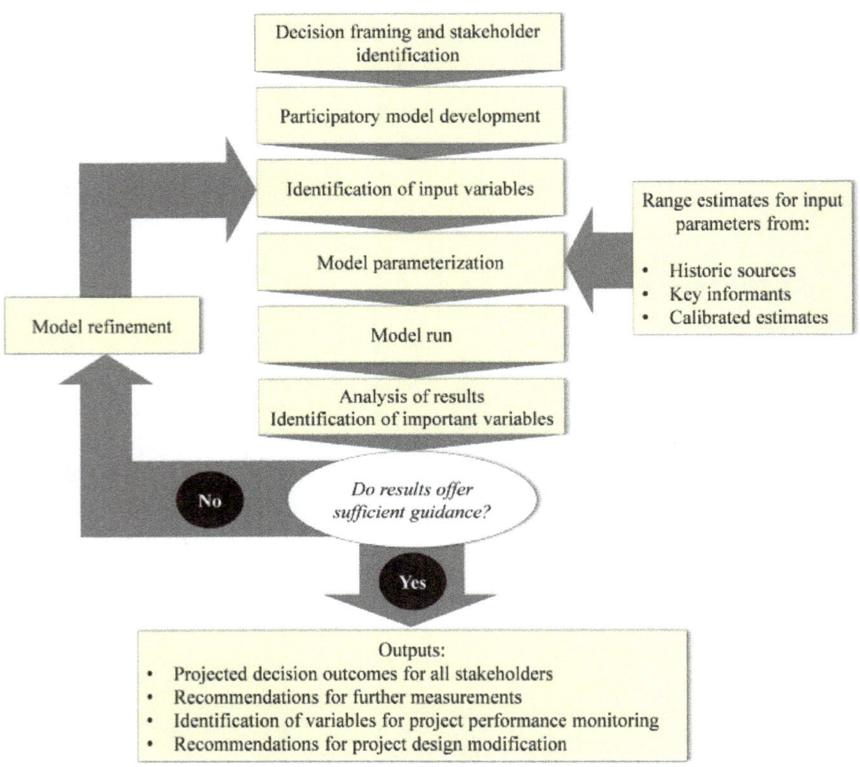

Figure 1: Sequence of activities in the decision modeling approach.

The expert group then engages in participatory model development, in which an impact pathway for the decision is constructed based on the experts' expectations of decision impacts. In this step, the group aims at including all relevant costs and benefits of the proposed intervention. They also thoroughly explore and list all plausible risks that may affect project success and outcomes. At this stage, high levels of mechanistic detail in considering particular costs, benefits, or risks are not needed and not desirable. It is also not important that the group has a detailed plan for measuring any of the variables that they introduce into the model. Instead, they should aim at capturing all relevant issues in the initial model. If any of these variables emerge as key uncertainties, they can be disaggregated at a later stage of the process. The key focus in the initial model building is on generating a conceptual model that represents the "big picture" of the decision at hand. This model is then translated into a set of equations that represent the group's understanding of the decision's impact pathway.

Once a comprehensive quantitative model has been developed, which basically represents the decision's impact pathway, members of the group of experts are requested to provide probability distributions of all input variables based on their uncertainty about the parameters used. At the simplest level, these can be provided as confidence intervals, i.e., as a set of two values of which the expert is 90% sure that they bracket the real value of the variable. They also select a probability distribution. This can be a normal distribution or a distribution with a different shape, if experts feel confident that this is more appropriate. To prepare experts for the estimation exercise, they undergo calibration training (Hubbard, 2014) before making estimates. During this training, they are instructed in a number of estimation techniques that measurably improve people's ability to estimate distributions (Hubbard, 2014). In consultation with the modeling team, estimates from the various experts are then consolidated into one single probability distribution for each variable.

Next, the model is run a large number of times as a probabilistic Monte Carlo simulation, each time fed with a different set of random draws from the defined distributions for each variable (Rosenstock et al., 2014). The result is a distribution of outcomes that are plausible given the experts' current state of uncertainty. In some cases, this outcome distribution may already suffice for deciding on a rational course of action for the decision (Peterman and Anderson, 1999). In most cases, however, decisions have a wide spread of potential outcomes, including negative ones. The width of the range of outcomes results from the uncertainty in the individual variables. Reduction of uncertainty of some influential variables will help reduce the range of overall outcomes, with a possibility of thus reducing the likelihood of negative outcomes. For such situations, mathematical procedures exist to explore the sensitivity of the model outputs to variation in the input variables and determine the value of information of every single variable (Hubbard, 2014). The value of information allows identification of those variables that most strongly affect uncertainty of the overall outcome and thus constrain the capacity to identify a preferable course of action. The variables that emerge with the highest information value can be interpreted as priorities for measurements to be undertaken to narrow the uncertainty around the decision.

Where particular variables emerge as priorities for measurement, additional measurements can be taken and the modeling procedure repeated (Figure 1). In some cases, simply decomposing a variable into a subset of variables that is easier to estimate may produce sufficient gain in certainty about decision outcomes. If the preferable course of action is not yet clear, the process can be used to guide further measurements and to successively reduce uncertainties about intervention outcomes, until sufficient confidence about which decision course is appropriate has been gained. Intervention designs can also be improved by directing attention to areas that pose greatest risk.

We next illustrate this approach with a case study we conducted in Wajir County in Northern Kenya. The procedures outlined above were used to simulate the outcomes of the proposed Habaswein-Wajir Water Supply Project, which aims to supply Wajir town with clean and reliable water.

THE HABASWEIN-WAJIR WATER SUPPLY PROJECT

Target Area

We analyzed the risks around the development of groundwater resources in the Habaswein Division of Wajir County to supply the town of Wajir with water, mainly for domestic and industrial use. Wajir (1.75°N; 40.1°E) is the largest town in Wajir County in the former Northeastern Province of Kenya (Figure 2). The climate is arid with an average annual rainfall of 250 mm. The official population estimates for Wajir were 42,207 in 1999 and 82,800 in 2009, reflecting a 6.7% increase per year, but these numbers have been contested and real population numbers are likely much higher. We consider that the true size of the population today (2015) is not well known and might be anywhere between 120,000 people, a conservative estimate based on the 2009 census and a 6.7% annual increase since then, and approximately 300,000, which was considered more realistic by local experts.

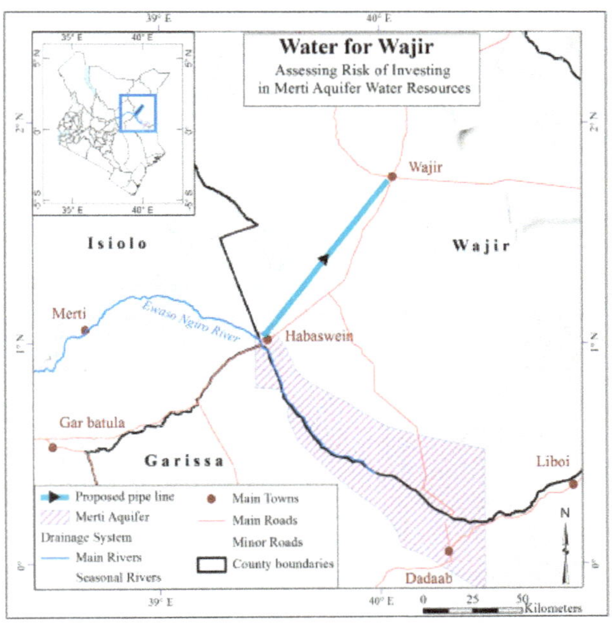

Figure 2: Geographic context of the proposed Habaswein-Wajir Water Supply Project.

The growing population complicates the continued provision of water through shallow wells, which have supplied the people of Wajir with water up to now. First of all, over-abstraction may lower the groundwater level so much that shallow wells can no longer function (Apaydın, 2012). Secondly, the biological quality of the water in these shallow wells is increasingly poor due to contamination with human and livestock excreta (Osman, 2012). Consequently, the national and county governments, who have identified the need for additional water of better quality, have developed a plan for a more sustainable solution to abstract water from the Merti aquifer, a rich body of groundwater located 110 km to the south (Mumma et al., 2011; De Leeuw et al., 2012).

The Planned Project

The plan for what is formally known as the Habaswein-Wajir Water Supply Project envisages the establishment of a well field with 12 deep wells in the vicinity of Habaswein (Figure 2), designed to allow for an abstraction of 2.2 million m^3 of fresh groundwater per year. This water is to be transported through a pressurized pipeline to Wajir, where it will be stored in reservoirs and distributed throughout the city via a reticulation system that is yet to be built. The project will also provide initial support to the water company that will operate and maintain this infrastructure and supply water to customers in Habaswein, Wajir and in settlements along the pipeline (Kanjara, Boji Yare, and Leheley).

In October 2010 the Ministry for Development of Northern Kenya and other Arid Lands (MDONK) applied through the Kenyan Treasury for a grant of € 37.7 million from the Dutch Facility for Infrastructure Development (ORIO). If approved, ORIO will donate 35% of the costs and provide a soft loan for the remainder. As a first step, ORIO approved a € 0.796 million project development phase grant in March 2012, and contracts for studies were established in April 2012. While several studies had been completed by the start of our research in November 2013, other studies were still pending, including social and environmental impact studies and a financial feasibility study. The research described in this article therefore overlapped with the decision-making process on the Habaswein-Wajir Water Supply Project.

The political approval process for the Habaswein-Wajir Water Supply Project has been complicated by a constitutional reform and devolution of authority in Kenya in 2012, which passed substantial decision-making power from central government to the county level (Kramon and Posner, 2011). While central government initiated and still formally leads the project, the devolution process has shifted the initiative toward the county government, which now plays an important role in decision-making. During the initial planning stages, the central government did not involve many of the stakeholders who stand to be affected by the project. These stakeholders remain poorly informed, and some groups have voiced concerns about the project (Ogalleh, 2015). These concerns are particularly strong among residents of Habaswein, who fear for the reliability of their water supply, which comes from boreholes tapping the Merti aquifer. Residents of Wajir town tend to be more in favor of the project, but some have also expressed reservations, caused for example by the prospect of having to pay a higher price for water in the future (Ogalleh, 2015).

Hydrogeological Setting

The most relevant water-related risks for the Merti Aquifer are drying up of wells in the proposed well field and deterioration of water quality as a result of increasing salinity of the abstracted water. While the newly proposed national water policy, which is currently under discussion in parliament, states the importance of ensuring inter-generational sustainability of groundwater use, inadequate information about basic aquifer parameters and limited practical experience with deep groundwater extraction hamper the application of this principle. In fact, lack of data and high uncertainty are typical features of many deep aquifers, particularly in Sub-Saharan Africa. Although the Merti aquifer is a vital source of fresh water in the region, relatively little is known about the extent of the aquifer, its hydrogeological parameters and groundwater recharge.

The Merti aquifer is defined by permeable water-bearing layers located within the Anza Rift (Bosworth and Morley, 1994). Generally, groundwater in the aquifer is confined and found at depths between 110 and 180 m below ground level (m bgl). Successful wells tap the more permeable zone of the Merti Formation commonly between 105 and 150 m bgl (GIBB Africa Ltd., 2004). Several attempts have been made to define the extent of the aquifer (e.g., Swarzenski and Mundorff, 1977; GIBB Africa Ltd., 2004). In these studies, the aquifer is believed to be limited by the presence of adjacent saline water bodies. Most studies agree that the main freshwater aquifer extends from Habaswein into Somalia at Liboi and beyond (see Figure 2). There is no comprehensive information that defines the aquifer's vertical extent. GIBB Africa Ltd. (2004) reports a likelihood that salinity of the groundwater increases with depth. The vertical extent of the freshwater aquifer is either limited by impervious layers, or the presence of saline groundwater.

Another source of substantial uncertainty is aquifer recharge. This has been a topic of debate among hydrogeologists, since knowledge of aquifer recharge is critical for designing a sustainable long-term groundwater use strategy. Mumma et al. (2011) report that groundwater abstraction comes from fossil water, with limited annual recharge (~3.3 million $m^3\ yr^{-1}$) and major recharge events (~30 million m^3) occurring at intervals of thousands to tens of thousands of years. GIBB Africa Ltd. (2004), however, suggests the possibility of a much higher recharge rate of 33 million $m^3\ yr^{-1}$.

Mumma et al. (2011) consider that most of the aquifer is currently under insignificant depletion stress, although very little data is available. The main groundwater abstractions take place in Habaswein and in the refugee camps of the Dadaab area (Figure 2). Influx of refugees into the Dadaab refugee camps has led to a sharp increase of groundwater abstraction from ~1 million m^3 in 2002 to ~3 million m^3 in 2011 (Mumma et al., 2011; Oord, 2011). In consequence, water quality in the Dadaab refugee camps has deteriorated over time, mainly due to increasing salinity, and also in Habaswein evidence exists of some salinization as a result of long term abstraction (Mumma et al., 2011). These observations strengthen the plausibility that a saline groundwater body underlies the freshwater aquifer.

Study Objectives

The objectives of the present study were two-fold: (1) to serve as a proof of concept that the Applied Information Economics approach, which has so far been applied predominately in single-decision-maker business decision settings, can improve decision-making processes in multiple stakeholder decision-making in a development context; (2) to comprehensively assess the Habaswein-Wajir Water Supply project, considering all relevant costs, benefits and risks, in order to provide holistic decision support on the preferable course of action. To account for the fact that most stakeholders perceived groundwater supply and quality as the main risks to the project, the hydrology of the aquifer was also evaluated with models that considered uncertainty about aquifer parameters.

METHODS

Participatory Model Development

A decision model was constructed using the framework outlined in the Decision Modeling section. The decision we modeled—whether to implement plans for the Habaswein-Wajir Water Supply project—was clearly defined and under consideration by national government and local authorities in Wajir County at the time of our research. Efforts for decision framing were therefore limited to clarification of design details of the proposed intervention. Stakeholders in the project, who were carefully identified through consultations with people familiar with the project and the target area, were then invited to an inception workshop. This workshop was held on November 4th 2013 and attended by approximately 30 stakeholder representatives, including potentially affected groups of water users, local government, water management authorities, the research community, and practical hydrogeologists with experience in the proposed project area. Stakeholder expectations of project outcomes and attitudes toward the project were explored through open-ended discussions, and a small team of eight experts was selected from among the workshop participants to contribute to the participatory model building process. To become part of the model building team, participants had to be available during the 2 days following the workshop and knowledgeable about the proposed project, the study area or the hydrology of the region. Care was also taken to ensure that the team included members from Wajir and Habaswein, who were well-respected by these communities and by local government.

The development of the quantitative impact model was initiated by brainstorming on all major benefits and costs expected to arise from the project, as well as on all the risks that might threaten project success. Relationships between these variables were then established through group discussions, aided by flow diagrams that were iteratively developed by the modeling team. In these diagrams, cost and benefit items were decomposed into their constituent parts, until members of the modeling team felt confident that all variables could be estimated. In addition, the group identified risks that might impact project outcomes and illustrated causal impact pathways in diagrams for each risk. The conceptual drawings were then translated into equations describing project outcomes and coded in R language (R Development Core Team, 2011). The resulting code is provided as supplementary material. Separate impact models were constructed for seven different stakeholders: (1) the donor, (2) the residents of Wajir, (3) the residents of Habaswein, (4) the residents of communities along the pipeline, (5) the water company that will run the water supply system, (6) downstream areas, and (7) upstream areas. All stakeholder outcomes were added up to produce a total estimate of the project's overall outcome prospects.

A list of all input variables to the model was assembled and annotated to facilitate the quantitative estimation process for the modeling team members. All team members were subjected to "calibration training" (Hubbard, 2014), a process that aims at enhancing the capacity of participants involved in model development to estimate probability distributions. The training involves instruction in techniques such as the exclusion of impossible values (Hubbard, 2014), the equivalent bet approach (Hubbard, 2014) and Klein's premortem (Klein, 2007). It also includes several iterations of estimating probability distributions for trivia questions, which aim at assessing the level of overconfidence or underconfidence displayed by participants. These errors of judgment manifest themselves in estimated ranges that are either too wide (underconfidence) or too narrow (overconfidence—the more common case), so that they do not accurately reflect the estimators' true level of confidence. Providing such feedback to estimators has been shown to greatly improve the accuracy of their estimates (Hubbard, 2014). Accuracy in this context should not be confused with precision. Instead of estimating narrow ranges or even single values for a variable (which would be very precise), the estimation process aims at obtaining an interval with a 90% chance of including the correct value. Where estimators have little knowledge about a variable, intervals should be wider than those provided by experts on the variable in question, but in both cases, estimators should be 90% certain that their intervals include the "correct" value.

Estimates of all variables were elicited from all experts, and five sets of estimates were returned. These were then consolidated by the main analyst into probability distributions that reflected the uncertainty of the entire team (Supplementary Table 3). Doing this required some additional discussion with experts about the ranges they chose, as well as some subjective credibility judgments by the analyst. For example, stakeholders from the target area were considered more credible sources of information for site-specific parameters, while hydrogeologists were given more credence for hydrological variables. A validation workshop was also convened with some of the experts in November 2013. Estimates on risks were collected for two scenarios:

Scenario 1: "All Risks"

In the "all risks" scenario, all risks to the project received full consideration. This can be considered as a forecast of intervention outcomes.

Scenario 2: "Cooperation"

In the "cooperation" scenario, political risks were assumed to have been overcome, and only biophysical and economic risks were considered. This scenario aimed at illustrating the general desirability of the pipeline project from the perspective of the different stakeholders and to inform stakeholders' expectations of project prospects.

Simulation Modeling

Once distributions had been defined for all uncertain variables (provided as Table 3 in the Supplementary Materials), these were used to run the decision model as a Monte Carlo simulation. In this simulation, the model was run 10,000 times for each scenario ("all risks" and "cooperation"), using random draws of all variables from the specified distributions. For some sets of variables that were likely to be correlated, values were drawn such that they met certain correlation requirements, expressed by prescribed coefficients of variation. These procedures were implemented in the MonteCaRlo package written in R (Luedeling, unpublished). This package produces estimates of the project's Net Present Value (NPV) for each stakeholder, which express the monetized net benefits of the project, while adjusting for time preference, i.e., the observation that most people assign lower values to benefits arising in the future than to immediate benefits. The analysis also conducts an assessment of the extent to which the various uncertainties affect the precision of project outcome estimates. In other words, we identified the key uncertainties in the decision-making process on the project. This was achieved through Partial Least Squares (PLS) regression (Wold, 1995; Wold et al., 2001; Luedeling and Gassner, 2012), which was used to correlate the various outcome variables (i.e., net present values for each stakeholder) to all input variables. A key output of the PLS procedure is the Variable-Importance-in-the-Projection (VIP) statistic, which provides a quantitative estimate of the degree to which uncertainty of each of the individual variables contributes to the uncertainty in the overall project outcome. A value of this statistic above 0.8 is considered to signify that the respective variable is important (Wold et al., 2001). The procedure also produces a coefficient indicating the direction of this effect, i.e., whether the variable is positively or negatively correlated with the outcome. We computed both statistics and combined them into variable-importance plots showing the VIP values and indicating the direction of the correlation by a color code. The main outputs of the modeling procedure are thus the projected outcome distributions for all stakeholders as well as the VIP plots.

When interpreting the VIP plots, it is important to remember that scores reflect the degree to which variation in the input variables is correlated with variation in the simulated outcomes. Variables with high variable importance scores are those variables that are most responsible for our inability to make precise forecasts of decision outcomes. In other words, if measurements can be undertaken to reduce uncertainty about some variables, they are most effective if they target variables with high variable importance scores, because these are the greatest sources of uncertainty about outcomes. It is also worth mentioning that even major risks to project success may not register high variable importance scores, if they are known with high precision.

Hydrogeological Modeling

The uncertainties related to the ability of the well-field to provide sustainable, high-quality water were studied in more detail in a hydrogeological assessment. For this, process-based hydrogeological models were combined with stochastic methods (Monte Carlo simulations) to calculate the expected drop in groundwater levels or drawdown (potentially leading to drying up of wells) and saline water intrusion, potentially leading to deteriorating water quality.

BAM International et al. (2012) estimated a total water demand of 6200 m^3 day^{-1} for the Habaswein-Wajir Water Supply Project. The proposed abstraction rate from the Habaswein-Dilmanyale aquifer is 5945 m^3 day^{-1}. They recommended a total number of 12 wells, with a minimum spacing of 700 m and a projected safe yield of 24 m^3 per hour per well. The model simulations were run for a well field of 12 wells, as well as 14 wells, to incorporate a scenario where the abstraction is distributed over a larger number of wells.

Wells Falling Dry

The widely used three-dimensional finite-difference groundwater model MODFLOW-2005 (Harbaugh, 2005) was applied to adequately represent the complexity of the aquifer. To determine the specific drawdown at the location of the wells, the MODFLOW model was extended with an analytical approach, using the Theis equation (Theis, 1935), which describes transient flow of water to a single pumping well in a confined aquifer. We used the principle of superposition to calculate the combined effect of multiple wells, and we introduced image wells to incorporate the influence of aquifer boundaries (Reilly et al., 1984). The Merti aquifer, which is mainly confined (Mumma et al., 2011), matches the conditions for application of the analytical approach described by Kruseman and de Ridder (1990). Conceptualizing the aquifer as homogenous is justified, since the effect of variation in aquifer properties is addressed by the Monte Carlo approach.

To evaluate whether calculated drawdown leads to drying of wells, we determined a drawdown threshold. At the time of writing, a well field design was not yet available. Therefore, the threshold was based on common well design limitations, aquifer characteristics, recommendations from BAM International et al. (2012) and the consideration that in the Merti Aquifer the static water level in wells is generally several meters above the top of the aquifer. Based on these figures, a threshold value for drawdown of 10–12 m was selected. With a Monte Carlo approach an ensemble of simulations was performed to establish the uncertainty range of the groundwater drawdown exceeding the threshold value. In these simulations the relevant aquifer parameters and geometry (i.e., conductivity, vertical anisotropy, storativity, effective porosity and aquifer thickness) were varied randomly within their probability distributions. A summary of the aquifer properties used in the simulation is given in the supplementary materials.

Salinity of the Abstracted Water

Groundwater abstraction from aquifers containing saline water can lead to deteriorating quality of the water supply. The most relevant process in this case is upconing, the upward movement of saline water (with a higher density) from deeper layers as a result of reduced pressure around the well. Given the low flow velocity of the groundwater (according to BAM International et al., 2012 in the order of several meters per year) and the distance to the saline fringes of the aquifer (more than 10 km), lateral intrusion of saline water into the well-field was considered unlikely within a 30-year time frame.

To establish the chance of the upward movement of saline water resulting in pumping of saline water, we extended the MODFLOW model with the SWI2 package, which allows simulation of a three-dimensional vertically integrated variable-density groundwater flow (Bakker et al., 2013). The SWI2 package incorporates density effects into MODFLOW-2005, without the need to solve a separate advective-dispersive transport equation. Furthermore, aquifers can be represented by a limited number of modeled layers. This results in substantial model run-time savings, making the SWI2 package very suitable for stochastic modeling, in comparison to more computationally intensive approaches, such as SEAWAT (Langevin et al., 2008).

A limitation of the numerical approach used in the SWI2 package is that it does not account for diffusion and dispersion (i.e., it takes a sharp-interface approach). These processes can lead to the formation of a transition zone along the fresh-saline groundwater interface (Paster and Dagan, 2008). The width of such a zone increases during the upconing process and can lead to increased salinity of the abstracted water, before the initial interface between fresh and saline water reaches the well (Diersch et al., 1984). The reported increase of salinity in wells over time in the Merti Aquifer indicates that such a transition zone may be present underneath the freshwater aquifer. For practical applications it is customary to limit the pumping rate, when sharp-interface calculations are used for design purposes (Diersch et al., 1984). A safety factor can be introduced to ensure that salinity of the pumped water does not increase. Following the analytical approach reviewed byDiersch et al. (1984), a safety factor of 0.4–0.6 was introduced to determine the threshold value for the upward movement of the fresh-saline water interface. Increased salinity in the abstracted water was assumed to occur when the fresh-saline interface has risen to 40–60% of the original depth below the well-field.

The probability of 30 years of pumping leading to increased salinity was determined by using the model described above, combined with Monte Carlo simulations. The mixing zone was simulated by introducing separate brackish and saline layers into the model. The depth of the saline water below the top of the aquifer, as well as the density of saline water, was varied in the Monte Carlo simulations, in addition to relevant aquifer properties and geometry. Aquifer properties used in the simulations are given in the supplementary materials (Supplementary Table 1).

RESULTS

The Decision Model

The modeling team members identified 14 separate types of benefits, 10 types of costs and 15 risk factors that they deemed important enough for inclusion in the model (Table 1). Risks were classified into five categories: some risk factors, such as the risk of political interference, could prevent the pipeline from being built, in which case the project would never produce any benefits. Project cancellation could also happen later, e.g., if wells run dry or water quality declines due to upconing of saline water. In these cases, benefit accrual was assumed to end as soon as the respective events occur. Benefits could also fail in some years only, e.g., if regional conflict disrupts operation of the pipeline temporarily. Such risks were captured by setting benefits to zero for individual years but allowing them to resume in the following year. In addition to risks that lead to permanent or temporary failure of the project to deliver benefits, there were also risk factors that merely reduced benefits. Such reductions could occur permanently, starting from the beginning of the project, if the water supply system is poorly designed. For other factors, reduction in benefits would vary from year to year, e.g., for poor maintenance and operation or illegal abstractions. A diagram showing how risks were considered is provided in the supplementary materials (Figure S3). The model had 102 variables for the "all risks" scenario, and 99 for the "cooperation" scenario (including risk, cost and benefit factors, as well as some technical model parameters).

Table 1: Relevant benefits, costs and risks of the pipeline project, as identified by members of the modeling team.

Benefits of the project:	Risks:
• Reduced infant mortality • Reduced costs for disease treatment • Higher productivity • Job creation • Higher investments • Reduced brain drain • Revenue from more demand for local products and services • Higher tax revenue • Reduced reliance on shallow wells • Sanitation benefits • Water during the dry season • General livelihood improvement • Revenue from water sales • Revenue from Payments for Ecosystem Services	Factors that lead to immediate project cancellation: • Negative feasibility report • Water yield too low • Inadequate benefit sharing • Political interference Factors that lead to cancellation later: • Wells run dry • Increasing water salinity • Oil development (raising the risks of wells running dry or turning saline) • Dam development (raising the risks of wells running dry or turning saline) Factors that cause temporary failure of project benefits: • Maintenance problems • Water price is too high • Regional conflict
Costs of the project:	
• Initial investment • Running costs and metering • Salaries • Repairs • Aquifer monitoring • Infrastructure maintenance • Pipeline security • Payments for ecosystem services • Costs for water purchases • Negative environmental impacts	Factors that reduce project performance from the beginning: • Poor project design Factors that temporarily reduce project performance: • Poor maintenance and operation • Illegal abstractions • Lack of cooperation

The general structure of the decision model is shown in Figure 3. It consists of all costs and benefits accruing to the various stakeholders and considers the impact of the risk factors on these costs and benefits. More detailed illustrations of model components that address the different items shown in gray in Figure 3 are provided in the supplementary materials (Figures S1–S3). The programming code in the supplementary materials (Section R Code used to run analyses) provides all equations of the model.

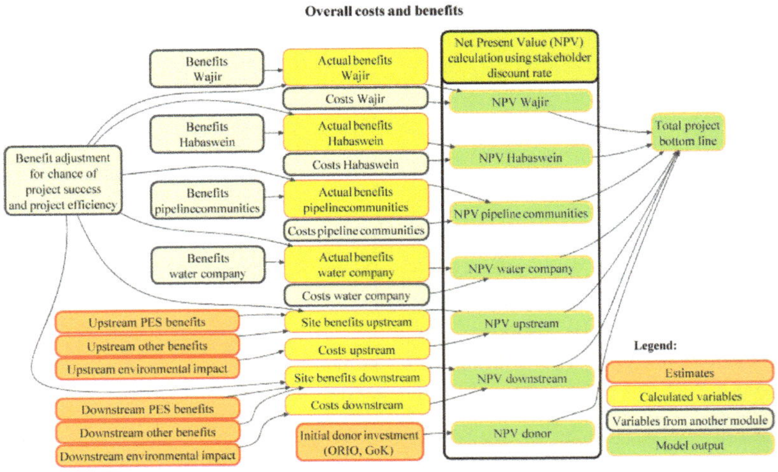

Figure 3: Overall structure of the decision model.

Projected Project Outcomes

Based on the 10,000 model runs for each of the two scenarios, probability distributions of likely outcomes for all stakeholders could be estimated (Figure 4). Members of the modeling team expected that water from the pipeline would reach between 80,000 and 300,000 people in Wajir, 20,000–100,000 in Habaswein, and 1500–50,000 along the pipeline (Supplementary Table 3). For all stakeholders, expected net present values varied widely, and in almost all cases, positive and negative outcomes were predicted, in both scenarios. In the "all risks" scenario (top panel of Figure4), chances were high (41.7%) that the pipeline would fail to materialize. Logically, financial outcomes for the donor consortium were negative for both scenarios, because the donors are not expected to benefit financially. They may of course still be interested in supporting the project, because of their interest in benefits to their target population. That such benefits would arise is not clear, however. In particular for the residents of Wajir, net benefits are not guaranteed, and negative outcomes emerged more frequently (34.0% of model runs) from the model runs than positive outcomes (24.3%; all remaining model runs produced neither net benefits nor net returns for Wajir, because the project failed). For Habaswein, the general prospects were more positive, with 51.3% positive outcomes and a 7.0% chance of net losses. Communities along the pipeline were also more likely to gain (43.0%) than to lose (15.3%). Effects on downstream and upstream areas were relatively small, but predominantly negative for downstream areas (95% chance of losses up to 80.4 million USD, and only a 2.9% chance of net gains up to 0.4 million USD), and balanced between positive and negative outcomes for upstream areas (49.7% chance of losses up to 0.5 million USD and 48.4% chance of gains up to 4.2 million USD). The water company faced substantial risks in the "all risks" scenario. Net losses emerged from model runs slightly more frequently (51.4%) than net gains (48.6%), though plausible gains were higher than plausible losses (as judged by the 10th and 90th percentiles, respectively, which were at a loss of 109.3 million USD and a gain of 447.7 million USD). Overall, negative outcomes from the project were slightly more probable (55.9%) than positive outcomes (44.1%), with total net present values likely between −192.3 and 302.2 million USD (the 10th and 90th percentiles of the distribution).

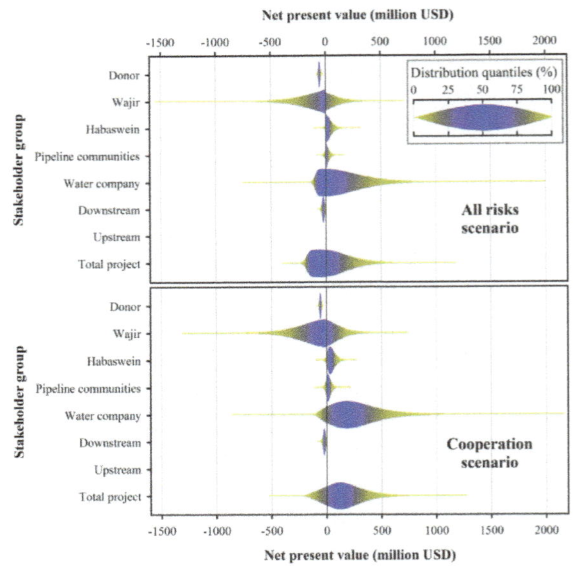

Figure 4: Projected probability distribution of net present value for all stakeholders and the project as a whole, considering a scenario that includes all project risks (top) and a scenario in which all political resistance has been overcome (bottom).

Assuming that political resistance can be overcome (the "cooperation" scenario) changed outcomes toward more positive prospects for most stakeholders (bottom panel of Figure 4). While the donor's outlook did not change, the chances of net gains for Habaswein, communities along the pipeline and the water company rose to 83.0, 70.3, and 79.0%, respectively. While a sizeable chance of net losses (21.0%) remained for the water company, there was a 75% chance of net gains of more than 26.3 million USD (and reaching 564.7 million USD at the 90th percentile). For the residents of Wajir, however, the chance of net losses from the project increased to 53.6%. These losses could be substantial, with a 25% chance of exceeding 176.6 million USD and a 10% chance of surpassing 348.2 million USD. Changes for downstream and upstream areas were relatively small. The overall outcome of the project looked more positive in the "cooperation" scenario, although a 25.9% chance of negative NPV remained. The most likely outcomes were between -115.9 million USD (the 10th percentile) and 404.8 million USD (the 90th percentile). Overall, simulation results showed that the potential net benefits, or net costs, of the water project could amount to a multiple of the sum that is to be invested.

Key Uncertainties

Analysis of the VIP scores revealed that the most influential variable introducing uncertainty in overall project outcome was the value attached to an additional surviving infant. This was the case for both the "all risks" scenario (Figure 5) and the "cooperation" scenario (Figure 6). VIP scores for this variable were considerably higher than for all other variables, flagging this as a key issue for further exploration. In both scenarios, the second-most important uncertainty was the risk of poor project design, followed by the number of additional surviving infants in Wajir, the largest community involved in the project. In the "all risks" scenario, political risks emerged as the next important issue, with the risks of political interference and inadequate benefit sharing also assigned high VIP scores. These risks were correlated with the degree to which Payments for Ecosystem Services (PES) are used to overcome political resistance, hence the PES rate was also found to correlate strongly with project outcomes. PES are transfer payments from the water users (i.e., Wajir or the water company) to stakeholders in upstream areas (i.e., upstream communities and Habaswein) as compensation for foregone opportunities or additional risks related to water abstraction. PES could also be used to compensate stakeholders downstream from Habaswein, but this was considered implausible by all experts. Other uncertainties of importance were again related to benefits from the project, namely the value of a disease treatment, the number of disease treatments that cleaner water would render unnecessary at Wajir and the number of additional surviving infants at Habaswein (and also for communities along the pipeline in the "cooperation" scenario). The discount rate used for the water company, and for the "cooperation" scenario also for the residents of Wajir, also featured in the list of important uncertainties. Further important uncertainties common to both scenarios were the risk of losses due to illegal abstractions, the risk of poor maintenance and operation, and salaries paid by the water company. In the "all risks" scenario, the initial investment by the donor, downstream environmental impacts and running costs of the water company also emerged as important uncertainties. Comparing the two scenarios showed that most of the key uncertainties were similar, even though projected outcome distributions differed substantially (insets in Figures 5, 6). Adding risks of political resistance to the projections strongly reduced the chance of positive project outcomes.

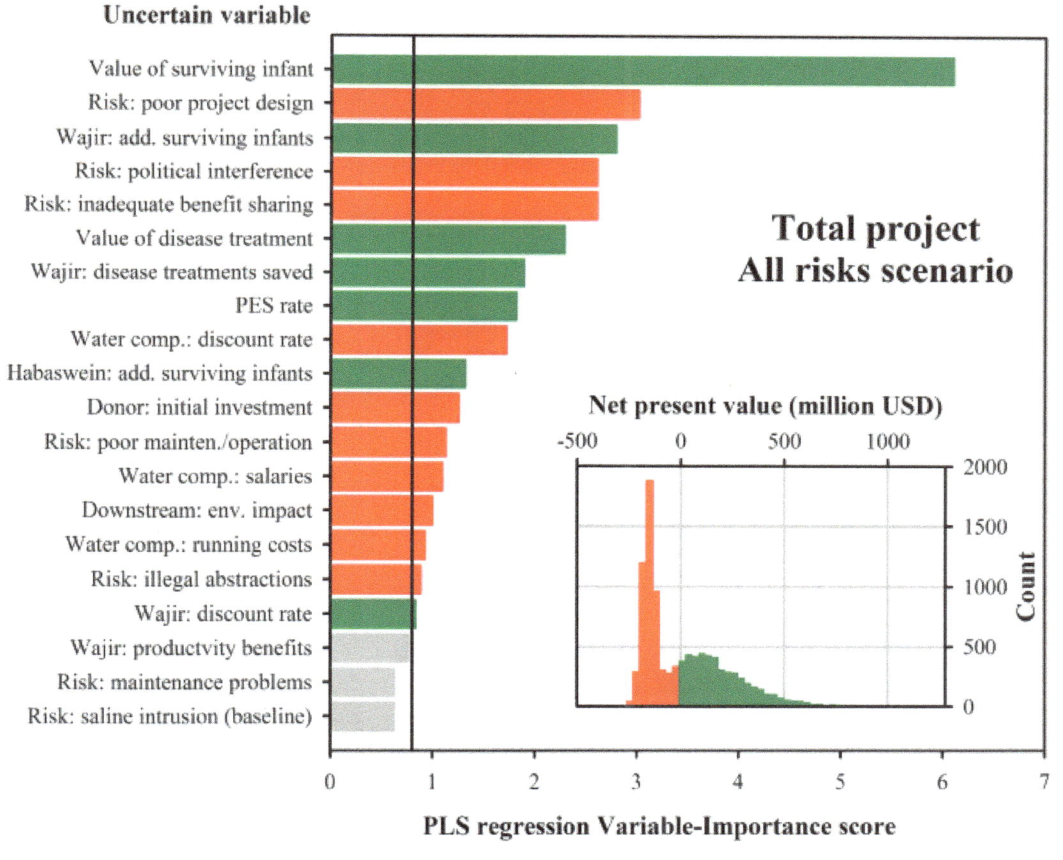

Figure 5: Key sources of uncertainty in projecting outcomes of the overall project for the "all risks" scenario, as identified by Partial Least Squares (PLS) regression. Bars show the Variable-Importance-in-the-Projection statistic, with high values indicating that uncertainty about the variable is a major source of uncertainty in model outputs. Green bars in the VIP plot indicate that the variable is positively correlated to outcomes, while red bars signify a negative correlation. The inset shows projected project outcomes as the distribution of net present value among 10,000 model runs of a Monte Carlo simulation. In this plot, red bars indicate negative values, and green bars signify positive outcomes.

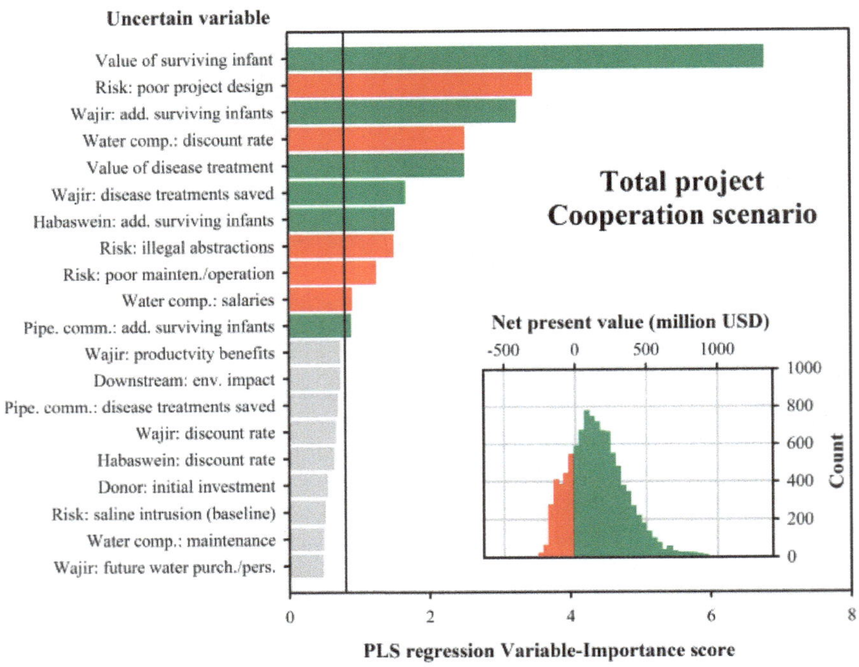

Figure 6: Key sources of uncertainty in projecting outcomes of the overall project for the "cooperation" scenario, as identified by Partial Least Squares (PLS) regression. For a detailed description of the plots, see legend to Figure 5.

Besides prospects for the entire project, we also evaluated likely outcomes for different stakeholders. For all major stakeholders, political risks, in particular the risks of political interference and inadequate benefit sharing, featured strongly among the key uncertainties. When assuming that these risks can be overcome, prospects for the residents of Wajir depended primarily on the economics of water sales, with future water purchases per person and future water price emerging as key uncertainties (Figure 7). The values and numbers of additional surviving infants and disease treatments came next in term of importance, followed by the water revenue trend for the water company, which closely reflects water price developments, and the risk of losses due to poor project design.

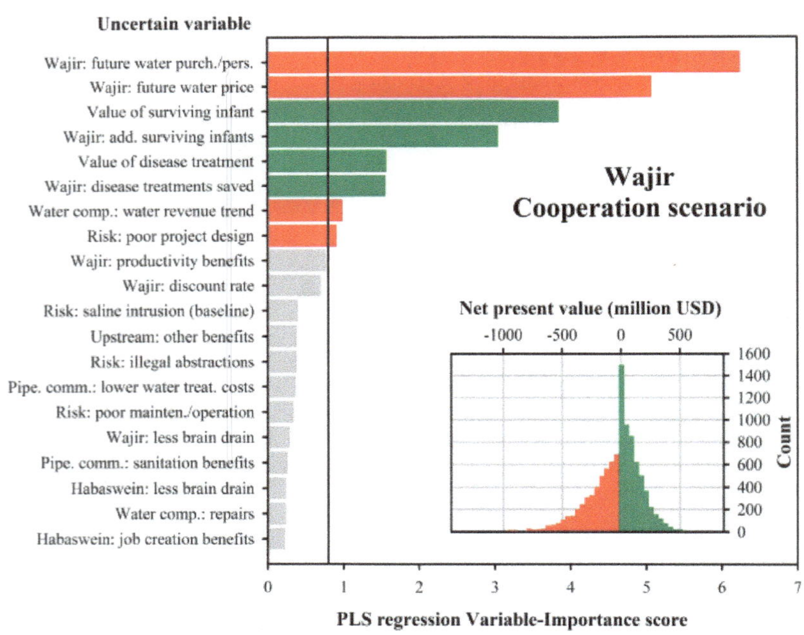

Figure 7: Key sources of uncertainty in projecting outcomes for the residents of Wajir for the "cooperation" scenario, as identified by Partial Least Squares (PLS) regression. For a detailed description of the plots, see legend to Figure 5.

Key uncertainties for the residents of Habaswein were the value and number of additional surviving infants, followed by the amount of water purchased per person in the future (Figure 8). Number and value of disease treatments rendered unnecessary, as well as the discount rate applied for the residents of Habaswein followed in terms of importance. The risk of losses due to poor project design, the future water price and the number of water users were also identified as key uncertainties.

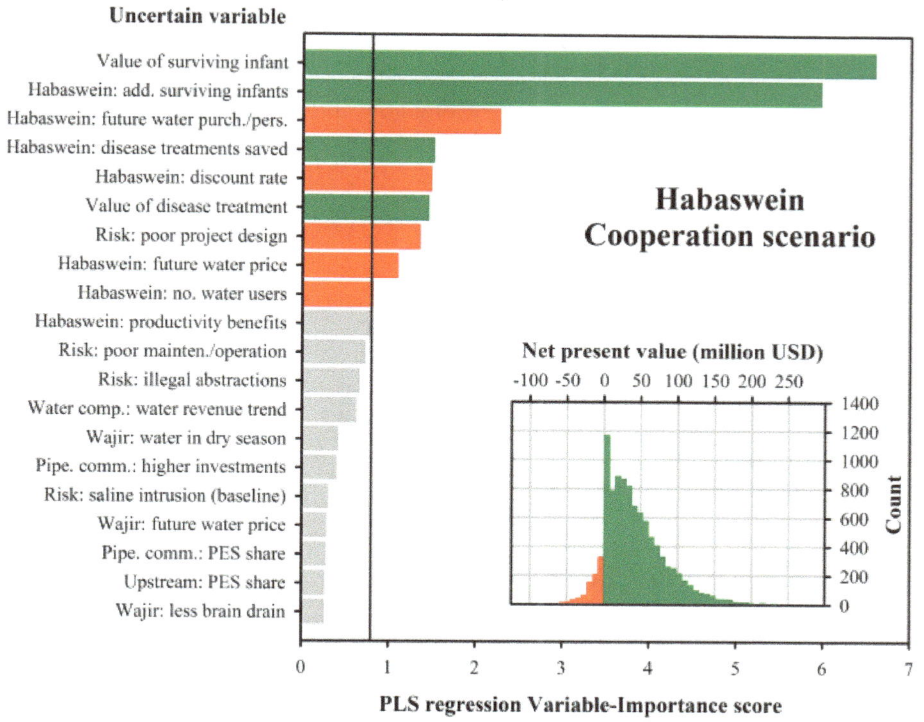

Figure 8: Key sources of uncertainty in projecting outcomes for the residents of Habaswein for the "cooperation" scenario, as identified by Partial Least Squares (PLS) regression. For a details description of the plots, see legend to Figure 5.

For the water company, key uncertainties revolved around the core of their business operations, placing particular importance on the future amount of water purchased per person in Wajir, as well as the future water price there (Figure 9). The discount rate applied by the company was also a major source of outcome uncertainty. Further uncertainties of importance were the risk of poor project design, the water revenue trend, water purchases per person in Habaswein and salaries paid by the water company. The risk of saline intrusion into the aquifer and losses from illegal abstractions were also key uncertainties. It is important to note that while certain variables like the water price and the amount of water purchased per person emerged as key uncertainties for both the water company and residents of Wajir, Habaswein and the communities along the pipeline, correlations between these variables and project outcomes were of opposite signs. While, e.g., a high water price was related to positive project outcomes for the water company, it led to negative outcomes for those purchasing the water.

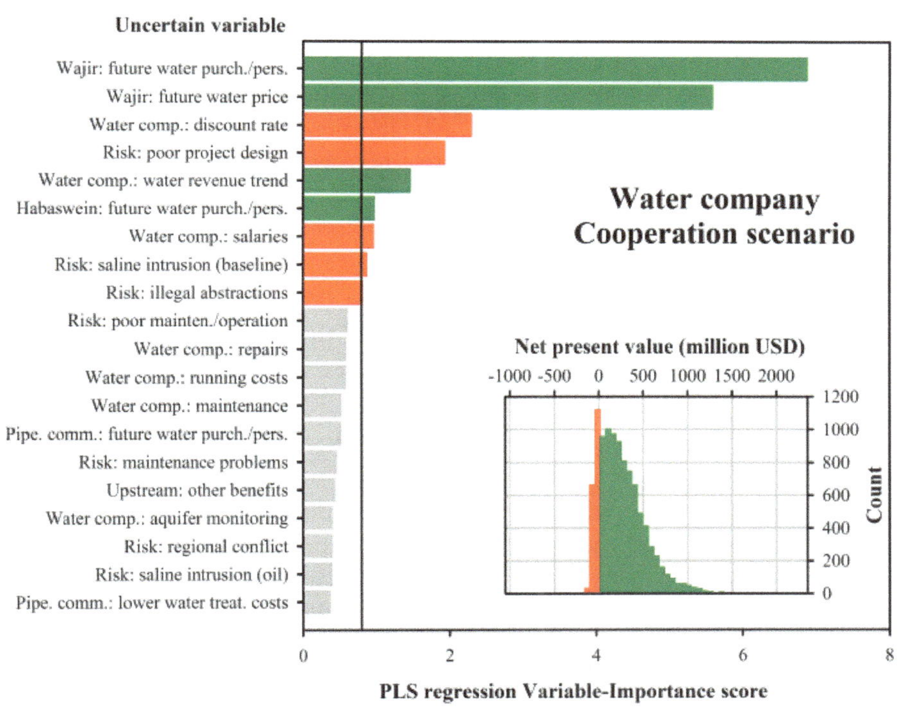

Figure 9: Key sources of uncertainty in projecting outcomes for the water company for the "cooperation" scenario, as identified by Partial Least Squares (PLS) regression. For a details description of the plots, see legend to Figure 5.

Hydrogeological Modeling

Based on Monte Carlo simulations using the Theis equation, plausible drawdowns after a 30-year period ranged from 5.4 m (10th percentile) to 13.7 m (90th percentile) with a median drawdown of 8.0 m in a well field using 12 wells (Figures 10A,B). In a well field using 14 wells, drawdown ranged from 5.2 m (10th percentile) to 13.3 m (90th percentile) with a median of 7.8 m. The difference between the 12 and 14 well design was relatively small (Figures 10C,D). Due to the small distance between wells, the total drawdown at a well was mainly caused by the combined effect of surrounding wells. From the probability-based calculations we concluded that the chance of wells running dry after 30 years was 15–30%, assuming that 10–12 m drawdown would cause wells to fall dry. The chance of failure due to deteriorating water quality was 25–55%, depending on the safety factor for critical uprising (0.4–0.6) and the number of wells (12–14).

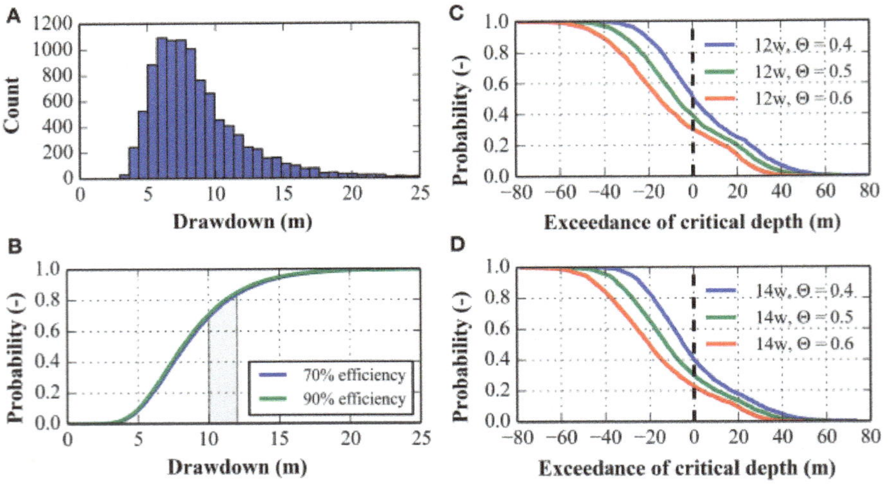

Figure 10: Results of the hydrogeological assessment. (A) Histogram of calculated drawdown values, (B) cumulative probability distribution of calculated drawdown, **(C)**cumulative probability distribution of the fresh-saline water interface exceeding the critical depth in a field of 12 wells, using different safety factors (Θ)—positive values represent exceedance of the threshold, **(D)** cumulative probability distribution of the fresh-saline water interface exceeding the critical depth in a field of 14 wells.

DISCUSSION

The Winners and the Losers

Contrary to the expectations of most workshop attendees, the greatest net losses were likely to accrue at Wajir, while residents of Habaswein appeared likely to gain. Stakeholders from Habaswein especially had previously voiced concerns about the proposed abstractions, which they felt could put their currently reliable water supply at risk. Residents of Wajir appeared to be more positively disposed toward the project, even though nuanced views were also recorded there (Ogalleh, 2015). Among the projected winners were also the communities along the pipeline, the water company and upstream areas, while downstream regions and—unsurprisingly—the donor were among the likely losers. The distribution of projected outcomes of the overall project looked substantially more favorable in the "cooperation" scenario than in the "all risks" scenario, with only Wajir having greater losses when political resistance could be overcome. This shows the benefits of including all stakeholders in project planning and ensuring fair benefit sharing among affected parties, which is likely to dissipate resistance against the intervention. Such resistance was very prevalent at the time of the study, but it became clear from site interviews that this was often based on very limited information (Ogalleh, 2015). Measures to build awareness and disseminate information might help overcome this situation. The predominantly negative outcomes for Wajir were an unexpected finding that is explained to a large degree by the expectation that residents there will have to pay higher prices for their future water supply than they are paying at present. This may lead to financial losses for many residents. It should be considered, however, that the current water supply system is unlikely to be sustainable in the long run because of rising population and increasing contamination of the shallow aquifer. Continuation of these trends may even lead to disease epidemics or famines, which can only be avoided if new sources of freshwater are secured. Preserving the status quo may therefore not be a viable option. In hindsight, it would have been prudent for the model building team to explicitly include into the model the risk of crises related to food security or public health in the event that no additional water sources become available.

Key Uncertainties and Implications for Measurements and Project Design

For almost all stakeholders, it remained unclear whether they would gain or lose from a pipeline. This was due to a number of key uncertainties that emerged from the variable importance calculations. One set of uncertainties concerned benefits expected to arise from the project. Probably the most controversial among these are related to reduced infant mortality due to the provision of cleaner water. At present, official statistics on infant mortality in Wajir list 9.8% as the countywide figure (www.wajirgovernment.com). Better sanitation and cleaner water would almost certainly reduce this number. While uncertainty about the number of additional infants that would survive because of the project could probably be reduced through some targeted study, the value judgment of what expenses are justified for saving the life of an infant is very controversial. A number of studies have used human behavior to derive such values, basing estimates of the value of a life on people's willingness to invest in measures to reduce their chance of death (Dardis, 1980;Hubbard, 2014). Premiums paid to workers in risky professions have also been used (Viscusi and Aldy, 2003). In most cases, such studies result in estimates in the millions of USD. Applying these techniques in Wajir, however, is quite difficult because the desire for a long life that many people may have is not backed by money, i.e., people do not have resources to invest in greater safety. Focusing specifically on Africa, Deaton et al. (2009) used a number of techniques and lines of evidence to argue that the perceived value of a life in Africa is very low, although they shy away from providing concrete numbers. It is also worth keeping in mind that the uncertain variable in question refers to the valuation of a life by the decision maker, rather than the people affected. The value a decision maker attaches to a human life must be decided on in the context of scarce resources and competing objectives. Many governments have undergone a formal process to define the value of a life, leading U.S. Agencies to work with figures between 2 and 20 million USD (Hubbard, 2014). Multiplying values of a Disability-Adjusted Life Year (DALY) defined in the Copenhagen Consensus, which aimed at evaluating the cost of terrorism (Sandler et al., 2008), by the average life expectancy in Kenya results in values between approximately 60,000 and 300,000 USD for the value of a statistical life. Even such values seemed high to members of the modeling team, because, even though many premature deaths in northeastern Kenya could be avoided with relatively little effort, this effort is often not made, because funds are lacking or needed elsewhere. The resulting range of estimates (200–100,000 USD) is thus considerably lower than what has been used in other contexts. This will make further reduction of this range difficult, unless it could rely on a formal definition of the value of a life by the decision-makers themselves.

Other uncertain variables are more amenable to reductions of uncertainty. The value of a disease treatment, for example, should be relatively easy to measure, and epidemiological studies should be able to provide good estimates of the number of treatments that would no longer be necessary with a reliable clean water supply. It should also be possible to reduce uncertainty about future water prices and the number of water users, which could be derived from a comprehensive financial feasibility study undertaken for the water company. Such a study has not yet been done, explaining the prominence of items related to water business economics among the uncertain variables. Risk factors surrounded by important uncertainties were primarily the benefit losses due to poor project design and the risk of political interference. These risks could be measured by investigating in detail the project plans and gaining clarity on the stance different stakeholders have taken toward the project, but they can also be influenced by improvements to the proposed intervention. Adequate investment in project planning, including thorough risk assessment and measures to safeguard against risks wherever possible would improve likely project outcomes. Similarly, a more inclusive process during the planning stage, which considers the interests of all stakeholder groups and ensures fair sharing of benefits, could help overcome political resistance. Even though such measures would naturally require additional investments during the planning phase, they would likely prove cost-effective given the potential of these risk factors to undermine project success.

Hydrogeology

The initial intuitive stance of stakeholders for or against the project appeared to be determined mainly by stakeholders' perceptions of hydrogeological risks. Wajir looked to gain water, hence supported the project, while Habaswein would be giving up part of "their" water, making them take an opposing stance. While such a focus on water is understandable in the arid northeast of Kenya, stakeholders seemed to overlook other important issues, such as the viability for the water company of the business case for selling piped water or the risk of political interference. These emerged as the greater risks, and also as those whose magnitudes were least well known. This is not to say that hydrogeological risks were not important. In fact, there was a substantial chance of saline water upconing (25–55%), and also the chance of wells running dry was not negligible (15–30%), as indicated by the probabilistic simulation over the entire 30-year time frame. The reason why hydrogeological risks did not feature strongly in the important variable lists for any stakeholders (except the water company in the "cooperation" scenario) was that a small body of research already existed on this topic, and more work was undertaken in the course of the present study. These previous studies provided greater certainty about project hydrogeology than about many other project aspects that had not been studied. Uncertainty about aquifer properties was still high, but much lower than for some of the other variables, for which virtually no data was available.

Negative hydrological effects could probably be overcome by drilling additional wells and risks could be mitigated by choosing an appropriate well field design, the costs of which would be justified in light of the large potential benefits of the project. Measurements about the thickness of the freshwater layer which would allow to more precisely assess the risk of upconing saline water and careful long-term monitoring of the aquifer are also absolutely recommended. It should be noted that there is currently no basis for ruling out adverse consequences for the long-term water supply stability at Habaswein. We therefore recommend more specific research on this subject.

Regarding the hydrogeological modeling undertaken in the present study, the use of simplified stochastic-deterministic models proved a practical way to assess the risks as a result of hydrogeological limitations in a data-scarce environment. Given current uncertainties about aquifer characteristics, uncertainty in the estimated risks was relatively low. These estimates highlighted a mismatch between stakeholder perception and model outputs: the risk of drying wells was the stakeholders' main hydrogeological concern, while the model showed a higher risk of upconing and deteriorating water quality. The probability-based approach greatly enhanced understanding of the aquifer's response to abstraction. The relative importance of parameters can easily be evaluated for different parameter ranges, which can help in planning field-research and support the technical design, even in data-scarce environments. Such considerations revealed a higher annual probability for the risk of saline water upconing (1.0–2.5% per year) than for wells running dry (0.5–1.2%). While the annual risks may seem low, they correspond to risks of 25–55% and 15–30% over the 30 year time frame considered by the model.

Outcomes of the Decision Modeling Process

The decision modeling process clearly enhanced stakeholder understanding of the proposed project, as stated by several workshop participants. Regardless of the results from the model runs, delineating the project's impact pathway for various stakeholders, reflecting on the risks that the project faces, and making estimates for uncertain variables improved the team members' capacity to evaluate the pipeline plans. Attendees of the larger stakeholder workshops, including an outreach workshop in Isiolo in September 2014, also gained such understanding. Although this did not go quite as far as for those directly involved in the model building, several workshop participants notably changed their attitudes toward the project. One senior water management professional stated during the final workshop that exposure to model results had made him more skeptical toward the project. On the other hand, an email survey of workshop attendees revealed that the views of several people became more favorable toward the pipeline. An evaluation survey of the local stakeholders in a recent workshop (Ogalleh, 2015) also found that exposure to decision analysis results made workshop participants more favorable toward the project. When informed about identified risks to the project during the workshop, participants also advised to explore alternative water supply options for Wajir, which might be less risky than the pipeline from Habaswein. Whether these alternatives, which included a greater focus on rainwater harvesting and tapping another aquifer to the north, are feasible is difficult to judge without further investigation, but we were encouraged by the reemergence of a debate about solutions to Wajir's water problems, which might eventually put more options on the table. We are thus confident that our approach has enriched the project planning process and raised the chances of positive outcomes of whichever project design is ultimately chosen.

Methodological Limitations

There is no guarantee that the model accurately captured all important aspects of the project. The development of the model was driven by stakeholders, and most of them had had no formal training in anticipating the impacts of development interventions. Nevertheless, the model that emerged appeared quite comprehensive and included variables that are easily overlooked when technical experts are charged with model development. Examples of such variables are the risk of political interference or the benefits arising from higher demand for local services. Project-related studies that had been commissioned before our modeling activities had focused exclusively on hydrogeological risks. They had not paid much attention to project economics or environmental, social and health impacts of the project, which our results highlighted as priorities. We are not suggesting that a stakeholder-driven modeling approach can replace technical studies, but it can complement them and provide guidance on what types of technical studies are most urgently needed.

Another concern about the approach was the ability of stakeholders to accurately estimate probability distributions for uncertain variables. While all those who provided estimates received calibration training, many variables remained difficult to fathom. Estimators were encouraged to provide wide ranges whenever they were highly uncertain, but they may still have been wrong in some cases. For some variables, estimated ranges by different experts did not overlap, which means that at least one of the experts was wrong. We are confident that the consolidated estimates that were distilled from the different expert assessments were accurate in most cases, but with the large number of uncertain variables in the model (102 for the "all risks" scenario and 99 for the "cooperation" scenario), chances are high that some of them may have been wrong. It is difficult to evaluate the implications of such erroneous estimates.

The difficulty of evaluating the implications of potentially missing variables and inaccurate estimates partly arises from the challenge of validating our model. Our decision model concerns a situation that has not arisen yet, so that no data for validation is available. But even in the future, after project impacts have unfolded, validation will remain difficult because of the probabilistic nature of our model outputs. There is a wide range of outcomes that could be realized without falsifying our model. It is of course possible to determine in retrospect how probable a realized project outcome would have been according to *ex-ante* projections, but this can only give a weak indication of model validity, and it is only available after the project has been implemented. In this context, however, it is important to recall the objective of the modeling process: to improve the decision-making process over what is currently practiced. At present, development decisions are often largely based on "gut feelings" or use of data that addresses only part of the decision problem. In contrast, we followed a structured analysis process, which helped stakeholders and decision-makers make their expectations and assumptions about the project's impact pathway explicit. Using such a model should thus constitute an improvement over current processes, even though—as every model—it remains an imperfect and possibly incomplete reflection of reality.

Implications for Development Planning

Many development decisions are made without structured *ex-ante* analysis of decision outcomes. We are therefore convinced that the methods used in this study, or similar approaches, can contribute greatly to improving decision-making processes and raising the returns on investments in development projects. Even with such methods, of course, making decisions in uncertain and data-scarce environments will remain risky and difficult. But holistic impact projections for the various decision alternatives can improve the robustness and transparency of decision-making processes. It can make use of both available data and expert knowledge, and it can provide guidance on what measurements are most urgently needed for improving a particular decision. The approach used here is a structured and unbiased method to prioritize those variables most urgently requiring additional measurement. Not all variables will likely be amenable to reduction of uncertainty, and other arguments such as feasibility of further measurement will often guide the selection of candidate variables for further research. However, initiating this process with a structured and unbiased approach opens opportunities for spending resources on variables that matter most. While more work undoubtedly needs to be done to adjust decision analysis methods to the context of multi-stakeholder decisions in the developing world, these methods open opportunities for the kind of holistic decision-support analysis that is needed for enhancing the outcomes of development initiatives.

CONCLUSIONS

The town of Wajir needs to improve its water supply to ensure acceptable standards of living for its growing number of residents. Results from our study indicate that better water will not be free, and may not even produce net benefits, but since the current situation may not remain viable in the future, there is little alternative to innovation in the water sector. In this context, the proposed project may be acceptable to the county government of Wajir, even if it leads to additional expenses for the residents of the county's capital.

Regardless of these considerations, certain measurements would increase certainty about project outcomes. The financial viability of operating the water supply system should be clarified, because it is essential for the sustainability of the investment. Information on project benefits should also be collected, including better appraisal of the number of additional infants that can be expected to survive and reductions in disease treatment costs due to better water quality. Regarding technical aspects of the project, careful project design emerged as a key priority, along with the necessity to implement a fair benefit sharing mechanism that ensures that all stakeholders receive benefits and reduces the risk of political interference. Our results indicate that this should be possible, because plausible negative impacts in downstream and upstream areas, in Habaswein and the communities along the pipeline are relatively small compared to the profits that could be reaped from running the system. Payments for Ecosystem Services and financial mechanisms to compensate those who expect net losses from the project may thus be a promising approach for breaking the resistance of important stakeholders. Investing in such measures may well pay off, since it greatly raises the chance of positive project outcomes. Regarding hydrogeological aspects of the project, a few deep test drills to determine the thickness of the freshwater aquifer should be undertaken before the project is approved, and well design should be optimized to reduce the risk of both saline water upconing and wells running dry. Lastly, it will be critical to carefully monitor the aquifer, so that adverse developments can be detected early and undesirable processes halted before they become irreversible.

Overall, the participatory modeling process and the probabilistic model runs based on uncertain information produced very encouraging results. Stakeholders took the study results very seriously and many changed their expectations about project outcomes and desirable measurements to be taken. This provided evidence that inclusion of decision-makers and other stakeholders in the research enhances the likelihood that study results will be given full consideration in development processes. It also showed that even with imperfect knowledge about a system, it is possible to provide useful guidance for project development, if uncertainties are fully considered in project impact projections. We conclude that the Applied Information Economics approach can effectively be applied beyond single-stakeholder business decisions. It can make a valuable contribution to supporting risky decisions in development, in the face of imperfect information and considering the potentially conflicting interests of multiple stakeholders.

CONFLICT OF INTEREST STATEMENT

The authors declare that the research was conducted in the absence of any commercial or financial relationships that could be construed as a potential conflict of interest.

ACKNOWLEDGMENTS

We acknowledge funding by UKAID (NERC UPGRO grant no NE/L00206X/1: Assessing Risks of Investment in Groundwater Resources in Sub-Saharan Africa) and support from the CGIAR Research Program on Water Land and Ecosystems (WLE). Furthermore, we are grateful to all members of the modeling team and to all participants of our workshops for invaluable insights into the Habaswein-Wajir Water Supply Project. We thank Lieselotte Tolk, Katherine Homewood, Alex Oduor, Evanson Njuguna, Benson Macharia and Elsabijn Koelman for their contribution during manuscript preparation. We are also grateful to Douglas Hubbard and Matthew Millar of Hubbard Decision Research for guidance on decision modeling.

SUPPLEMENTARY MATERIAL

The Supplementary Material for this article can be found online at:http://www.frontiersin.org/journal/10.3389/fenvs.2015.00016/abstract

REFERENCES

1. Apaydın, A. (2012). Dual impact on the groundwater aquifer in the Kazan Plain (Ankara, Turkey): sand–gravel mining and over-abstraction. *Environ. Earth Sci.* 65, 241–255. doi: 10.1007/s12665-011-1087-8
2. Bakker, M., Schaars, F., Hughes, J. D., Langevin, C. D., and Dausman, A. M. (2013). *Documentation of the Seawater Intrusion (SWI2) Package for, MODFLOW.* Reston, VA: USGS.
3. BAM International, G. I., BAM, Balteau, and Vitens Evides, International. (2012). *Development Phase Wajir/ Habaswein Water Supply Project: Hydrological Report (Revision 1).* Ministry Of State For Development Of Northern Kenya And Other Arid Lands, The Hague.
4. Bosworth, W., and Morley, C. K. (1994). Structural and stratigraphic evolution of the Anza rift, Kenya. *Tectonophysics* 236, 93–115.
5. Clapp, A., DauSchmidt, N., Millar, M., Hubbard, D., and Shepherd, K. (2013). *A Survey and Analysis of the Data Requirements for Stakeholders in African Agriculture.* World Agroforestry Centre (ICRAF), Nairobi.
6. Dardis, R. (1980). The value of a life: new evidence from the marketplace. *Am. Econ. Rev.* 70, 1077–1082.
7. Deaton, A., Fortson, J., and Tortora, R. (2009). *Life (Evaluation), HIV/AIDS, and Death in Africa.* Cambridge, MA: National Bureau of Economic Research.
8. De Leeuw, J., Said, M. Y., Kifugo, S., Musyimi, Z., Mutiga, J. K., and Peden, D. (2012). Benefits of riverine

water discharge into the Lorian Swamp, Kenya. *Water* 4, 1009–1024. doi: 10.3390/w4041009

9. Diersch, H.-J., Prochnow, D., and Thiele, M. (1984). Finite-element analysis of dispersion-affected saltwater upconing below a pumping well. *Appl. Math. Model.* 8, 305–312.

10. GIBB Africa Ltd. (2004). *UNICEF Kenya Country Office - Study of the Merti Aquifer - Technical Report ISsue 2.0.*

11. Harbaugh, A. W. (2005). *MODFLOW-2005, the US Geological Survey Modular Ground-Water Model: The Ground-Water Flow Process*. Reston, VA: US Department of the Interior, US Geological Survey.

12. Hubbard, D. W. (2014). *How to Measure Anything - Finding the Value of Intangibles in Business*. Hoboken, NJ: Wiley and Sons.

13. Jay, S., Jones, C., Slinn, P., and Wood, C. (2007). Environmental impact assessment: retrospect and prospect. *Environ. Impact Assess. Rev.* 27, 287–300. doi: 10.1016/j.eiar.2006.12.001

14. Klein, G. (2007). Performing a project premortem. *Harv. Bus. Rev.* 85, 18. Available online at:https://hbr.org/2007/09/performing-a-project-premortem

15. Kramon, E., and Posner, D. N. (2011). Kenya's new constitution. *J. Democracy* 22, 89–103. doi: 10.1353/jod.2011.0026

16. Kruseman, G. P., and de Ridder, N. A. (1990). *Analysis and Evaluation of Pumping Test Data*. Wageningen: International Institute for Land Reclamation and Improvement Publication 47.

17. Langevin, C., Thorne Jr, D., Dausman, A., Sukop, M., and Guo, W. (2008). *SEAWAT Version 4: A Computer Program for Simulation of Multi-Species Solute and Heat Transport*. Reston, VA: US Geological Survey.

18. Luedeling, E., and Gassner, A. (2012). Partial least squares regression for analyzing walnut phenology in California. *Agric. Forest Meteorol.* 158, 43–52. doi: 10.1016/j.agrformet.2011.10.020

19. Mumma, A., Lane, M., Kairu, E., Tuinhof, A., and Hirji, R. (2011). *Kenya: Groundwater Governance Case Study*. Washington, DC: The World Bank Group. Available online at: http://documents.worldbank.org/curated/en/2011/06/16583819/kenya-groundwater-governance-case-study

20. Ogalleh, S. (2015). *ARIGA Social Survey Report*. Centre for Training and Integrated Research in ASAL Development (CETRAD), Nanyuki.

21. Oord, A. L. (2011). Droogte in de Hoorn van Afrika Ontwikkeling van de Merti aquifer in noordoost Kenia. *Stromingen* 16, 5–13. Available online at: http://edepot.wur.nl/314538

22. Osman, A. D. (2012). *Groundwater Quality in Wajir (Kenya) Shallow Aquifer: An Examination of the Association Between Water Quality and Water-Borne Diseases in Children*. Ph.D. thesis, La Trobe University, Bundoora. Available online at:http://hdl.handle.net/1959.9/308716

23. Paster, A., and Dagan, G. (2008). Mixing at the interface between two fluids in aquifer well upconing steady flow. *Water Resour. Res.* 44:W05408. doi: 10.1029/2007WR006510

24. Peterman, R. M., and Anderson, J. L. (1999). Decision analysis: a method for taking uncertainties into account in risk-based decision making. *Hum. Ecol. Risk Assess.* 5, 231–244.

25. R Development Core Team. (2011). *R: A Language and Environment for Statistical Computing*. Vienna: R Foundation for Statistical Computing.

26. Regan, H. M., Ben-Haim, Y., Langford, B., Wilson, W. G., Lundberg, P., Andelman, S. J., et al. (2005). Robust decision-making under severe uncertainty for conservation management. *Ecol. Appl.* 15, 1471–1477. doi: 10.1890/03-5419

27. Reilly, T. E., Franke, O. L., and Bennett, G. D. (1984). *The Principle of Superposition and its Application in Ground-Water Hydraulics*. Reston, VA: US Geological Survey.

28. Rosenstock, T. S., Mpanda, M., Rioux, J., Aynekulu, E., Kimaro, A. A., Neufeldt, H., et al. (2014). Targeting conservation agriculture in the context of livelihoods and landscapes. *Agric. Ecosyst. Environ.* 187, 47–51. doi: 10.1016/j.agee.2013.11.011

29. Sandler, T., Arce, D. G., and Enders, W. (2008). *Copenhagen Consensus 2008 Challenge Paper: Terrorism*. Copenhagen: Copenhagen Consensus Center. Retrieved July, 29: 2011.

30. Swarzenski, W. V., and Mundorff, M. J. (1977). *Geohydrology of North Eastern Province*. Washington, DC: US Geological Survey Water-Supply Paper 1757-N.

31. Theis, C. V. (1935). *The Relation between the Lowering of the Piezometric Surface and the Rate and Duration of Discharge of a Well Using Ground Water Storage*. Washington, DC: US Department of the Interior, Geological Survey, Water Resources Division, Ground Water Branch.

32. Viscusi, W. K., and Aldy, J. E. (2003). The value of a statistical life: a critical review of market estimates throughout the world. *J. Risk Uncertain.* 27, 5–76. doi: 10.1023/A:1025598106257

33. Wold, S. (1995). "PLS for multivariate linear modeling," in *Chemometric Methods in Molecular Design: Methods and Principles in Medicinal Chemistry*, ed H. van der Waterbeemd (Weinheim: Verlag-Chemie), 195–218.

34. Wold, S., Sjostrom, M., and Eriksson, L. (2001). PLS-regression: a basic tool of chemometrics. *Chemom. Intell. Lab. Syst.* 58, 109–130. doi: 10.1016/S0169-7439(01)00155-1

CHAPTER 3

GRAPHENE OXIDE: A PROMISING NANOMATERIAL FOR ENERGY AND ENVIRONMENTAL APPLICATIONS

Fen Li, Xue Jiang, Jijun Zhao, Shengbai Zhang

Keywords: Graphene oxide, Hydrogen energy, Lithium battery, Supercapacitor, Gas capture, Water purification

INTRODUCTION

As the global concerns in the development of human civilization, the scientific and technological issues of energy utilization and environment protection are currently facing challenges. Nowadays, enormous energy demands of the world are mainly met by the nonrenewable and environmental unfriendly fossil fuels. To replace the conventional energy platform, pursuit of the renewable and clean energy sources and carriers, including hydrogen storage, lithium batteries, and supercapacitors, has become extremely urgent. Additionally, long-term industrial and agricultural activities induce serious environmental pollution (such as greenhouse and toxic gases, heavy metal ions and organic species) in air and water to deteriorate the ecological balance and the daily human health.

GO is a monolayer of graphite oxide, which can be obtained by exfoliating graphite oxide into layered sheets through sonicating or mechanical stirring [1]. The graphene-based lattice and existence of various oxygen-containing groups (mainly epoxy and hydroxyl groups) enable GO abundant fascinating properties. First, the functional groups on GO surface act as effective anchoring sites to immobilize various active species. Furthermore, GO possesses tunable electronic properties. Typically, GO is insulating due to the large portion of sp^3 hybridized carbon atoms bonded with the oxygen-containing groups, which results in a sheet resistance of ~$10^{12}\ \Omega\ sq^{-1}$ or higher [2]. However, after reduction, the sheet resistance of reduced GO (namely, RGO) can be degraded by several orders of magnitude, hence transforming the material into a semiconductor or even into a graphene-like semimetal [3]. It has been demonstrated that band gap of GO can be tailored by controlling coverage, arrangement, and relative ratio of the epoxy and hydroxyl groups [4], [5], [6], [7] and [8].

Besides, GO also displays excellent optical and mechanical properties for a wide landscape of applications. The optical transmittance of GO films can be continuously tuned by varying the film thickness or the extent of reduction [9]. Generally, a suspension of GO films in water is dark brown to light yellow, depending on the concentration, whereas that of reduced graphene oxide (RGO) thin films (with a thickness less than 30 nm) is semitransparent [10]. The optical absorption of GO is dominated by the π–π transitions, which typically give rise to an absorption peak between 225 and 275 nm (4.5–5.5 eV). During reduction, the strength of optical absorption increases while the plasmon peak shifts to ~270 nm, reflecting an increased π-electron concentration and structural ordering [11]. Usually, the mechanical properties of GO rely on the details of sample, such as the oxidation degree (especially coverage of the epoxy and hydroxyl groups) and thickness [12], [13], [14] and [15]. The reported Young's modulus and intrinsic strength of GO sheets show a wide range of distributions of 6–42 GPa and 76–293 MPa, respectively [16]. More details about the fundamental physical properties of GO can be found in a recent book and a review article [1] and [17].

Citation: Fen Li, Xue Jiang, Jijun Zhao, Shengbai Zhang, Graphene oxide: A promising nanomaterial for energy and environmental applications, Nano Energy, Volume 16, September 2015, Pages 488-515, ISSN 2211-2855, http://dx.doi.org/10.1016/j.nanoen.2015.07.014.

Despite the aforementioned fascinating properties, there are still some drawbacks of GO for practical applications. The combination of structural defects, poor dispersion, restacking and multilayer thickness can affect the electrical properties and high surface area of GO materials [18]. The insulating nature of regular GO also limits its applications in electronic devices and energy storage. Furthermore, the residual defects and holes degrade the electronic quality of RGO [19] and [20]. Fortunately, the oxygenated groups can largely expand the structural/chemical diversity of GO by further chemical modification or functionalization, which offer an effective way to tailor the physical and chemical properties of GO to expected extents. As a consequence, GO and GO-based composites have shown great potentials in the applications of energy storage/conversion and environment protection. Figure 1 shows the numbers of journal publications searched by ISI with some relevant keywords. One can see that there have been tremendous efforts in developing GO-based materials for various kinds of Li batteries and supercapacitors, whereas there are also certain activities on hydrogen generation/storage as well as purification of water and air by using GO-based materials. This review aims to highlight the recent progresses on the energy and environmental applications of GO/RGO-based materials and their composites.

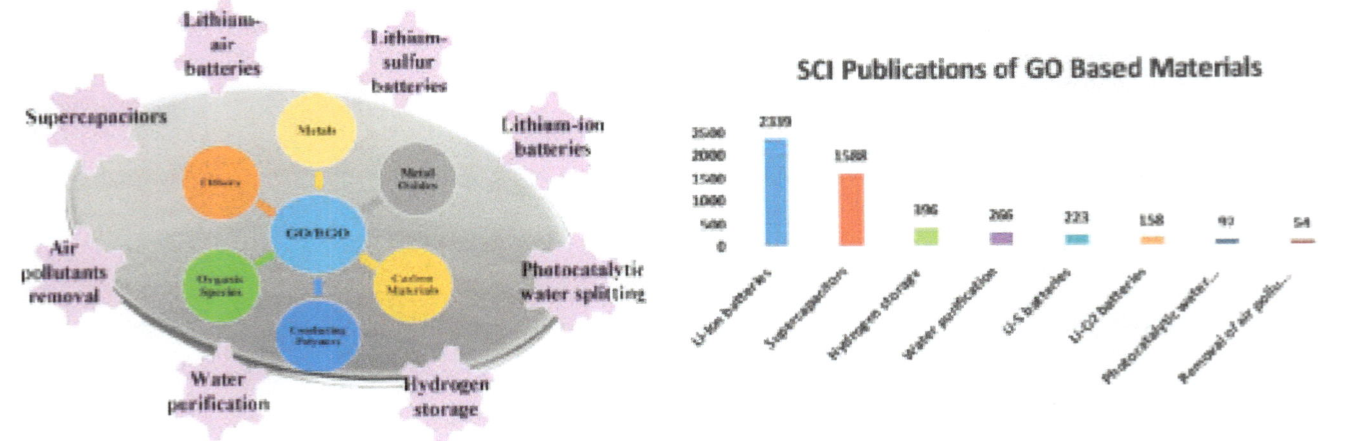

Figure 1: A schematic showing the GO/RGO-based hybrid materials for energy and environmental applications along with the SCI-indexed journal publications until now (January, 2015), searched from ISI by the keyword of graphene oxide combined with another one listed in the outer circle (left panel) or the X axis (right panel).

APPLICATIONS OF GO IN ENERGY CONVERSION AND STORAGE

Photocatalytic water splitting

Hydrogen is considered as a clean and renewable energy carrier in the future. Both production and storage of hydrogen are crucial for the utilization of hydrogen energy. One promising strategy of hydrogen production is water splitting by semiconductor photocatalyst under visible-light irradiation [21] and [22]. In the practical applications, this technique is limited by the inability to utilize visible light, insufficient quantum efficiency, amount of charge trapping centers, recombination of electron–hole pair, improper energy level offsets, and small contact area between catalyst and water, etc. Thus, designing novel catalysts to meet all the industrial requirements remains a big challenge.

Considering the superior electron mobility, high surface area, feasible assembly and tunable electronic band structures [19] and [23], GO and GO-based materials have been developed as excellent catalysts for light-driven hydrogen generation from water splitting[24], [25], [26], [27], [28], [29] and [30]. Yeh and coworker [25] demonstrated photocatalytic activity of GO with a band gap of 2.4–4.3 eV for H_2 evolution under mercury light irradiation, even in absence of the Pt cocatalyst. Since the GO sheets with high oxidation degree had larger band gap and limited absorption of light [28], appropriate reduction level of GO sheets was facilitated for the photoreactions to generate H_2[30].

Photocatalytic activity of GO elucidated by DFT calculations. Jiang et al. [31] investigated the electronic and optical properties of GO with different coverage and relative ratio of the epoxy and hydroxyl groups using DFT calculations. They found that the flexibility of controlling the amount of –O– and –OH functional groups in a GO sheet allows efficient tailoring of the band gap along with band positions. GO materials with 40–50% (33–67%) coverage and OH:O ratio of 2:1 (1:1) are suitable for both reduction and oxidation reactions for water splitting (Figure 2a). Among these systems, the GO composition with 50% coverage and OH:O (1:1) ratio is a very promising material for visible-light-driven photocatalyst. Simulated optical absorption spectra (Figure 2b) further confirms that those carefully chosen GO can harvest the major portion of solar light efficiently.

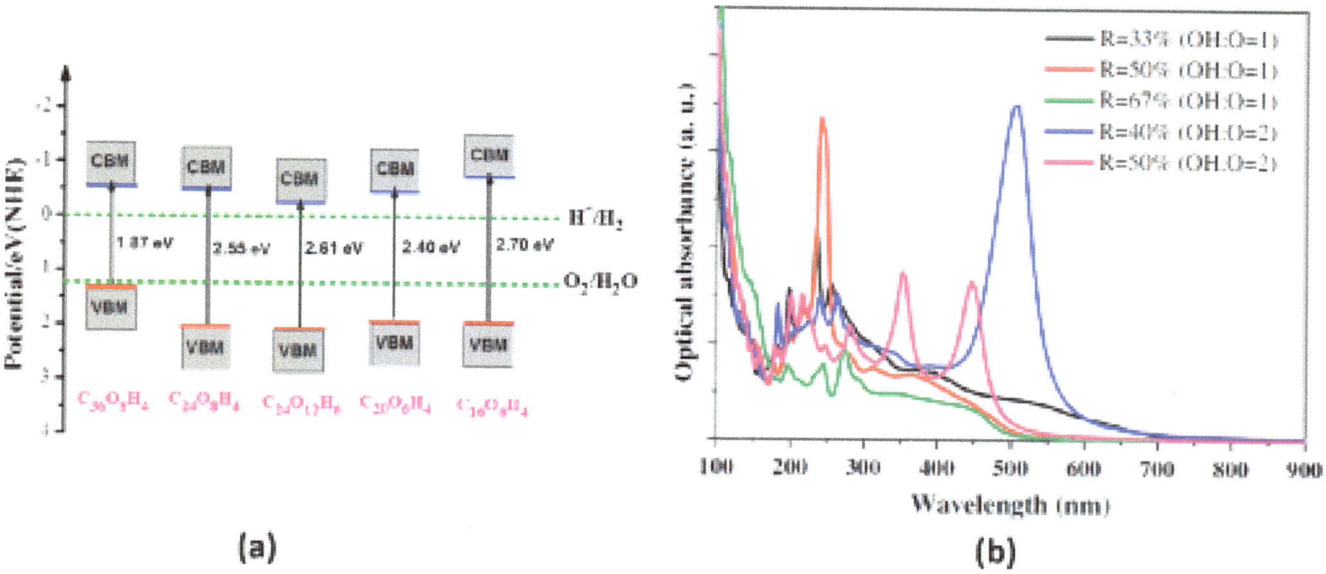

Figure 2: (a) Site levels of VBM and CBM for OH:O=1 and OH:O=2 graphene oxide with different coverage rate. The dot lines are standard water redox potentials. The reference potential is the vacuum level. (b) Calculated optical absorption curves for GO with OH:O=1:1 and OH:O=2:1 under different coverage rates. Reprinted with permission from Ref. [31]. Copyright (2013) Elsevier B.V.

In view of the p-type conductivity which hinders the hole transfer for water oxidation and suppresses O_2 evolution, Yeh and coworkers [24] introduced amino and amide groups on the GO surface and demonstrated that the ammonia-modified GO exhibits n-type conductivity and is able to catalyze the H_2 and O_2 evolutions simultaneously. Furthermore, the p–n diodes configuration of nitrogen-doped graphene oxide quantum dots (NGO-QDs) can fulfill the simultaneous H_2 and O_2 evolutions from pure water under visible light irradiation [29]. Besides, metal deposition can enhance the photocatalytic activity of GO. For example, Agegnehu et al. [32] found that the hydrogen generation rate of Ni/GO composite from aqueous methanol solution under UV–visible light illumination is enhanced by approximately four to seven times compared to that of the bare GO.

GO-semiconductor composites replacing noble metals as catalyst. To date, large numbers of efficient GO-based composites for photocatalytic water splitting have been synthesized, including GO-semiconductor binary systems and even more complicated ternary composites [19], [33], [34], [35], [36], [37], [38], [39], [40], [41], [42], [43], [44], [45],[46], [47], [48] and [49]. The improvement of H_2 production rate is mainly attributed to the role of GO as an electron acceptor and transporter to separate the photogenerated electron and hole pairs. For instance, H_2 production rate and quantum efficiency of the GO–$Zn_{0.8}Cd_{0.2}S$ system are 1824 mol h^{-1} g^{-1} and 23.4% under 420 nm light, respectively [37], which are much better than the Pt–$Zn_{0.8}Cd_{0.2}S$ photocatalyst (Figure 3). Another system with high quantum efficiency is the GO–CdS nanocomposite, which reaches a high H_2 production rate of 1.12 mol h^{-1} under visible light irradiation, corresponding to an apparent quantum efficiency of 22.5% at wavelength of 420 nm. It was demonstrated that the electron and hole can be efficiently separated by transferring photoinduced electrons from CdS to GO, and recombination of the electron and hole pairs in the excited semiconductor material is simultaneously suppressed [38]. First-principles investigation [39] indicated that the excited electrons in CdS are injected into GO and transport along graphene layer through π□ orbital under visible light irradiation to achieve electron and hole separation, consistent with experiment [38]. Similarly, binary composites of GO/TiO_2 [40] and [41], GO/AgBr [42], GO/g–C_3N_4[43], GO/3C–SiC [44], GO/Cu_2O [45], GO/$Sr_2Ta_2O_{7-x}$[46] also showed higher photocatalytic H_2 evolution than the bare semiconductors.

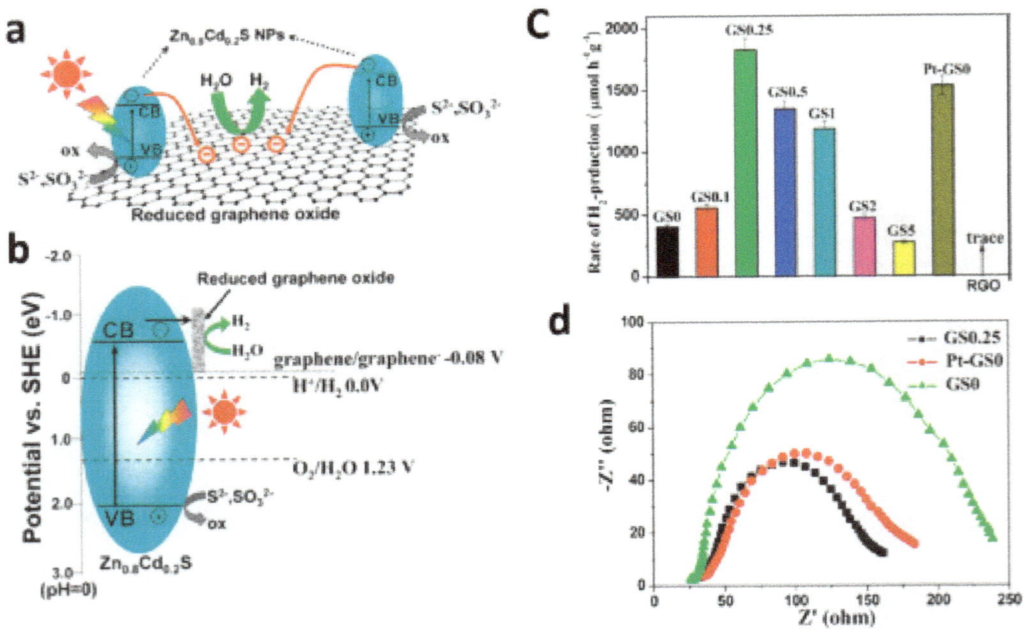

Figure 3: (a) Schematic illustration for charge transfer and separation in the GO–$Zn_{0.8}Cd_{0.2}S$ system. (b) Mechanism for photocatalytic H_2 production under simulated solar irradiation. (c) Comparison of the photocatalytic H_2-production activity under simulated solar irradiation over GS0, GS0.1, GS0.25, GS0.5, GS1, GS2, GS5, Pt-GS0, and RGO samples. (d) Nyquist plots of GS0, GS0 loaded with 1 wt% Pt, and GS 0.25 electrodes under solar irradiation. Reprinted with permission from Ref. [37]. Copyright (2012) American Chemical Society.

Beyond the GO-based binary composites, ternary GO–CdS@TaON composite with core–shell structure (containing 1 wt% CdS nanocrystals) showed a high rate of hydrogen production at 306 mol h^{-1} with an apparent quantum efficiency of 15% under 420 nm monochromatic light [47]. Other ternary composites such as CdS/Al_2O_3/GO [48], CdS/ZnO/GO [48], and TiO_2/MoS_2/GO [49] also exhibited satisfactory photocatalytic behavior.

GO in the Z-scheme photocatalysis system. Alternatively, GO can be used as a solid-state electron mediator for water splitting in the Z-scheme photocatalysis system [22] to overcome the problems with electron recombination and transportability. Iwase's group[35] constructed a model system of using GO as solid electron mediator, $BiVO_4$ as O_2photocatalyst, and Ru/$SrTiO_3$:Rh as H_2 photocatalyst, respectively. Such system exhibited higher H_2 production rate than the GO/$BiVO_4$ and Ru/$SrTiO_3$:Rh. The underlying mechanism of electron transfer can be illustrated as the following picture. First, GO serves as an electron mediator to transfer electrons from the conduction band of $BiVO_4$ to the Ru/$SrTiO_3$:Rh. Meanwhile, electrons in Ru/$SrTiO_3$:Rh reduce water into H_2 on the Ru cocatalyst, while holes in $BiVO_4$ oxidize water into O_2 simultaneously, accomplishing a complete water splitting cycle.

Open issues. As promising photocatalysts of water splitting, GO and GO-based composites not only show the ability of separating the photogenerated electron–hole pair, but also exhibit the capability for photocatalytic H_2 evolution by itself. However, research on GO-based materials for H_2 generation from light driven water splitting is still at its early stage and requires further attentions. Explanation of the photocatalytic activity by GO content in these composites is still controversial. In this regard, theoretical calculations are highly desirable to provide critical insights on the mechanism.

Hydrogen storage

Currently, the hydrogen storage strategies can be categorized into chemical storage (hydrogen in the form of hydrides) and physical storage (hydrogen in the form of H_2molecules). Both of them have their own deficiencies to store necessary amounts of hydrogen under technologically useful conditions. As for chemical storage, the hydrides usually possess high hydrogen contents, but concomitant with the unsatisfactory performance for hydrogen release. Meanwhile, physical storage suffers from the weak adsorption energy of H_2 molecules, which leads to low storage capacity at ambient conditions. Besides, the host sorbents of physical storage should possess large surface areas and be light in weight.

Physical storage of hydrogen

On the pristine two-dimensional (2D) graphene nanosheets, the weak adsorption energy of H_2 molecule (only 1.2 kJ mol^{-1}) [50] usually leads to very low storage capacity of hydrogen (<2 wt%) [51]. However, Guo et al. fabricated a hierarchical graphene material with the micropore of ~0.8 nm, mesopore of ~4 nm and macropore above 50 nm, which achieved a high H_2 storage capacity over 4.0 wt% at atmospheric air pressure [52]. For the GO layers, it was found that modulating the interlayer distance [53] and pore size [54]of multilayered graphene oxide to optimal values can improve the maximum hydrogen storage capacity.

Metal decoration on GO surface. More importantly, the functional groups on GO surface offer feasible ways to further composite with other species to improve the H_2 binding energy. In principle, if the obstacle of metal aggregation [55] can be conquered, metal (especially transition metal) decoration on carbon sorbents is a promising route to enhance H_2 binding energy and thus storage capacity [56]. For example, first-principles calculations by Wang and co-workers predicted that Ti atoms can be stably anchored by the hydroxyl groups on GO surface and simultaneously retain sufficient chemical activity to adsorb H_2 molecules [57]. Each Ti atom is able to bind multiple H_2 with a desired average binding energy (14–41 kJ mol^{-1}), corresponding to a theoretical gravimetric density of 4.9 wt% or volume density of 64 g L^{-1} (see Figure 4a and b). In addition to Ti, it was theoretically proposed that Mg can reduce the hydroxyl groups on GO and thus result in –(C–O)*x*–Mg (*x*=1 or 2) [58]. H_2 molecules can be polarized by strong electric field jointly produced by the anchored Mg and O on GO surface, and therefore leading to a hydrogen storage capacity up to 5.6 wt% (at 200 K without any pressure) with an average binding energy of 28 kJ mol^{-1}.

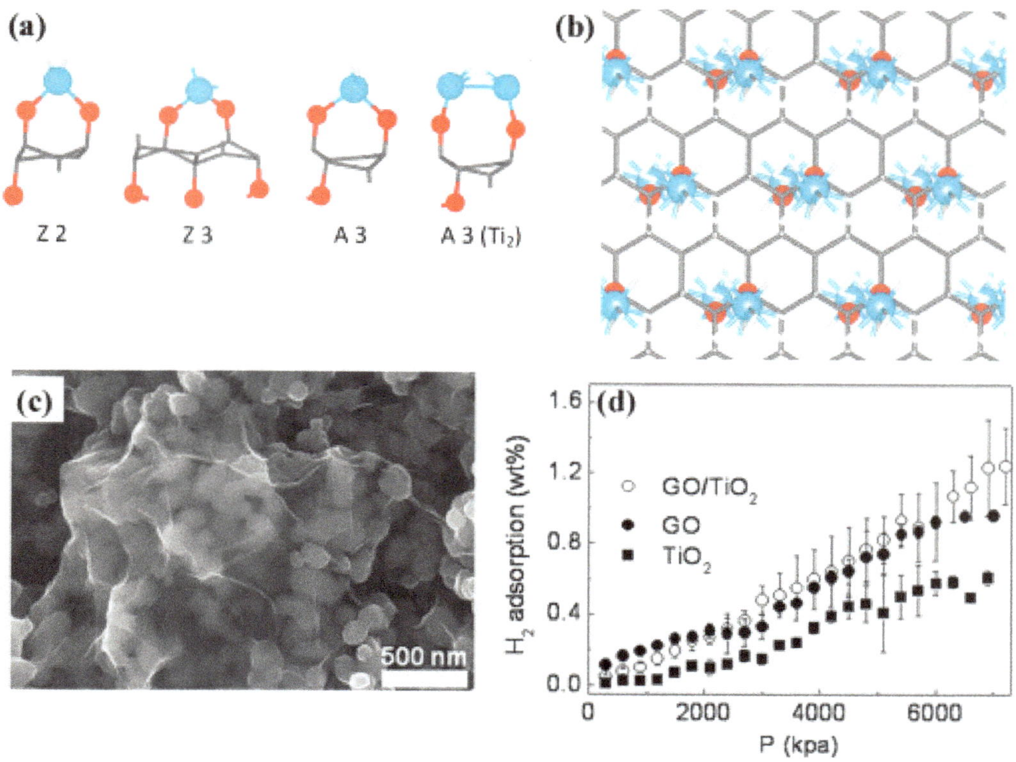

Figure 4: (a) Stable structural motifs for Ti decoration on GO: three possible sites for Ti atom and one site for Ti_2; (b) 2D periodic structural model for Ti-decorated GO with multiple H_2 adsorption (gravimetric density: 4.9 wt%). Reprinted with permission from Ref. [57]. Copyright (2009) American Chemical Society. (c) SEM image of GO/TiO_2; (d) High-pressure H_2 adsorption of TiO_2, GO and GO/TiO_2. Reprinted with permission from Ref. [60]. Copyright (2012) Elsevier Ltd.

These theoretical predictions have been partly confirmed by experiments. For instance, Pd decoration on graphite oxide can significantly increase the hydrogen storage capability by a factor of 3.3 in comparison with that of pristine GO [59]. Moreover, Figure 4c and d show that the GO wrapped TMO composites without any additional agents, which exhibited enhanced hydrogen storage capacities, i.e., 1.36 wt% for GO/V_2O_5 or 1.26 wt% for GO/TiO_2, compared to that of 0.16 wt% for V_2O_5 or 0.58 wt% for TiO_2 alone, respectively [60].

GO-based three-dimensional (3D) porous pillared materials possessing tunable porosity, accessible surface area and versatile electronic properties for hydrogen adsorption. Such materials can be formed by assembling other reactive agents with the functional groups on GO. Froudakis et al. [61] proposed a new class of pillared graphene oxide in conceptual by introducing carbon nanotubes (CNTs) in-between the GO layer and implemented with Li decoration to increase H_2 accommodation. The Li decorated pillared GO model with a pore dimension of *d*=23 Å and an O/C ratio of 1/8, achieved a gravimetric capacity greater than 10 wt% and a volumetric capacity of 55 g L^{-1} at 77 K and 100 bar, respectively. Later, experiment by Aboutalebi et al. [62] demonstrated the feasibility of self-aligned GO/MWCNT hybrid frameworks, which reached an appreciable hydrogen uptake of 2.6 wt% at room temperature with proper interlayer distance.

As shown in Figure 5a, graphene oxide frameworks (GOFs) with 3D porosity was synthesized by intercalating the boronic acid (which can react with the hydroxyl groups) into GO layers, which showed superior hydrogen adsorption behavior [63], [64] and [65]. By adopting various linear boronic acid pillaring units, the interlayer spacing between graphene planes can be tuned to an optimum value for H_2 adsorption on both surfaces, which is twice of typical porous carbon materials and comparable to MOFs. The outstanding hydrogen storage properties are attributed to the porous spaces, and the enhanced hydrogen adsorption is mainly caused by the benzenediboronic acid pillars between graphene sheets [65]. By modulating the intercalation via three kinds of diaminoalkanes between GO layers, Kim et al. [66] found an optimum GO interlayer distance of 6.3 Å with maximum H_2 uptake, similar to the predicted distance from thermally modulated GO [53].

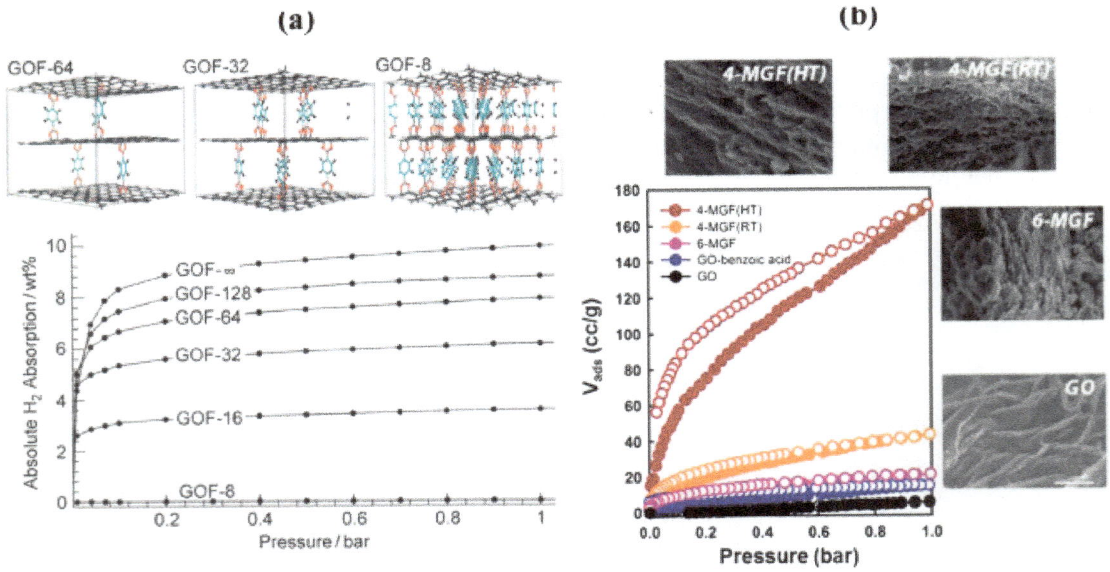

Figure 5: (a) Structures of three representative GOFs with *n*=64, 32, 8 (Upper) and hydrogen adsorption for several GOFs as function of hydrogen gas pressure (Lower). Reprinted with permission from Ref. [63]. Copyright (2010) WILEY-VCH Verlag GmbH & Co. KGaA, Weinheim. (b) SEM images and H_2 adsorption–desorption isotherms of GO, GO–benzoic acid, and the three MGFs. Reprinted with permission from Ref. [67]. Copyright (2013) The Royal Society of Chemistry.

Moreover, tunable porosity and surface area are also beneficial for accommodation of hydrogen gas [67] and [68]. The metallomacrocycle–graphene frameworks (Figure 5b) possess tunable porosity from microporous to hierarchical micro- and mesoporous, which allows the usually unstable square planar Ni(III) species to be stabilized in the solid state [67]. The H_2 adsorption–desorption isotherms shows that the highest hydrogen uptake of the composites achieved 1.54 wt% at 77 K and 1 bar. Recently, by the C–C coupling reaction of RGO and iodobenzene, the porous graphene frameworks displayed a high BET surface area of 825 m^2 g^{-a} and thus achieved a H_2 storage capacity of 1.2 wt% at 77 K and 1 atm [68].

Intercalation of MOFs into GO. MOF Intercalation can enlarge the space for accommodating H_2 molecules, and also increase the number of effective sites to adsorb H_2. Petit and Bandosz [69] observed enhancement of hydrogen uptake in copper-based GO/MOFs composites, therefore confirming the formation of new small pores. Later, Liu et al. [70] synthesized Cu–BTC/GO composite with nanosized Cu–BTC (copper–benzene-1, 3,5-tricarboxylate), which significantly improved hydrogen storage capacity compared to the pristine Cu–BTC (from 2.81 wt% of Cu–BTC to 3.58 wt% of Cu–BTC/GO at 77 K and 42 atm). Recently, Pt was incorporated into the GO/HKUST-1 composite to adsorb and dissociate H_2 molecules, which yielded nearly twofold enhancement with regard to HKUST-1 itself [71].

Open issues. Overall, GO and GO-based composites exhibit promising hydrogen storage properties, especially the 3D pillared porous GO materials show remarkable potential for the on-broad applications. More efforts should be devoted to designing new 3D architectures on the basis of GO sheets and enhancing the interaction between H_2molecules and GO-based sorbent materials.

Chemical storage of hydrogen

Noble metal catalysts supported by GO/RGO. As for hydrogen storage in the form of hydrates, one major obstacle is to lower the working temperature for release of hydrogen. Because of the large surface area and excellent chemical stability, GO and RGO can serve as appropriate supports to disperse and stabilize the metal nanoparticles (NPs) for catalyzing and dissociating H_2 molecules. Noble metal catalysts supported by GO/RGO have been also investigated [72], [73], [74], [75] and [76]. A facile route to synthesize Ru/RGO NPs via methylamine borane (MeAB) as a reducing agent was reported by Cao's group [72]. Their experiments provided the ever reported lowest activation energy of 11.7 kJ mol^{-1} during the hydrolytic dehydrogenation of amine borane (AB). To further improve the catalytic activity, they synthesized the RGO supported core–shell NPs of Ag@M (M=Co, Ni, Fe) [73] and Ru@Ni [74] for the hydrolysis of amine borane (AB) and MeAB. Additionally, Pt–CeO_2 NPs supported by RGO exhibited synergistic effect to stabilize the active centers and increase the catalytic activities [75]. Nevertheless, the turnover frequency (TOF) value of the hybrid Pt–CeO_2/RGO NPs was only 48.0, much lower than the nanocatalysts with core–shell structure [73], [74],[75] and [76].

Non-noble metal composites supported by RGO. These composites also showed good catalytic activity of hydrolytic dehydrogenation of AB. Xi et al. [77] demonstrated the catalytic activity of RGO/Pd nanocomposite in hydrogen generation of AB hydrolysis with a hydrolysis completion time (12.5 min) and an activation energy (51±1 kJ mol^{-1}). Metin and coworkers [78] further exploited a better catalytic performance of RGO/Pd nanocomposite for AB hydrolysis with lower activation energy of 40±2 kJ mol^{-1} and higher TOF than Xi's results [77]. In addition, Zhou et al. [79] found that addition of polyethyleneimine can significantly affect the morphology and size of the resulting Fe–Ni NPs, and therefore modulate the catalytic activity of AB hydrolysis. Very recently,

RGO supported Cu NPs were used as catalyst for hydrolytic dehydrogenation of AB, which exhibited the highest TOF value of 3.61 among all of the Cu nanocatalysts ever reported for this reaction [80].

GO facilitated AB (MAB) dehydrogenation. For hydrogen release through thermo-dehydrogenation, Tang et al. [81] reported a recyclable dehydrogenation of AB with a GO-based hybrid nanostructure. The hydroxyl groups on GO surface act as proton donors to react with AB and yield $H_3NBH_2^+$ cation, which is the key to facilitating AB dehydrogenation (see Figure 6a–c). Using first-principles calculations, Li and co-workers[82] predicted a novel composite by GO and lithium amine borane (LiAB), in which the hydroxyl groups on GO surface may interact with LiAB via one molar equivalent of H_2released (Figure 6d). Compared to the pure LiAB, GO–LiAB hybrid shows better dehydrogenation performance with reduced reaction barriers. Most strikingly, the dehydrogenated products of GO–LiAB complex with lithium well dispersed on the GO support are still capable of adsorbing H_2 molecules with appreciable H_2 binding energy (7.9 kJ mol^{-1}) and maximum capacity of 5 wt% (Figure 6e). In such a way, chemisorption and physisorption can be appropriately combined for superior hydrogen storage, which might be a promising direction for future development of hydrogen storage materials.

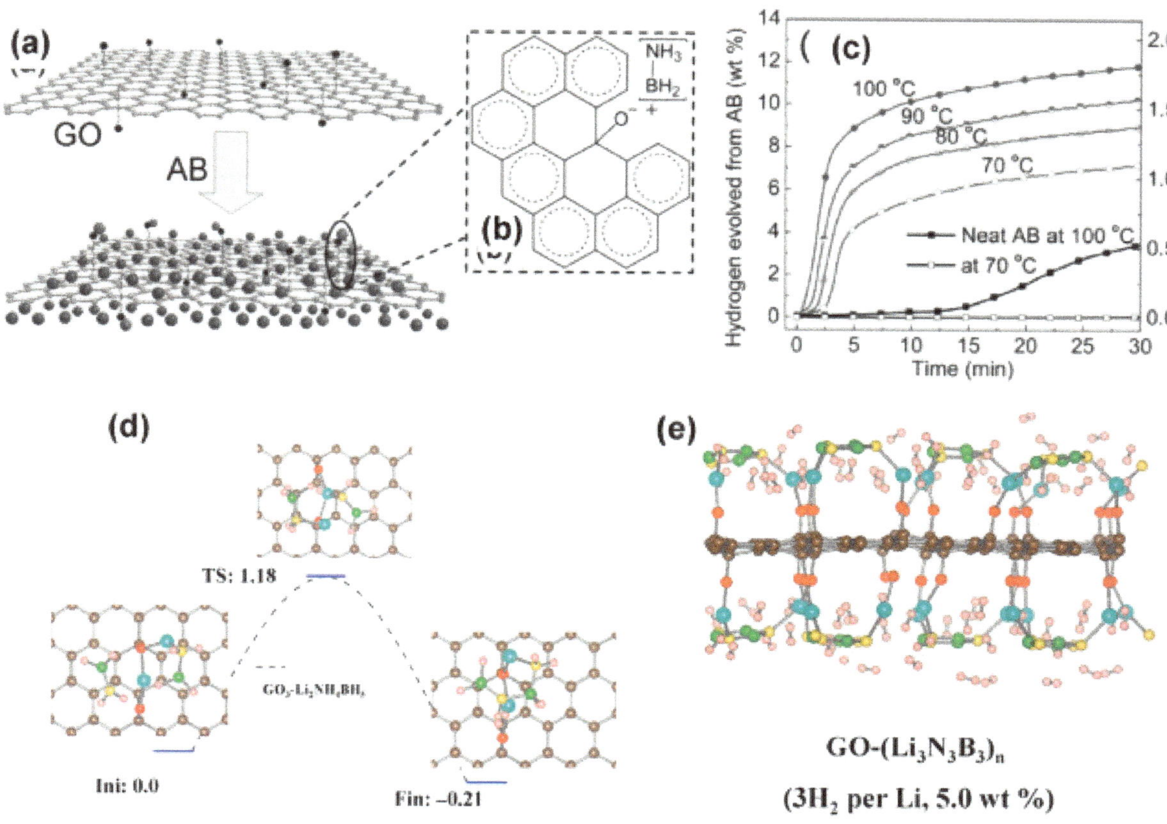

Figure 6: (a) Schematic representation of the mechanism for the formation of GO–AB hybrid nanostructure; (b) Detailed illustration of interaction between AB cation and negatively charged oxygen of GO; (c) Isothermal TPD results of hydrogen released from AB in GAB30 at 70, 80, 90, and 100 °C. Reprinted with permission from Ref. [81]. Copyright (2012) American Chemical Society. (d) Energy profiles (eV) for dehydrogenation processes of GO_3–$Li_2NH_4BH_5$ complexes; (e) Hydrogen storage capacities in GO–$(Li_3N_3B_3)_{(n)}$. Reprinted with permission from Ref. [82]. Copyright (2013) The Royal Society of Chemistry.

Lithium batteries

To meet the growing demand of portable electronic products and electric vehicles, lithium batteries (Li–ion, Li–S and Li–O_2) hold great promise for future energy storage and utilization owing to their high specific energy and energy density. In principle, GO with heterogeneous chemical and electronic structures can be used as electrode material for lithium batteries. The functional groups on GO may serve as sites for chemical modification or functionalization, which in turn immobilize various active species and render hybrid architectures for electrode materials. However, applications of pristine GO in lithium batteries suffer from its insulating nature [83]. It is thus necessary to improve the conductivity while keeping the inherent merits of GO.

Lithium ion batteries

As the most popular rechargeable batteries for critical applications, lithium-ion batteries (LIBs) with conventional electrode materials suffer from its low theoretical capacity limits. In this regard, elaborately designed GO/RGO-based composites as both anode and cathode materials exhibit superior electrochemical performances, as summarized inTable 1.

Table 1: Electrochemical properties of LIBs with electrode using GO/RGO composites.

	System	Current density (mA cm^{-2}or mA g^{-1})	Charge/discharge rate (C)	Specific capacity (mA h g^{-1})	Capacity retention (%)	Cycles	Ref.
Anode	Graphite/GO		0.5	690		60	[84]
	Al reduced GO	100		540		30	[325]
	CNT/GO	200		1500		100	[85]
	Mn_3O_4/RGO	400		730		40	[87]
	MnO/ZnO–RGO	100		660		100	[88]
	Fe_2O_3 NPs/RGO	302	0.3	881		90	[89]
	Fe_2O_3/RGO	100		1027		50	[90]
	SnO_2/RGO	100		718		200	[91]
	Free standing CuO NPs/RGO	67		736.8		50	[92]
	Zn_2GeO_4/GO	200		1150		100	[93]
	$Li_4Ti_5O_{12}$/RGO		10	154		200	[94]
	Si nanowires @G/RGO	2,100		1600	80	100	[100]
	Si nanoparticle/RGO	24,000	10	700	87	152	[101]
	Si nanowires/RGO	1,200	1/3	2300	91.8	100	[102]
Cathode	GO	50		360.4		60	[103]
	RGO	137		125		30	[104]
	$Li_3V_2(PO_4)_3$/RGO		0.5	177	96	50	[105]
	$Li_3V_2(PO_4)_3$/RMGO		0.1	186	99.9	50	[106]
	V_2O_5nanorod/RGO	100		287	89	20	[107]
	V_2O_5nanowire/RGO		0.2	225	56	60	[108]
	PPy/RGO		20	55	100	200	[109]

Anode

Direct combination of GO with commercial graphite. Here, GO acts as the binder-free anode material in LIBs to increase the reversible capacity to 690 mA h g^{-1} at the rate of 0.5 C (1 C=372 mA g^{-1}), along with excellent cycle performance and rate capability. In the combined material, the original graphite maintains the conductivity and GO mainly contributes to the enhanced capacity as lithium accommodator [59] and [84]. Indeed, the conductive RGO itself can replace the conventional graphite in anode. More importantly, the residual functional groups on RGO provide wide range of the porous structures, which are beneficial for the electrochemical performance of LIBs [85] and [86].

GO/RGO serves as excellent host materials. To disperse anode materials, concomitant with the synergistic effect of improving the configuration and stability of the anode, it has been demonstrated that incorporation of RGO into either metal oxides or sulfides induces remarkably enhanced battery performance with regard to pure oxides or sulfides. The pulverization process of transition metal oxide (TMO) induced rapid capacity degradation can be effectively suppressed in the RGO/TMO hybrids [87], [88], [89], [90], [91], [92],[93] and [94]. For instance, two-step synthesized Mn_3O_4/RGO hybrid renders a high specific capacity up to ~900 mA h g^{-1} (near the theoretical capacity) as well as good rate capability and cycling stability [87]. Very recently, the synergistic coupling of a ternary metal oxide (Zn_2GeO_4) and GO layers yields extraordinary high specific capacity, superior rate capability and long cycle life [93].

Moreover, RGO can offer enough buffer space for large volume expansion of the metal sulfide during insertion and extraction of the Li ions and alleviate the insulating problems of polysulfide [95], [96] and [97]. The FeS@RGO nanocomposite with robust sheet-wrapped structure exhibited better electrochemical performance than the isolated FeS NPs because of the smaller particle sizes and the synergetic effects between FeS and RGO sheets, i.e., increased conductivity, shortened lithium ion diffusion path, and effective prevention of the polysulfide dissolution [95].

Application of GO/RGO in Si based anode. Compared to carbon, silicon exhibited higher theoretical capacity of ~4200 mA h g^{-1} and appropriately lower working potential [98]. However, serious pulverization of bulk silicon during cycling limits its cycle life [99]. Compositing silicon with GO/RGO is an effective strategy to improve the cycling performance [77],

[85], [100], [101] and [102]. Hybrid of leaf-like GO and Si NPs as anode of LIBs achieved rather high capacity retention over 100 cycles [61], although the surface groups and large surface area resulted in irreversibility on the initial cycling. Moreover, RGO as the overcoats of Si nanowire@graphene (SiNW@G) nanocables exhibited great capability to accommodate the volume change of the embedded SiNW@G nanocables (see Figure 7) [100]. The SiNW@G@RGO composite maintains the structural and electrical integrity of the electrode and yields a high specific capacity of 1600 mA h g^{-1} at a current density of 2.1 A g^{-1}. Afterwards, it was demonstrated that Si nanowires directly grow on flexible RGO sheets can also maintain the structural integrity and provide a continuous conductive network of the electrode, which resulted in over 100 cycles serving as anode in half cells at a high lithium storage capacity of 2300 mA h g^{-1}[102].

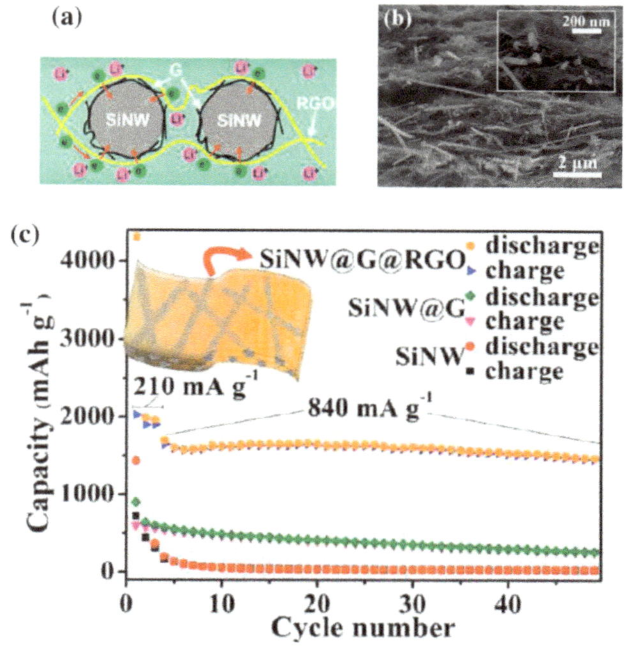

Figure 7: (a) Schematic of SiNW@G/RGO composite; (b) Cross-section SEM image; (c) Comparison of capacity retention of different electrodes. All electrodes were cycled at a charge/discharge rate of 210 mA g^{-1} for the first three cycles and then 840 mA g^{-1} for the subsequent cycles. Reprinted with permission from Ref.[100]. Copyright (2013) American Chemical Society.

Cathode

In addition to anode, inclusion of GO/RGO in cathode materials also shows significant advantages compared to the conventional polymer cathodes and lithium-TMOs cathodes in LIBs. The epoxide-enriched GO can be lithiated/delithiated as rechargeable cathode with high capacity and good stability [103] and [104], in which the hydroxyl groups were identified as the lithiation-active species. More efforts have been devoted to the RGO/TMO hybrid cathodes in LIBs [105], [106], [107] and [108]. Enwrapping $Li_3V_2(PO_4)_3$ on RGO sheets as cathode material can facilitate the charge transfer but concomitant with low initial discharge capacity (177 mA h g^{-1} at 0.1 C) [105]. Even with modification of carbon [105] and cetyltrimethyl ammonium bromide (CTAB) [106] in the $Li_3V_2(PO_4)_3$/RGO composite, the storage capacity only slightly increased to about 186 mA h g^{-1}. In comparison, vanadium oxides/RGO composite [107] and [108] exhibits better lithium storage performance than the $Li_3V_2(PO_4)_3$/RGO, as shown in Table 1. Moreover, the PPy/RGO hybrid shows significantly improved electrochemical properties, i.e., high rate capability and excellent cycling stability for being both cathode and anode materials in LIBs [109]. Finally, GO utilized as a protective coat can effectively adhere to aluminum and thus inhibit corrosion of the aluminum current collector used in LIBs [110].

Open issues. Limited by the horizon of theoretical capacity, even the improved lithium storage capacity of LIBs after modification of both anode and cathode materials is still inadequate for the increasing demands of portable electronics and electric vehicles, which inspires further explorations of other kinds of batteries with higher capacity.

Lithium sulfur batteries

Compared to the traditional cathode materials based on transition metal oxides or phosphates in LIBs, Li–S batteries possess higher specific capacity up to about 1675 mA h g^{-1}[111]. However, they suffer several operational problems in the commercial applications, such as insulating nature of sulfur [112], dissolution of polysulfides in electrolyte [113], and volume expansion of sulfur during discharge [114],[115] and [116]. Incorporation of GO/RGO into the sulfur cathode of Li–S batteries is an effective way to deter these operational issues. The oxygenated functional groups on the basal planes and edges as well as the structural defects of GO/RGO provide strong anchoring points [117] and [118], while the 2D sheet-like structure promotes the formation of self-assembled films for improving the electrochemical performance of Li–S batteries. The reactive functional groups on GO also exhibit strong adsorption of sulfur or polysulfides and effectively prevent loss of the active materials (the lithium polysulfides dissolved in the electrolyte during cycling), therefore leading to high reversible capacities[119], [120], [121] and [122] (see Table 2). However, the stable charge–discharge process of S/GO composite can only sustain limited number of cycles (e.g., 50 cycles[119]), followed by a dramatic capacity fading.

Table 2: Electrochemical performances of Li–S batteries with GO/RGO-based electrodes.

System	Charge/discharge rate (C)	Specific capacity (mA h g^{-1})	Capacity retention (%)	Cycles	Ref.
S/GO	0.1	1550	58	50	[119]
S/GO	0.1	430	81	19	[121]
S/GO core shell	0.5	308	89	210	[122]
PEG coated S/GO	0.2	750	85	100	[123]
CATB coated S/GO	6	1356	59	1500	[124]
Amylopectin wrapped S/GO	5 C/16	441	68	175	[125]
Li_2S/GO	0.2	782	88	150	[126]
Saccule-like S/RGO	1	724.5	85.8	60	[128]
S/RGO	0.1	1260	71	100	[129]
Li_2S/RGO	0.1	982	32	100	[130]
TG@S/RGO	0.95	1064	62.6	200	[131]
CMK-3@RGO/S	0.1	1333	48.7	100	[132]
CMK-3@RGO/S	0.5	1147.7	64	100	[133]
NMCNT@RGO/S	2	1164.5	69.3	200	[134]
PEG@RGO/S	0.1	1021	72	100	[135]
Pyridinic-N RGO/S	0.1	1356.8	62.4	100	[136]
Pyrrolic-N RGO/S	0.1	1298.6	58.9		

Addition of effective assistant components into GO/S composites. It has been demonstrated to further improve the electrochemical performance of GO/S composites[123], [124], [125], [126] and [127]. For instance, the surfactant of polyethylene glycol (PEG) modified S/GO composites (Figure 8a) exhibit high capacity retention after long cycles, indicating an effective shuttle inhibition of polysulfide [123]. Zhou et al. [125]modified the GO/S nanocomposite with amylopectin to suppress the Li polysulfides escaped from the open channels among GO layers, and thus largely improved cyclability. Moreover, Cui et al. [126] demonstrated that wrapping of GO onto the Li_2S surface via lithium–oxygen interactions can alleviate dissolution of intermediate polysulfides into electrolyte and therefore achieve a high discharge capacity of 782 mA h g^{-1} of Li_2S (~1122 mA h g^{-1} of S) with stable cycling performance over 150 charge–discharge cycles.

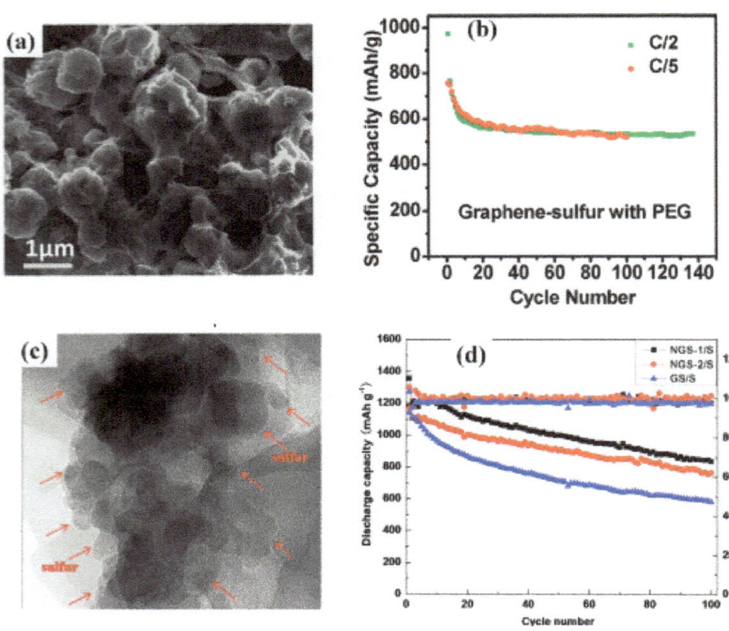

Figure 8: (a) SEM characterization of graphene sulfur composite at low magnifications; (b) Cycling performance of the same composite as in (a) at rates of ~C/5 and ~C/2. Reprinted with permission from Ref. [123], Copyright (2011) American Chemical Society. (c) TEM image of NGS-1/S composites; (d) Cycle performance of GS/S and NGS/S electrodes at 0.1 C (Inset shows the corresponding Coulombic efficiency). Reprinted with permission from Ref [136], Copyright (2014) Elsevier B.V.

S/RGO-based sulfur cathode. Compared to the S/GO composites, the sulfur cathode prevails due to better electrical conductivity [127], [128], [129] and [130]. RGO provides a conductive network surrounding the sulfur particles, which facilitates both electron conductivity and Li^+ ion transportation [127]. The saccule-like S@RGO composites shows better electrochemical performance as compared to the common RGO sheets since that the unique saccule-like RGO can provide a more effective electrically conductive network and offer sufficient space to accommodate the stress and volumetric expansion of sulfur during charge–discharge process [128]. Moreover, the Li_2S/RGO nanocomposite with a unique 3D pocket structure showed a high initial capacity of 982 mA h g^{-1}[130]. But the remaining problem of polysulfide dissolution (even with the presence of functional groups on RGO) still led to noticeable capacity fading over time[128], [129] and [130].

Ternary RGO/S-based cathode materials. Further carbon decoration of the RGO/S cathode materials can enhance the Li^+ ion transportation and electrical transport properties of the cathode. The thermally exfoliated G@S/RGO nanocomposite possesses superior electrical conductivity and thus resulted in a high specific capacity of 1064 mA h g^{-1} at the high rate of 1.6 A g^{-1}[131]. Moreover, synergistic coupling of RGO with mesoporous CMK-3/sulfur composite provides sufficient buffering space and extra diffusion channel and consequently improved the overall electrochemical performance[132] and [133]. Enhanced electrochemical performance was also achieved in the MWCNTs@S/RGO cathode [134]. Nevertheless, adding carbon additive into cathode is still insufficient to restrain the polysulfides and alleviate the capacity fading. The synergistic coupling of RGO and PEG with sulfur can keep the sulfur well confined within the PEG–S–RGO nanostructure and therefore effectively suppress dissolution of the polysulfides, accommodate the volume change and enhance the electrical conductivity of sulfur [135]. In addition, nitrogen doped RGO shows enhanced electrical conductivity and stronger polysulfide confinement compared to the common RGO/S composite, as depicted in Figure 8c and d [136]. The pyridinic-N enriched RGO/S exhibits better discharge capacity (1356.8 mA h g^{-1}) and capacity retention (847.4 mA h g^{-1}) than the Pyrrolic-N RGO/S (1298.6 mA h g^{-1} and 764.8 mA h g^{-1}, respectively).

Open issues. GO/RGO-based materials facilitate the electrochemical performance of Li–S batteries such as rate capacity and cycle stability significantly, which makes the Li–S batteries rather promising in future commercial utilization. Further efforts are still needed to avoid the polysulfide-induced capacity fading. In this regard, the detailed interaction mechanism between polysulfides and GO/RGO composites has to be elucidated both experimentally and theoretically.

Lithium air batteries

Compared to other kinds of Li batteries, Li–O_2 batteries depict super promising prospects due to its extremely high theoretical energy density (up to 13,000 W h kg^{-1}, comparable to petroleum) [137]. However, development of Li–O_2 batteries also suffers from some major challenges such as block of the diffusion path induced by discharged products (Li_2O_2 and Li_2O) [138], corrosion by atmospheric moisture [139], and attack by highly reactive reduced O_2^-[140]. Therefore, the ideal cathode materials of Li–O_2 batteries require good electrical conductivity, outstanding O_2 reduction performance, high structural stability, and suitable path for fast oxygen diffusion.

Hierarchical GO-based materials. With highly interconnected 3D channels for rapid oxygen diffusion, these materials have been used to improve the electrochemical performance of Li–O_2 batteries [141] and [142]. With the assistance of binder materials, the hierarchical GO exhibited an exceptionally high capacity of 15,000 mA h g^{-1} in Li–O_2battery, which is attributed to the formation of isolated Li_2O_2 nanoparticles and alleviation of air blocking in the cathode induced by defects and functional groups [141]. Besides, GO gel acted as a special carbon source and provided the framework of 3D gel to form a highly efficient cathode material for Li–O_2 battery, which achieved a high capacity of 11,060 mA h g^{-1} at the current density of 0.2 mA cm^{-2} (280 mA g^{-1}) [142]. A recent study of carbon materials (RGO and thermally exfoliated graphite) in a cell with an all-solid-state positive electrode under true operando conditions, however, suggested that the functional groups of epoxy and carbonates generated during the discharge reaction can limit the rechargeability of Li–O_2 cells [140].

Synergistic effect of GO/CNT cathode materials. The leaf-like GO and CNT in the cathode of Li–O_2 batteries has a discharge capacity of 6000 mA h g_{GO}^{-1} with a cutoff voltage of 2.0 V and stable cyclic lifetime over 150, which is much better than the 2250 mA g_{GO}^{-1} for GO-based, 2500 mA g_{CNT}^{-1} for CNT-based, and 3000 mA $g_{CNT/GO}^{-1}$for CNT/GO mixture-based Li–O_2 batteries [85]. Moreover, the nitrogen doped GO as cathode catalyst substantially improved the electrochemical performance in the oxygen reduction reaction (ORR) process [143]. Specifically, the graphitic nitrogen content affected the electrocatalytic activity and the pyridinic nitrogen content improved the onset potential for ORR.

GO/RGO as an ideal host for metal catalysts dispersion. GO/RGO can fulfill the requirement of high dispersion and low aggregation of metal catalysts to reduce the high overpotential. The Ru and RuO_2-based nanomaterials hybrid with RGO exhibited superior electrocatalytic activity for the oxygen evolution reaction (OER) reaction in Li–O_2 cells, leading to a high capacity of 5000 mA h g^{-1}[144]. However, the stable cycling of the Ru-based materials only sustained up to 30 cycles due to degradation of the Li–metal anode and pore clogging. In addition, Co and N-doped Co oxides [145] and Mn oxides[146] and [147] supported on RGO also showed high catalytic activity for the ORR of Li–O_2 batteries. Nevertheless, there was also a rapid degradation of capacity during the charge–discharge cycles in the RGO/α-MnO_2 composite, as illustrated in Figure 9.

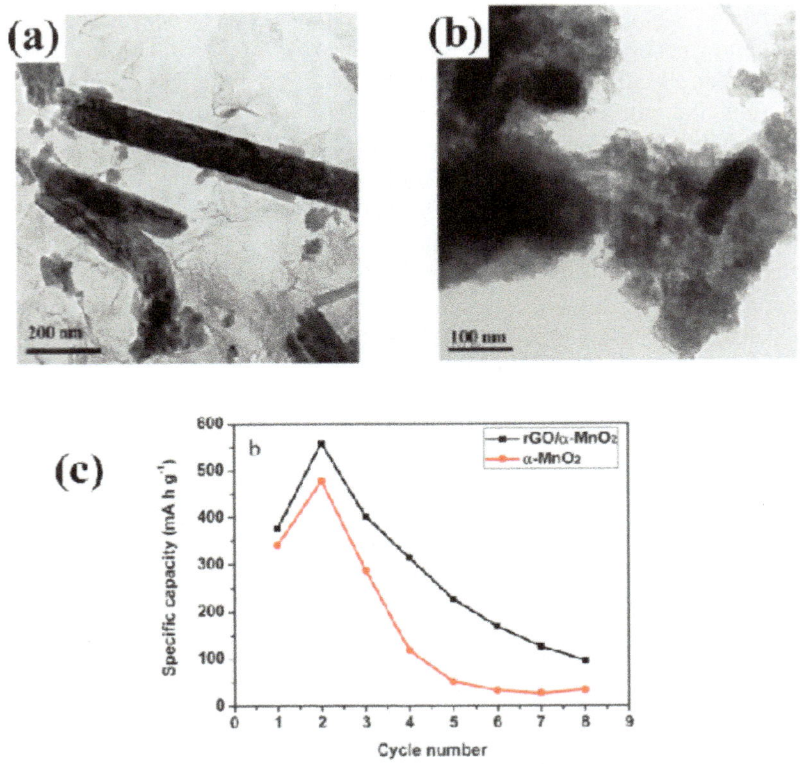

Figure 9: TEM images of RGO/α-MnO_2 composite cathodes (a) before the cyclic test and (b) after 20 cycles. (c) Cyclic performance of neat α-MnO_2 and RGO/α-MnO_2 composite cathodes at the current density of 100 mA g^{-1}. Reprinted with permission from Ref. [147]. Copyright (2013) The Royal Society of Chemistry.

Open issues. Despite the remarkable progresses summarized above, there are still daunting challenges in the Li–O_2 batteries, including the practical discharge capacities far below the theoretical value and the complex catalytic mechanisms. These difficulties indicate the necessity of continuous exploration to improve the electrochemical performance in the future, especially, to achieve a high capacity and a high rate in the Li–O_2 batteries simultaneously.

Supercapacitors

Compared to the lithium batteries, supercapacitors (SCs) as an alternative energy storage device possess unique merits of high power densities, short charge–discharge durations, bare memory effect, long cycle lifetime and environmentally benign [148] (seeTable 3). According to the charge–discharge mechanism, SCs can be classified into two categories: (1) electrochemical double-layer capacitors (EDLCs) where the electrical charge is accumulated at the electrode–electrolyte interface [148]; (2) pseudocapacitors where the fast and reversible Faradaic reactions take place at the electrode material[149].

Table 3: Electrochemical performances of GO/RGO-based supercapacitors.

Systems	Scan rate (mV s^{-1})	Current density (mA g^{-1})	*Cs*(F g^{-1})	Capacity retention (%)	Cycles	Ref.
GO		50	189	93	5,000	[150]
RGO		100	276	74.3		[152]
RGO		100	238	97	10,000	[153]
N doped RGO	5		258	87	1,000	[157]
3D RGO		10,000	210	100	2,000	[160]
Non stocking RGO		1,000	236.8	103.5	15,000	[161]
Template assisted RGO		500	456	134	10,000	[162]
MWCNT@GO nanoribbons	50		252.4	87	1,000	[163]
SSCNTs/RGO	50		244	96.4	1,000	[164]
MnO_2/GO		200	197.2	84.1	1,000	[170]
Mn_3O_4/GO	5		344	87	3,000	[171]

MnO/GO		500	51.5	82	15,000	[172]
Co_3O_4 hollow spheres/RGO		1,000	163.8	93	1,000	[173]
Co_3O_4/RGO		1,000	636	95	1,000	[174]
Co_3O_4/RGO		2,000	416	95.6	1,000	[175]
Co_3O_4 NSs/RGO		1,200	187	94	1,000	[176]
MnO_2/RGO	10		327.5	88.3	1,000	[177]
MnO_2/RGO		100	217	84	3,600	[178]
NiO/RGO		380	428	90.2	5,000	[179]
Glu/CP/NiO/RGO		1,000	1077	99	1,500	[180]
ZnO/RGO	50		107.9	95.8	200	[181]
Fe_3O_4/RGO		1,000	843	100	10,000	[182]
Fe_3O_4/RGO		1,000	195	100	1,000	[183]
Fe_3O_4/RGO		900	151	85	1,000	[184]
CuO_2/RGO		100	33	72.7	5,000	[185]
CuO_2 NSs/RGO	2		163.7	50.4	1,000	[186]
TiO_2/RGO		125	225	86.5	2,000	[187]
RGO/TiO_2Nanorod/RGO	5		114.5	85	4,000	[188]
MnO_2/C fiber/RGO		100	393	98.5	2,000	[189]
NiO/CNT/RGO		1,000	1180	95	2,000	[190]
WS_2/RGO	2		350	10	1,000	[191]
NiS/GO		1,000	800	87.5	1,000	[192]
Ni-Al-LDHs/RGO		1,000	2712.7	98.9	5,000	[194]
Ni-Al-LDHs/CNT/RGO		5 (mA cm^{-3})	1562	96.5	1,000	[195]
PANI/GO		200	555	92	2,000	[197]
PANI/GO		200	746	73	500	[198]
PANI/GO		300	525	91	200	[199]
PANI nanofibers/GO	1		1136.4	89	1,000	[200]
PANI/GO		900	797	118	500	[201]
PANI/NH_2-RGO	2		500	100	680	[202]
PANI–fRGO		100	590	100	200	[203]
PANI–RGO		450	431	74	500	[204]
PPy/GO-fiber		300	510	70	1,000	[205]
PPy/GO		1,000	633	94	1,000	[206]
PPy/GO		500	356	78	1,000	[207]
PPy/GO		500	728	93	1,000	[208]
PPy/GO	20		960	100	300	[209]
Benzimidazole/GO		100	370	90	5,000	[211]
Ion gel/GO (all-solid state SCs)		1,000	190	80	5,000	[215]
PPD/Au/GO		2 (mA cm^{-3})	238	81	500	[216]
TiO_2/PANI/GO		1,000	430	95	1,000	[217]
CNT/PANI/GO		5,000	413	81	1,000	[218]
CNF/PANI/GO	10		450.2	90.2	1,000	[219]
CNT–CNF/PPy/RGO		500	82.4%	93	2,000	[220]

GO used in SCs. In principle, GO cannot be directly used in SCs due to its intrinsically poor electrical conductivity, although there have been some investigations that demonstrated feasible usage of GO in SCs [150] and [151]. To improve the electrical conductivity of GO and to retain the pseudo-capacitive behavior, partially reduced GO has been widely adopted [152], [153], [154], [155], [156], [157], [158], [159], [160],[161] and [162]. The residual oxygenated groups on RGO can induce large pseudocapacitance in SCs [152]. Moreover, the co-contribution of double layer capacitance and pseudocapacitance from the oxygenated groups were observed in the SCs with water-soluble RGO as the electrode materials [153]. The reduction level of the RGO sheets (especially the variation of oxygen-containing groups) plays a significant role in controlling the intrinsic properties such as the interlayer spacing, oxygen content, BET specific surface area, and thus affects the overall performance of SCs[154] and [155].

Furthermore, it was shown that pH value also affects the performance of supercapacitors[156]. RGO sheets in the acidic and neutral media, which display both electrochemical double-layer (EDL) and pseudocapacitive behavior, have more oxygenated groups and lower surface areas but broader pore size distributions compared with those in the basic medium (mainly EDL). Meanwhile, chemically nitrogen doping of RGO can also tune the electronic properties to enhance the supercapacitor performance [157], [158] and [159]. For instance, the nitrogen doped RGO exhibited much higher specific capacitance than pure RGO [159]. In particular, the content of N atoms in N-RGO has remarkable influence on the capacitance, as displayed in Figure 10a.

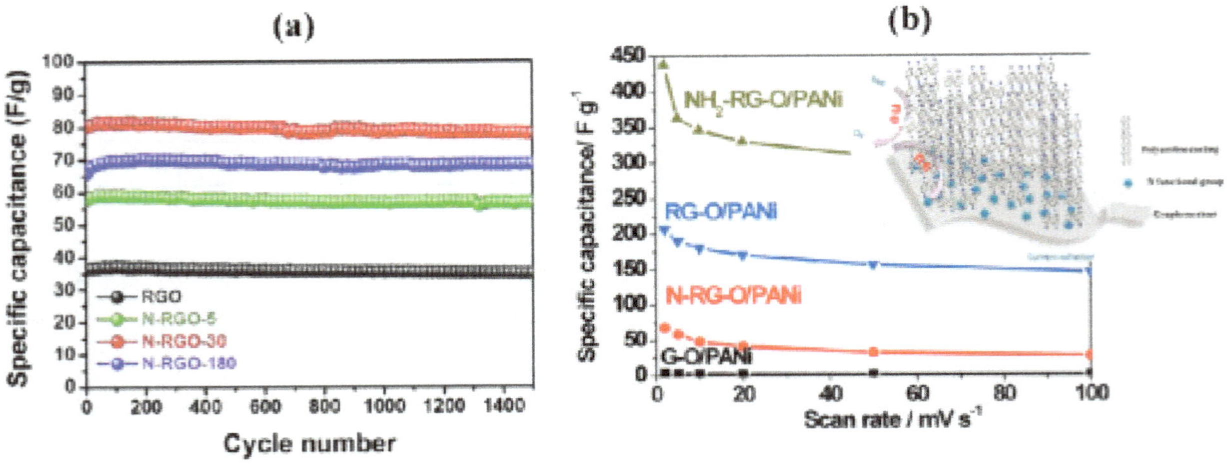

Figure 10: (a) The cyclic retentions of the supercapacitors based on N-RGOs at 0.5 A g^{-1}. Reprinted with permission from Ref. [159]. Copyright (2014) Elsevier Ltd. (b) Supercapacitance performance of the PANI/GO-based composite. Reprinted with permission from Ref. [202]. Copyright (2012) American Chemical Society.

Non-staking GO/RGO for SCs. Considering the negative influence of intercalated water molecules on restacking of RGO at the early dried stage of GO, Lee et al. [160]functionalized GO sheets with melamine resin (MR) monomers to prevent the hydrogen bonding with water molecules upon drying and consequently obtained the porous restacking-inhibited GO sheets with a specific capacitance of 210 F g^{-1} and superior capacitance retention for 20,000 cycles. Moreover, the anti-solvent (hexane) without any interaction with various oxygenated groups on GO was used to replace the water solution for fabricating the non-stacked RGO, which achieved high specific capacitance up to 236 F g^{-1} at the current density of 1 A g^{-1} and long cycle life [161]. Meanwhile, incorporation of $Mg(OH)_2$ nanosheets as the spacer in-between GO sheets effectively inhibited the restacking behavior during heat-treatment of GO, but preserved the stable oxygen-containing groups to enhance the pseudocapacitance at the same time [162].

Coupling of carbon materials with GO/RGO. It has been demonstrated to be a feasible strategy to enhance the supercapacitance [163], [164], [165], [166] and [167]. For example, the core–shell MWCNT@GO nanoribbons exhibited a high specific capacitance of 252 F g^{-1}, higher than 194 F g^{-1} for the SWCNT/GO electrode [163]. The CNT can be used as the spacer to inhibit restacking of RGO and to increase the supercapacitance [165] and [166]. Meanwhile, addition of electrochemically reduced GO resulted in a unique wrinkled yet porous structure with higher available surface area for accumulating the electrolyte ions to produce excellent electrochemical double-layers[167].

Incorporation of GO/RGO with transition metal oxides (TMOs). This can break some restrictions of TMOs electrode, such as low working voltage, poor stability and unsatisfactory high-rate capabilities [168] and [169], therefore improving the overall performance of SCs [170], [171] and [172]. For example, GO induced better electrochemical stability (84.1% retention of supercapacitance) in the needle-like MnO_2/GO composite electrode compared to the pure nano-MnO_2 (69% retention) even with a slightly smaller supercapacitance compensation of 197 F g^{-1} (211 F g^{-1} for the pure nano-MnO_2) [170]. Furthermore, the RGO/TMOs composites with controllable reduction level possess better electrical conductivity than the GO/TMOs composites, making them more suitable for SC applications [173], [174], [175], [176], [177], [178],[179], [180], [181], [182], [183], [184], [185], [186], [187] and [188]. For instance, the one-step synthesized Co_3O_4 nanosheets-RGO (Co_3O_4 NSs-RGO) hybrid with good redox activity exhibited a supercapacitance of 187 F g^{-1} at 1.2 A g^{-1} and maintained 91–94% capacitance even after 1000 cycles [176]. Similarly, the MnO_2/RGO composite as

SC electrode material displayed good flexibility and strength [178]. In addition, it facilitates easy access of electrolyte ions into the large surface area of the pseudocapacitive hybrid paper electrode. In a 3D flower-like hierarchical NiO/RGO composite, the NiO NPs not only inhibited the RGO restacking but also functioned as catalyst for GO reduction, which led to a high supercapacitance up to 428 F g^{-1} at the discharge current density of 0.38 A g^{-1}[179]. Most strikingly, the morphology of TMOs NPs was found as a key factor to determine the overall electrochemical performance of SCs [180] and [181]. Other TMOs/RGO hybrid composites like Fe_3O_4/RGO [182], [183] and [184], CuO/RGO[185] and [186] and TiO_2/RGO [187] and [188] also exhibited great electrochemical performance of SCs. Especially, the one-step solvothermal synthesized Fe_3O_4/RGO composite achieved an extremely high specific capacitance of 843 F g^{-1} at the discharge current density of 1 A g^{-1} in concomitant with very stable cycling performance (without any lose in supercapacitance after 1000 charge–discharge cycles) [182].

Addition of ternary carbon material into the GO/RGO–MOs can further improve supercapacitor performance [189] and [190]. In this case, the low solubility and aggregation tendency of CNTs restricted the growth of $Ni(OH)_2$ particles and therefore improved the activity of the NiO–CNT/RGO electrode to reach a specific capacitance of 1180 F g^{-1} at 1 A g^{-1}, along with excellent capacity retention of 95% after 2000 cycles (Figure 11a and b) [190].

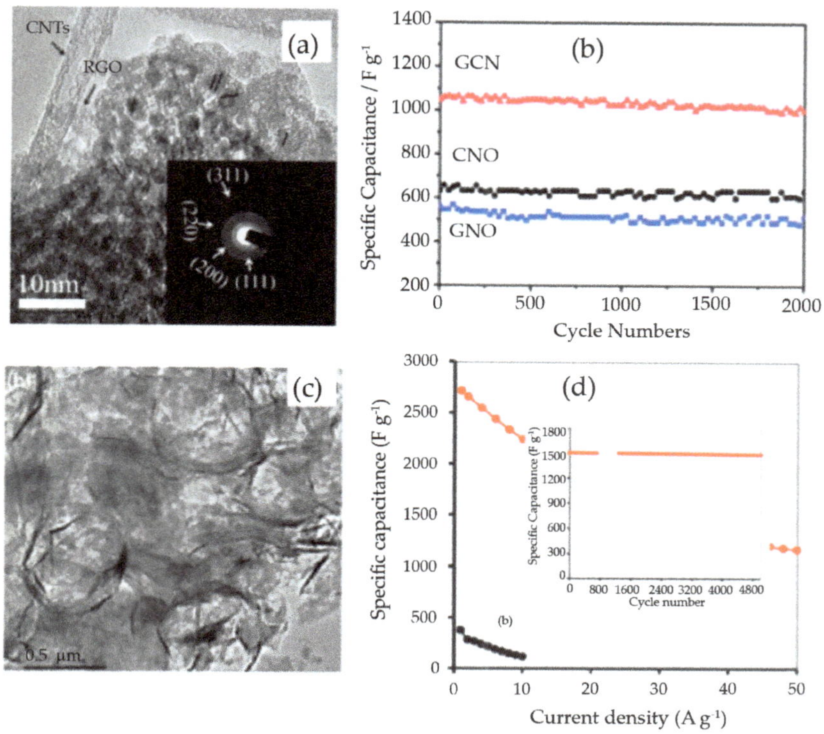

Figure 11: (a) TEM image of the GCN composites; (b) Cyclic performance of CNO, GNO, and GCN electrodes at 4 A g^{-1}. Reprinted with permission from Ref. [190]. Copyright (2014) The Royal Society of Chemistry. (c) High-resolution image of the 3D –ARGON/NiAl-LDH; (d) Specific capacitances of the 3D –ARGON/NiAl-LDH (red) and pure NiAl-LDH (black) capacitor cells in different discharge current density (Inset shows the cyclic performance of the 3D-ARGON/NiAl-LDH at a current density of 30 A g^{-1}). Reprinted with permission from Ref. [194]. Copyright (2013) Elsevier Ltd. (For interpretation of the references to color in this figure legend, the reader is referred to the web version of this article.)

Synergistic coupling of metal sulfide with GO/RGO. This has been demonstrated to be an effective strategy to improve the supercapacitor performance [191] and [192]. The NiS/GO nanocomposite achieved a high specific capacitance of 800 F g^{-1} at 1 A g^{-1} as well as a long life over 1000 cycles [192]. The hybrid of layered double hydroxides (LDHs) and GO/RGO composites have also been investigated for their extraordinary SCs performances [193], [194] and [195]. Figure 11c shows the high-resolution image of the 3D-RGO nanocup/Ni-Al LDHs composite. This composite exhibits an extremely high specific capacitance of 2712 F g^{-1} at the current density of 1 A g^{-1} and kept 98.9% of the initial capacitance after 5000 cycles at the current density of 30 A g^{-1} (Figure 11d) [194].

GO/RGO with conducting polymers. Compared to the TMOs, conducting polymers [196] exhibit superior conductivity, higher supercapacitance, lower cost; but the relatively poor cycling performance hinders their practical applications. To overcome this shortcoming, GO/RGO are uniformly dispersed in the conducting polymers, e.g., polyaniline (PANI)[166], [167], [168], [169], [170], [171], [197], [198], [199], [200], [201], [202],[203] and [204] and polypyrrole (PPy) [172], [173], [174], [175], [176], [205], [206], [207],[208] and [209], and thus form stable hybrid composites for SC electrodes. The compositing mechanisms between GO/RGO and conducting polymers (including electrostatic interaction, hydrogen bonding and π–π stacking) influence the supercapacitor performance [198]. For instance, owing to the strong interaction between GO and PANI, the PANI/GO composite reached an extremely high supercapacitance of 1136 F g^{-1}, compared to 484 F g^{-1} for the PANI alone at the scan rate of 1 mV s^{-1}[200]. Moreover,

GO can serve as a weak electrolyte as well as act as an effective charge balancing dopant within the PPy, both being beneficial for the SC applications [207].

Chemistry of GO/RGO-conducting polymer surfaces. The surface chemistry has demonstrated significance for controlling the overall electrochemical performance of SCs. Among four kinds of surface functionalized graphene/PANI composites, i.e., GO, RGO, nitrogen-doped RGO (N-RGO), and amine-modified RGO (NH_2-RGO) in Figure 10b, the NH_2-RGO/PANI composite shows largest specific capacitance of 500 F g^{-1} at 2 mV s^{-1} over 680 cycles [202]. Furthermore, it was found that ethanol can facilitate uniform coating of (poly 3,4-ethylenedioxythiophene), PANi, and PPy on RGO sheets[210]. Another polymer of benzimidazole cross-linked with GO formed a porous composite and therefore showed good cyclic stability of SCs [211]. Moreover, RGO/polyamide-66 (PA66) nanofiber (as electrodes) [212] and GO/PPy composite (as organic current collector) also exhibited great electrochemical performance in SCs [213]. In comparison with the pure ion gel and the conventional supercapacitor, the GO-doped ion gel as electrolyte in the all-solid-state supercapacitor possess less internal resistance, higher capacitance performance, and better cycle stability [214] and [215].

Based on the GO/RGO-polymer composites, other active materials such as gold [216], TMOs [217] and [218], and carbon materials [219], [220] and [221] have been further introduced to enhance the SCs performances. Especially, the hierarchical free-standing CNF/GO/PANI composite possesses 3D open diffusion channels for electrolyte ions and thus yields a high supercapacitance of 450 F g^{-1} at the scan rate of 10 mV s^{-1} and good cyclic stability up to 1000 times [220].

Open issues. The GO/RGO-based composites exhibit superior supercapacitive performance by providing the synergistic effects of EDLs and pseudocapacitive behavior. For the commercialization of high performance supercapacitors, however, there are still significant challenges. Further investigations not only improving the performance of electrodes but also illustrating the intrinsic synergistic effects are expected for materials design and applications.

APPLICATIONS OF GO IN ENVIRONMENTAL PROTECTION

Management of harmful gases

Air pollution by industrial release of greenhouse and toxic gases is one of the biggest threatens to environment. In principle, removal of air pollutants can be divided into three routes: (1) direct reduction of gases emission, (2) gases capture and storage, (3) ultimate utilization. The oxygen-containing functional groups on the basal plane and sheet edges make GO capable to covalently and non-covalently interact with various molecules. The unique electronic structures of GO also endow effective catalyst for converting undesirable gases during the industrial processing. Therefore, GO-based composites are promise materials for capture and conversion of harmful gases such as CO_2, CO, NO_2, and NH_3.

Capture of harmful gases

Capture of greenhouse gas CO2. The few-layer GO sheets show superior adsorption behavior, especially under the assistance of water molecules [222], [223],[224] and [225]. Molecular dynamics simulation demonstrated that the functional groups of GO is able to enhance CO_2 adsorption [223]. The presence of water can maintain the integration of CO_2 intercalated GO structure [224] and affect CO_2 migration by repulsive interaction between CO_2 and the oxygenated groups attached on the GO sheets [225]. Compared to the MWCNTs-LDHs [226], GO–LDHs [227] exhibit better efficiency of CO_2adsorption under low GO content, while the poor network forming ability limited the CO_2adsorption capacity even with further increased GO loading (see Figure 12a and b). For the layered double oxides (LDOs), it has been demonstrated that incorporation of GO can alleviate the capacity decay of CO_2[228]. Addition of GO (20 wt% content) into the hybrid monolith aerogels of chitosan affected the morphological characteristics and dramatically enhanced the BET surface area for strong interactions with CO_2; thus the CO_2 storage capacity increased from 1.92 to 4.15 mol kg^{-1}[229]. Besides, GO facilitated the dispersion of the nanoscale Cu–MOF, which provided more active sites for CO_2adsorption and thus led to a high CO_2 adsorption capacity of 8.26 mmol g^{-1} at 273 K under 1 atm [70].

Removal of ammonia. In addition to the greenhouse gases, GO/RGO-based composites have excellent adsorption capability of ammonia [230], [231], [232], [233], [234], [235],[236], [237] and [238]. Bandosz's group found that the oxygenated groups, remained sulfonic groups, and water molecules play important role in the ammonia adsorption of graphite oxides [230], [231], [232] and [233]. Further incorporation of additional active materials such as polyoxometalate [234], Al_{13}[235] and MnO_2[236] into graphite oxides can dramatically enhance the ammonia accommodation. Moreover, first-principles calculations showed that the diverse active sites on GO (hydroxyl and epoxy functional groups and their neighboring carbon atoms) are effective for facilitating the charge transfer between NH_3 and GO [237]. Later, the co-contribution of epoxide groups and carbon vacancies to NH_3 dissociation has been confirmed by in situ IR microspectroscopy experiments combined with DFT calculations [238].

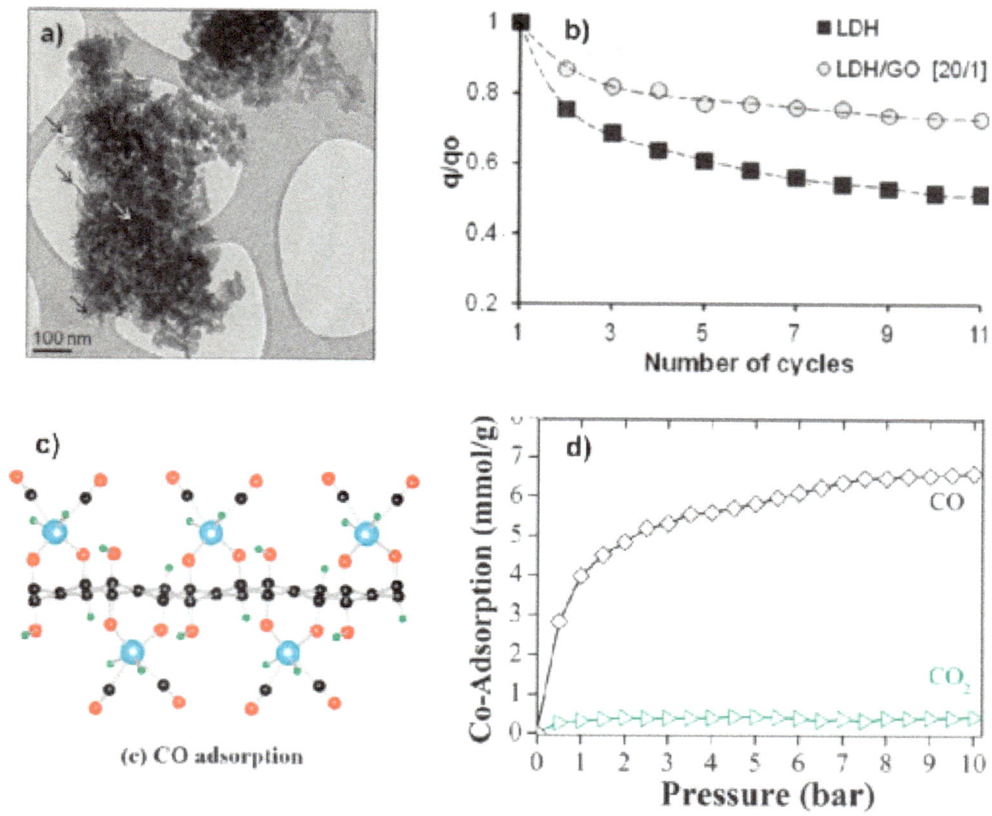

Figure 12: (a) HRTEM image of the LDH/GO sample; (b) Normalized CO_2 adsorption capacity over 11 adsorption–desorption cycles (adsorption 573 K; desorption 673 K). Reprinted with permission from Ref. [227]. Copyright (2012) American Chemical Society. (c) Structural model of Ti decorated on both sides of GO surface with CO molecular adsorption; (d) CO/CO_2 co-adsorption isotherm of Ti decorated GO at 298 K by GCMC simulation. Reprinted with permission from Ref. [245]. Copyright (2011) The Royal Society of Chemistry.

Removal of other harmful gases. It has shown that other *harmful gases* like formaldehyde [239], acetone [240], H_2S [241] and [242], SO_2 [243] and [244], CO [245], nitrogen oxides [244] and [246] can be also effectively removed via GO/RGO-based composites. DFT calculations predicted that the hydroxyl and carbonyl functional groups as well as the nearby carbon atoms induced strong interactions with NO_x (x=1, 2, 3) molecular species, which eventually led to chemisorption of these gaseous molecules[246]. For the extremely dangerous acetone gas, GO foam displayed higher affinity than that of RGO foams and other carbon materials [240]. In addition, it has been demonstrated that a small amount of GO integrated with MOF-5 can increase the available Zn sites in MOF-5 and also create pore space with strong dispersive force for H_2S capture [241]. A theoretical study by Huang et al. predicted that the carbonyl groups were the active sites for H_2S dissociation among various functional groups on GO, while the H_2O molecules behaved as the challenging gas to block the potential active sites for H_2S decomposition [242]. Moreover, in the graphite oxide/Zr$(OH)_4$ composite, SO_2retention is not only accounted by physical adsorption in small pores but also contributed by strong attraction of terminal –OH groups of hydrous zirconia [243].

The atomistic mechanism for selective gas adsorption of GO-metal hybrid materials has been revealed by DFT calculations and grand canonical Monte Carlo (GCMC) simulations by Zhao's group [245]. As shown in Figure 12c, the titanium-decorated GO (Ti–GO) monolayer sheet is an ideal sorbent for carbon monoxide capture with a large binding energy of about 70 kJ mol^{-1}, due to stronger hybridization between the empty d orbitals of Ti and the occupied p orbitals of CO. GCMC simulations further demonstrate strong selectivity of Ti–GO sheet for CO adsorption in a mixture with other gases like CH_4, CO_2, N_2 (Figure 12d).

Conversion of harmful gases

GO/RGO-based composites as chemical catalysts for CO2conversion. Besides gas capture, GO/RGO-based composites have attracted substantial attentions on converting the air pollutants to some useful energy resource for full utilization. With the assistance of tetrabutylammonium bromide (Bu_4NBr) as co-catalyst to convert CO_2 into propylene oxide, the GO–Bu_4NBr composite yielded 96% of propylene carbonate under relatively mild conditions [247]. The porous GO foams served as not only oxidant but also catalyst for oxidization process from SO_2 to SO_3[248]. SO_2 can also be inserted into GO as the oxidized intermediate, which led to functionalization of graphene oxide with sulfur and further decoration by various types of organic moieties [249]. In addition to acting as catalyst itself, GO/RGO may serve as excellent host to support efficient catalyst for conversion of CO [250], CO_2[251] and NH_3[252]. For instance, the oxygenated groups on GO can accelerate formation of Co nanocrystals, leading to 3D reticular structure, which can be seen in the TEM image of Figure 13a. This structure displays good catalytic performance in the Fischer–Tropsch CO_2 hydrogenation process (see Figure 13b) [251].

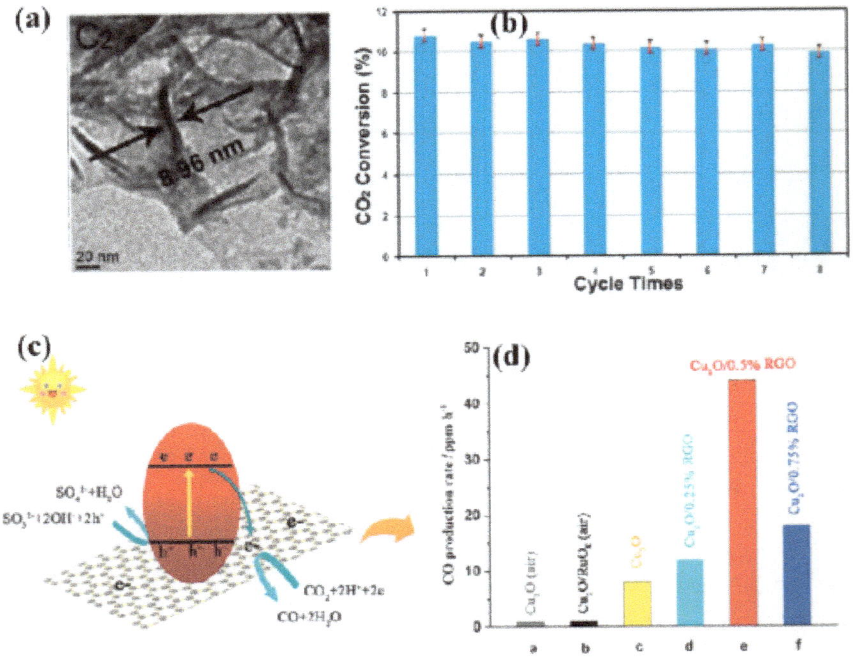

Figure 13: (a) TEM image and (b) catalytic stability of the catalyst made from Co/RGO composite. Reprinted with permission from Ref. [251]. Copyright (2013) The Royal Society of Chemistry. (c) Schematic illustration of the charge transfer in Cu_2O/RGO composites; (d) CO yield in the 20th hour over different photocatalysts. Reprinted with permission from Ref. [253]. Copyright (2014) Wiley-VCH Verlag GmbH & Co. KGaA, Weinheim.

Photocatalytical conversion of CO2by GO/RGO-based composites. Benefited from its unique electronic structures, GO/RGO-based composites behave as superior photocatalysts for CO_2 conversion [253], [254], [255], [256], [257] and [258]. The Pt modified RGO and Pt modified TiO_2 nanotubes were combined as cathode and photo-anode catalysts, respectively, which exhibited high efficiency of converting CO_2 into chemical materials (CH_3OH, C_2H_5OH, HCOOH and CH_3COOH) [254]. Considering the high cost of Pt catalyst, alternative non-noble metal catalysts have been explored for CO_2 conversion [253], [255], [256], [257] and [258]. For Cu_2O, RGO coating also induced retarded electron–hole recombination, efficient charge transfer, and protection, which dramatically enhanced the catalytic activity and the system stability for photoreduction from CO_2 to CO (Figure 13c and d) [253]. By modulating the oxygenated groups, the modified GO reached a methanol conversion rate of 0.172 μmol g^{-1} cat^{-1} h^{-1} (under 300-Watt halogen lamp irradiation) [255]. RGO incorporation also caused charge anti-recombination of TiO_2[257] and CdS [258] and therefore enhanced the photocatalytic activity for reducing CO_2 to useful CH_4, which is even superior to the optimized Pt–CdS nanorod photocatalyst [258]. Except for CO_2, colloidal GO can enhance the photocatalytic activity of nanocrystalline TiO_2 for gas-phase oxidation of ethanol and benzene by air oxygen, since the GO sheets were capable to accept photoelectrons from titania nanocrystals and to bound to TiO_2 surface [259].

Water purification

Adsorption of containments in wastewater

Parallel to air pollution, water pollution is another worldwide environmental concern. The effective strategies of water purification can be categorized into pollutants adsorption and conversion. For the pollutants (mainly heavy metal ions and organic dyes) in wastewater that strongly threaten human, animals and plants, GO/RGO-based composites typically show strong binding with these pollutant species, as summarized inTable 4.

Table 4: Adsorption capacities of heavy metal ions and organic containments from wastewater.

Adsorbent	Adsorbate	Adsorption capacity (mg g^{-1})	Condition	Ref.
2,6-Diamino pyridine–RGO	Cr(VI)	393.7	C_0 500 mg L^{-1}	[267]
PPY–GO	Cr(VI)	497.1	PH 3.0	[268]
PANI–GO	Cr(VI)	1149.4	PH 3.0; *T* 298 K	[269]
M–GO	Co(II)	12.9	PH 6.8; *T* 303 K	[265]
GO	Cu(II)	46.6	$C_e \geq 21.8$ mg L^{-1}	[260]
GO aerogel	Cu(II)	19.1	PH 6.3; *T* 298 K	[261]
GO	Pb(II)	842	PH 6.0; *T* 293 K	[264]
EDTA–GO	Pb(II)	479±46	PH 6.8; *T* 298 K	[326]

MC–GO	Pb(II)	76.9	PH 5.0	[271]
PPy–RGO	Hg(II)	980	PH 3.0; *T* 293 K	[270]
CGGO	Cu(II)	~120	PH 6.0; *T* 298 K	[272]
	Pb(II)	99		
GO	Au(III)	108.3	PH 6.0; *T* 298 K	[263]
	Pd(II)	80.8		
	Pt(IV)	71.4		
GO	Cd(II)	106.3	PH 6.0; *T* 303 K	[262]
	Co(II)	68.2		
Chitosan–GO	Au(III)	1076.6	PH 3.0–5.0; *T*298 K	[273]
	Pd(II)	216.9	PH 3.0–4.0; *T*298 K	
GO	Methylene blue	714	PH 6.0; *T* 298 K	[274]
RGO	Methylene blue	158	*T* 283 K	[278]
MCGO	Methylene blue	179.6	PH 10.0	[288]
Fe_3O_4/SiO_2–GO	Methylene blue	97	*T* 298 K	[280]
CA–GO	Methylene blue	181.8	PH 5.4; *T* 298 K	[290]
MCCG	Methylene blue	84.3		[289]
$Na_2S_2O_4$ in situ reduced GO	Acridine orange	3300	PH 4.0–5.0	[279]
GEPMs	Amaranth	800		[291]
Fe_3O_4@mTiO_2–GO	Congo red	89.9	PH 6.0; *T* 303 K	[281]
EGO	Methylene blue	17.3	PH 10.0	[275]
	Methyl violet	2.47	PH 6.0	
	Rhodamine B	1.24	PH 6.0	
RGO	Orange G	5.98		
GO sponge	Methylene blue	397	PH 7.0; *T* 343 K	[276]
	Methyl violet	467		
GO	Rhodamine 6G	23.3	*T* 298 K	[277]
	Dopamine	40		
Fe_3O_4–RGO	Rhodamine B	~50		[282]
	Rhodamine 6G	~30		
	Acid blue 92	~90		
	Orange(II)	~90		
	Malachite green	~50		
	New coccine	~45		
Fe_3O_4–RGO	Methylene blue	167.2	PH 3.0–11.0	[283]
	Neutral red	171.3	PH 3.0–7.0	
PDDA–GO	Ponceau S	188.7	PH 6.0; *T* 298 K	[292]
	Trypan blue	50.0		
Magnetic Fe_2O_4–RGO	Rhodamine B	22.5	PH 7.0; *T* 298 K	[284]
	Methylene blue	34.7		
Poly(acrylamide) –RGO	Pb(II)	1000	PH 6.0; *T* 298 K	[293]
	Methylene blue	1530		
IO–GO(IO–RGO)	Pb(II)	588.2 (454.5)	PH 6.5; *T* 303 K	[285]
	1-Naphthylamine	228.4 (243.1)		
IO–RGO	Pb(II)	~250	PH 7.0; *T* 301 K	[286]
	Cd(II)	~210		
	TBBPA	~35		

Magnetic GO	Cd(II)	91.3	PH 6.0; *T* 298 K	[287]
	Methylene blue	64.2		
	Orange G	20.8		

GO for heavy metal ion adsorption. Even with the oxygen-containing functional groups on GO/RGO surface acting as the active sites, GO exhibits only very low adsorption capacity of Cu^{2+} ions because of the aggregation triggered by the interaction between GO and Cu^{2+} ions [260]. Similar result for adsorption of Cu^{2+} ions was observed in GO aerogels [261]. Later, it was reported that GO nanosheets showed increased adsorption ability for Cd(II), Co(II), Au(III), Pd(II), and Pt(IV) ions than Cu^{2+}, due to their good structural stability [262] and [263]. Most strikingly, the few-layered GO materials (which showed no aggregation even after five months) exhibited excellent adsorption capacities of about 842 mg g^{-1} for Pb(II) ions at 293 K [264]. However, further attempts of applying the magnetite-GO composite for Co(II) ions removal only led to very low adsorption capacity of 12.9 mg g^{-1}[265].

GO with organic compounds. This can provide more feasible anchoring sites for heavy metal ions [266], [267], [268], [269] and [270]. The UV-activated 2,6-diamino pyridine–RGO composite created an extra pyridinic-nitrogen lone pair, which facilitated the removal efficiency of Cr(VI) by not only adsorption but also partial reduction [267]. The synergistic combination of conducting polymers and GO exhibits high removal capacity of heavy metal ions [268], [269] and [270]. Especially, the hierarchical PANI/GO [269], which exhibited extremely high removal capacity of 1149.4 mg g^{-1} for Cr(VI), simultaneous with partial reduction from Cr(VI) to Cr(III) by positive nitrogen groups and the assistance of p electrons on the carbocyclic six-member ring. Other chitosan modified GO, e.g., magnetic chitosan-GO for Pb(II) [271], chitosan–gelatin–GO for Cu(II) and Pd(II) [272] and chitosan–GO for Ag(II) and Pd(II) [273], also showed strong attraction to the heavy metal ions.

GO for organic dyes adsorption. On the other hand, GO/RGO itself renders strong ability to remove another kind of harmful containments in wastewater, i.e., organic dyes. Zhang et al. [274] reported that GO exhibited high removal efficiency (>99%) of methylene blue (MB) due to the electrostatic interaction and partially contributed by the π–π stacking interaction between GO and MB. In addition, competitive adsorption of various organic dyes on GO surface was also found [275], [276] and [277]. After removing some functional groups, the partially reduced GO still kept strong affinity for MB adsorption owing to the C=O or C=N double bonds and C–C/C=C bonds induced conjugate interaction [278]. $Na_2S_2O_4$ in situ reduction of GO can convert carbonyl groups into hydroxyl groups, which acted as the key sites for a sharply adsorption enhancement of acridine orange from 1400 to 3300 mg g^{-1}[279].

GO/RGO-iron oxides composites for adsorption of containments in wastewater. For GO/RGO-based composites, it has been demonstrated that addition of iron oxides can effectively enhance the containment removal in wastewater under external magnetic field [280], [281], [282], [283], [284], [285], [286], [287], [288] and [289]. For example, the Fe_3O_4–RGO hybrid greatly enhanced the thermal stability of GO and also provided additional magnetic field for facilitating the removal of MB and neutral red (NR) [283]. Similarly, the magnetic chitosan–GO composite functioned as efficient adsorbents in removal of MB, which was benefited from the surface chemistry of GO, the hydrophobicity of β-cyclodextrin, the abundant amino and hydroxyl functional groups of chitosan, and the magnetism of Fe_3O_4, simultaneously [289]. In addition, it was found that the iron oxide decorated GO only exhibited adsorption ability for Pb(II) but not for 1-naphthol and 1-naphthylamine, whereas the iron oxide-RGO composite behaved in an opposite way [285]. A recent study of magnetic iron oxide/RGO nanohybrid revealed effective adsorption ability for both heavy ions and organic dyes from the wastewater[286].

Besides, there have been other attempts of utilizing the calcium alginate-GO hybrid for MB adsorption [290] and the 3D porous GO-polyethylenimine composite for amaranth adsorption [291], which achieved maximum adsorption capacities of 181.8 and 800 mg g^{-1}, respectively. Furthermore, the poly(diallyldimethylammonium chloride)–GO hybrid exhibited high removal efficiency for both ponceau S (PS) and trypan blue (TB) owing to the strong π–π stacking and anion–cation interactions [292]. The poly(acrylamide)–RGO nanocomposite showed enormous attraction for both Pb(II) ions and MB simultaneously (1000 mg g^{-1} for Pb(II) and 1530 mg g^{-1} for MB, respectively) [293].

Conversion of containments in wastewater

Chemical reduction of containments. In general, the pollutant conversion technology in water purification via GO-based materials can be categorized into chemical- and photo-reduction. For chemical reduction of Cr(VI) to low toxic Cr(III) species at low pH, the ethylenediamine coated RGO (ED-RGO) exhibited a relatively high conversion rate with the aid of π electrons on the carbocyclic six-membered ring of ED-RGO [266]. The conversion mechanism and PH affected conversion ratio of Cr(VI) ions have been displayed in Figure 14a and b, respectively. In addition, the zigzag edges of RGO acted as the catalytic active sites, while the basal plane of RGO served as the conductor for electron transfer during the catalytic reduction of nitrobenzene by Na_2S [294]. Other GO-based composites like Co_3O_4 coated GO and Co_3O_4 nanorods coated RGO also displayed high chemical catalytic ability for the Orange II decomposition [295] and [296]and the methylene blue degradation [297], respectively.

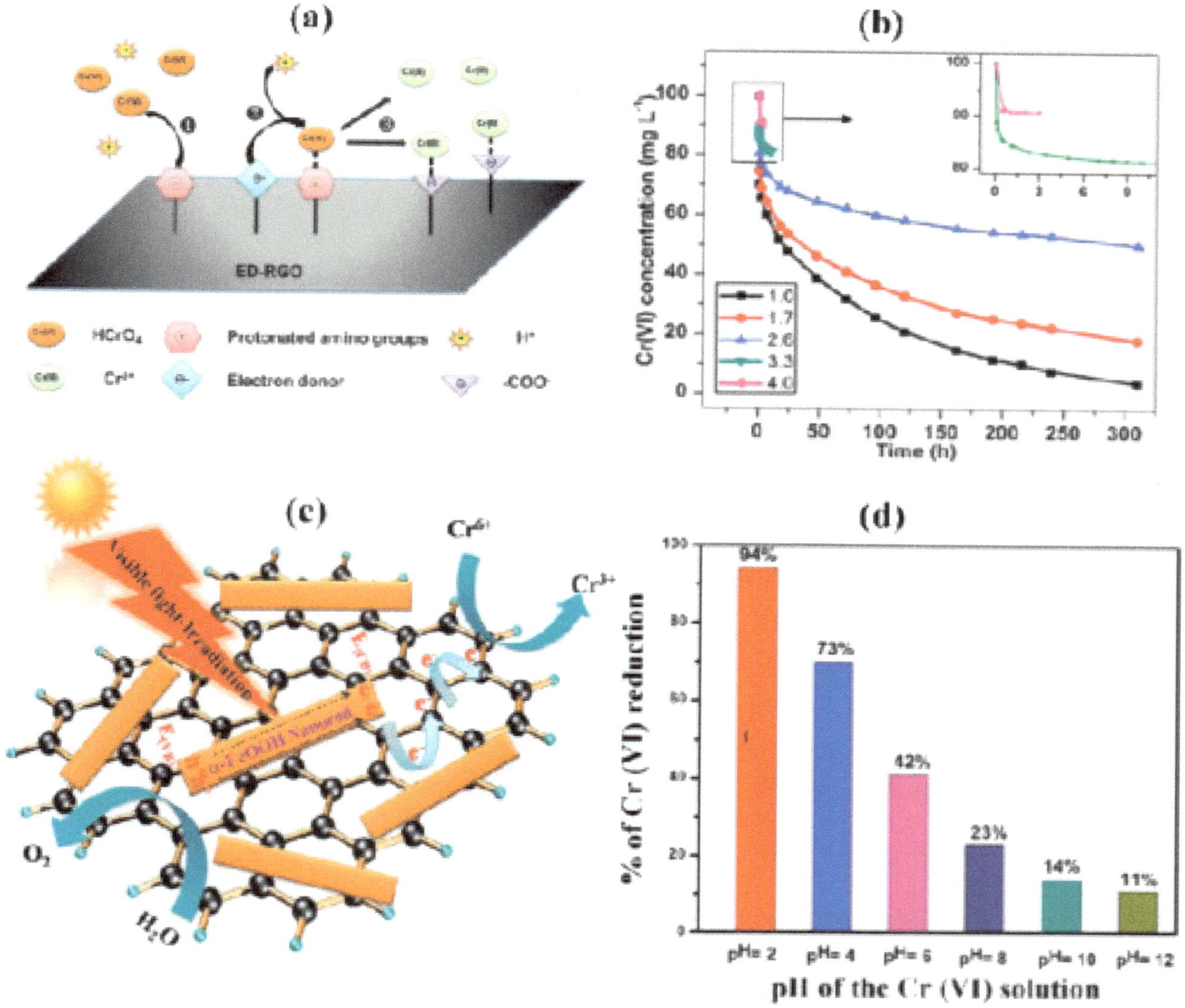

Figure 14: (a) Proposed mechanism of Cr(VI) removal by ED-RGO; (b) Effect of the solution pH on Cr(VI) removal in 50 mL 100 mg L^{-1} solution. Reprinted with permission from Ref. [266]. Copyright (2012) The Royal Society of Chemistry. (c) Mechanism of photocatalytic reduction of Cr(VI) over 3GFeOOH; (d) Effect of pH on the initial rate of Cr(VI) photo-reduction. Reprinted with permission from Ref. [300]. Copyright (2012) The Royal Society of Chemistry.

Photoreduction of heavy metal ions. On the other hand, GO-based materials as photocatalysts show high conversion rate for both toxic Cr(VI) ions and organic containments. It has been demonstrated that additions of RGO into TiO_2 NPs [298], CdS[299] and α-FeOOH nanorod [300] can significantly enhance the photocatalytic activity of Cr(VI) reduction. In the α-FeOOH nanorod/RGO composite, the extended π-conjugated flat 2D layer of graphene played a crucial role in channelizing the photoexcited electrons on its surface, which led to minimization of the electron–hole recombination for enhancing the photocatalytic activity of the α-FeOOH nanorod (Figure 14c and d) [300].

Photoreduction of RhB. Due to the extraordinarily slow rate of RhB degradation by RGO alone under visible-light irradiation [301], many compounds have been incorporated with RGO to improve the photocatalytic performance for RhB degradation. The CNT-pillared RGO exhibited excellent visible light photocatalytic activity of RhB degradation owing to its unique porous structure and the exceptional electron mobility of graphene [302]. The synergistic combination of RGO with other photocatalysts such as metal oxides [303],[304] and [305], Ag_3PO_4[306], $BiVO_4$[307] and Ag@AgCl [308] resulted in outstanding photocatalytic performance for the degradation of rhodamine dyes. For example, Figure 15a shows the TEM micrograph of TiO_2–RGO nanocomposites. This composite exhibits high conversion rates of 98.8 wt% after 80 min under visible light (Figure 15b) [303]. Similarly, the ZnO nanowire–RGO [304] also achieved high conversion rates of 98% within 10 min under ultraviolet irradiation. Moreover, the excellent electron-accepting and -transporting properties of RGO obviously enhanced the photocatalytic activity of RHB in the Ag_3PO_4–RGO composite [306].

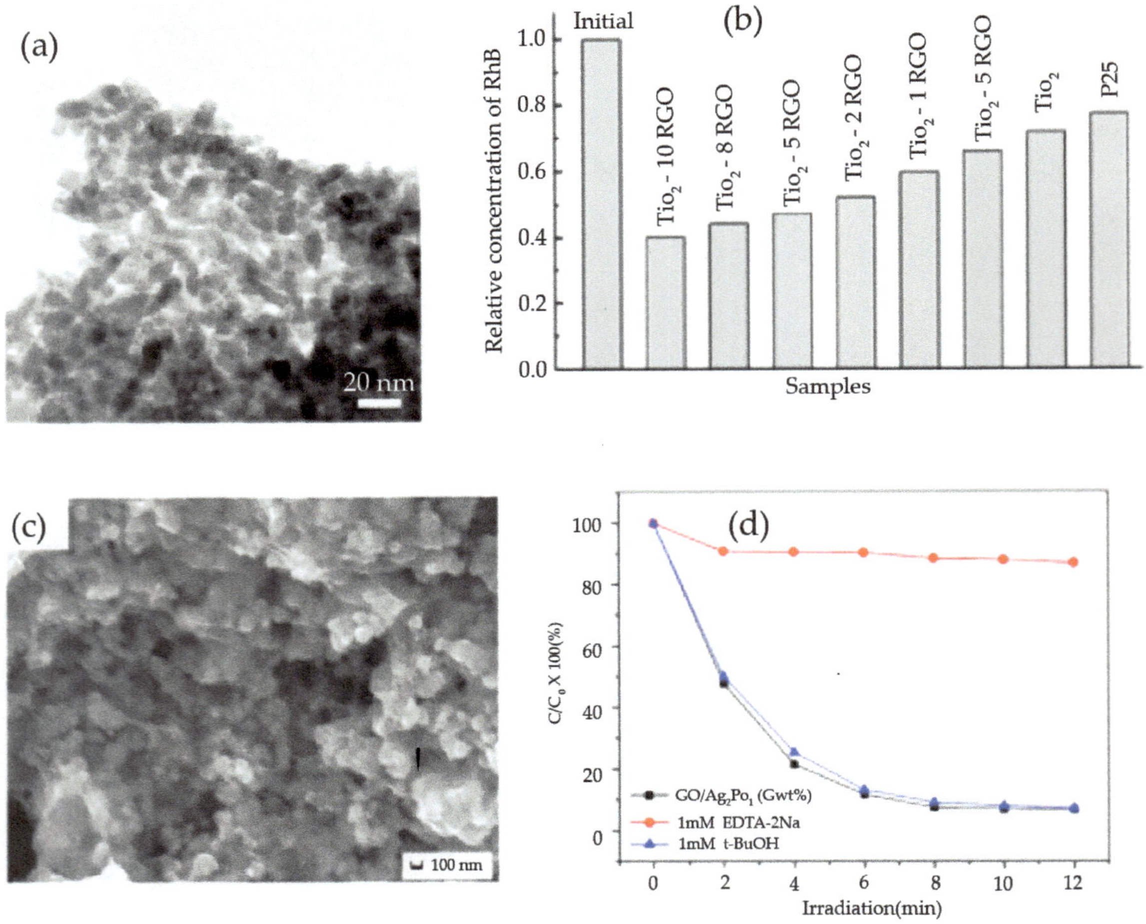

Figure 15: (a) Typical TEM micrograph of TiO_2–RGO nanocomposites; (b) Bar plot showing the remaining relative concentration of RhB after dark absorption by TiO_2–RGO nanocomposites. Reprinted with permission from Ref. [303]. Copyright (2012) American Chemical Society. (c) SEM image of GO/Ag_3PO_4(6 wt%) composite; (d) Plots of photogenerated active species trapped in the system of photodegradation of RhB by GO/Ag_3PO_4 (6 wt%) under visible light irradiation. Reprinted with permission from Ref. [322]. Copyright (2014) Elsevier Ltd and Techna Group S.r.l.

Photoreduction of other organic dyes. Other organic dyes such as MB [309], [310], [311],[312] and [313], methyl orange (MO) [314] and [315] and crystal violet (Cv) [316] have also been effectively photodegraded via GO/RGO-based photocatalysts. Recently, the Z-scheme photocatalytic system of the Ag@AgCl/RGO composite displayed much higher photocatalytic activity of MB degradation than the normal Ag@AgCl/RGO composite because of the two parallel photochemical reactions by the electron–hole pairs of the low energy level recombined in space by metallic Ag as a solid-state electron mediator and the remaining electron–hole pairs of the high energy level [309]. Furthermore, the GO nanosheets not only controlled the fabrication of Ag/AgCl nanostructures, but also acted as a promoter for the photocatalytic performances of MO photodegradation [314]. In the nanoscale hybrid of ZnO–GO composite, GO also induced high e^-–h^+ pair separation under light illumination (~95% Cv degraded within 80 min) for high rate photodegradation of Cv dye [316].

Besides dyes, other kinds of organic pollutants can be also photodegraded by the GO/RGO-based composites. The typical herbicide 2,4-dichlorophenoxyacetic acid (2,4-D) was nearly 100% photodegraded by Ag/RGO–TiO_2 composite under simulated solar light irradiation with good cyclic stability over ten times [317]. Similarly, the epoxy-GO*x*hybrid showed photocatalytic activity of phenol degradation [318] and the CdS nanowires/RGO nanocomposites displayed selectively photocatalytic reduction of aromatic nitro organics in water under visible light irradiation [319].

Various organic containments can be simultaneously removed via GO-based composites [320], [321], [322], [323] and [324]. Both GO/Ag_3PO_4[320] and GO/TiO_2/Ag nanocomposites [321] showed substantial photocatalytic activities in degradation of organic dye and phenol under visible light irradiation. The GO/Ag_3PO_4 composite also exhibited efficient photodegradation of RhB and bisphenol A (Figure 15c and d) [322]. Additionally, combination of RGO with β-$SnWO_4$ increases the degradation efficiency of MO (from 55% to 90%) and RhB (from 60% to 91%), as compared to β-$SnWO_4$ alone[323].

SUMMARY AND PERSPECTIVE

Benefited from the oxygenated groups, GO (including RGO) can be chemically modified or functionalized by cooperating with other active species. The GO/RGO composites therefore possess unique surface chemistry and architectures, such as three dimensional network, large surface area, tunable electrical conductivity, satisfactory chemical/electrochemical stability, high flexibility and excellent elasticity. To promote further investigations of such kind of advanced materials, this review comprehensively summarizes recent progresses on the energy storage and conversion (hydrogen storage, photocatalytic water splitting, lithium batteries, and supercapacitors) as well as environmental protections (air pollutants removal and water purification) using GO/RGO and the related composite materials.

By varying the concentrations of surface functional groups, band gap and work function of GO sheets can be tuned to suitable values as efficient catalysts for photocatalytic water splitting, reduction of harmful gases or heavy metal ions, and degradation of organic containments. Also, the high surface area and abundant functional groups of GO offer enough space and active sites for adsorption of gaseous molecules and various species in solutions, such as hydrogen storage, capture of CO_2, CO, NO_2, and NH_3gases for air purification, removal of heavy metal ions and organic containments for water purification. In principle, applications of pristine GO as electrode materials in lithium batteries and supercapacitors are restricted by its insulating nature. However, appropriate reduction can improve the electrical conductivity while keeping the inherent merits of GO. By further compositing with other active materials, such as carbon materials, metals, metal oxides, conducting polymers and organic species, GO/RGO-based materials show greatly improved performance in the energy storage devices.

As for the environmental applications, metal-decorated GO exhibit satisfactory capability of gas capture and detection, especially the harmful gases like CO_2, CO, NO_2, and NH_3; meanwhile even GO itself can act as catalyst for oxidation of the harmful gases. In the polluted water, various GO-based composites not only display strong affinity for the adsorption of heavy metal ion and organic containments but also function as efficient chemical- and photo-catalysts in converting the toxic metal ions and organic containments into the harmless products.

Despite the great advances in the GO/RGO-based materials for energy utilization and air/water purification described above, development of mature technologies using GO and GO composite materials still remains a long-term target in the future and requires more attentions from different aspects. From the fundamental point of view, it would be crucial to elucidate the interaction mechanism between GO/RGO and the additional material in a composite and to further understand the possible synergistic effects. Such knowledge would provide useful guidance to design GO-based composites with optimal performance. On the technological side, integration of GO/RGO-based materials into a realistic device for energy or environmental applications remains a grand challenge. For instance, the restacking issue of 2D GO sheets must be taken into consideration. The thermal/chemical stability of GO-based materials (especially those hybrid ones) under operational conditions is also a critical concern.

Nevertheless, by combining the superior physical/ chemical properties of GO/RGO themselves and the versatile nanomaterials that can couple with GO/RGO, GO/RGO-based materials have a bright future in the energy and environmental applications. We anticipate this active field will continue growing rapidly, leading eventually to a variety of mature materials and devices that would benefit the society.

ACKNOWLEDGEMENTS

This work was supported by the National Natural Science Foundation of China(11134005, 11404050), China Postdoctoral Science Foundation (2014M551065 and2015T80241), the Fundamental Research Funds for the Central Universities of China(DUT14RC(3)041), the Liaoning BaiQianWan Talents Program, and SBZ was supported by the US National Science FoundationDMR-1305293.

REFERENCES

1. J. Zhao, L. Liu, F. Li, Graphene Oxide: Physics and Applications, Springer, 2015.
2. H.A. Becerril, J. Mao, Z. Liu, R.M. Stoltenberg, Z. Bao, Y. Chen, ACS Nano 2 (2008) 463–470.
3. D. Chen, H. Feng, J. Li, Chem. Rev. 112 (2012) 6027–6053.
4. D.W. Boukhvalov, M.I. Katsnelson, J. Am. Chem. Soc. 130 (2008) 10697–10701.
5. R. Lahaye, H. Jeong, C. Park, Y. Lee, Phys. Rev. B 79 (2009) 125435.
6. J.-A. Yan, L. Xian, M.Y. Chou, Phys. Rev. Lett. 103 (2009) 086802.
7. J.-A. Yan, M.Y. Chou, Phys. Rev. B 82 (2010) 125403.
8. L. Liu, L. Wang, J. Gao, J. Zhao, X. Gao, Z. Chen, Carbon 50 (2012) 1690–1698.
9. G. Eda, G. Fanchini, M. Chhowalla, Nat. Nanotechnol. 3 (2008) 270–274.
10. G. Eda, M. Chhowalla, Adv. Mater. 22 (2010) 2392–2415.
11. J. Robertson, Mater. Sci. Eng., R 37 (2002) 129–281.
12. H. Chen, M.B. Müller, K.J. Gilmore, G.G. Wallace, D. Li, Adv. Mater. 20 (2008) 3557–3561.
13. C. Chen, Q.-H. Yang, Y. Yang, W. Lv, Y. Wen, P.-X. Hou, M. Wang, H.-M. Cheng, Adv. Mater. 21 (2009) 3007–3011.
14. Q. Zheng, Y. Geng, S. Wang, Z. Li, J.-K. Kim, Carbon 48 (2010) 4315–4322.
15. L. Liu, J. Zhang, J. Zhao, F. Liu, Nanoscale 4 (2012) 5910–5916.
16. Y. Gao, L.-Q. Liu, S.-Z. Zu, K. Peng, D. Zhou, B.-H. Han, Z. Zhang, ACS Nano 5 (2011) 2134–2141.
17. F. Perrozzi, S. Prezioso, L. Ottaviano, J. Phys. Condens. Matter 27 (2015) 013002.
18. K.S. Novoselov, V. Fal, L. Colombo, P. Gellert, M. Schwab, K. Kim, Nature 490 (2012) 192–200.
19. K.P. Loh, Q. Bao, G. Eda, M. Chhowalla, Nat. Chem. 2 (2010) 1015–1024.
20. C. Gómez-Navarro, J.C. Meyer, R.S. Sundaram, A. Chuvilin, S. Kurasch, M. Burghard, K. Kern, U. Kaiser, Nano Lett. 10 (2010) 1144–1148.
21. Fujishima, Nature 238 (1972) 37–38.
22. Kudo, Y. Miseki, Chem. Soc. Rev. 38 (2009) 253–278.
23. S. Park, R.S. Ruoff, Nat. Nanotechnol. 4 (2009) 217–224.

24. T.-F. Yeh, S.-J. Chen, C.-S. Yeh, H. Teng, J. Phys. Chem. C 117 (2013) 6516–6524.
25. T.F. Yeh, J.M. Syu, C. Cheng, T.H. Chang, H. Teng, Adv. Funct. Mater. 20 (2010) 2255–2262.
26. K. Krishnamoorthy, R. Mohan, S.-J. Kim, Appl. Phys. Lett. 98 (2011) 244101.
27. T.-F. Yeh, H. Teng, ECS Trans. 41 (2012) 7–26.
28. T.-F. Yeh, F.-F. Chan, C.-T. Hsieh, H. Teng, J. Phys. Chem. C 115 (2011) 22587–22597.
29. T.-F. Yeh, C.-Y. Teng, S.-J. Chen, H. Teng, Adv. Mater. 26 (2014) 3297–3303.
30. Y. Matsumoto, M. Koinuma, S. Ida, S. Hayami, T. Taniguchi, K. Hatakeyama, H. Tateishi, Y. Watanabe, S. Amano, J. Phys. Chem. C 115 (2011) 19280–19286.
31. X. Jiang, J. Nisar, B. Pathak, J. Zhao, R. Ahuja, J. Catal. 299 (2013) 204–209.
32. A.K. Agegnehu, C.-J. Pan, J. Rick, J.-F. Lee, W.-N. Su, B.- J. Hwang, J. Mater. Chem. 22 (2012) 13849–13854.
33. X. An, C.Y. Jimmy, RSC Adv. 1 (2011) 1426–1434.
34. Q. Xiang, J. Yu, J. Phys. Chem. Lett. 4 (2013) 753–759.
35. Iwase, Y.H. Ng, Y. Ishiguro, A. Kudo, R. Amal, J. Am. Chem. Soc. 133 (2011) 11054–11057.
36. N. Zhang, Y. Zhang, Y.-J. Xu, Nanoscale 4 (2012) 5792–5813.
37. J. Zhang, J. Yu, M. Jaroniec, J.R. Gong, Nano Lett. 12 (2012) 4584–4589.
38. Q. Li, B. Guo, J. Yu, J. Ran, B. Zhang, H. Yan, J.R. Gong, J. Am. Chem. Soc. 133 (2011) 10878–10884.
39. Dong, X. Li, P. Jin, W. Zhao, J. Chu, J. Qi, J. Phys. Chem. C 116 (2012) 15833–15838.
40. Q. Xiang, J. Yu, M. Jaroniec, Nanoscale 3 (2011) 3670–3678.
41. X.-J. Lv, W.-F. Fu, H.-X. Chang, H. Zhang, J.-S. Cheng G.-J. Zhang, Y. Song, C.-Y. Hu, J.-H. Li, J. Mater. Chem. 22 (2012) 1539–1546.
42. J. Wang, C. An, J. Liu, G. Xi, W. Jiang, S. Wang, Q.-H. Zhang, J. Mater. Chem. A 1 (2013) 2827–2832.
43. Q. Xiang, J. Yu, M. Jaroniec, J. Phys. Chem. C 115 (2011) 7355–7363.
44. J. Yang, X. Zeng, L. Chen, W. Yuan, Appl. Phys. Lett. 102 (2013).
45. P.D. Tran, S.K. Batabyal, S.S. Pramana, J. Barber, L.H. Wong, S.C.J. Loo, Nanoscale 4 (2012) 3875–3878.
46. Mukherji, B. Seger, G.Q. Lu, L. Wang, ACS Nano 5 (2011) 3483–3492.
47. J. Hou, Z. Wang, W. Kan, S. Jiao, H. Zhu, R. Kumar, J. Mater. Chem. 22 (2012) 7291–7299.
48. Z. Khan, T.R. Chetia, A.K. Vardhaman, D. Barpuzary, C. V. Sastri, M. Qureshi, RSC Adv. 2 (2012) 12122–12128.
49. Q. Xiang, J. Yu, M. Jaroniec, J. Am. Chem. Soc. 134 (2012) 6575–6578.
50. S. Patchkovskii, S.T. John, S.N. Yurchenko, L. Zhechkov, T. Heine, G. Seifert, Proc. Natl. Acad. Sci. U.S.A. 102 (2005) 10439–10444.
51. G. Srinivas, Y. Zhu, R. Piner, N. Skipper, M. Ellerby, R. Ruoff, Carbon 48 (2010) 630–635.
52. C.X. Guo, Y. Wang, C.M. Li, ACS Sustainable Chem. Eng. 1 (2013) 14–18.
53. B.H. Kim, W.G. Hong, H.Y. Yu, Y.-K. Han, S.M. Lee, S. J. Chang, H.R. Moon, Y. Jun, H.J. Kim, Phys. Chem. Chem. Phys. 14 (2012) 1480–1484.
54. J.M. Kim, W.G. Hong, S.M. Lee, S.J. Chang, Y. Jun, B.H. Kim, H.J. Kim, Int. J. Hydrogen Energy 39 (2014) 3799–3804.
55. Q. Sun, Q. Wang, P. Jena, Y. Kawazoe, J. Am. Chem. Soc. 127 (2005) 14582–14583.
56. Y. Zhao, Y.-H. Kim, A.C. Dillon, M.J. Heben, S.B. Zhang, Phys. Rev. Lett. 94 (2005) 155504.
57. L. Wang, K. Lee, Y.-Y. Sun, M. Lucking, Z. Chen, J.J. Zhao, S. B. Zhang, Acs Nano 3 (2009) 2995–3000.
58. Chen, J. Zhang, B. Zhang, H. Ming Duan, J. Phys. Chem. C 117 (2013) 4337–4344.
59. L. Wang, F.H. Yang, R.T. Yang, M.A. Miller, Ind. Eng. Chem. Res. 48 (2009) 2920–2926.
60. W.G. Hong, B.H. Kim, S.M. Lee, H.Y. Yu, Y.J. Yun, Y. Jun, J. B. Lee, H.J. Kim, Int. J. Hydrogen Energy 37 (2012) 7594–7599.
61. Tylianakis, G.M. Psofogiannakis, G.E. Froudakis, J. Phys. Chem. Lett. 1 (2010) 2459–2464.
62. S.H. Aboutalebi, S. Aminorroaya-Yamini, I. Nevirkovets, K. Konstantinov, H.K. Liu, Adv. Energy Mater. 2 (2012) 1439–1446.
63. J.W. Burress, S. Gadipelli, J. Ford, J.M. Simmons, W. Zhou, T. Yildirim, Angew. Chem. Int. Ed. 49 (2010) 8902–8904.
64. G. Srinivas, J.W. Burress, J. Ford, T. Yildirim, J. Mater. Chem. 21 (2011) 11323–11329.
65. Y. Chan, J.M. Hill, Nanotechnology 22 (2011) 305403.
66. B.H. Kim, W.G. Hong, H.R. Moon, S.M. Lee, J.M. Kim, S. Kang, Y. Jun, H.J. Kim, Int. J. Hydrogen Energy 37 (2012) 14217–14222.
67. T.K. Kim, J.Y. Cheon, K. Yoo, J.W. Kim, S.-m. Hyun, H.S. Shin, S.H. Joo, H.R. Moon, J. Mater. Chem. A 1 (2013) 8432–8437.
68. R. Kumar, V.M. Suresh, T.K. Maji, C. Rao, Chem. Commun. 50 (2014) 2015–2017.
69. C. Petit, J. Burress, T.J. Bandosz, Carbon 49 (2011) 563–572.
70. S. Liu, L. Sun, F. Xu, J. Zhang, C. Jiao, F. Li, Z. Li, S. Wang, Z. Wang, X. Jiang, H. Zhou, L. Yang, C. Schick, Energy Environ. Sci. 6 (2013) 818–823.
71. H. Zhou, X. Liu, J. Zhang, X. Yan, Y. Liu, A. Yuan, Int. J. Hydrogen Energy 39 (2014) 2160–2167.
72. N. Cao, W. Luo, G. Cheng, Int. J. Hydrogen Energy 38 (2013) 11964–11972.
73. L. Yang, W. Luo, G. Cheng, ACS Appl. Mater. Interfaces 5 (2013) 8231–8240.
74. N. Cao, J. Su, W. Luo, G. Cheng, Int. J. Hydrogen Energy 39 (2014) 426–435.

75. X. Wang, D. Liu, S. Song, H. Zhang, Chem. Eur. J. 19 (2013) 8082–8086.
76. J. Wang, Y.-L. Qin, X. Liu, X.-B. Zhang, J. Mater. Chem. 22 (2012) 12468–12470.
77. P. Xi, F. Chen, G. Xie, C. Ma, H. Liu, C. Shao, J. Wang, Z. Xu, X. Xu, Z. Zeng, Nanoscale 4 (2012) 5597–5601.
78. B. Kılıç, S. Şencanlı, Ö. Metin, J. Mol. Catal. A: Chem. 361– 362 (2012) 104–110.
79. X. Zhou, Z. Chen, D. Yan, H. Lu, J. Mater. Chem. 22 (2012) 13506–13516.
80. Y. Yang, Z.-H. Lu, Y. Hu, Z. Zhang, W. Shi, X. Chen, T. Wang, RSC Adv. 4 (2014) 13749–13752.
81. Z. Tang, H. Chen, X. Chen, L. Wu, X. Yu, J. Am. Chem. Soc. 134 (2012) 5464–5467.
82. Li, J. Gao, J. Zhang, F. Xu, J. Zhao, L. Sun, J. Mater. Chem. A 1 (2013) 8016–8022.
83. V.S. Channu, R. Bobba, R. Holze, Colloids Surf., A 436 (2013) 245–251.
84. J. Zhang, H. Cao, X. Tang, W. Fan, G. Peng, M. Qu, J. Power Sources 241 (2013) 619–626.
85. Z. Guo, J. Wang, F. Wang, D. Zhou, Y. Xia, Y. Wang, Adv. Funct. Mater. 23 (2013) 4840–4846.
86. S.-L. Kuo, W.-R. Liu, C.-P. Kuo, N.-L. Wu, H.-C. Wu, J. Power Sources 244 (2013) 552–556.
87. H. Wang, L.-F. Cui, Y. Yang, H. Sanchez Casalongue, J. T. Robinson, Y. Liang, Y. Cui, H. Dai, J. Am. Chem. Soc. 132 (2010) 13978–13980.
88. Jiang, L.W. Yang, Y. Tian, P. Yang, S.W. Hu, K. Huang, X. L. Wei, J.X. Zhong, Ceram. Int. 40 (2014) 4297–4304.
89. J. Zhu, T. Zhu, X. Zhou, Y. Zhang, X.W. Lou, X. Chen, H. Zhang, H.H. Hng, Q. Yan, Nanoscale 3 (2011) 1084–1089.
90. X. Zhu, Y. Zhu, S. Murali, M.D. Stoller, R.S. Ruoff, ACS Nano 5 (2011) 3333–3338.
91. L. Wang, D. Wang, Z. Dong, F. Zhang, J. Jin, Nano Lett. 13 (2013) 1711–1716.
92. Y. Liu, W. Wang, L. Gu, Y. Wang, Y. Ying, Y. Mao, L. Sun, X. Peng, ACS Appl. Mater. Interfaces 5 (2013) 9850–9855.
93. Zou, X. Hu, L. Qie, Y. Jiang, X. Xiong, Y. Qiao, Y. Huang, Nanoscale 6 (2014) 924–930.
94. Ni, W.-L. Song, L.-Z. Fan, Electrochem. Commun. 40 (2014) 1–4.
95. L. Fei, Q. Lin, B. Yuan, G. Chen, P. Xie, Y. Li, Y. Xu, S. Deng, S. Smirnov, H. Luo, ACS Appl. Mater. Interfaces 5 (2013) 5330–5335.
96. B. Qiu, X. Zhao, D. Xia, J. Alloys Compd. 579 (2013) 372–376.
97. D. Chen, H. Quan, X. Luo, S. Luo, Scr. Mater. 76 (2014) 1–4.
98. C.K. Chan, H. Peng, G. Liu, K. McIlwrath, X.F. Zhang, R. A. Huggins, Y. Cui, Nat. Nanotechnol. 3 (2008) 31–35.
99. U. Kasavajjula, C. Wang, A.J. Appleby, J. Power Sources 163 (2007) 1003–1039.
100. B. Wang, X. Li, X. Zhang, B. Luo, M. Jin, M. Liang, S.A. Dayeh, S.T. Picraux, L. Zhi, ACS Nano 7 (2013) 1437–1445.
101. Chang, X. Huang, G. Zhou, S. Cui, P.B. Hallac, J. Jiang, P. T. Hurley, J. Chen, Adv. Mater. 26 (2014) 758–764.
102. J.-G. Ren, C. Wang, Q.-H. Wu, X. Liu, Y. Yang, L. He, W. Zhang, Nanoscale 6 (2014) 3353–3360.
103. D.-W. Wang, C. Sun, G. Zhou, F. Li, L. Wen, B.C. Donose, G. Q. Lu, H.-M. Cheng, I.R. Gentle, J. Mater. Chem. A 1 (2013) 3607–3612.
104. S.H. Ha, Y.S. Jeong, Y.J. Lee, ACS Appl. Mater. Interfaces 5 (2013) 12295–12303.
105. Wu, J. Yang, Mater. Res. Bull. 48 (2013) 435–439.
106. B. Pei, Z. Jiang, W. Zhang, Z. Yang, A. Manthiram, J. Power Sources 239 (2013) 475–482.
107. D. Chen, H. Quan, S. Luo, X. Luo, F. Deng, H. Jiang, Physica E 56 (2014) 231–237.
108. D. Pham-Cong, K. Ahn, S.W. Hong, S.Y. Jeong, J.H. Choi, C. H. Doh, J.S. Jin, E.D. Jeong, C.R. Cho, Curr. Appl. Phys. 14 (2014) 215–221.
109. Y. Yang, C. Wang, B. Yue, S. Gambhir, C.O. Too, G.G. Wallace, Adv. Energy Mater. 2 (2012) 266–272.
110. S.J. Richard Prabakar, Y.-H. Hwang, E.G. Bae, D.K. Lee, M. Pyo, Carbon 52 (2013) 128–136.
111. P.G. Bruce, S.A. Freunberger, L.J. Hardwick, J.-M. Tarascon, Nat. Mater. 11 (2012) 19–29.
112. H. Yamin, E. Peled, J. Power Sources 9 (1983) 281–287.
113. Y.V. Mikhaylik, J.R. Akridge, J. Electrochem. Soc. 151 (2004) A1969–A1976.
114. R. Elazari, G. Salitra, Y. Talyosef, J. Grinblat, C. ScordilisKelley, A. Xiao, J. Affinito, D. Aurbach, J. Electrochem. Soc. 157 (2010) A1131–A1138.
115. S.-E. Cheon, K.-S. Ko, J.-H. Cho, S.-W. Kim, E.-Y. Chin, H.- T. Kim, J. Electrochem. Soc. 150 (2003) A800–A805.
116. B.H. Jeon, J.H. Yeon, K.M. Kim, I.J. Chung, J. Power Sources 109 (2002) 89–97.
117. Kim, L.J. Cote, J. Huang, Acc. Chem. Res. 45 (2012) 1356–1364.
118. I.V. Lightcap, P.V. Kamat, Acc. Chem. Res. 46 (2013) 2235–2243.
119. Ji, M. Rao, H. Zheng, L. Zhang, Y. Li, W. Duan, J. Guo, E. J. Cairns, Y. Zhang, J. Am. Chem. Soc. 133 (2011) 18522–18525.
120. Zhang, L. Ji, P.-A. Glans, Y. Zhang, J. Zhu, J. Guo, Phys. Chem. Chem. Phys. 14 (2012) 13670–13675.
121. J.W. Kim, J.D. Ocon, D.-W. Park, J. Lee, J, Energy Chem. 22 (2013) 336–340.
122. Xiao, M. Huang, S. Zeng, D. Han, S. Wang, L. Sun, Y. Meng, RSC Adv. 3 (2013) 4914–4916.
123. H. Wang, Y. Yang, Y. Liang, J.T. Robinson, Y. Li, A. Jackson, Y. Cui, H. Dai, Nano Lett. 11 (2011) 2644–2647.
124. M.-K. Song, Y. Zhang, E.J. Cairns, Nano Lett. 13 (2013) 5891–5899.
125. W. Zhou, H. Chen, Y. Yu, D. Wang, Z. Cui, F.J. DiSalvo, H. D. Abruña, ACS Nano 7 (2013) 8801–8808.

126. Z.W. Seh, H. Wang, N. Liu, G. Zheng, W. Li, H. Yao, Y. Cui, Chem. Sci. 5 (2014) 1396–1400.

127. X. Gao, J. Li, D. Guan, C. Yuan, ACS Appl. Mater. Interfaces 6 (2014) 4154–4159.

128. F.-f. Zhang, X.-b. Zhang, Y.-h. Dong, L.-m. Wang, J. Mater. Chem. 22 (2012) 11452–11454.

129. X. Wang, Z. Wang, L. Chen, J. Power Sources 242 (2013) 65–69.

130. K. Han, J. Shen, C.M. Hayner, H. Ye, M.C. Kung, H.H. Kung, J. Power Sources 251 (2014) 331–337.

131. Li, M. Zheng, H. Lu, Z. Hu, C. Shen, X. Chang, G. Ji, J. Cao, Y. Shi, Chem. Commun. 48 (2012) 4106–4108.

132. X.Y. Zhao, J.P. Tu, Y. Lu, J.B. Cai, Y.J. Zhang, X.L. Wang, C. D. Gu, Electrochim. Acta 113 (2013) 256–262.

133. X. Zhou, J. Xie, J. Yang, Y. Zou, J. Tang, S. Wang, L. Ma, Q. Liao, J. Power Sources 243 (2013) 993–1000.

134. J. Xie, J. Yang, X. Zhou, Y. Zou, J. Tang, S. Wang, F. Chen, J. Power Sources 253 (2014) 55–63.

135. L.Q. Lu, L.J. Lu, Y. Wang, J. Mater. Chem. A 1 (2013) 9173–9181.

136. X. Wang, Z. Zhang, Y. Qu, Y. Lai, J. Li, J. Power Sources 256 (2014) 361–368.

137. S.D. Beattie, D.M. Manolescu, S.L. Blair, J. Electrochem. Soc. 156 (2009) A44–A47.

138. J. Christensen, P. Albertus, R.S. Sanchez-Carrera, T. Lohmann, B. Kozinsky, R. Liedtke, J. Ahmed, A. Kojic, J. Electrochem. Soc. 159 (2011) R1–R30.

139. Kraytsberg, Y. Ein-Eli, J. Power Sources 196 (2011) 886–893.

140. D.M. Itkis, D.A. Semenenko, E.Y. Kataev, A.I. Belova, V. S. Neudachina, A.P. Sirotina, M. Hävecker, D. Teschner, A. Knop-Gericke, P. Dudin, A. Barinov, E.A. Goodilin, Y. Shao-Horn, L.V. Yashina, Nano Lett. 13 (2013) 4697–4701.

141. J. Xiao, D. Mei, X. Li, W. Xu, D. Wang, G.L. Graff, W. D. Bennett, Z. Nie, L.V. Saraf, I.A. Aksay, J. Liu, J.- G. Zhang, Nano Lett. 11 (2011) 5071–5078.

142. Z.-L. Wang, D. Xu, J.-J. Xu, L.-L. Zhang, X.-B. Zhang, Adv. Funct. Mater. 22 (2012) 3699–3705.

143. L. Lai, J.R. Potts, D. Zhan, L. Wang, C.K. Poh, C. Tang, H. Gong, Z. Shen, J. Lin, R.S. Ruoff, Energy Environ. Sci. 5 (2012) 7936–7942.

144. H.-G. Jung, Y.S. Jeong, J.-B. Park, Y.-K. Sun, B. Scrosati, Y. J. Lee, ACS Nano 7 (2013) 3532–3539.

145. Y. Liang, Y. Li, H. Wang, J. Zhou, J. Wang, T. Regier, H. Dai, Nat. Mater. 10 (2011) 780–786.

146. Li, P. Xu, B. Zhang, H. Tsai, J. Wang, H.-L. Wang, G. Wu, Chem. Commun. 49 (2013) 10838–10840.

147. Y. Yu, B. Zhang, Y.-B. He, Z.-D. Huang, S.-W. Oh, J.-K. Kim, J. Mater. Chem. A 1 (2013) 1163–1170.

148. Simon, Y. Gogotsi, Nat. Mater. 7 (2008) 845–854.

149. Ghosh, E.J. Ra, M. Jin, H.-K. Jeong, T.H. Kim, C. Biswas, Y. H. Lee, Adv. Funct. Mater. 21 (2011) 2541–2547. 150 B. Xu, S. Yue, Z. Sui, X. Zhang, S. Hou, G. Cao, Y. Yang, Energy Environ. Sci. 4 (2011) 2826–2830.

150. M. Zabihinpour, H.R. Ghenaatian, Synth. Met. 175 (2013) 62–67.

151. Z. Lin, Y. Liu, Y. Yao, O.J. Hildreth, Z. Li, K. Moon C.-p. Wong, J. Phys. Chem. C 115 (2011) 7120–7125.

152. D. Zhang, X. Zhang, Y. Chen, C. Wang, Y. Ma, Electrochim. Acta 69 (2012) 364–370.

153. W. Zhang, Y. Zhang, Y. Tian, Z. Yang, Q. Xiao, X. Guo, L. Jing, Y. Zhao, Y. Yan, J. Feng, K. Sun, ACS Appl. Mater. Interfaces 6 (2014) 2248–2254.

154. B. Zhao, P. Liu, Y. Jiang, D. Pan, H. Tao, J. Song, T. Fang, W. Xu, J. Power Sources 198 (2012) 423–427.

155. Y. Bai, R.B. Rakhi, W. Chen, H.N. Alshareef, J. Power Sources 233 (2013) 313–319.

156. K. Gopalakrishnan, K. Moses, A. Govindaraj, C.N.R. Rao, Solid State Commun. 175–176 (2013) 43–50.

157. H. Nolan, B. Mendoza-Sanchez, N. Ashok Kumar, N. McEvoy, S. O'Brien, V. Nicolosi, G.S. Duesberg, Phys. Chem. Chem. Phys. 16 (2014) 2280–2284.

158. J. Yang, M.R. Jo, M. Kang, Y.S. Huh, H. Jung, Y.-M. Kang, Carbon 73 (2014) 106–113.

159. J.H. Lee, N. Park, B.G. Kim, D.S. Jung, K. Im, J. Hur, J. W. Choi, ACS Nano 7 (2013) 9366–9374.

160. Y. Yoon, K. Lee, C. Baik, H. Yoo, M. Min, Y. Park, S.M. Lee, H. Lee, Adv. Mater. 25 (2013) 4437–4444.

161. J. Yan, Q. Wang, T. Wei, L. Jiang, M. Zhang, X. Jing, Z. Fan, ACS Nano 8 (2014) 4720–4729.

162. L.-Y. Lin, M.-H. Yeh, J.-T. Tsai, Y.-H. Huang, C.-L. Sun K.-C. Ho, J. Mater. Chem. A 1 (2013) 11237–11245.

163. F. Zeng, Y. Kuang, N. Zhang, Z. Huang, Y. Pan, Z. Hou, H. Zhou, C. Yan, O.G. Schmidt, J. Power Sources 247 (2014) 396–401.

164. M. Beidaghi, C. Wang, Adv. Funct. Mater. 22 (2012) 4501–4510.

165. J. Liu, D. Galpaya, M. Notarianni, C. Yan, N. Motta, Appl. Phys. Lett. 103 (2013) 063108.

166. Y. Cao, M. Zhu, P. Li, R. Zhang, X. Li, Q. Gong, K. Wang, M. Zhong, D. Wu, F. Lin, H. Zhu, Phys. Chem. Chem. Phys. 15 (2013) 19550–19556.

167. Liu, S.B. Lee, J. Am. Chem. Soc. 130 (2008) 2942–2943.

168. Y. Sun, Q. Wu, G. Shi, Energy Environ. Sci. 4 (2011) 1113–1132.

169. Chen, J. Zhu, X. Wu, Q. Han, X. Wang, ACS Nano 4 (2010) 2822–2830.

170. G.S. Gund, D.P. Dubal, B.H. Patil, S.S. Shinde, C.D. Lokhande, Electrochim. Acta 92 (2013) 205–215.

171. D. Antiohos, K. Pingmuang, M.S. Romano, S. Beirne, T. Romeo, P. Aitchison, A. Minett, G. Wallace, S. Phanichphant, J. Chen, Electrochim. Acta 101 (2013) 99–108.

172. W. Zhou, J. Liu, T. Chen, K.S. Tan, X. Jia, Z. Luo, C. Cong, H. Yang, C.M. Li, T. Yu, Phys. Chem. Chem. Phys. 13 (2011) 14462–14465.

173. L.-J. Xie, J.-F. Wu, C.-M. Chen, C.-M. Zhang, L. Wan, J.- L. Wang, Q.-Q. Kong, C.-X. Lv, K.-X. Li, G.-H. Sun, J. Power Sources 242 (2013) 148–156.

174. C. Xiang, M. Li, M. Zhi, A. Manivannan, N. Wu, J. Power Sources 226 (2013) 65–70.
175. C. Yuan, L. Zhang, L. Hou, G. Pang, W.-C. Oh, RSC Adv. 4 (2014) 14408–14413.
176. M. Kim, Y. Hwang, J. Kim, J. Power Sources 239 (2013) 225–233.
177. Sumboja, C.Y. Foo, X. Wang, P.S. Lee, Adv. Mater. 25 (2013) 2809–2815.
178. W. Li, Y. Bu, H. Jin, J. Wang, W. Zhang, S. Wang, J. Wang, Energy Fuels 27 (2013) 6304–6310.
179. G. Lee, Y. Cheng, C.V. Varanasi, J. Liu, J. Phys. Chem. C 118 (2014) 2281–2286.
180. Prakash, D. Bahadur, ACS Appl. Mater. Interfaces 6 (2014) 1394–1405.
181. W. Shi, J. Zhu, D.H. Sim, Y.Y. Tay, Z. Lu, X. Zhang, Y. Sharma, M. Srinivasan, H. Zhang, H.H. Hng, Q. Yan, J. Mater. Chem. 21 (2011) 3422–3427.
182. J.P. Cheng, Q.L. Shou, J.S. Wu, F. Liu, V.P. Dravid, X.B. Zhang, J. Electroanal. Chem. 698 (2013) 1–8.
183. W.-H. Khoh, J.-D. Hong, Colloids Surf., A 436 (2013) 104–112.
184. B. Li, H. Cao, G. Yin, Y. Lu, J. Yin, J. Mater. Chem. 21 (2011) 10645–10648.
185. Y. Liu, Y. Ying, Y. Mao, L. Gu, Y. Wang, X. Peng, Nanoscale 5 (2013) 9134–9140.
186. C. Xiang, M. Li, M. Zhi, A. Manivannan, N. Wu, J. Mater. Chem. 22 (2012) 19161–19167.
187. Ramadoss, G.-S. Kim, S.J. Kim, CrystEngComm 15 (2013) 10222–10229.
188. M. Sawangphruk, P. Srimuk, P. Chiochan, A. Krittayavathananon, S. Luanwuthi, J. Limtrakul, Carbon 60 (2013) 109–116.
189. Y. Bai, M. Du, J. Chang, J. Sun, L. Gao, J. Mater. Chem. A 2 (2014) 3834–3840.
190. Ratha, C.S. Rout, ACS Appl. Mater. Interfaces 5 (2013) 11427–11433.
191. Wang, H. Wang, S. Zhang, C. Mao, J. Song, H. Niu, B. Jin, Y. Tian, Appl. Surf. Sci. 282 (2013) 704–708.
192. S. Huang, G.-N. Zhu, C. Zhang, W.W. Tjiu, Y.-Y. Xia, T. Liu, ACS Appl. Mater. Interfaces 4 (2012) 2242–2249.
193. Y. Lin, L. Ruiyi, L. Zaijun, L. Junkang, F. Yinjun, W. Guangli, G. Zhiguo, Electrochim. Acta 95 (2013) 146–154.
194. W. Yang, Z. Gao, J. Wang, J. Ma, M. Zhang, L. Liu, ACS Appl. Mater. Interfaces 5 (2013) 5443–5454.
195. K.R. Prasad, N. Munichandraiah, J. Power Sources 112 (2002) 443–451.
196. J. Xu, K. Wang, S.-Z. Zu, B.-H. Han, Z. Wei, ACS Nano 4 (2010) 5019–5026.
197. H. Wang, Q. Hao, X. Yang, L. Lu, X. Wang, ACS Appl. Mater. Interfaces 2 (2010) 821–828.
198. Y. Liu, R. Deng, Z. Wang, H. Liu, J. Mater. Chem. 22 (2012) 13619–13624.
199. Q. Zhang, Y. Li, Y. Feng, W. Feng, Electrochim. Acta 90 (2013) 95–100.
200. Z. Luo, L. Zhu, H. Zhang, H. Tang, Mater. Chem. Phys. 139 (2013) 572–579.
201. L. Lai, H. Yang, L. Wang, B.K. Teh, J. Zhong, H. Chou, L. Chen, W. Chen, Z. Shen, R.S. Ruoff, J. Lin, ACS Nano 6 (2012) 5941–5951.
202. L. Wang, Y. Ye, X. Lu, Z. Wen, Z. Li, H. Hou, Y. Song, Sci. Rep. 3 (2013).
203. M. Kim, C. Lee, J. Jang, Adv. Funct. Mater. 24 (2014) 2489–2499.
204. L.L. Zhang, S. Zhao, X.N. Tian, X.S. Zhao, Langmuir 26 (2010) 17624–17628.
205. J. Li, H. Xie, Mater. Lett. 78 (2012) 106–109.
206. C. Zhu, J. Zhai, D. Wen, S. Dong, J. Mater. Chem. 22 (2012) 6300–6306.
207. J. Li, H. Xie, Y. Li, J. Power Sources 241 (2013) 388–395.
208. Z. Chen, D. Yu, W. Xiong, P. Liu, Y. Liu, L. Dai, Langmuir 30 (2014) 3567–3571.
209. J. Zhang, X.S. Zhao, J. Phys. Chem. C 116 (2012) 5420–5426.
210. Y. Cui, Q.-Y. Cheng, H. Wu, Z. Wei, B.-H. Han, Nanoscale 5 (2013) 8367–8374.
211. Y.-S. Wang, S.-M. Li, S.-T. Hsiao, W.-H. Liao, P.-H. Chen, S.- Y. Yang, H.-W. Tien, C.-C.M. Ma, C.-C. Hu, Carbon 73 (2014) 87–98.
212. B.-H. Wee, J.-D. Hong, Langmuir 30 (2014) 5267–5275.
213. X. Yang, F. Zhang, L. Zhang, T. Zhang, Y. Huang, Y. Chen, Adv. Funct. Mater. 23 (2013) 3353–3360.
214. X. Yang, L. Zhang, F. Zhang, T. Zhang, Y. Huang, Y. Chen, Carbon 72 (2014) 381–386.
215. X. Han, S.-j. Liu, Y. Yuan, Y. Wang, L.-j. Hu, J. Alloys Compd. 543 (2012) 200–205.
216. H. Su, T. Wang, S. Zhang, J. Song, C. Mao, H. Niu, B. Jin, J. Wu, Y. Tian, Solid State Sci. 14 (2012) 677–681.
217. B. Mu, W. Zhang, S. Shao, A. Wang, Phys. Chem. Chem. Phys. 16 (2014) 7872–7880.
218. G. Ning, T. Li, J. Yan, C. Xu, T. Wei, Z. Fan, Carbon 54 (2013) 241–248.
219. D. Xu, Q. Xu, K. Wang, J. Chen, Z. Chen, ACS Appl. Mater. Interfaces 6 (2014) 200–209.
220. C. Yang, J. Shen, C. Wang, H. Fei, H. Bao, G. Wang, J. Mater. Chem. A 2 (2014) 1458–1464.
221. H.W. Kim, H.W. Yoon, S.-M. Yoon, B.M. Yoo, B.K. Ahn, Y. H. Cho, H.J. Shin, H. Yang, U. Paik, S. Kwon, J.-Y. Choi, H. B. Park, Science 342 (2013) 91–95.
222. D. Kim, D.W. Kim, H.-K. Lim, J. Jeon, H. Kim, H.-T. Jung, H. Lee, J. Phys. Chem. C 118 (2014) 11142–11148.
223. S. Eigler, C. Dotzer, A. Hirsch, M. Enzelberger, P. Müller, Chem. Mater. 24 (2012) 1276–1282.
224. Yumura, A. Yamasaki, Phys. Chem. Chem. Phys. 16 (2014) 9656–9666.
225. Garcia-Gallastegui, D. Iruretagoyena, M. Mokhtar A.M. Asiri, S.N. Basahel, S.A. Al-Thabaiti, A.O. Alyoubi, D. Chadwick, M.S.P. Shaffer, J. Mater. Chem. 22 (2012) 13932–13940.

226. Garcia-Gallastegui, D. Iruretagoyena, V. Gouvea, M. Mokhtar, A.M. Asiri, S.N. Basahel, S.A. Al-Thabaiti, A. O. Alyoubi, D. Chadwick, M.S.P. Shaffer, Chem. Mater. 24 (2012) 4531–4539.

227. D. Iruretagoyena, M.P. Shaffer, D. Chadwick, Adsorption 20 (2014) 321–330.

228. A.A. Alhwaige, T. Agag, H. Ishida, S. Qutubuddin, RSC Adv. 3 (2013) 16011–16020.

229. M. Seredych, T.J. Bandosz, J. Phys. Chem. C 111 (2007) 15596–15604.

230. Petit, M. Seredych, T.J. Bandosz, J. Mater. Chem. 19 (2009) 9176–9185.

231. M. Seredych, T.J. Bandosz, Langmuir 26 (2010) 5491–5498.

232. M. Seredych, J.A. Rossin, T.J. Bandosz, Carbon 49 (2011) 4392–4402.

233. Petit, T.J. Bandosz, J. Phys. Chem. C 113 (2009) 3800–3809.

234. M. Seredych, T.J. Bandosz, Colloids Surf., A 353 (2010) 30–36.

235. M. Seredych, T.J. Bandosz, Microporous Mesoporous Mater. 150 (2012) 55–63.

236. S. Tang, Z. Cao, J. Phys. Chem. C 116 (2012) 8778–8791.

237. E.C. Mattson, K. Pande, M. Unger, S. Cui, G. Lu, M. Gajdardziska-Josifovska, M. Weinert, J. Chen, C. J. Hirschmugl, J. Phys. Chem. C 117 (2013) 10698–10707.

238. J. Górka, A. Zawislak, J. Choma, M. Jaroniec, Carbon 46 (2008) 1159–1161.

239. Y. He, N. Zhang, F. Wu, F. Xu, Y. Liu, J. Gao, Mater. Res. Bull. 48 (2013) 3553–3558.

240. Z.-H. Huang, G. Liu, F. Kang, ACS Appl. Mater. Interfaces 4 (2012) 4942–4947.

241. L. Huang, M. Seredych, T.J. Bandosz, A.C.T. van Duin, X. Lu, K.E. Gubbins, J. Chem. Phys. 139 (2013) 194707.

242. M. Seredych, T.J. Bandosz, J. Phys. Chem. C 114 (2010) 14552–14560.

243. Chen, K. Xu, X. Ji, L. Miao, J. Jiang, Phys. Chem. Chem. Phys. 16 (2014) 11031–11036.

244. L. Wang, J. Zhao, L. Wang, T. Yan, Y.-Y. Sun, S.B. Zhang, Phys. Chem. Chem. Phys. 13 (2011) 21126–21131.

245. S. Tang, Z. Cao, J. Chem. Phys. 134 (2011) 044710.

246. D.-H. Lan, F.-M. Yang, S.-L. Luo, C.-T. Au, S.-F. Yin, Carbon 73 (2014) 351–360.

247. Y. Long, C. Zhang, X. Wang, J. Gao, W. Wang, Y. Liu, J. Mater. Chem. 21 (2011) 13934–13941.

248. Humeres, N.A. Debacher, A. Smaniotto, K.M. de Castro, L. O.B. Benetoli, E.P. de Souza, R.d.F.P.M. Moreira, C.N. Lopes, W.H. Schreiner, M. Canle, J.A. Santaballa, Langmuir 30 (2014) 4301–4309.

249. T.-T. Jia, C.-H. Lu, Y.-F. Zhang, W.-k. Chen, J. Nanopart. Res. 16 (2014) 1–11.

250. He, N. Niu, F. Qu, S. Wei, Y. Chen, S. Gai, P. Gao, Y. Wang, P. Yang, Nanoscale 5 (2013) 8507–8516.

251. R. Nie, J. Shi, S. Xia, L. Shen, P. Chen, Z. Hou, F.-S. Xiao, J. Mater. Chem. 22 (2012) 18115–18118.

252. X. An, K. Li, J. Tang, ChemSusChem 7 (2014) 1086–1093.

253. J. Cheng, M. Zhang, G. Wu, X. Wang, J. Zhou, K. Cen, Environ. Sci. Technol. 48 (2014) 7076–7084.

254. H.-C. Hsu, I. Shown, H.-Y. Wei, Y.-C. Chang, H.-Y. Du Y.-G. Lin, C.-A. Tseng, C.-H. Wang, L.-C. Chen, Y.-C. Lin, K.-H. Chen, Nanoscale 5 (2013) 262–268.

255. P. Kumar, A. Kumar, B. Sreedhar, B. Sain, S.S. Ray, S.L. Jain, Chem. Eur. J. 20 (2014) 6154–6161.

256. L.-L. Tan, W.-J. Ong, S.-P. Chai, A. Mohamed, Nanoscale Res. Lett. 8 (2013) 465.

257. J. Yu, J. Jin, B. Cheng, M. Jaroniec, J. Mater. Chem. A 2 (2014) 3407–3416.

258. N.S. Andryushina, O.L. Stroyuk, Appl. Catal., B 148–149 (2014) 543–549.

259. S.-T. Yang, Y. Chang, H. Wang, G. Liu, S. Chen, Y. Wang, Y. Liu, A. Cao, J. Colloid Interface Sci. 351 (2010) 122–127.

260. X. Mi, G. Huang, W. Xie, W. Wang, Y. Liu, J. Gao, Carbon 50 (2012) 4856–4864.

261. Zhao, J. Li, X. Ren, C. Chen, X. Wang, Environ. Sci. Technol. 45 (2011) 10454–10462.

262. L. Liu, S. Liu, Q. Zhang, C. Li, C. Bao, X. Liu, P. Xiao, J. Chem. Eng. Data 58 (2013) 209–216.

263. Zhao, X. Ren, X. Gao, X. Tan, J. Li, C. Chen, Y. Huang, X. Wang, Dalton Trans. 40 (2011) 10945–10952.

264. M. Liu, C. Chen, J. Hu, X. Wu, X. Wang, J. Phys. Chem. C 115 (2011) 25234–25240.

265. H.-L. Ma, Y. Zhang, Q.-H. Hu, D. Yan, Z.-Z. Yu, M. Zhai, J. Mater. Chem. 22 (2012) 5914–5916.

266. D. Dinda, A. Gupta, S.K. Saha, J. Mater. Chem. A 1 (2013) 11221–11228.

267. S. Li, X. Lu, Y. Xue, J. Lei, T. Zheng, C. Wang, PLoS ONE 7 (2012) e43328.

268. S. Zhang, M. Zeng, W. Xu, J. Li, J. Li, J. Xu, X. Wang, Dalton Trans. 42 (2013) 7854–7858.

269. Chandra, K.S. Kim, Chem. Commun. 47 (2011) 3942–3944.

270. L. Fan, C. Luo, M. Sun, X. Li, H. Qiu, Colloids Surf., B 103 (2013) 523–529.

271. N. Zhang, H. Qiu, Y. Si, W. Wang, J. Gao, Carbon 49 (2011) 827–837.

272. L. Liu, C. Li, C. Bao, Q. Jia, P. Xiao, X. Liu, Q. Zhang, Talanta 93 (2012) 350–357.

273. S.-T. Yang, S. Chen, Y. Chang, A. Cao, Y. Liu, H. Wang, J. Colloid Interface Sci. 359 (2011) 24–29.

274. G.K. Ramesha, A. Vijaya, H.B. Muralidhara Kumara, S. Sampath, J. Colloid Interface Sci. 361 (2011) 270–277.

275. F. Liu, S. Chung, G. Oh, T.S. Seo, ACS Appl. Mater. Interfaces 4 (2012) 922–927.

276. Ren, D.D. Kulkarni, R. Kodiyath, W. Xu, I. Choi, V. V. Tsukruk, ACS Appl. Mater. Interfaces 6 (2014) 2459–2470.

277. L. Chen, J. Yang, X. Zeng, L. Zhang, W. Yuan, Mater. Express 3 (2013) 281–290.

278. L. Sun, H. Yu, B. Fugetsu, J. Hazard. Mater. 203–204 (2012) 101–110.
279. Y. Yao, S. Miao, S. Yu, L. Ping, Ma, H. Sun, S. Wang, J. Colloid Interface Sci. 379 (2012) 20–26.
280. L. Li, X. Li, H. Duan, X. Wang, C. Luo, Dalton Trans. 43 (2014) 8431–8438.
281. Z. Geng, Y. Lin, X. Yu, Q. Shen, L. Ma, Z. Li, N. Pan, X. Wang, J. Mater. Chem. 22 (2012) 3527–3535.
282. G. Xie, P. Xi, H. Liu, F. Chen, L. Huang, Y. Shi, F. Hou, Z. Zeng, C. Shao, J. Wang, J. Mater. Chem. 22 (2012) 1033–1039.
283. S. Bai, X. Shen, X. Zhong, Y. Liu, G. Zhu, X. Xu, K. Chen, Carbon 50 (2012) 2337–2346.
284. Yang, C. Chen, J. Li, G. Zhao, X. Ren, X. Wang, RSC Adv. 2 (2012) 8821–8826.
285. S. Thakur, N. Karak, Mater. Chem. Phys. 144 (2014) 425–432.
286. J.-H. Deng, X.-R. Zhang, G.-M. Zeng, J.-L. Gong, Q.-Y. Niu, J. Liang, Chem. Eng. J. 226 (2013) 189–200.
287. L. Fan, C. Luo, M. Sun, X. Li, F. Lu, H. Qiu, Bioresour. Technol. 114 (2012) 703–706.
288. L. Fan, C. Luo, M. Sun, H. Qiu, X. Li, Colloids Surf., B 103 (2013) 601–607.
289. Li, Q. Du, T. Liu, J. Sun, Y. Wang, S. Wu, Z. Wang, Y. Xia, L. Xia, Carbohydr. Polym. 95 (2013) 501–507.
290. Z.-Y. Sui, Y. Cui, J.-H. Zhu, B.-H. Han, ACS Appl. Mater. Interfaces 5 (2013) 9172–9179.
291. Wang, Z. Liu, X. Ye, K. Hu, H. Zhong, J. Yu, M. Jin, Z. Guo, Appl. Surf. Sci. 308 (2014) 82–90.
292. Yang, Y. Xie, L. Pang, M. Li, X. Song, J. Wen, H. Zhao, Langmuir 29 (2013) 10727–10736.
293. H. Fu, D. Zhu, Environ. Sci. Technol. 47 (2013) 4204–4210.
294. P. Shi, R. Su, S. Zhu, M. Zhu, D. Li, S. Xu, J. Hazard. Mater. 229–230 (2012) 331–339.
295. P. Shi, X. Dai, H. Zheng, D. Li, W. Yao, C. Hu, Chem. Eng. J. 240 (2014) 264–270.
296. Zhang, J. Hao, W. Yang, B. Lu, X. Ke, B. Zhang, J. Tang, ACS Appl. Mater. Interfaces 5 (2013) 3809–3815.
297. Y. Zhao, D. Zhao, C. Chen, X. Wang, J. Colloid Interface Sci. 405 (2013) 211–217.
298. R.C. Pawar, C.S. Lee, Mater. Chem. Phys. 141 (2013) 686–693.
299. D.K. Padhi, K. Parida, J. Mater. Chem. A 2 (2014) 10300–10312.
300. Z. Xiong, L.L. Zhang, J. Ma, X.S. Zhao, Chem. Commun. 46 (2010) 6099–6101.
301. L.L. Zhang, Z. Xiong, X.S. Zhao, ACS Nano 4 (2010) 7030–7036.
302. M. Sher Shah, A. Park, K. Zhang, J. Park, P. Yoo, ACS Appl. Mater. Interfaces 4 (2012) 3893–3901.
303. C. Zhang, J. Zhang, Y. Su, M. Xu, Z. Yang, Y. Zhang, Physica E 56 (2014) 251–255.
304. H. Liu, T. Liu, X. Dong, Y. Lv, Z. Zhu, Mater. Lett. 126 (2014) 36–38.
305. B. Chai, J. Li, Q. Xu, Ind. Eng. Chem. Res. 53 (2014) 8744–8752.
306. S. Dong, Y. Cui, Y. Wang, Y. Li, L. Hu, J. Sun, J. Sun, Chem. Eng. J. 249 (2014) 102–110.
307. C.J. Miller, H. Yu, T.D. Waite, Colloids Surf., A 435 (2013) 147–153.
308. Y. Min, G. He, Q. Xu, Y. Chen, J. Mater. Chem. A 2 (2014) 1294–1301.
309. M. Sookhakian, Y.M. Amin, W.J. Basirun, Appl. Surf. Sci. 283 (2013) 668–677.
310. P.T.N. Nguyen, C. Salim, W. Kurniawan, H. Hinode, Catal. Today 230 (2014) 166–173.
311. S.N. Ariffin, H.N. Lim, F.A. Jumeri, M. Zobir, A.H. Abdullah, M. Ahmad, N.A. Ibrahim, N.M. Huang, P.S. Teo, K. Muthoosamy, I. Harrison, Ceram. Int. 40 (2014) 6927–6936.
312. Huang, Y.H. Li, S. Lin, C. Liang, X. Xu, Y.F. Zhou, D.Y. Fan, H.J. Yang, P.L. Lang, R. Zhang, Y.G. Wang, M. Lei, Mater. Lett. 124 (2014) 223–226.
313. M. Zhu, P. Chen, M. Liu, Langmuir 29 (2013) 9259–9268.
314. P. Ma, W. Jiang, F. Wang, F. Li, P. Shen, M. Chen, Y. Wang, J. Liu, P. Li, J. Alloys Compd. 578 (2013) 501–506.
315. S. Ameen, M. Shaheer Akhtar, H.-K. Seo, H. Shik Shin, Mater. Lett. 100 (2013) 261–265.
316. Y. Tang, S. Luo, Y. Teng, C. Liu, X. Xu, X. Zhang, L. Chen, J. Hazard. Mater. 241–242 (2012) 323–330.
317. M. Sangermano, P. Calza, M.A. Lopez-Manchado, J. Mater. Sci. 48 (2013) 5204–5208.
318. S. Liu, Z. Chen, N. Zhang, Z.-R. Tang, Y.-J. Xu, J. Phys. Chem. C 117 (2013) 8251–8261.
319. Liu, J. Liu, D.D. Sun, Catal. Sci. Technol. 2 (2012) 2525–2532.
320. Liu, H. Bai, J. Liu, D.D. Sun, J. Hazard. Mater. 261 (2013) 214–223.
321. 322 C. Wang, J. Zhu, X. Wu, H. Xu, Y. Song, J. Yan, Y. Song, H. Ji, K. Wang, H. Li, Ceram. Int. 40 (2014) 8061–8070.
322. H. Liu, Z. Chen, Z. Jin, Y. Su, Y. Wang, Dalton Trans. 43 (2014) 7491–7498.
323. S. Thangavel, G. Venugopal, S.-J. Kim, Mater. Chem. Phys. 145 (2014) 108–115.
324. D. Wan, C. Yang, T. Lin, Y. Tang, M. Zhou, Y. Zhong, F. Huang, J. Lin, ACS Nano 6 (2012) 9068–9078.
325. C.J. Madadrang, H.Y. Kim, G. Gao, N. Wang, J. Zhu, H. Feng, M. Gorring, M.L. Kasner, S. Hou, ACS Appl. Mater. Interfaces 4 (2012) 1186–1193.

CHAPTER 4

GRAPHENE-BASED MICROBOTS FOR TOXIC HEAVY METAL REMOVAL AND RECOVERY FROM WATER

Diana Vilela, Jemish Parmar, Yongfei Zeng, Yanli Zhao, and Samuel Sánchez

Keywords: Catalytic microswimmers, environmental application, Graphene, heavy metals, metal recovery, microbots, wastewater treatment

Pollution in water from heavy metals such as arsenic, mercury, cadmium, chromium, and lead, originates from various human industrial activities such as electroplating, mining, fabrication of batteries and microelectronics. It is a potential hazard to living systems; hence, it is essential to develop efficient and inexpensive materials and technologies to remove and recycle them from polluted water. Various methods are in use to remove and recover heavy metals such as chemical precipitation, adsorption, ion exchange, and membrane filtration among which adsorption is considered an economical and effective strategy.(1)

Recent developments in nanotechnology have further increased the effectiveness of adsorbent materials providing innovative systems for improving environmental remediation.(2, 3) Lately, numerous reports described the utilization of graphene and its composites as good adsorbents for the removal of dyes and heavy metal ions from aqueous solutions.(4-6) Furthermore, in the last couple of years, catalytic self-propelled micro- and nanomotors have demonstrated diverse applications within the environmental field. The synergy between active motion of micromotors,(7-10) which enhances micromixing (and mass transfer in the solution(11, 12) and surface multifunctionalities(13, 14) opens many possibilities of artificial swimmers as water remediation tools.(11, 15, 16)For instance, they are very efficient in the degradation of organic pollutants,(11, 17-20)chemical warfare agents,(21, 22) and the capture of organics(23-29) from water. They can analyze the water quality from the fluid where they swim such as the detection of heavy metals,(30, 31) pH,(32) or other analytes.(33-35) However, reusable micromotors for the capture and release of heavy metals in defined locations have not been yet reported.

Here, we present the removal and recovery of heavy metals (lead) from contaminated water by using graphene oxide (GOx) based tubular micromotors, dubbed microbots, propelled by a catalytic reaction. The high adsorption of Pb (II) ions on the graphene oxide (GOx) nanosheets of the microbots surface is a spontaneous process due to the strong surface complexation between the Pb (II) ions and the abundant oxygen moieties on the GOx. Moreover, due to the magnetic properties of these microbots, they can be easily removed from the water using a magnet after successful lead decontamination. The adsorbed Pb(II) ions on the microbots can be recovered via acid pH adjustment allowing them to be recycled and reused for further decontamination processes.

Citation: Diana Vilela, Jemish Parmar, Yongfei Zeng, Yanli Zhao, and Samuel Sánchez, Graphene-Based Microbots for Toxic Heavy Metal Removal and Recovery from Water, DOI: 10.1021/acs.nanolett.6b00768.

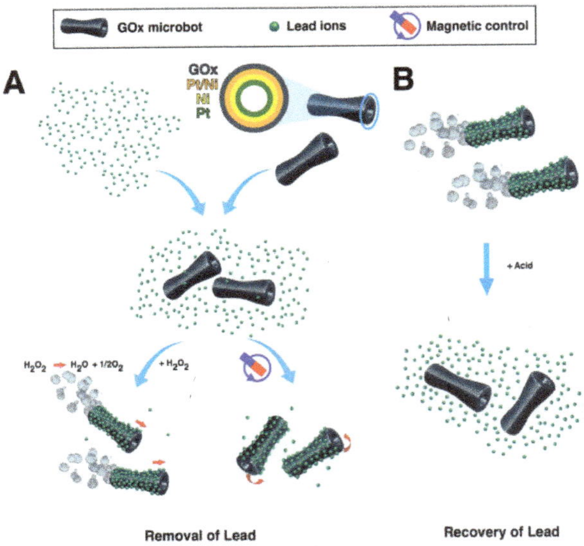

Figure 1: Scheme of GOx-microbots based approach for lead decontamination and recovery. (A) Decontamination of polluted water using GOx-microbots fabricated by electrodeposition of nanolayers of graphene oxide (GOx), Pt/Ni layer, Ni magnetic layer, and Pt catalytic inner layer. The decontamination strategy for lead ions can be carried out by two different techniques: self-propulsion of the GOx-microbots in the presence of H_2O_2 or by using an external rotating magnetic field. (B) Recovery of lead ions from the GOx-microbots in the presence of acidic media.

The detailed fabrication of conical, self-propelled tubular motors by sequential electrochemical deposition of nanolayers on the inner wall of a polycarbonate membrane is described in the Supporting Information (SI). As shown in Figure 1A, the microbot structure consists of an outer graphene layer and a platinum inner layer. The platinum layer decomposes hydrogen peroxide into water and oxygen microbubbles, and the ejection of microbubbles from one side of microbot provides enough force for its self-propulsion. Between the Pt and GOx, layers of Pt/Ni and Ni were deposited to control and guide microbots motion by externally applied magnetic field.

Taking advantage of the self-propulsion and magnetic characteristics of the microbots, in combination with the GOx adsorption properties to attach Pb (II) ions, two approaches for the water cleaning of lead were carried out (Figure 1A). In addition, the recovery of lead after its removal from wastewater is also carried out, as it is displayed in Figure 1B.

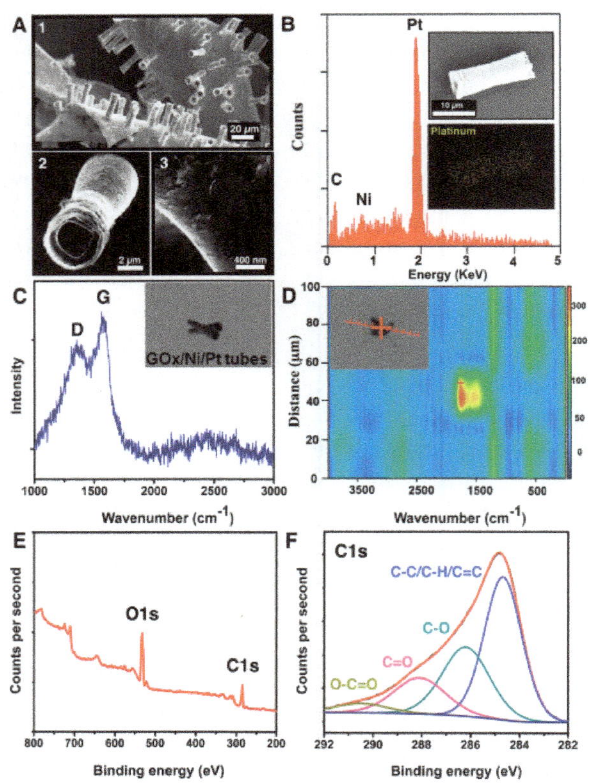

Figure 2: Characterization of GOx-microbots: (A) SEM images illustrating: (1) GOx-microbots attached to the gold layer, (2) structure of a single GOx-microbot, and (3) close look of the surface of a GOx-microbot. (B) The EDX spectrum of GOx-microbots (inset: EDX mapping

of the GOx-microbot for Pt). (C) Raman spectrum from the surface of GOx-microbots, showing characteristic D and G band of graphene oxide. (Inset is an illustration of the GOx-microbots analyzed by Raman spectroscopy.) (D) Raman spectroscopy displays the Raman scan map of the GOx-microbots, confirming the presence of graphene on the full surface of microbots (see inset: image of the GOx/Ni/Pt tube and scanning path (red line) of the analysis). (E) XPS spectra survey of GOx-microbots, showing O 1s and C 1s peaks. (F) High-resolution C 1s XPS spectra of GOx microbots displaying various functional groups identified on the surface.

GOx-microbots were characterized by scanning electron microscopy (SEM), energy dispersive X-ray spectroscopy (EDX), Raman spectroscopy, and high resolution X-ray photoelectron spectroscopy (XPS), as it is observed in Figure 2. Figure 2A,1 illustrates the tubular morphology and size polydispersity of the microbots after their fabrication. Most of the microbots show biconical morphology and a low polydispersity indicating that fabrication was homogeneous. Figure 2A,2 displays the hollow inside layer of a representative microbot. Their average outer diameter is 4.6 ± 0.1 μm (n = 10) and their inner diameter 2.5 ± 0.1 μm (n = 10). The surface of microbots is not homogeneous, as it is shown in Figure 2A,3 which can be due to the high number of defects produced by the electrochemical deposition of graphene nanosheets on the polycarbonate membrane during the fabrication of these microbots. Figure 2B illustrates the EDX analysis and reveals the components of the microbots including carbon, nickel, and platinum where platinum is the major component. Raman spectroscopy analysis and mapping for GOx-microbots are shown in Figure 2C and D, respectively. Figure 2C shows Raman spectrum of the microbot surface where the characteristic peaks of D and G band are observed at 1350 and 1570 cm^{-1}, respectively. Usually, the D-mode is caused by disordered structure of graphene, while band G arises from the stretching of the C–C bond in graphitic materials and is common to all sp^2 carbon systems. The band intensity ratio of I_D/I_G suggests the presence of oxidated carbon in GOx-microbots. To further confirm the complete coverage of GO layer on microbots, the GOx-microbots were characterized by confocal Raman technique. As shown in the bright field image (Figure 2D inset), a trimer of GOx-microbots formed by physical attachment was chosen for the Raman mapping. By mapping integrated intensity values of D and G bands in Raman spectrum of GOx-microbots, a 2-D Raman line mapping image was acquired and is presented in Figure 2D. A homogeneous distribution of high intensity signal in red and yellow colors resembles the width of the trimmer with similar dimensions. The confocal Raman line mapping image confirmed the successful and uniform coverage of GO on the surface of microbots. XPS spectra survey further reveals the presence of carbon (C 1s peak) and oxygen (O 1s peak) elements on the surface of GOx-microbots, where the O 1s signal is higher compared to the C 1s signal (Figure 2E) characteristic for GOx. The peaks of high-resolution C 1s spectra (Figure 2F) and O 1s spectra correspond to the binding energy of various functional groups such as C–C/C–H/C=C, C–OH, C=O, and O–C=O, revealing the nature of the covalent bonds of oxygen atoms and carbon atoms. The ratio of percentage atomic concentration of C–C/C–H/C=C functional groups to all carbon–oxygen functional groups is 0.94, which denotes the degree oxidation of GOx. Abundant carbonyl and carboxyl groups are present on the GOx-microbots which are considered very important for the adsorption of heavy metals on the surface of graphene oxide.(36)

To prove the capability of the self-propelled GOx-microbots for the purification of lead contaminated water by adsorption, a concentration of 1.5% (v/v) of H_2O_2 and 0.1% (w/v) of sodium dodecyl sulfate (SDS) were used systematically in the all experiments as the optimal conditions for the swimming of the GOx-microbots. Average velocities of the microbots in these conditions were around 500 μm s^{-1}. A swarm of approximately 2 × 10^{5}GOx-microbots was deployed in the lead contaminated water (1 ppm) for adsorptive removal. Figure 3A shows snapshots from characteristic microbots swimming in lead-contaminated water at different time periods. Bubble tails released from microbots indicate the trajectories and displacement of microbots at initial time, 15, 30, and 60 min, respectively.

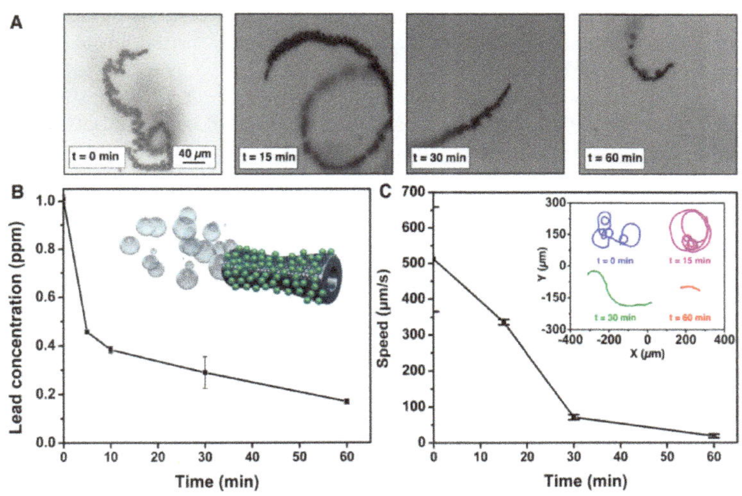

Figure 3: Kinetics of Pb(II) decontamination and speed for the GOx-microbots. (A) Optical snapshots from videos of GOx-microbots moving at different times during the decontamination process. (B) Pb(II) ion concentration at different time intervals during decontamination by GOx-microbots (inset: illustration of a GOx-microbot with adsorbed lead (green dots) on the surface after the decontamination process). (C) Speed of GOx-microbots at different times (0, 15, 30, and 60 min, inset: trajectories of the GOx-microbots for 5 s). Experimental conditions: 1 ppm as initial Pb(II) concentration, 1.5% (v/v) of H_2O_2 and 0.1% (w/v) SDS.

Figure 3B shows the decrease in lead concentration over 60 min in the presence of motile GOx-microbots. The lead concentration was measured using inductively coupled plasma optical emission spectrometry (ICP-OES) which is able to detect traces of metals. In the first 10 min, the GOx-microbots were able to decrease the lead concentration from 1 ppm to lower than 0.4 ppm, and in 1 h the GOx-microbots were able to remove more than the 80% of lead from the contaminated water. The GOx-microbots were allowed to swim for 24 h observing that after this time the concentration of lead was slightly higher (0.29 ± 0.02 ppm) than it was in 1 h (0.17 ± 0.01 ppm). This could be attributed to the fact that with time, the process of adsorption reaches to a desorption–adsorption equilibrium where desorption phenomenon could also occur. Therefore, 1 h was selected as the optimal time for the lead remediation from polluted water. Figure 3C (inset) and Video 1 (SI) displays the tracking of the average speed of the microbots for 5 s at different times (0, 15, 30, and 60 min). When the microbots were initially added to the lead contaminated solution and 1.5% H_2O_2 (v/v), high velocities and frequent reorientation of trajectories were observed. After 15 min, the speed of the microbots slightly decreased, and their trajectories were usually circular or straight. After 30 and 60 min, microbots swam at slower speed compared to their initial values and with less frequent reorientation in the direction of swimming. This swimming behavior was due to the fact that the H_2O_2 was being consumed continuously from the solution during the catalytic reaction on the inner platinum surface. To demonstrate that the presence of lead does not affect the motor speeds, new H_2O_2 was added to the slow or nonswimming microbots that had previously undergone 24 h of catalytic reaction. Addition of the fresh H_2O_2 reestablished microbot swimming with similar velocities and movements as were observed at the beginning of the experiments.

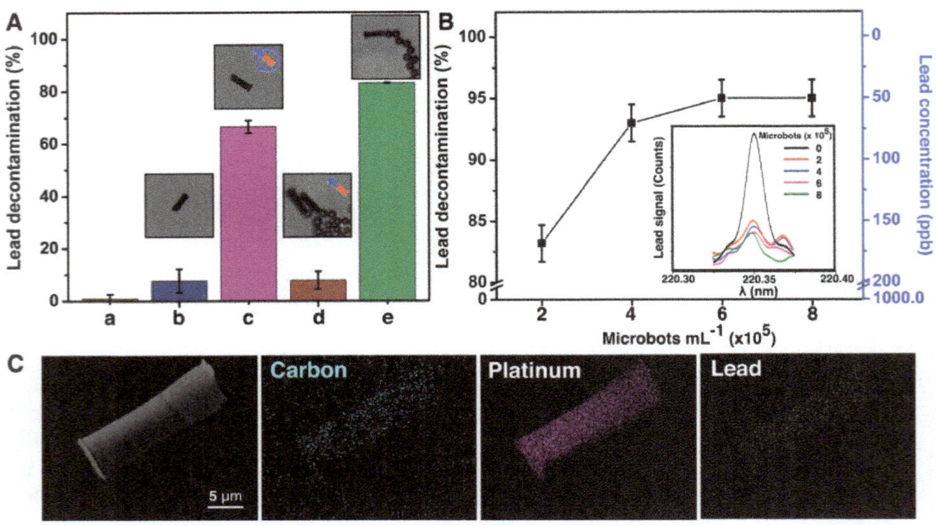

Figure 4: Pb (II) ions decontamination by GOx-microbots and characterization of microbots after decontamination. (A) Decontamination of Pb(II) ions in different systems: In the presence of (a) H_2O_2 (1.5% v/v) and of SDS (0.1% w/v) after 24 h, (b) SDS (0.1% w/v) and nonmotile GOx- microbots after 1 h, (c) SDS (0.1% w/v) and GOx-microbots stirred by external magnets after 1 h, (d) H_2O_2 (1.5% v/v), SDS (0.1% w/v) and docked GOx-microbots after 1 h (immobilized by stationary magnetic field), and (e) in the presence of H_2O_2 (1.5% v/v) and SDS (0.1% w/v) motile GOx-microbots after 1 h. (B) Decontamination of Pb(II) ions for different concentrations of GOx-microbots after 1 h in the presence of H_2O_2 (1.5% v/v) and of SDS (0.1% w/v). (Inset: ICP-OES signal of lead concentration after 1 h of decontamination process for increasing amount of motors.) (C) EDX mapping: the SEM image of the analyzed GOx-microbot after the decontamination process showing, carbon distribution, platinum distribution, and Pb(II) distribution.

Control experiments were carried out to demonstrate that the decontamination process was due to the adsorption on the GOx-microbots (Figure 4). Figure 4A,a shows that without the use of microbots, the lead concentration does not decrease when in contact with the fuel H_2O_2 and SDS after 24 h. Figure 4A,b shows minor decontamination of lead when a fixed amount of GOx-microbots were left in contact with the lead contaminated solution for 1 h. This was performed without adding H_2O_2 in the solution, so that GOx-microbots could not swim, leading to a low decontamination of lead. The GOx-microbots were then stirred by rotation of an external magnetic field generated by the magnetic stirrer, as it is represented in the inset in Figure 4A,c. Here, microbots were able to remove 66.6 ± 2.4% lead from water. When the GOx-microbots are rotated at high speed in the solution, the diffusion of lead ions on the microbot surface is enhanced due to induced microconvection. This increases the chances of contact between contaminant and microbots. Figure 4A,d displays the decontamination when the bubbled GOx-microbots left in polluted solution, but their swimming was inhibited by the presence of a strong magnet, which immobilizes them in a fixed location. However, when these GOx-microbots were left free to swim (Figure 4A,e), the decontamination process is 10 times more effective as it varies from 7.7 ± 4.5% (nonmotile) to 83.2 ± 1.0% (motile). These results reveal the high significance of the synergy between the presence of graphene oxide and self-propulsion of the GOx-microbots.

Because the lead decontamination is an adsorption process based on the adhesion of lead onto a GOx surface, when the number of GOx-microbots in the contaminated water was increased, the concentration of lead in the solution decreased as displayed in Figure 4B. We doubled the number of microbots and up to four times the standardized amount of microbots (2 × 10^5 microbots), resulting in an increase of lead capture from 83.2% to more than 95% (remaining lead concentration was

<50 ppb). The plateau at 6 × 10^5 microbots indicates the lead detection limit of the analytical system used for the analysis.

EDX mapping was used to verify the decontamination by GOx-microbots, which showed the presence of Pb on their surfaces (Figure 4C). Lead was not found in EDX mapping performed previous to the decontaminating experiment (Figure 1B), which gives direct evidence of adsorption of lead on the surface of the microbots. The effective adsorption of lead on graphene surface of the GOx-microbots is based on the strong interactions produced between graphene oxide and Pb (II) ions due to the formation of electron donor–acceptor complexes. The presence of oxygen moieties and delocalized π-electron systems in graphene oxide act as Lewis base and attach to the Pb(II) ions which act as Lewis acid. In addition, this process is strongly dependent on the pH and temperature and independent of ionic strength.(37) We also observed that the structure of the GOx-microbots was not damaged by the decontamination process maintaining its composition (Figure 4C, SEM image).

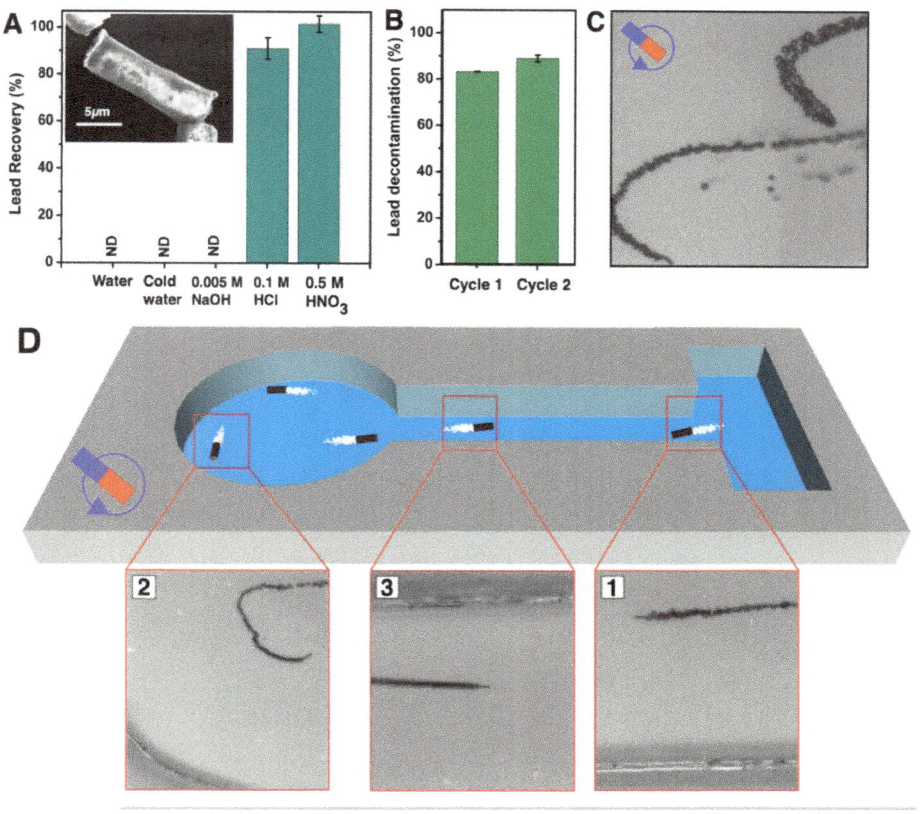

Figure 5: Lead recovery, reusability and magnetic control of GOx-micromotors. (A) Recovery of Pb(II) ions from GOx-microbots after different desorption treatments for 1 h, which were previously employed in decontamination process. (Inset shows SEM image illustrating GOx-microbot tube after the decontamination and desorption (with 0.5 M HNO_3) process. (B) Reusability of GOx-microbots, representing lead decontamination in the first and second cycle. (C) Optical snapshot from a video of GOx-microbot controlled by magnetic guidance after the lead recovery process. (D) GOx-microbot controlled by magnetic guidance swimming in a PDMS microchannel as a prototype system. Experimental conditions for B and C: 1.5% (v/v) of H_2O_2 and 0.1% (w/v) of SDS.

We further demonstrate the recovery of lead after its removal from polluted water. Figure 5A shows the efficiency of lead recovery after treating GOx-microbots in different chemical conditions to induce the desorption of Pb(II) from their surfaces (see SI). Different aliquots of previously used GOx-microbots for lead removal were extracted using a magnet, dissolved in 3 mL of different media (Figure 5A), and stirred for 1 h. After the removal of GOx-microbots from the solution by magnetic control, the remaining solution was analyzed for lead by ICP-OES analysis. When the GOx-microbots were in the presence of water, cold water (temp 4 °C), and basic pH (11), the Pb(II) ion concentration was not detectable at the detection limit of the instrument (50 ppb). However, when the GOx-microbots were in the presence of acidic media, i.e., HCl (pH = 1) and HNO_3 (pH = 0.3), a desorption event reached recoveries up to 91.2 ± 4.6% and 101.0 ± 3.5%, respectively. The structure of the GOx-microbots was not significantly affected as it is observed in the inset of Figure 5A which shows a SEM image of GOx-microbot after the decontamination and recovery processes. After cleaning and washing GOx-microbots, the reusability of GOx-microbots was studied. The GOx-microbots were added into the lead contaminated water as the first decontamination assays. Figure 5B represents that the GOx microbots retained their lead removal efficiently in the second cycle after removing and recovering lead in the first cycle. Activation of graphene oxide by treatment with the oxidant acids such as nitric acids oxidize the graphene surface and increase the number of oxygen moieties, which mainly interact with Pb(II) ions.

As a proof-of-concept, we tested the capabilities of GOx-microbots to perform various tasks inside a PDMS microchannel by external magnetic guidance. Because GOx-microbots contain Ni layer, they are ferromagnetic in nature which allows controlling trajectories by external magnetic field once they are self-propelled in solution. Figure 5C and Video 2 (SI) show the trajectory control of GOx-microbots by using the external magnetic field after the process of Pb(II) decontamination on the surface. Figure 5D andVideo 3 (SI) illustrate a GOx-microbot guided in the microfluidic channel from the lead contaminated water containing reservoir after decontamination to the other location where lead can be recovered and concentrated for recycling. The ability to magnetically control the microbots makes it possible to develop and program an automated system to guide swarm of microbots to accomplish the assigned tasks.

In conclusion, we have demonstrated graphene oxide based microbots for very efficient removal of toxic heavy metal (Pb) from contaminated water through an adsorption process, the recovery of Pb(II) ions, and the subsequent reusability of GOx-microbots. GOx-microbots can be deployed in contaminated water to swim randomly and easily collected using magnets once the water purification process has been completed. As a proof of the concept, magnetic control of GOx-microbots swimming inside a microfluidic system was demonstrated. GOx-microbots can be useful as new devices for future decontamination of heavy metals from industrial wastewater due to their efficiency for decontamination, their easy removal from the solution and the possibility of lead recovery and their reusability. The use of active systems and graphene nanomaterials can pave the way for new functionalities of self-propelled micronanomotors, from drug delivery, sensing, and energy to new environmental applications.

ACKNOWLEDGMENT

The results leading to the publication have received financial support from the European Research Council for the European Union's Seventh Framework Programme (FP7/2007-2013)/ERC Grant Agreement 311529 (Lab-in-a-tube and Nanorobotics biosensors) and the Alexander von Humboldt Foundation (DV). Authors thank L. Soler for help with XPS analysis.

REFERENCES

1. Fu, F.; Wang, Q. J. Environ. Manage. 2011, 92, 407–418.
2. Shannon, M. A.; Bohn, P. W.; Elimelech, M.; Georgiadis, J. G.; Marinas, B. J.; Mayes, A. M. Nature 2008, 452 (7185), 301–310.
3. Ali, I. Chem. Rev. 2012, 112 (10), 5073–5091.
4. Zhao, G.; Li, J.; Ren, X.; Chen, C.; Wang, X. Environ. Sci. Technol. 2011, 45 (24), 10454–10462.
5. Cong, H.-P.; Ren, X.-C.; Wang, P.; Yu, S.-H. ACS Nano 2012, 6 (3), 2693–2703.
6. Huang, Z.-H.; Zheng, X.; Lv, W.; Wang, M.; Yang, Q.-H.; Kang, F. Langmuir 2011, 27 (12), 7558–7562.
7. Wang, W.; Duan, W.; Ahmed, S.; Sen, A.; Mallouk, T. E. Acc. Chem. Res. 2015, 48 (7), 1938–1946.
8. Sanchez, S.; Soler, L.; Katuri, J.´ Angew. Chem., Int. Ed. 2015, 54 (5), 1414–1444.
9. Paxton, W. F.; Kistler, K. C.; Olmeda, C. C.; Sen, A.; St. Angelo, S. K.; Cao, Y.; Mallouk, T. E.; Lammert, P. E.; Crespi, V. H. J. Am. Chem. Soc. 2004, 126 (41), 13424–13431.
10. Wang, H.; Pumera, M. Chem. Rev. 2015, 115 (16), 8704–8735.
11. Soler, L.; Magdanz, V.; Fomin, V.
12. Orozco, J.; Jurado-Sanchez, B.; Wagner, G.; Gao, W.; Vazquez-´ Duhalt, R.; Sattayasamitsathit, S.; Galarnyk, M.; Cortes, A.; Saintillan,´ D.; Wang, J. Langmuir 2014 , 30 (18), 5082 –5087.
13. Guix, M.; Mayorga-Martinez, C. C.; Merkoci, A.¸ Chem. Rev. 2014 , 114 (12), 6285 –6322.
14. Wang, J.Nanomachines: fundamentals and applications; John Wiley & Sons, 2013.
15. Soler, L.; Sanchez, S.´ Nanoscale 2014 , 6 (13), 7175 –7182.
16. Gao, W.; Wang, J. ACS Nano 2014 , 8 (4), 3170 –3180.
17. Li, T.; Li, L.; Song, W.; Wang, L.; Shao, G.; Zhang, G. ECS J. Solid State Sci. Technol. 2015 , 4 (10), S3016 –S3019.
18. Wani, O. M.; Safdar, M.; Kinnunen, N.; Janis, J.¨ Chem. - Eur. J. 2016 , 22 (4), 1244 –1247.
19. Mushtaq, F.; Guerrero, M.; Sakar, M. S.; Hoop, M.; Lindo, A. M.; Sort, J.; Chen, X.; Nelson, B. J.; Pellicer, E.; Pane, S. J. Mater. Chem. A 2015 , 3 (47), 23670 –23676.
20. Teo, W. Z.; Zboril, R.; Medrik, I.; Pumera, M. Chem. - Eur. J. 2016 , 22, 4789.
21. Orozco, J.; Cheng, G.; Vilela, D.; Sattayasamitsathit, S.; VazquezDuhalt, R.; Valdes-Ramírez, G.; Pak, O. S.; Escarpa, A.; Kan, C.; Wang, J.´ Angew. Chem., Int. Ed. 2013 , 52 (50), 13276 –13279.
22. Li, J.; Singh, V. V.; Sattayasamitsathit, S.; Orozco, J.; Kaufmann, K.; Dong, R.; Gao, W.; Jurado-Sanchez, B.; Fedorak, Y.; Wang, J. ACS Nano 2014 , 8 (11), 11118 –11125.
23. Guix, M.; Orozco, J.; García, M.; Gao, W.; Sattayasamitsathit, S.; Merkoci, A.; Escarpa, A.; Wang, J.¸ ACS Nano 2012 , 6 (5), 4445 –4451.
24. Seah, T. H.; Zhao, G.; Pumera, M. ChemPlusChem 2013 , 78 (5), 395 –397.
25. Gao, W.; Feng, X.; Pei, A.; Gu, Y.; Li, J.; Wang, J. Nanoscale 2013 , 5 (11), 4696 –4700.
26. Singh, V. V.; Martin, A.; Kaufmann, K.; D. S. de Oliveira, S.; Wang, J. Chem. Mater. 2015 , 27 (23), 8162 –8169.
27. Jurado-Sanchez, B.; Sattayasamitsathit, S.; Gao, W.; Santos, L.;´ Fedorak, Y.; Singh, V. V.; Orozco, J.; Galarnyk, M.; Wang, J. Small 2015 , 11 (4), 499 –506.
28. Orozco, J.; Mercante, L. A.; Pol, R.; Merkoci, A. J. Mater. Chem. A 2016 , 4, 3371.
29. Mou, F.; Pan, D.; Chen, C.; Gao, Y.; Xu, L.; Guan, J. Adv. Funct. Mater. 2015 , 25 (39), 6173 –6181.
30. Orozco, J.; García-Gradilla, V.; D 'Agostino, M.; Gao, W.; Cortes¸ A.; Wang, J. ACS Nano 2013 , 7 (1), 818 –824.
31. Moo, J. G. S.; Wang, H.; Zhao, G.; Pumera, M. Chem. - Eur. J. 2014 , 20 (15), 4292 –4296.
32. Dey, K. K.; Bhandari, S.; Bandyopadhyay, D.; Basu, S.; Chattopadhyay, A. Small 2013 , 9 (11), 1916 –1920.

33. Vilela, D.; Orozco, J.; Cheng, G.; Sattayasamitsathit, S.; Galarnyk, M.; Kan, C.; Wang, J.; Escarpa, A. Lab Chip 2014 , 14 (18), 3505 −3509.

34. Esteban-Fernandez de A´´ vila, B.; Lopez-Ramirez, M. A.; Baez, D.´ F.; Jodra, A.; Singh, V. V.; Kaufmann, K.; Wang, J. ACS Sens. 2016 , DOI: 10.1021/acssensors.5b00300 .

35. Moreno-Guzman, M.; Jodra, A.; Lopez, M.-A´ ´ .; Escarpa, A. Anal. Chem. 2015 , 87 (24), 12380 −12386.

36. Moo, J. G. S.; Khezri, B.; Webster, R. D.; Pumera, M. ChemPhysChem 2014 , 15 (14), 2922 −2929.

37. Zhao, G.; Ren, X.; Gao, X.; Tan, X.; Li, J.; Chen, C.; Huang, Y.; Wang, X. Dalton Trans. 2011 , 40 (41), 10945 −10952.

CHAPTER 5

BRACKISH GROUNDWATER MEMBRANE SYSTEM DESIGN FOR SUSTAINABLE IRRIGATION

Beni Lew, Lolita Trachtengertz,Shany Ratsin,Gideon Oronand Amos Bick

Keywords: Analytical hierarchical process, irrigation, multi-dimension scaling, nanofiltration, reverse-osmosis

INTRODUCTION

Reuse of brackish water, mainly for agricultural irrigation, simultaneously solves water shortage problem and allows for overcoming environmental pollution problems (Qadir et al., 2007;Ghermandi and Mesalem, 2009). It is recognized that water for irrigation should be without particles with average size larger than 50–100 micron and, have a low salt content to avoid certain ion toxicities and increase of soil salinity (Ben-Gal et al., 2008; Edelstein et al., 2009). Salt concentration is generally tested by electrical conductivity (EC) or total dissolved solids (TDS) levels (Bui, 2013). In addition, the sodium to water hardness (calcium and magnesium) ratio should be managed to keep sodium absorption ratio (SAR) at values lower than 3–4, if possible. Together with that, precipitation potential of the water should be to some extent negative in order to avoid the precipitation of calcium and magnesium on the productive system. Also, iron and manganese ions need to be managed to prevent staining harms.

Other parameters, such as boron, fluoride, and heavy metals should also be low to ease the possibility of ion toxicity. Besides, irrigation water needs to be without any disease-causing microorganisms, viruses, bacteria, fungi, nematodes, cysts, etc. (Yermiyahu et al., 2007; Bustan et al., 2013).

Hence, irrigation water has to be treated properly to get rid of all the undesirable physical, chemical, and biological contaminants that are able to: (i) reduce the choice of crops (Levy and Tai, 2013); (ii) cut down crop yield (Rasouli et al., 2013); (iii) damage crop quality (Bernstein et al., 2011); (iv) injure soil appropriateness (Liu et al., 2013); and (v) harm the irrigation tools (Connor et al., 2012).

Water treatment regarding irrigation purposes is quite new to the agricultural business. The advanced technology is based on membrane technology (Oron et al., 2008; Bick et al., 2012). Mainly, Nano Filtration (NF) and Reverse-Osmosis (RO) can usually be implemented for removal of organic matter, particles, turbidity, pathogenic micro-organisms, and selective ions without the use of disinfectants (Mrayed et al., 2011; Riera et al., 2013).

Membrane system presents several advantages: modularity, automation, low maintenance costs, and low chemicals consumption (Miller, 2006; Birnhack et al., 2010). However, there are several drawbacks: high energy consumption, permeate quality (lack of vital ions for plant growth, e.g., magnesium), high costs relative to other sources of water and in particular chemical and biological fouling of the membranes (Oren et al., 2012).

For a known feed water source, the selection between membrane and any other alternative for water treatment is usually made on technical, economical, and/or political criteria. Yet, environmental concern have demonstrated its importance to integrate sustainability factors into the judgment process by comparing the "green" impacts of the different alternatives (Vince et al., 2008; Lew et al., 2009).

Downstream supervision on desalination plants, such as brine dilution, treatment, and/or removal may diminish the ecological impacts. However, reducing the upstream sources through means of optimization may be more successfully in minimizing the environmental impacts. The actual performances

Citation: Lew B, Trachtengertz L, Ratsin S, Oron G and Bick A (2014) Brackish groundwater membrane system design for sustainable irrigation. Front. Environ. Sci. 2:56. doi: 10.3389/fenvs.2014.00056.

of the membrane course of action outcome from model choices, which usually are designed to minimize the total water cost, taking in account technical constraints and project necessities but, with no consideration for the environmental impact (Bartels et al., 2008; Garcia et al., 2013; Wei et al., 2013). The environmental performances of the membrane system could be enhanced by revaluating these previously defined outline selections (Gur-Reznik et al., 2011;Chakrabortty et al., 2013; Zheng et al., 2013; Zhou et al., 2013). It is thus anticipated the introduction of the environmental criteria directly in the near the beginning design phases, in order to classify environmental responsible process configurations (Vince et al., 2008).

Concerning agricultural applications, NF membranes showed up to be a new opportunity for salty water desalination (Hilal et al., 2005): (i) high rejection rate for divalent ions, capability of knocking down TDS significantly and may constitute a potentially cost-effective alternative for irrigation with brackish water; (ii) higher tolerance (in general) for fouling conditions, as compared to RO (Gutman et al., 2012; Sotto et al., 2013); (iii) removal of specific pollutants, so that the concentrate stream can be used without increased membrane fouling; and (iv) operation at lower applied pressures, as compared to RO, saving energy and operating costs (Nederlof et al., 2005; Al-Amoudi and Lovitt, 2007; Zhao et al., 2012).

NF-RO configuration can be a promising solution, since (i) the NF stage can separate the agriculturally advantageous ions, such as magnesium from the non-desirable ions, such as sodium, which are finally removed from the water by the following RO stage; and (ii) the advantageous ions (in the NF brine flow) can be blended with the free sodium water produced in the RO stage (RO permeate stream) to create a nutrient-enriched, low-salt water (Mrayed et al., 2011).

The Arava Valley in Israel was used for this study, and the modeling and design of a desalination plant featuring both NF and RO membranes is discussed. The objectives of this study are: (i) to examine the designs of NF and RO membranes using typical brackish water from Hatzeva-Idan aquifer in the Arava Valley (Israel), which is characterized to some extent by saline water with a typical TDS concentration of 1200 ppm; (ii) to simulate RO and NF technologies for desalination of brackish water (using IMSDesign software, Verhuelsdonk et al., 2010; Karabelas et al., 2012); (iii) to compare the design configuration using Analytic Hierarchy Process (AHP) model (Caputo et al., 2013); and (iv) to rank the various alternatives using multi-dimension scaling (MDS) technique (Lespinats et al., 2009) in order to find the optimal design configuration for technology selection.

MATERIALS AND METHOD

Management Modeling and the Analytical Hierarchical Process (AHP)

The amount of data and information collected and retained by organizations and businesses is persistently growing, due to advances in data gathering, computerization of transactions, and breakthroughs in storage tools. In order to take out practical information from such large datasets, it is crucial to be able to spot patterns, trends and interactions in the data and visualize their global configuration to ease decision making. Decision making techniques used for the ranking of various options on the basis of more than one attribute are strictly dependent on the attributes setting and thus can be completely different for different settings (Saaty, 1980).

The analytical hierarchical process (AHP) is a strong and flexible decision-supporting process that helps in setting priorities and making the best decision when mutually qualitative and quantitative aspects are measured (Tzfati et al., 2011; Zhang et al., 2013). The AHP is designed for individual evaluation of a set of options based on multiple attributes, and is approved in a hierarchical building (Kayastha et al., 2013). The estimate of the options is founded on a pair-wise judgment (Bozoki et al., 2011). The pair-wise comparisons are translated from linguistic/verbal expressions in numerical values (integers 1–9) using the original Saaty's Scale (Saaty, 2003) for the comparative judgments (Table 1).

Table 1: Fundamental Saaty's Scale for pair-wise comparison (Saaty, 1980).

Numerical values	Verbal term	Explanation
1	Equally important	Two elements have equal importance
3	Moderately more important	Experience or judgment slightly favors one element
5	Strongly more important	Experience or judgment strongly favors one element
7	Very strongly more important	Dominance of one element proved in practice
9	Extremely more important	The highest order dominance of one element over another
2,4,6,8	Important intermediate values	Compromise is needed

Multi Dimension Scaling (MDS) and CoPlot

Many research questions dealing with information require the analysis of complex multivariate data (Gomez-Silvan et al., 2013). Briefly, most multivariate approaches can be roughly classified as dependence methods (e.g., multiple regression, discriminant analysis, multivariate analysis of variance), that are usually used to assess the relationship between dependent and independent variables; or as interdependence methods (e.g., principal component analysis, factor analysis, cluster analysis), that are typically used to estimate the mutual association among all variables with no difference made among the variable types (Schilli et al., 2010). Multi dimensional scaling (MDS) is an interdependence methods that facilitates the examination of multivariate data by reducing multidimensional data into a two-dimensional structure that attempts to expose the "out of sight structure" in a data set by creating a pictorial or mapping image of the data. The MDS map graphically represents the proximities (or similarities) between objects (i.e., observations or events). Similarities between the observations in the data set are transformed into distances on a map such that observations with great similarity are closer together than less similar observations.

In this way, a single picture illustrates the relationships among all the observations. MDS, initially developed in the

1960s, has been used to evaluate the relationships among observations, to identify clusters of similar observations and to locate outliers (Bick et al., 2011). However, MDS maps have two key limitations: (i) MDS does not simultaneously map the variables and the observations; and (ii) the MDS map has no orientation, thereby limiting the map's interpretability.

CoPlot, a variant of multi-dimensional scaling, addresses both these limitations and locates each decision-making unit in a two-dimensional space in which the location of each observation is strong-minded by every variables simultaneously (hence, its name) (Lipshitz and Raveh, 1994). The graphical put on view technique exhibits observations as points and variables as arrows, relative to the same center-of-gravity. CoPlot is rooted in the integration of mapping concepts, using a variant of regression analysis that superimposes two graphs sequentially. Additionally, CoPlot maps the observations and variables in a manner that preserves their relationships, allowing richer interpretations of the data. Importantly, CoPlot allows analysis of a dataset where the number of variables is greater than the number of observations. Also, CoPlot map can be used to detect outliers and errors in the data, assessment of the relationships within the data and for selection of key variables for subsequent analysis (Adler and Raveh, 2008).

Coplot's output is a visual display of its findings [Given an input matrix Yn × v of v variable values for each of n observations (see for example Table 2)] and it is based on two graphs that are superimposed on each other (Bravata et al., 2008). The first graph maps the n observations into a two-dimensional space: those observations that are perceived to be very similar to each other are placed near each other on the map, and those observations that are perceived to be very different from each other are placed far away from each other on the map. The second graph (Figure 1) consists of v arrows, representing the variables and it shows the direction of the gradient for each arrow.

Table 2: Production technologies for irrigation.

Design code	Membrane type	Number of passes	Additional data
NF-L-H	ESNA1-LF-LD 86% rejection 10,500 gpd	1	Nanofiltration Low rejection High flux
NF-M-M	ESNA1-LF2-LD 91% rejection 8200 gpd	1	Nanofiltration Medium rejection Medium flux
RO-H-H	CPA5-MAX 99.7% rejection 12,000 gpd	1	Reverse-Osmosis High rejection High flux
RO-M-H	ESPA4-MAX 99.2% rejection 13,200 gpd	1	Reverse-Osmosis Medium rejection Extreme high flux
NF-RO-1P	NF-M-M at the first stage RO-M-H at the second stage	1	
NF-RO-2P	NF-M-M at pass I RO-M-H at pass II	2	Desalination of Nanofiltration permeate (the brine of the second pass is recycled to the feed)

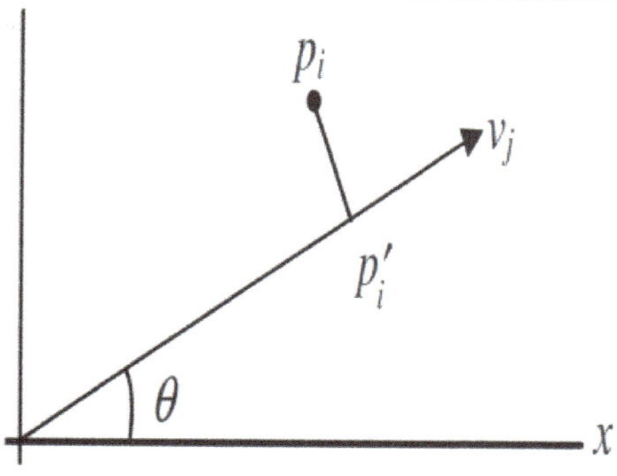

Figure 1: CoPlot: Adding the variable vectors. The point p_i corresponds to the coordinates for observation i = 1... n. The vector v_j is for the variable j = 1 ... m. The x-axis is rotated through an angle θ to give a point p'_i, which is the projection of p_i onto the vector v_j. The correlation between the n new projected scores and the original n values for variable j are computed, and the choice for θ is the one that maximizes this correlation.

The CoPlot analysis consists of four stages, two preliminary adaptations of the data matrix and two subsequent stages that compute two maps sequentially that are then superimposed to produce a single map. The goal of the first stage is to normalize the variables, which is needed in order to allow the variables to relate to each other, although each variable has a different unit and scale. This is done in the usual way, the difference between the elements of the matrix Y_{ij} (which are scores) and the deviations from column means$\bar{Y}_j$ is divided by their standard deviations (D_j), given the normalized Z_{ij}, according to Equation (1).

$$Z_{ij} = (Y_{ij} - \bar{Y}_j)/D_j \tag{1}$$

In the second stage, a measure of dissimilarity ($S_{ik} \geq 0$) between each pair of observations (rows of Zn × v) is chosen and a symmetric n × n matrix is produced from all the different pairs of observations. To measure S_{ik}, the sum of absolute deviations (generally defined as city-block distance) as a measure of dissimilarity is used (Equation 2).

$$S_{ik} = \sum_{j=1}^{v} |Z_{ij} - Z_{kj}| \tag{2}$$

In the third stage the matrix S_{ik} is mapped by a MDS method. The algorithm maps the matrix S_{ik}into an Euclidean space, of two dimensions in our case, such that "similar" observations (with a small dissimilarity between them) are close to each other on the map, while "different" ones are also distant on the map. Formally, the requirement is as follows: consider two observations, i and k, which are mapped at a d_{ik} distance from each other. This distance has to reflect the dissimilarity S_{ik}(which is actually a relative measurement), taking in consideration that the important constraints are $S_{ik} < S_{lm}$ if $d_{ik} < d_{lm}$.

CoPlot procedure uses the Guttman's smallest space analysis (SSA) with the "coefficient of alienation" (Θ) as a measure of "goodness of fit" (Guttman, 1968; Raveh, 2000). The intuition for Θ comes directly from the above MDS requirement of dissimilarity measures and the map distances: A success of satisfying it implies that the product of the differences between the dissimilarity measures and the map distances are positive. In a normalized form a new variable is defined, μ_{ca}(Equation 3).

$$\mu_{ca} = \frac{\sum_{i,k,l,m}(S_{ik} - S_{lm})(d_{ik} - d_{lm})}{\sum_{i,k,l,m}|S_{ik} - S_{lm}|\,|d_{ik} - d_{lm}|} \quad (3)$$

μ_{ca} can attain a maximal value of 1 (Raveh, 2000) and is used to define Θ according to Equation (4).

$$\Theta = \sqrt{1 - \mu_{ca}^2} \quad (4)$$

The details of the SSA algorithm are beyond the scope of this paper and were presented in the literature (Guttman, 1968). The SSA algorithm is an extensively used method in social sciences and several examples along with intuitive descriptions can be found (Raveh, 2000). The outcome of this stage is a two-dimensional map of n observations and the CoPlot user can color code observations with any definite variable that has up to 16 different values.

The map generated so far is a classical MDS map, however without orientation or meaningful axes. In the fourth stage of the CoPlot method, v arrows are drawn on the Euclidean space previously obtained. Each variable j is represented by an arrow j, emerging from the center of gravity of the n points. The graphical display technique plots observations as points and variables as arrows, relative to the same arbitrary center-of-gravity. Observations are mapped such that similar entities are closely located on the plot, signifying that they belong to a group possessing comparable characteristics and behavior. The location of the center of gravity is in the middle of the plot in order to introduce all the observations and it does not affect the arrows direction. The direction of each arrow is selected so that the correlation between the actual values of the variable j and their projections on the arrow is maximal (the arrows' length is undefined). Therefore, observations with a high j value will be located in the part of the space which the arrow points to, while observations with a low j value will be located at the other side of the map. The magnitude of the j maximal correlations measures the "goodness of fit" of the R_j regressions. Higher is the correlation better is the arrow representation of the variables and, those having low correlations should be eliminated.

Moreover, arrows linked with highly correlated variables will point in about the same direction and vice versa. As a result, the cosines of angles between these arrows are approximately proportional to the correlations between their associated variables (Raveh, 2000).

The "goodness of fit" measured for each correlation (R_j) is obtained as follows: For each possible variable vector, CoPlot projects the points onto the vector, thereby yielding n projected values. These projected values can now be compared with the observed values. The axis that is chosen is the one that maximizes the correlation between the projected values and the observed values. Figure 1 depicts how this is performed. The point p_i corresponds to the coordinates for observation i = 1… n. The vector v_j is for the variable j = 1… m. The x-axis is rotated through an angle θ to give a point p'_i, which is the projection of p_i onto the vector v_j. The correlation between the n new projected scores and the original n values for variable j are computed and, the choice for "goodness of fit" is the one that maximizes this correlation. This maximization can be achieved numerically by calculating all 360° possibilities for θ, which is performed separately for each variable vector.

These variable vectors have four useful properties; first, vectors for highly correlated variables point in the same direction, vectors for highly negatively correlated variables are oriented along the same axis but in opposing directions and, vectors for variables that are not correlated are orthogonal to each other. Second, each vector emanates from the center of gravity, which serves as the origin. An observation located at or near the origin is an average observation (it has an average value in all variables). Third, the length of each vector is proportional to the correlation (namely R_j) between the original data for that variable and the projections of the observations onto the vector. Finally, the cosines of angles between the arrows are approximately proportional to the correlations between their associated Variables. Therefore, the correlational structure among the variables can be studied in a single graphical output (Raveh, 2000).

In practical terms, the user imports data, selects variables and observations for inclusion in the analysis, creates the CoPlot map, evaluates Θ parameters, selects the map to view (observations only, variables only, or both observations and variables) and, then selects variables for color coding the observations for greater interpretation. Qualitative variables can be selected for color coding and may either be included in the computation of the map or can be excluded from the computation of the map but still used for color coding. For example, if a variable was found to have low R_j, it might be excluded from the computation of the map but could still be used to color code variables to facilitate the interpretation of the data.

CoPlot produces two "goodness of fit" measurements, one that describes how well the CoPlot map represents the observations and another that describes how well the CoPlot map represents the variables. The first measure is a "coefficient of alienation" (Θ), which indicates the relative loss of information that arises when the multidimensional data are transformed into two dimensions. The lower the Θ value, the smaller the information loss in the process of reducing the original data set to a two-dimensional map. In other words, the lower the Θ, the more precise the representation of the MDS model to the proximities. A general rule-of-thumb states that the map is statistically significant if $\Theta \leq 0.15$ (Guttman, 1968).

In general, as the number of variables increases, Θ also increases. In such case, Θ measures the discrepancy between every pair of points and the original matrix of "similarities" that comprises distances between points, so that this index provides a comparison between two matrices, the matrix similarities (which are inputs) and the matrix of the distances on the map (which are outputs) obtained by the algorithm. When these two matrices (inputs and outputs) are identical, Θ is zero (precise).

The second "goodness of fit" measure is obtained at the stage of calculating the correlation between the original data, for each variable and, the projection of each observation onto that vector in the CoPlot map. In general, the methodology maximizes the correlations (actually the normalized cross-products) of the vector of inputs (the actual distances from each point to every other point) and the outputs (the coordinates of the vectors that go into the map); in other words, the R_j measurements are the correlational measure that relates the input with the output. The closer these are, in a correlational sense, higher the fit. Individual correlations are obtained for each of the variables separately. If a vector has a correlation of 1 it means that there is a perfect fit with the original variable data. In general, as the number of (poor) variables decreases the average correlation increases and, average of correlations with values of 0.7 or greater provides maps that fit the data (Bravata et al., 2008).

CASE STUDIES

The Arava Valley in Southern Israel is an example of highly efficient agriculture and greenhouse technologies (Villarreal-Guerrero et al., 2012) in a region of extreme water scarcity (Hillel et al., 2013). The water quality of the Arava Valley (Hatzeva wells) is as follows: Total Dissolved Solids-1577 ppm; Barium-0.2 ppm; Calcium-150 pm; Potassium-12.5 ppm; Magnesium-82.5 ppm; Sodium-225 ppm; Chloride-359 ppm; Bicarbonates-208 ppm; Nitrate-9.6 ppm; Sulfates-505 ppm (Ghermandi and Mesalem, 2009).

Simulation of Arava feed water treatment can be done by RO process design software's that were developed by membrane constructors such as ROSA from Filmtec or IMSDesign from Hydranautics (Penate and García-Rodríguez, 2011). Besides defining constructor good practices for membrane operation and shortcuts methods for pressure vessel modeling, the simulation allowed to verify flexible RO and NF configurations for different commercial membranes (Alghoul et al., 2009).

In this work, four types of 8″ commercial membranes were selected: (i) Nanofiltration low rejection and high flux (ESNA1-LF-LD, 86% rejection, 10,500 gpd); (ii) Nanofiltration medium rejection and medium flux (ESNA1-LF2-LD, 91% rejection, 8200 gpd); (iii) Reverse Osmosis high rejection and high flux (CPA5-MAX, 99.7 rejection, 12,000 gpd); and (iv) Reverse Osmosis medium rejection and extreme high flux (ESPA4-MAX, 99.2 rejection, 13,200 gpd). The simulation procedure includes six configurations (Table 2): (i) one pass with ESNA1-LF-LD membranes (code NF-L-H); (ii) one pass with ESNA1-LF2-LD membranes (code NF-M-M); (iii) one pass with CPA5-MAX membranes (code RO-H-H); (iv) one pass with ESPA4-MAX membranes (code RO-M-H); (v) one pass with ESNA1-LF2-LD membranes at the first stage and ESPA4-MAX membranes at the second stage (code NF-RO-1P); and (vi) two pass with ESNA1-LF2-LD membranes at the first pass and ESPA4-MAX membranes at the second pass (code NF-RO-2P).

The output of the simulation using IMSDesign software of the six different membrane configurations from Table 2 is presented in Table 3 (the design is based on six elements per vessel, production of 960 m^3/day, and operation at a permeate flux of 20 l/m^2-h). The pilot plant flow (960 m^3/day) can be increases without limitations, like any other membrane plants based on NF membranes. The 20 l/m^2-h permeated flux is based on pilot plant best performance as indicated by the model (increasing the flux in the system may cause blocking and fouling problems) The comparative performance of configuration technologies by the various alternative methods is very complex and is highly depending on various site-specific operational and economic attributes. General attributes must be fulfilled by an objective function (Gursoy et al., 2013): (i) minimization of energy consumption; (ii) minimization of brine production; (iii) minimization of brine concentration; and (iv) minimization of membrane types and passes. In order to support adequate selection the decision-maker has to define a utility function (Z_T) (Equation 5) that takes in account all the previous attributes.

$$Z_T = \begin{pmatrix} \text{Min} \\ \text{Energy} \\ \text{Consumption} \end{pmatrix} + \begin{pmatrix} \text{Min} \\ \text{Brine} \\ \text{Production} \end{pmatrix} + \begin{pmatrix} \text{Min} \\ \text{Brine} \\ \text{Concentration} \end{pmatrix} + \begin{pmatrix} \text{Min} \\ \text{Number} \\ \text{of Passes} \end{pmatrix} \quad (5)$$

Table 3: Output technologies for irrigation (IMSDesign software).

Design code	Energy kwhr/m^3	Brine percent	Brine salinity ppm	Number of passes
NF-L-H	0.26	18.6	5377	1
NF-M-M	0.20	22.5	4524	1
RO-H-H	0.52	16.1	6111	1
RO-M-H	0.33	17.0	5817	1
NF-RO-1P	0.31	19.5	5145	1
NF-RO-2P	0.28	16.3	5280	2

To use the AHP, the decision-maker must specify his requirements (based on previous experience) for achieving the overall goal.

Model Implementation and Discussion

From the matrix obtained in Table 3, the geometric mean (w_i, approximately the product of the elements in each row regarding to a matrix of n rows and n columns) and the normalized geometric mean (p_i) are determined according to Equations (6) and (7), respectively.

$$w_i = \left(\prod_{j=1}^{n} a_{ij}\right)^{1/n} \quad i = 1, \ldots, n \quad (6)$$

$$p_i = \left(\prod_{i=1}^{n} a_{ij}\right)^{1/n} / \sum_{i=1}^{n}\left(\prod_{j=1}^{n} a_{ij}\right)^{1/n} \quad i = 1, \ldots, n \quad (7)$$

where, a_{ij} is an evaluated value, i is the row index (alternative, i = 1,…, n) and j is the column index (quality attribute, j = 1,…, n). In Table 4 its shown that for each decision attribute chosen, the importance of the design options. It designates how the alternatives are preferred over others with respect to each objective as well as to the whole objective (Cay and Uyan, 2013).

Table 4: Variability in importance across design options.

Attribute	Design option
Min energy consumprion kwhr/m^3	NF-M-M>NF-L-H>NF-RO-2P>NF-RO-1P> RO-M-H> RO-H-H
Min brine production percent	RO-H-H>NF-RO-2P>RO-M-H>NF-L-H> NF-RO-1P>NF-M-M
Min brine salinity ppm	NF-M-M>NF-RO-1P>NF-RO-2P>NF-L-H> RO-M-H> RO-H-H

The notations > and ≅ symbolize the option preceding the sign in "preferable to" and "equal to" the one after the sign, respectively.

Table 5 compares minimization of energy consumption related to each of the options. One can see that according to the comparison the configurations NF-M-M is favorable (0.381). A comparable comparison was run for each of the attributes. The resultant set of weights for each of the option (in this case technology) with respect to each attributes is presented in Table 6.

Table 5: AHP pair-wise evaluation of min energy consumption (Number are based on Saaty's Scale and an expert subjective point of view).

Design	NF-L-H	NF-M-M	RO-H-H	RO-M-H	NF-RO-1P	NF-RO-2P	Geometric mean (*)	Normalization (**)
NF-L-H	1	1/2	5	4	3	2	1.98	0.252
NF-M-M	2	1	6	5	4	3	2.99	0.381
RO-H-H	1/5	1/6	1	1/2	1/3	1/4	0.33	0.042
RO-M-H	1/4	1/5	2	1	1/2	1/3	0.51	0.064
NF-RO-1P	1/3	1/4	3	2	1	1/2	0.79	0.101
NF-RO-2P	1/2	1/3	4	3	2	1	1.26	0.160
Total							7.87	

*For example, the geometric mean of NF-M-M is $(2 \cdot 1 \cdot 6 \cdot 5 \cdot 4 \cdot 3)^{1/6} = 2.99$

**For example, the normalization of NF-M-M is 2.99/7.87 = 0.38.

Table 6: AHP pair-wise results of attribute weights.

Design	Min energy	Min brine production	Min brine salinity	Min number of passes
NF-L-H	0.252	0.101	0.101	0.196
NF-M-M	0.381	0.042	0.381	0.196
RO-H-H	0.042	0.381	0.042	0.196
RO-M-H	0.064	0.160	0.064	0.196
NF-RO-1P	0.101	0.064	0.252	0.196
NF-RO-2P	0.160	0.252	0.160	0.022

Discrepancies in the response assessments can occur and are related to human errors along the process. Let's assume that assumption A is preferred over assumption B and, assumption B is preferred over assumption C. Then, it can be assumed that assumption A should be preferred over assumption C by a wide margin. If throughout the pair-wise comparison A is vaguely preferred over C, then a contradiction is taken place. In this work, the statistics and implementation of a judgment matrix (Yang et al., 2013) was implemented.

Based on the results of the correlations between variables, the CoPlot diagram presented in Figure2 shows "min brine salinity" and "min energy" relatively close to each other. "Min brine production" is pointing to the opposite direction and "min passes" is lying in-between. Consequently "min brine salinity"(or "min energy") is identified as offering diminutive information and can be taken away from the analysis with no effect (or little effect) on the results. The results show that: NF-RO-2P is disconnected from the other units because it has the highest "min passes" value; RO-H-H is separated from the other units because it has the highest "min brine production" value; and NF-M-M is separated from the other units because it has the highest "min brine salinity "and "min energy" values. NF-RO-2P, RO-H-H and NF-M-M could be considered outliers (see Figure 2). Finally, NF-L-H is average in all variables and in conclusion appears quite near to the center-of-gravity, the point from which all arrows begin. It is being claimed in this paper that MDS can be used to present the result graphically, the lower value (min objective function) a unit gets, the more effective the design is assumed over that specific attribute and it is clearly shown (Figure 2) that the best configuration is achieved with NF-L-H.

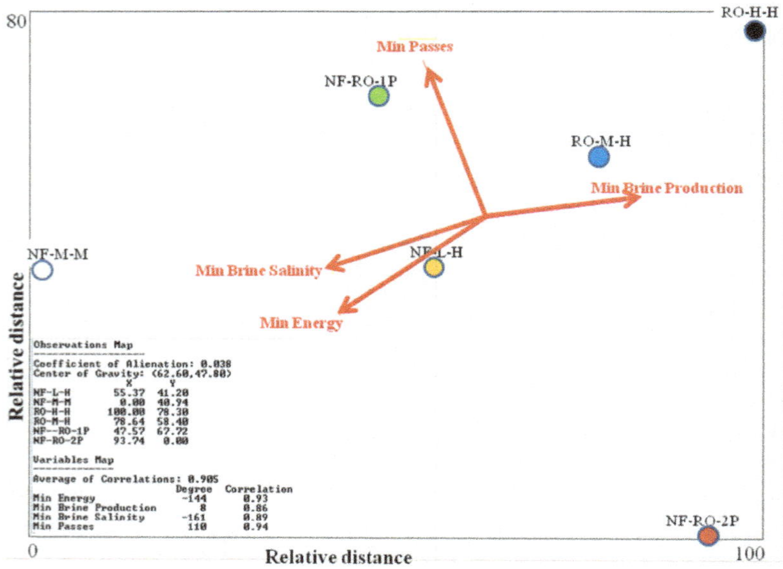

Figure 2: MDS map, generated by the proposed method (CoPlot): NF and RO configuration performance (at a permeate flux of 20 l/m²-h, six elements per vessel). Design: (i) one pass with ESNA1-LF-LD membranes (code NF-L-H); (ii) one pass with ESNA1-LF2-LD membranes

(code NF-M-M); (iii) one pass with CPA5-MAX membranes (code RO-H-H); (iv) one pass with ESPA4-MAX membranes (code RO-M-H); (v) one pass with ESNA1-LF2-LD membranes at the first stage and ESPA4-MAX membranes at the second stage (code NF-RO-1P); and (vi) two pass with ESNA1-LF2-LD membranes at the first pass and ESPA4-MAX membranes at the second pass (code NF-RO-2P). Each variable is represented by an arrow, emerging from the center of gravity of the n points. The direction of each arrow is chosen so that the correlation between the actual values of the variable and their projections on the arrow is maximal. (MDS Statistics: "Coefficient of alienation" = 0.038, "Average of Correlations" = 0.905). The "goodness of fit" measures (concerning the "Coefficient of alienation" and the variables correlations) is satisfied both for the configurations and the variables.

The "goodness of fit" in the Coplot system is calculated by two types of measures: the "coefficient of alienation" (Θ) and the extend of the v maximal correlations that measure the "goodness of fit" of the j regressions. Smaller "coefficient of alienation" indicates better the output and, values lower than 0.15 are assumed precise. In this study (Figure 2) a coefficient of 0.038 is considered as an excellent figure. The "goodness of fit" of the j regressions helps to decide whether one eliminates or adds variables: Variables that do not fit into the graphical display, meaning, have low correlations, should be removed. Higher the variable's correlation, the better the variable's arrow represents common direction and order of the projections of the n points along the axis.

Based on this test case, the correlations of MDS data are high: average 0.905 ("min energy" = 0.93; "min brine production" = 0.86; "min brine salinity" = 0.89; "min passes" = 0.94). The two "goodness of fit" measurements ("coefficient of alienation" for the first step and four correlations for each one of the variables for the second step) allow the researchers to point out that according to this map the optimal solution is Nanofiltration technology (low rejection and high flux).

There are some limitations of these approaches: (i) Some problems are too complex resulting in a very complicated pair wise process that cannot be conducted by AHP and MDS, (ii) Not all individuals are capable in using and understanding the process; there is a need for experts, and (iii) The focus is on immediate technology impacts and not on the gain in substantial knowledge expertise and future improvements.

For the use of NF technology a reform of water pricing may be needed, which is most often driven by budget government pressure, rise of the costs of the services provided and, government aspiration to reduce subsidies. The World Bank has encouraged governments to employ a policy of cost recovery for many years, on the belief that users should cover O&M and some of the capital costs (Bustan et al., 2013).

The Public-Private Partnerships (PPP) can be viewed as the governance policy for minimizing transaction costs and; coordinating and enforcing relations between partners engaged in agricultural goods and services production. The Built Own and Operate (BOO) projects promotes an optimal policy tactic to enable social and economic development, bringing together competence, plasticity, and competence for the private sector with the responsibility, long-term outlook and social interest of the public sector (Johannessen et al., 2014).

The use of membrane technology in agriculture can lead to additional gains: (i) understanding the entire supply-chain instead of the bottlenecks, (ii) improved quality and quantity of sales' product, (iii) new market penetration, and (iv) positive social effects. Agricultural research concerning NF technology responds to multiple objectives. Clearness of objectives in determining water charges is vital and there are limits of valuing as a practical tool for irrigation demand. The authorities must take into consideration the environmental benefits by including incentives for appropriate water allocation and demand management.

CONCLUSIONS

Membrane technology selection for water treatment is often a subjective task: Combining quantitative methods into the evaluation procedure permits the decision makers to recognize the most suitable option in an objective and efficient way. In this study two methodologies were used and discussed to analyse multiple variable data: AHP and MDS (CoPlot).

The AHP method establishes an evaluation model for design of membrane technology based on IMSDesign software. Analysis results using Saaty's Scale indicates that the decision-makers may select the most appropriate option based on the following attributes: minimization of energy consumption, minimization of brine production, minimization of brine salinity, and minimization of design passes (investment cost).

Using the AHP method for assortment of optimal technology provides a systematically decision making a framework with several characteristics: (i) the performance of different technologies can be evaluated using multiple attributes—both quantitative and qualitative—rather than profitability alone; (ii) ratings allow to evaluate the use of the technologies for the end user; (iii) AHP method provides an effective way for managing process documentation; and (iv) the proposed method forms the basis for a intermittent process of planning and managing technology, so that the priorities of the technologies can easily be modified and updated.

AHP database is graphically introduced by the CoPlot technique, a multivariate statistical method that is remarkably robust and provides new insights about membrane configurations and design, giving a strong picture of what needs to be done next. The CoPlot technique provides a powerful analytic tool for brackish groundwater membrane system: the optimal configuration using AHP and MDS design is based on NF membrane (low rejection and high flux).

CONFLICT OF INTEREST STATEMENT

The authors declare that the research was conducted in the absence of any commercial or financial relationships that could be construed as a potential conflict of interest.

ACKNOWLEDGMENT

The research was funded by the Chief Scientist, Israeli Ministry of Agriculture and Rural Development (Project n. 459-4437-12).

REFERENCES

1. Adler, N., and Raveh, A. (2008). Presenting DEA graphically. *Omega* 36, 715–729. doi: 10.1016/j.omega.2006.02.006

2. Al-Amoudi, A., and Lovitt, R. W. (2007). Fouling strategies and the cleaning system of NF membranes and factors affecting cleaning efficiency. *J. Membr. Sci.* 303, 4–28. doi: 10.1016/j.memsci.2007.06.002

3. Alghoul, M. A., Poovanaesvaran, P., Sopian, K., and Sulaiman, M. Y. (2009). Review of brackish water reverse osmosis (BWRO) system designs. *Renew. Sust. Energ. Rev.* 13, 2661–2667. doi: 10.1016/j.rser.2009.03.013

4. Bartels, C., Hirose, M., and Fujioka, H. (2008). Performance advancement in the spiral wound RO/NF element design.*Desalination* 221, 207–214. doi: 10.1016/j.desal.2007.01.077

5. Ben-Gal, A., Ityel, E., Dudley, L., Cohen, S., Yermiyahu, U., Presnov, E., et al. (2008). Effect of irrigation water salinity on transpiration and on leaching requirements: a case study for bell peppers. *Agric. Water Manage.* 95, 587–597. doi: 10.1016/j.agwat.2007.12.008

6. Bernstein, N., Ioffe, M., Luria, G., Bruner, M., Nishri, Y., Philosoph-Hadad, S., et al. (2011). Effects of K and N nutrition on function and production of *Ranunculus asiaticus*. *Pedosphere* 21, 288–301. doi: 10.1016/S1002-0160(11)60129-X

7. Bick, A., Gillerman, L., Manor, Y., and Oron, G. (2012). Economic assessment of an integrated membrane system for secondary effluent polishing for unrestricted reuse. *Water* 4, 219–236. doi: 10.3390/w4010219

8. Bick, A., Yang, F., Shandalov, S., Raveh, A., and Oron, G. (2011). Multi-dimension scaling as an exploratory tool in the analysis of an immersed membrane bioreactor. *Membr. Water Treat.* 2, 105–119. doi: 10.12989/mwt.2011.2.2.105

9. Birnhack, L., Shlesinger, N., and Lahav, O. (2010). A cost effective method for improving the quality of inland desalinated brackish water destined for agricultural irrigation. *Desalination* 262, 152–160. doi: 10.1016/j.desal.2010.05.061

10. Bozoki, S., Fulop, J., and Koczkodaj, W. W. (2011). An LP-based inconsistency monitoring of pairwise comparison matrices.*Math. Comput. Model.* 54, 789–793. doi: 10.1016/j.mcm.2011.03.026

11. Bravata, D. M., Shojania, K. G., Olkin, I., and Raveh, A. (2008). CoPlot: a tool for visualizing multivariate data in medicine. *Stat. Med.* 27, 2234–2247. doi: 10.1002/sim.3078

12. Bui, E. N. (2013). Soil salinity: a neglected factor in plant ecology and biogeography. *J. Arid Environ.* 92, 14–25. doi: 10.1016/j.jaridenv.2012.12.014

13. Bustan, A., Avni, A., Yermiyahu, U., Ben-Gal, A., Riov, J., Erel, R., et al. (2013). Interactions between fruit load and macroelement concentrations in fertigated olive (Olea europaea L.). trees under arid saline conditions. *Sci. Hortic.* 152, 44–55. doi: 10.1016/j.scienta.2013.01.013

14. Caputo, A. C., Pelagagge, P. M., and Salini, P. (2013). AHP-based methodology for selecting safety devices of industrial machinery. *Saf. Sci.* 53, 202–218. doi: 10.1016/j.ssci.2012.10.006

15. Cay, T., and Uyan, M. (2013). Evaluation of reallocation criteria in land consolidation studies using the Analytic Hierarchy Process (AHP). *Land Use Policy* 30, 541–548. doi: 10.1016/j.landusepol.2012.04.023

16. Chakrabortty, S., Roy, M., and Pal, P. (2013). Removal of fluoride from contaminated groundwater by cross flow nanofiltration: transport modelling and economic evaluation. *Desalination* 313, 115–124. doi: 10.1016/j.desal.2012.12.021

17. Connor, J. D., Schwabe, K., King, D., and Knapp, K. (2012). Irrigated agriculture and climate change: the influence of water supply variability and salinity on adaptation. *Ecol. Econ.* 77, 149–157. doi: 10.1016/j.ecolecon.2012.02.021

18. Edelstein, M., Plaut, Z., Dudai, N., and Ben-Hur, M. (2009). Vetiver (*Vetiveria zizanioides*) responses to fertilization and salinity under irrigation conditions. *J. Environ. Manage.* 91, 215–221. doi: 10.1016/j.jenvman.2009.08.006

19. Garcia, N., Moreno, J., Cartmell, E., Rodriguez-Roda, I., and Judd, S. (2013). The cost and performance of an MF-RO/NF plant for trace metal removal. *Desalination* 309, 181–186. doi: 10.1016/j.desal.2012.10.017

20. Ghermandi, A., and Mesalem, R. (2009). The advantages of NF desalination of brackish water for sustainable irrigation: the case of the Arava Valley in Israel. *Desalination Water Treat.* 10, 101–107. doi: 10.5004/dwt.2009.824

21. Gomez-Silvan, C., Arevalo, J., Perez, J., Gonzalez-Lopez, J., and Rodelas, B. (2013). Linking hydrolytic activities to variables influencing a submerged membrane bioreactor (MBR) treating urban wastewater under real operating conditions. *Water Res.* 47, 66–78. doi: 10.1016/j.watres.2012.09.032

22. Gur-Reznik, S., Koren-Menashe, I., Heller-Grossman, L., Rufel, O., and Dosoretz, C. G. (2011). Influence of seasonal and operating conditions on the rejection of pharmaceutical active compounds by RO and NF membranes. *Desalination* 277, 250–256. doi: 10.1016/j.desal.2011.04.029

23. Gursoy, B. B., Mason, O., and Sergeev, S. (2013). The analytic hierarchy process, max algebra and multi-objective optimisation.*Linear Algebra Appl.* 438, 2911–2928. doi: 10.1016/j.laa.2012.11.020

24. Gutman, J., Fox, S., and Gilron, J. (2012). Interactions between biofilms and NF/RO flux and their implications for control-a review of recent developments. *J. Membr. Sci.* 421–422, 1–7. doi: 10.1016/j.memsci.2012.06.032

25. Guttman, L. (1968). A general non-metric technique for finding the smallest space for a configuration of points. *Psychometrica*33, 479–506.

26. Hilal, N., Al-Zoubi, H., Darwish, N. A., and Mohammad, A. W. (2005). Characterisation of nanofiltration membranes using atomic force microscopy. *Desalination* 177, 187–199. doi: 10.1016/j.desal.2004.12.008

27. Hillel, D., Belhassen, Y., and Shani, A. (2013). What makes a gastronomic destination attractive? Evidence from the Israeli Negev. *Tourism Manage.* 36, 200–209. doi: 10.1016/j.tourman.2012.12.006

28. Johannessen, A., Rosemarin, A., Thomalla, F., Swartling, A. G., Stenström, T. A., Vulturius, G., et al. (2014). Strategies for building resilience to hazards in water, sanitation and hygiene (WASH) systems: the role of public private partnerships. *Int. J. Disaster Risk Reduct.* 10, 102–115. doi: 10.1016/j.ijdrr.2014.07.002

29. Karabelas, A. J., Koutsou, C. P., Gragopoulos, J., Isaias, N. P., and Al Rammah, A. S. (2012). A novel system for continuous monitoring of salt rejection characteristics of individual membrane elements in desalination plants. *Sep. Pur. Technol.* 88, 29–38. doi: 10.1016/j.seppur.2011.12.002

30. Kayastha, P., Dhital, M. R., and De Smedt, F. (2013). Application of the analytical hierarchy process (AHP). for landslide susceptibility mapping: a case study from the Tinau watershed, west Nepal. *Comput. Geosci.* 52, 398–408. doi: 10.1016/j.cageo.2012.11.003

31. Lespinats, S., Fertil, B., Villemain, P., and Herault, J. (2009). RankVisu: mapping from the neighborhood network.*Neurocomputing* 72, 2964–2978. doi: 10.1016/j.neucom.2009.04.008

32. Levy, D., and Tai, G. C. C. (2013). Differential response of potatoes (Solanum tuberosum L.) to salinity in an arid environment and field performance of the seed tubers grown with fresh water in the following season. *Agric. Water Manage.* 116, 122–127. doi: 10.1016/j.agwat.2012.06.022

33. Lew, B., Cochva, M., and Lahav, O. (2009). Potential effects of desalinated water quality on the operation stability of wastewater treatment plants. *Sci. Total Environ.* 407, 2404–2410. doi: 10.1016/j.scitotenv.2008.12.023

34. Lipshitz, G., and Raveh, A. (1994). Applications of the CoPlot method in the study of socioeconomic differences among cities: a basis for a differential development policy. *Urban Stud.* 31, 123–135.

35. Liu, Y., Li, X., Xing, Z., Zhao, X., and Pan, Y. (2013). Responses of soil microbial biomass and community composition to biological soil crusts in the revegetated areas of the Tengger Desert, *Appl. Soil Ecol.* 65, 52–59. doi: 10.1016/j.apsoil.2013.01.005

36. Miller, G. W. (2006). Integrated concepts in water reuse: managing global water needs. *Desalination* 187, 65–75. doi: 10.1016/j.desal.2005.04.068

37. Mrayed, S. M., Sanciolo, P., Zou, I., and Leslie, G. (2011). An alternative membrane treatment process to produce low-salt and high-nutrient recycled water suitable for irrigation purposes. *Desalination* 274, 144–149. doi: 10.1016/j.desal.2011.02.003

38. Nederlof, M. M., van Paassen, J. A. M., and Jong, R. (2005). Nanofiltration concentrate disposal: experiences in the Netherlands. *Desalination* 178, 303–312. doi: 10.1016/j.desal.2004.11.041

39. Oren, S., Birnhack, L., Lehmann, O., and Lahav, O. (2012). A different approach for brackish-water desalination, comprising acidification of the feed-water and CO_2(aq). reuse for alkalinity, Ca^{2+} and Mg^{2+} supply in the post treatment stage. *Sep. Purif. Technol.* 89, 252–260. doi: 10.1016/j.seppur.2012.01.027

40. Oron, G., Gillerman, L., Bick, A., Manor, Y., Buriakovsky, N., and Hagin, J. (2008). Membrane technology for sustainable treated wastewater reuse: agricultural, environmental and hydrological considerations. *Water Sci. Technol.* 57, 1383–1388. doi: 10.2166/wst.2008.243

41. Penate, B., and García-Rodríguez, L. (2011). Reverse osmosis hybrid membrane inter-stage design: a comparative performance assessment. *Desalination* 281, 354–363. doi: 10.1016/j.desal.2011.08.010

42. Qadir, M., Sharma, B. R., Bruggeman, A., Choukr-Allah, R., and Karajeh, F. (2007). Non-conventional water resources and opportunities for water augmentation to achieve food security in water scarce countries. *Agric. Water Manage.* 87, 2–22. doi: 10.1016/j.agwat.2006.03.018

43. Rasouli, F., Pouya, A. K., and Karimian, N. (2013). Wheat yield and physico-chemical properties of a sodic soil from semi-arid area of Iran as affected by applied gypsum. *Geoderma* 193–194, 246–255. doi: 10.1016/j.geoderma.2012.10.001

44. Raveh, A. (2000). CoPlot: a Graphic display method for geometrical representations of MCDM. *Eur. J. Oper. Res.* 125, 670–678. doi: 10.1016/S0377-2217(99)00276-3

45. Riera, F. A., Suarez, A., and Muro, C. (2013). Nanofiltration of UHT flash cooler condensates from a dairy factory: Characterisation and water reuse potential. *Desalination* 309, 52–63. doi: 10.1016/j.desal.2012.09.016

46. Saaty, T. L. (1980). *The Analytic Hierarchy Process*. New York, NY: McGraw-Hill.

47. Saaty, T. L. (2003). Decision-making with the AHP: why is the principal eigenvector necessary. *Eur. J. Oper. Res.* 145, 85–91. doi: 10.1016/S0377-2217(02)00227-8

48. Schilli, C., Lischeid, G., and Rinklebe, J. (2010). Which processes prevail?: analyzing long-term soil solution monitoring data using nonlinear statistics. *Geoderma* 158, 412–420. doi: 10.1016/j.geoderma.2010.06.014

49. Sotto, A., Arsuaga, J. M., and Van der Bruggen, B. (2013). Sorption of phenolic compounds on NF/RO membrane surfaces: Influence on membrane performance. *Desalination* 309, 64–73. doi: 10.1016/j.desal.2012.09.023

50. Tzfati, E., Sein, M., Rubinov, A., Raveh, A., and Bick, A. (2011). Pre-treatment of wastewater: optimal coagulant selection using partial order scaling analysis (POSA). *J. Hazard Mater.* 190, 51–59. doi: 10.1016/j.jhazmat.2011.02.023

51. Yermiyahu, U., Tal, A., Ben-Gal, A., Bar-Tal, J., and Tarchitzky, O. (2007). Lahav, rethinking desalinated water quality and agriculture. *Science* 318, 920–921. doi: 10.1126/science.1146339

52. Verhuelsdonk, M., Attenborough, T., Lex, O., and Altmann, T. (2010). Design and optimization of seawater reverse osmosis desalination plants using special simulation software. *Desalination* 250, 729–733. doi: 10.1016/j.desal.2008.11.031

53. Villarreal-Guerrero, F., Kacira, M., Fitz-Rodríguez, E., Kubota, C., Giacomelli, G. A., Linker, R., et al. (2012).

Comparison of three evapotranspiration models for a greenhouse cooling strategy with natural ventilation and variable high pressure fogging.*Sci. Hortic*. 134, 210–221. doi: 10.1016/j.scienta.2011.10.016

54. Vince, F., Marechal, F., Aoustin, E., and Breant, P. (2008). Multi-objective optimization of RO desalination plants. *Desalination*222, 96–118. doi: 10.1016/j.desal.2007.02.064

55. Wei, J., Qiu, C., Wang, Y. N., Wang, R., and Tang, C. Y. (2013). Comparison of NF-like and RO-like thin film composite osmotically-driven membranes-Implications for membrane selection and process optimization, J. *Membrane Sci*. 427, 460–471. doi: 10.1016/j.memsci.2012.08.053

56. Yang, X., Yan, L., and Zeng, L. (2013). How to handle uncertainties in AHP: The cloud delphi hierarchical analysis. *Inform. Sci*. 222, 384–404. doi: 10.1016/j.ins.2012.08.019

57. Zhang, R., Zhang, X., Yang, J., and Yuan, H. (2013). Wetland ecosystem stability evaluation by using analytical hierarchy process (AHP) approach in Yinchuan Plain, China. *Math. Comput. Model*. 57, 366–374. doi: 10.1016/j.mcm.2012.06.014

58. Zhao, S., Zou, L., and Mulcahy, D. (2012). Brackish water desalination by a hybrid forward osmosis–nanofiltration system using divalent draw solute. *Desalination* 284, 175–181. doi: 10.1016/j.desal.2011.08.053

59. Zheng, Y., Yu, S., Shuai, S., Zhou, Q., Cheng, Q., Liu, M., et al. (2013). Color removal and COD reduction of biologically treated textile effluent through submerged filtration using hollow fiber nanofiltration membrane. *Desalination* 314, 89–95. doi: 10.1016/j.desal.2013.01.004

60. Zhou, F., Wang, C., and Wei, J. (2013). Separation of acetic acid from monosaccharides by NF and RO membranes: performance comparison. *J. Membr. Sci*. 429, 243–251. doi: 10.1016/j.memsci.2012.11.043

CHAPTER 6 NANO-ENABLED MEMBRANES TECHNOLOGY

P.S. Goh, A.F. Ismail, N. Hilal

Keywords: Nano-enabled membrane, Desalination, Sustainability

INTRODUCTION: SUSTAINABILITY OF MEMBRANE DESALINATION

Continuing population growth and urbanization as well as rapid industrialization are constantly putting pressure on global water demand and resulting in an alarming scenario where accessible fresh water supply is no longer matching the patterns of these developments. Desalination has long served as a feasible option to provide safe drinking water in many deserted areas, coastal regions or remote locations [1]. Despite its versatility in tackling water scarcity, desalination has been unfavorably addressed as the most energy-intensive water treatment technology with undesirable consequences in terms of high economic cost and energy requirements [2]. While the applications of thermal distillation desalination technologies have been majorly restricted by their high power consumption and consequently rather prohibitive, membrane-based desalination has been widely promoted as the key and mainstream desalination technology. With their outstanding attributes for desalination, membrane desalination technology holds undoubtful potential to bridge and connect the dots to battle key issues regarding the current and future water gaps. For decades, the desalination plant designs as well as the guidance and evaluation on the performances of the conventionally used membrane desalination, particularly for seawater reverse osmosis (SWRO), have been well established to conform a complete range of industry standards. In the past decade, forward osmosis desalination has also been extensively studied to offer an alternative for leveraging low cost thermal energy 3 and 4. Nevertheless, despite the overwhelming features of these desalination technologies to render the impressive economical operational expenditures of the desalination plant, the impacts of environments and resource depletion associated to the pretreatment and recovery of the membrane-based desalination still deserve growing attentions 5 and 6. Although the impact of these issues can be alleviated with good equipment design, operation, practice and maintenance, there are always urgent needs to improve and strengthen the sustainability of this technology [7].

Nanotechnology, which literally refers to the manipulations of materials and processes that are engineered to the molecular scale of 1–100 nm, is an emerging field that poses immense ability to hierarchically construct materials from the bottom up approach to significantly improve the performance as well as add new functionality to the existing products. The evolutions and advancements explored in this field have urged seawater desalination progress as a more economical and sustainable option [8]. In this context, the paradigm shift made by the new approaches and technological innovation based on the new frontier in enabling nanoscience and nanotechnology, although not one size fits all, has advocated the progress of maintaining the sustainability in terms of reduced membrane materials, chemical and energy usage that eventually leads to minimized sources of environmental issues [9]. As membrane transports are inherently nanoscale where the transport of water and salts across the membrane primarily depends on the chemical properties as well as physical structures of the membranes at nano- to microscales, the application of nanotechnology knowledge in the fabrication of so-called nano-enabled membranes is anticipated to counter the limitations of current materials and

Citation: P.S. Goh, A.F. Ismail, N. Hilal, Nano-enabled membranes technology, Desalination, Volume 380, 15 February 2016, Pages 100-104, ISSN 0011-9164, http://dx.doi.org/10.1016/j.desal.2015.06.002.

processes as well as optimize the performance of processes used in desalination. Most importantly, the breakthroughs in this area have also delivered the astonishing tangible outcome: high performance nano-enabled membranes that combine both high productivity and high rejection to meet the hiking research and industry expectations [10]. The unprecedented quantum leaps evidence the versatility of nanomaterials and their nanocomposite to provide the alternative route for sustainable development. This serves the main reason for the industries and stakeholders to be optimistic with the capability of these new generation technology to make a large difference for modern, affordable and environmentally sound remedy for water shortage crisis. Undeniably, the transformation to the era of nanotechnology has the potential to bring the capacity of membrane science and engineering a big step forward for the desalination technology to flourish [11]. However, to precisely gauge the contribution of the nano-enabled membrane towards the advancement and sustainability, one should comprehend how the rapidly proceeding nanotechnology research and commercialization play their roles in providing solutions to the current desalination technology. Along the same vein, glancing back the past efforts towards maintaining the sustainability of the membrane-based desalination technology, few arisen questions must be carefully addressed:

i) Where are we now to contribute remarkable changes to the current sustainable development in membrane desalination?
ii) What can be done to extend the limit of current innovative nano-enabled membrane technology to further harness the economic and performance benefits while co-existing with the desired sustainability?
iii) Is nanotechnology the final piece of jigsaw which serves as the game changer that manipulates the niche and future mainstream of membrane desalination technology?

With the emphasis placed on the applications of nanomaterials in the current trends of desalination research, this review looks at the innovations and strategies of nanotechnologies that have been showcased in the state-of-the-art membrane-based desalination technologies. By referring the correlated questions raised above, the sustainable approaches and significant implications of nano-enabled membrane technology to shape the future of desalination research and industrial competitiveness are highlighted. There are positive signs across the desalination communities that we are maneuvering into the right direction with a growing momentum. In this sense, whether it is a long-term technology or a short path from present technologies, we are certain that when we are ready to leave behind some conservative mindsets in industry and open our arms to new ideas and embrace them, it will be the genuine revolutionary shifts in this field.

INNOVATIVE NANOMATERIALS: REVOLUTIONARY SOLUTIONS FOR SUSTAINABLE MEMBRANE DESALINATION?

As nanotechnology continues to precede steadily, the primary focus of the research agenda has progressively advanced from the fundamental discovery and characterizations to look into the opportunities of presenting nanotechnology and the resultant nanomaterials as the major key to address the challenges of improving global sustainability, particularly in a wide spectrum of pivotal areas involving energy and environment 12, 13 and 14. The application of the conventionally used RO membranes have been severely restricted by the unsatisfactory water recovery and high pressure operation while generating substantial amounts of liquid wastes. As the strongest drivers of the desalination market, industry and municipal suppliers are eagerly looking for the new breakthroughs from the research community to fill the gaps. Therefore, the development of next generation of low-pressure membranes with low energy consumption and heightened salt rejection performance is needed to provide a viable route for sustainable membrane desalination.

To date, a number of advances have been made in the areas of material science for the development of desalination membranes based on nanoparticles and nanocomposites. The plethora of different types of nanomaterials has conferred almost unlimited technological benefits in terms of the adaptation and optimization of new generation membrane for the existing desalination plants. Exciting findings and breakthroughs have been constantly made through the incorporation of conventional nanomaterials such as metal oxides, silica and zeolites to introduce new degrees of freedom in the desalination membrane design 15, 16, 17, 18, 19, 20, 21, 22 and 23. Carbon allotropes such as carbon nanotubes (CNTs) 24, 25, 26 and 27, graphene-family nanomaterials (GFNs) 28,29, 30 and 31 and nanodiamonds [32] are emerging nanomaterials that pose huge potential to cope with the sustainability issues of current membrane desalination technology. The frictionless CNT channels with precisely controlled pore entries could simultaneously facilitate the fast transport of water molecules and offer exceptional salt ion rejection capability. Kim et al. demonstrated that thin film RO nanocomposite membranes containing well-dispersed functionalized CNTs possess tremendously high water flux and salt rejection that is comparable to that of without CNTs, which can be attributed to the hydrophobic nanochannels of CNTs [33]. Similarly, the highly accessible nanopores of the graphene monolayers could also promote ultra-fast water flow that surpasses that of conventional polymeric RO membranes. Hu and Mi reported graphene oxide membrane flux that up to 276 LMH/MPa, which was 10 times higher than that of most commercial nanofiltration membranes [34]. Additionally, the abundant surface functional groups have endowed the graphene oxide sheets with hydrophilic nature that reinforced membrane durability associated with their antifouling properties [35]. Recently, biomimetic and bioinspired membranes have been addressed as forefront materials with efficient avenues for membrane desalination 36, 37 and 38. A considerable number of recent studies have evidenced the fascinating properties of the nanocomposite membranes consist of aquaporins (AQPs) 39, 40 and 41. Xie et al. [40]demonstrated that the incorporation of AQP into polymer vesicles through surface imprinting technique could overcome the fragility of biomimetic bilayers while retaining the excellent NaCl and $MgCl_2$ salt rejection and rapid transport capability of AQP. With the capacity and strength to make things with a wide characteristic diversity and less resource consumption, the manipulation of nanomaterials in membrane desalination technology has opened a great potential to design and create a completely new range of novel and innovative membrane materials that have been carefully customized to meet the desired requirements.

A convergence between nanotechnology and membrane technology has led to revolutionary advances in water desalination. With the significantly improved water productivity and separation performance, the desalination plant operators can take the full advantages featured by these nano-enabled membranes to significantly reduce the capital and operating expenses. To name a few conditions: i) fewer pressure vessel and membrane modules are required to retain the same desalination capacity; ii) lower feed pressure will be sufficient to generate same permeate flux; iii) the same number of membrane modules can effectively increase the water recovery and iv) greater fouling resistance to promote extended membrane life span 42 and 43 — all thanks to the impactful contributions of the nano-enabling synthesis and fabrication strategies. In this context, energy consumption and the capital involved in the pretreatment and membrane cleaning can be drastically minimized. By taking membrane desalination technology into the consideration, reaching sustainability implies tackling the exciting transformation with minimum undesired consequences towards the environment, which means less energy wastage, less materials and chemicals usage. Integrating the newly designed membrane elements into the retrofitted desalination system might offer the best solutions of delivering maximum sustainability benefits and optimal saving of desalination cost. The cradle-to-grave environmental and commercial cost is expected to be much cheaper to surpass any in the historical experience. The economical potentials have been concretely evidenced by the growing rhythm of public and private spending on nanoscience and nanotechnology research and development.

NANO-ENABLED MEMBRANE DESALINATION: TRANSFORMING CHALLENGES INTO OPPORTUNITIES

After all these years, the research community and industries might no longer seeing the bottlenecks in desalination technology as a threat, but a welcoming challenge where they can find the turning point for green employment. A subtle shift is essential to find solutions and build a shared vision for the raising sustainability awareness and challenges. Nevertheless, several cross-cutting issues need to be addressed in order to promote growth in this field. This in turn indicates that there are issues ranging from fundamental sciences to engineering applications to be unraveled by the scientific community. As summarized in Fig. 1, nano-enabling membrane technology is a double-edged sword. It reconciles the underlying issues of membrane desalination technology and carries the benefits by sustainably promoting improved desalination system and providing greater return across the sectors; but at the same time may also call attentions to multifaceted social and environmental pitfalls with unintended consequences. Although these two situations are polar opposites, responding positively or at least neutrally will trigger a completely different set of outcomes. The barriers must be overcome to make it possible to open up the value chains and realize their full economic potentials.

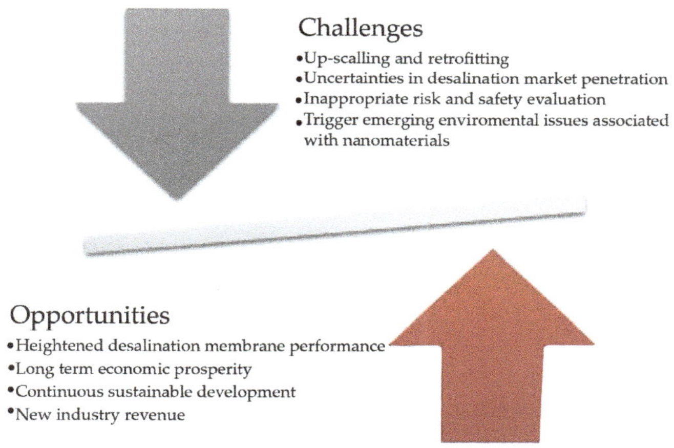

Figure 1: Challenges and opportunities in nano-enabled membrane desalination. Transformation of challenges to opportunities is the crucial strategy to ensure the commercial success and industry engagement of the technology.

As doubts and negativities cannot possibly create wealth and good fortune, it is clear that some membrane manufacturers and desalination plant operators will be keeping their faith as the 'first mover' to respond better to the opportunities that arise with their availability to the sources of innovations and their promptitude in capitalizing the emerging opportunities. Wise investment will lead to multiple future benefits but the bottom line question for prospective investors is whether the innovation and invention based on nanomaterials can profitably access and penetrate the desalination market over a long period. The answer to this question depends on how the stakeholders are involved in these activities to estimate the potential advances measured against today's technologies based on the substantial input for formulating and reviewing the values of the changes. Although it is not easy to get to this point, but it has reached at the junction where the technology makes it possible to switch from the traditional route. It is anticipated that the commercial success of the nano-enabled membrane technologies would catalyze the long term economic prosperity by enabling the continuation of sustainable development while also creating industries with valuable new opportunities. If such development is bridged with a proper platform for innovative technologies and management practices, it is quite certain that the technology and investments can be scaled up to generate paramount economic and environmental impacts for years to come.

The excitement and uncertainty harnessed from the nano-innovations have stimulated an increasing scientific and public debate. To ensure that the desalination industry sectors utilizing nanotechnologies are going to flourish and expand, a favorable environment must be adopted to allow the new commercial entities to be able to take the necessary risks to bring these new innovations into successful marketable products. At this moment, another rising key challenge is to ensure the availability of resources or test methods to get rid of laborious and problematic safety assessment of the nanomaterials that are emerging at an ever increasing pace [44]. Inappropriate risk and safety evaluation and counter measures can have serious knock-on effects to the future development where they may lead to disoriented stakeholder's confidence as well as a missed opportunity for the industries to simultaneously tackle issues and generate wealth [45]. While developing the infrastructure and facilities to promote the research innovation along this line within the research community or beyond to other key stakeholders such as industry and regulatory bodies, it is essential to establish regulatory framework to allow good practice and communication in order to cater the value chain for environmental health and safety [46]. It is of utmost priority to assess and address any potential adverse effects for the upscale and deployment of nano-enabled desalination technologies.

FUTURE OUTLOOK AND CONCLUDING REMARKS

Over the last decade, there has been a remarkable progress in the applications of nanotechnology in the areas of desalination to provide solutions in waste treatment and environmental remediation. Application of engineered nanomaterials in the membrane fabrication has enabled the exploration of a new generation of nano-enabled desalination membranes. Putting the pieces together, from where we stand today, the positive transformation of nano-enabled membranes holds good potential to realize the sustainable development of current membrane desalination industries like what has already been imagined for years. Nanomaterials are arguably a win–win option to mutually enhance the performance of the desalination membrane and sustain a clean environment. However, large-scale implementation of nano-enabled desalination technology by industry might take decades. The implementation of sustainable desalination strategy should unify and assimilate the factors of economic, social and environment when the scientists and engineers pursue and expand the key research areas that involve broad and interdisciplinary research goals.

By looking closer to the details underlying the current and forthcoming issues concerning the sustainability of desalination technology, the major constraint at the current stage is probably the response and reaction of stakeholders towards the risks and opportunities presented. As illustrated in Fig. 2, the critical drivers of the holistic development cannot be divorced from: i) formulation of an integrated research and development to allow effective transfer of research results into sustainable innovations; ii) adoption of positive industrial attitude towards the acceptance of environmentally benign products and processes; iii) dissemination and exploitation of the safety of the nano-enabled membrane desalination technology along the entire life-cycles and iv) establishment of long term roadmap to provide a cohesive structure for addressing hurdles proactively as the industry moves towards sustainable operation. Although considerable effort is still needed to fill the gaps and reduce disparities between the pipe-dream and the reality, with the accelerating knowledge and technological transfer from academic to industries, it is envisaged that in the next five to ten years, the fundamental science and applied engineering knowledge in nanotechnology R&D and infrastructure development will help to obtain ultimate solutions to develop and commercialize the next generation of sustainable membrane products as well as desalination technologies.

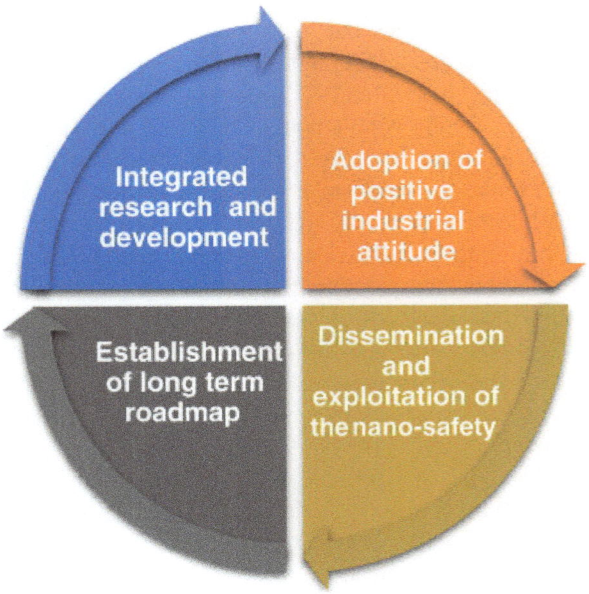

Figure 2: The critical drivers of the holistic and sustainable development of nano-enabled membrane desalination technology.

REFERENCES

1. March H, Saur D, Rico-Amors AM: The end of scarcity? Water desalination as the new cornucopia for Mediterranean Spain. J Hydrol 2014, 519:2642–51.
2. Schallenberg-Rodríguez J, Veza JM, Blanco-Marigorta A: Energy efficiency and desalination in the Canary Islands. Renew Sustain Energy Rev 2014, 40:741–8.
3. Shaffer DL, et al: Forward osmosis: where are we now? Desalination 2014 http://dx. doi.org/10.1016/j.desal.2014.10.031.
4. McGovern RK, Lienhard V JH: On the potential of forward osmosis to energetically outperform reverse osmosis desalination. J Membr Sci 2014, 469:245–50.
5. Baten R, Stummeyer K: How sustainable can desalination be? Desalin Water Treat 2013, 51:44–52.
6. Miller S, et al: Energy and environmental issues in desalination. Desalination 2014 http://dx.doi.org/10.1016/j.desal.2014.11.034.
7. Haddad BM: A case for an ecological–economic research program for desalination. Desalination 2013, 324:72–8.
8. Ghobeity A, Mitsos A: Optimal design and operation of desalination systems: new challenges and recent advances. Curr Opin Chem Eng 2014, 6:61–8.
9. Goh PS, Ismail AF: Review: is interplay between nanomaterial and membrane technology the way forward for desalination? J Chem Technol Biotechnol 2015 http://dx.doi.org/10.1002/jctb.4531.

10. Wang P, Ma J, Shi F, Ma Y, Wang Z, Zhao X: Behaviors and effects of differing dimensional nanomaterials in water filtration membranes through the classical phase inversion process: a review. Ind Eng Chem Res 2013, 52:10355–63.
11. Bhadra M, Mitra S: Advances in nanostructured membranes for water desalination. Nanotechnology Applications for Clean Water Second Edition. Elsevier Inc.; 2014 109–22.
12. Hussein AK: Applications of nanotechnology in renewable energies—a comprehensive overview and understanding. Renew Sustain Energy Rev 2015, 42: 460–76.
13. Qu X, Brame J, Li Q, Alvarez PJJ: Nanotechnology for a safe and sustainable water supply: enabling integrated water treatment and reuse. Acc Chem Res 2013, 46: 834–43.
14. Trujillo-Reyes J, Peralta-Videa JR, Gardea-Torresdey JL: Supported and unsupported nanomaterials for water and soil remediation: are they a useful solution for worldwide pollution? J Hazard Mater 2014, 280:487–503.
15. Emadzadeh D, Lau WJ, Matsuura T, Rahbari-Sisakht M, Ismail AF: A novel thin film composite forward osmosis membrane prepared from PSf–TiO2 nanocomposite substrate for water desalination. Chem Eng J 2014, 237:70–80.
16. Dong H, Zhao L, Zhang L, Chen H, Gao C, Winston Ho WS: High-flux reverse osmosis membranes incorporated with NaY zeolite nanoparticles for brackish water desalination. J Membr Sci 2015 http://dx.doi.org/10.1016/j.memsci.2014.11.054.
17. Duan J, Pan Y, Pacheco F, Litwiller E, Lai Z, Pinnau I: High-performance polyamide thin-film-nanocomposite reverse osmosis membranes containing hydrophobic zeolitic imidazolate framework-8. J Membr Sci 2015, 476:303–10.
18. Li Q, Wang Y, Song J, Guan Y, Yu H, Pan X, Wu F, Zhang M: Influence of silica nanospheres on the separation performance of thin film composite. Appl Surf Sci 2015, 324:757–64.
19. Mollahosseini A, Rahimpour A: Interfacially polymerized thin film nanofiltration membranes on TiO2 coated polysulfone substrate. J Ind Eng Chem 2014, 20:1261–8.
20. An W, Zhou X, Liu X, Chai PW, Kuznicki T, Kuznicki SM: Natural zeolite clinoptilolitephosphate composite membranes for water desalination by pervaporation. J Membr Sci 2014, 470:431–8.
21. 21 Peyki A, Rahimpour A, Jahanshahi M: Preparation and characterization of thin film composite reverse osmosis membranes incorporated with hydrophilic SiO2 nanoparticles. Desalination 2014 http://dx.doi.org/10.1016/j.desal.2014.05.025.
22. Niksefat N, Jahanshahi M, Rahimpour A: The effect of SiO2 nanoparticles on morphology and performance of thin film composite membranes for forward osmosis application. Desalination 2014, 343:140–6.
23. Shaban M, et al: Titanium dioxide nanotube embedded mixed matrix PES membrane characterization and membrane performance. Chem Eng Res Des 2014 http://dx.doi.org/10.1016/j.cherd.2014.11.008.
24. Das R, Ali ME, Abd Hamid SB, Ramakrishna S, Chowdhury ZZ: Carbon nanotube membranes for water purification: a bright future in water desalination. Desalination 2014, 336:97–109.
25. Goh PS, Ismail AF, Ng BC: Carbon nanotubes for desalination: performance evaluation and current hurdles. Desalination 2013, 308:2–14.
26. Zhao H, Qiu S, Wu L, Zhang L, Chen H, Gao C: Improving the performance of polyamide reverse osmosis membrane by incorporation of modified multi-walled carbon nanotubes. J Membr Sci 2014, 450:249–56.
27. Das R, Abd Hamid SB, Ali ME, Ismail AF, Annuar MSM, Ramakrishna S: Multifunctional carbon nanotubes in water treatment: the present, past and future. Desalination 2014, 354:160–79.
28. Goh PS, Ismail AF: Graphene-based nanomaterial: the state-of-the-art material for cutting edge desalination technology. Desalination 2015, 356:115–28.
29. Mahmoud KA, Mansoor B, Mansour A, Khraisheh M: Functional graphene nanosheets: the next generation membranes for water desalination. Desalination 2015, 356:208–25.
30. Sun C, Boutilier MSH, Au H, Poesio P, Bai B, Karnik R, Hadjiconstantinou NG: Mechanisms of molecular permeation through nanoporous graphene membranes. Langmuir 2014, 30:675–82.
31. Akin I, Zor E, Bingol H, Ersoz M: Green synthesis of reduced graphene oxide/ polyaniline composite and its application for salt rejection by polysulfonebased composite membranes. J Phys Chem B 2014, 118:5707–16.
32. Bhadra M, Roy S, Mitra S: Nanodiamond immobilized membranes for enhanced desalination via membrane distillation. Desalination 2014, 341:115–9.
33. Kim HJ, Choi K, Baek Y, Kim D-G, Shim J, Yoon J, Lee J-C: High-performance reverse osmosis CNT/polyamide nanocomposite membrane by controlled interfacial interactions. ACS Appl Mater Interfaces 2014, 6:2819–29.
34. Hu M, Mi B: Enabling graphene oxide nanosheets as water separation membranes. Environ Sci Technol 2013, 47:3715–23.
35. Choi W, Choi J, Bang J, Lee J: Layer-by-layer assembly of graphene oxide nanosheets on polyamide membranes for durable reverse-osmosis applications. ACS Appl Mater Interfaces 2013, 5:12510–9.
36. Zhao J, Zhao X, Jiang Z, Li Z, Fan X, Zhu J, Wu H, Su Y, Yang D, Pan F, Shi J: Biomimetic and bioinspired membranes: preparation and application. Prog Polym Sci 2014, 39: 1668–720.
37. Yang CY, Zhao Y, Wang R, Helix-Nielson C, Fane AG: Desalination by biomimetic aquaporin membranes: review of status and prospects. Desalination 2013, 308:34–40.

38. Shen Y-X, Saboe PO, Sines IT, Erbakan M, Kumar M: Biomimetic membranes: a review. J Membr Sci 2014, 454:359–81.

39. Sun G, Chung T-S, Jeyaseelan K, Armugam A: A layer-by-layer self-assembly approach to developing an aquaporin-embedded mixed matrix membrane. RSC Adv 2013, 3:473–81.

40. Xie W, He F, Wang B, Chung T-S, Jeyaseelan K, Armugam A, Tong YW: An aquaporinbased vesicle-embedded polymeric membrane for low energy water filtration. J Mater Chem A 2013, 1:7592–600.

41. Li X, Wang R, Wicaksana F, Tang C, Torres J, Fane AG: Preparation of high performance nanofiltration (NF) membranes incorporated with aquaporin Z. J Membr Sci 2014, 450:181–8.

42. de Lannoy C-F, Jassby D, Gloe K, Gordon AD, Wiesner MR: Aquatic biofouling prevention by electrically charged nanocomposite polymer thin film membranes. Environ Sci Technol 2013, 47:2760–8.

43. Buonomenna MG: Nano-enhanced reverse osmosis membranes. Desalination 2013, 314:73–88.

44. Kühnel D, Nickel C: The OECD expert meeting on ecotoxicology and environmental fate-towards the development of improved OECD guidelines for the testing of nanomaterials. Sci Total Environ 2014, 472:347–53.

45. Qu X, Alvarez PJJ, Li Q: Applications of nanotechnology in water and wastewater treatment. Water Res 2013, 47:3931–46.

46. Duic N, Urbanie K, Huisingh D: Components and structures of the pillars of sustainability. J Clean Prod 2014 http://dx.doi.org/10.1016/j.jclepro.2014.11.030.

CHAPTER 7

COAGULATION-MICROFILTRATION WATER TREATMENT

Q. Ding, H. Yamamura, N. Murata, N. Aoki, H. Yonekawa, A. Hafuka, Y. Watanabe

Keywords: Coagulation-microfiltration, Irreversible membrane fouling, Meso-particles, Particle number, Zeta potential

INTRODUCTION

Low-pressure membrane filtration methods facilitate simpler and more precise solid-liquid separation than sand filtration, and their introduction is therefore being considered at many water treatment plants. When introducing membrane filtration to water treatment processes, it is common to combine it with a preliminary coagulation process to remove dissolved organic matter and control membrane fouling (Gao et al., 2011, Huang et al., 2009, Matilainen et al., 2010 and Singer, 1994). Also, because existing sand filter tanks can effectively be used as membrane submersion tanks, processing in which pre-coagulation and membrane filtration are combined is more often used in full-scale water treatment plants (as opposed to laboratories) than are ozone or activated carbon methods (Lebeau et al., 1998). However, although coagulation-membrane filtration processes are being considered as replacements for conventional water treatment processes, there are as yet no clear guidelines for optimal operation procedures when coagulation and membrane filtration are combined.

With sand filtration methods, it is important to optimize conditions such that turbidity and organic concentration are controlled. Optimal pH and coagulant dose are determined using jar tests. In most cases circumneutral conditions with a large dose of coagulants are used, to form large flocs (Amirtharajah and Mills, 1982). With membrane filtration methods, on the other hand, there is no need to sediment the flocs; instead they need only be larger than the membrane pore size. Thus these methods are more efficient than sand filtration because of the lower dose of coagulant required and smaller amount of coagulant sludge produced (Choi and Dempsey, 2004). In addition, floc formation and sedimentation basins are not required, so the size of processing facilities can be greatly reduced (Lebeau et al., 1998). A range of factors affect the process of membrane filtration and should be optimized, including the rate of removal of organic matter, turbidity, and the rate of membrane fouling. However, the conditions under which membrane fouling is best controlled differ from the optimal coagulation conditions obtained by jar tests (Choi and Dempsey, 2004, Kimura et al., 2008 and Lee et al., 2000). In fact, optimal conditions are currently identified through a process of trial and error, by conducting membrane filtration tests under various coagulation conditions over an extended period of time. In order for the membrane filtration process to become more widespread, a method to replace the jar test that can simply and accurately explore the optimal coagulation conditions for pre-coagulation processing is crucial.

There are two types of membrane fouling: reversible and irreversible (Kimura et al., 2004). In reversible membrane fouling, there is a correlation between the fractal dimension of coagulation particles of 0.5 μm or greater (hereafter, micro-particles) and the rate of fouling. The mechanism that has been proposed to explain this is that floc strength increases with fractal dimension, and that any trans-membrane pressure (TMP) increase is thus inhibited by the formation of a coarse cake layer (Cho et al., 2005). It is however difficult to explain irreversible membrane fouling based on micro-particle behavior. This process is instead thought to be the result of microscopic particles with a particle size of 0.5 μm or less, which are close to or smaller than the membrane pore size

Citation: Q. Ding, H. Yamamura, N. Murata, N. Aoki, H. Yonekawa, A. Hafuka, Y. Watanabe, Coagulation-microfiltration water treatment, Water Research, Volume 101, 15 September 2016, Pages 127-136, ISSN 0043-1354, http://dx.doi.org/10.1016/j.watres.2016.05.076.

(hereafter, meso-particles). Wiesner et al. (1989) calculated the balance between back-transportation speed and advection for each particle size, and proposed that particles around 0.1 μm in size are particularly likely to adhere to the membrane and are thus the primary culprits in membrane fouling. Furthermore, because physicochemical interactions between particles and the membrane surface are heavily involved in irreversible membrane fouling, we believe that in addition to rheological properties, physicochemical behaviors such as the effects of electrostatic force are instrumental in understanding the fouling mechanism of meso-particles.

Dynamic light scattering (DLS), light scattering, atomic force microscopes, and scanning electron microscopes have all been used in the investigation of the physicochemical behavior of microscopic particles (Buffle et al., 1998, Lead et al., 2006, Li et al., 1997,Perret et al., 1991 and Tombacz et al., 1999). Because residual post-coagulation meso-particles are highly dispersed and therefore scatter light weakly, it is however extremely difficult to observe their particle characteristics underwater. Even with DLS, which is used in nanoparticle characterization studies, the lower concentration limit for measurement is around 10 mg/L, and residual post-coagulation meso-particles, at a concentration of around 10,000 particles/mL, cannot be observed (Table S1). We have successfully detected, for the first time and with high sensitivity, underwater particles between 20 nm and 0.5 μm in size, by combining high-output short-wavelength laser light with an optical microscope, thus facilitating the measurement of zeta potential (ZP) and particle number (PN, lower bound 50,000 particles/mL) of meso-particles that remain after coagulation.

Thepurposeofthisstudywastodescribethephysicochemical behavior of meso-particles, which has to date been difficult to ascertain, and to identify the particular characteristics that affect irreversible membrane fouling. Meso-particles were operationally defined in this study as particles between 20 nm and 0.5 μm after fractionation of coagulated water, described in 2.2. Jar tests in detail. Using a polyaluminum chloride (PACl) with a basicity of 54.5% as coagulant, we varied the coagulation pH and PACl dose, measured the ZP and PN of the particles generated at coagulation, and compared the values for micro-particles and meso-particles. We also selected six sets of conditions from among the coagulation conditions tested, to conduct coagulation-MF tests using ceramic membranes and polyvinylidene fluoride (PVDF) membranes. This allowed us to identify the meso-particle characteristics that affect the rate of irreversible membrane fouling, and therefore propose a method for determining optimal coagulation conditions, based on meso-particle characteristics.

MATERIALS AND METHODS

Water samples and membranes

Water sampled from two representative bodies of surface water used as sources of drinking water in Japan (water A and water B) were used as raw water. Four hundred liters of water A and nine hundred liters of water B were collected, and large particles were removed using a 10 μm stainless steel cartridge filter. The characteristics of the two types of water are listed in Table 1. Water A was from a river which is about 59.5 km long and drains an area of 200 km^2. The level of organic matter in water A was very low compared to water B, as the total organic carbon (TOC) of water A was approximately one fourth of that of water B in Table 1. Water B was sampled from a water path running from a eutrophic lake. This lake's catchment area is approximately 220 km^2, which has recently been polluted by human sewage and agricultural fertilization. In addition, Table 1showed that specific ultraviolet absorbance (SUVA) of water B was smaller than that of water A, which implies that water B contained much more hydrophilic organic matters including biopolymers (Kimura et al., 2014 and Yamamura et al., 2014). We used two kinds of membranes for the filtration experiments: PVDF hollow fiber membrane (Asahi Kasei Chemicals, Japan) and ceramic monolith membrane (METAWATER, Japan). The nominal pore size of both membranes was 0.1 μm. For the PVDF membrane, 39 cm^2 of tiny-scale membrane modules were assembled with 10 fibers, each 10 cm in length. For the ceramic membrane, 53 of 55 holes in the 10 cm piece were filled with glue, leaving two holes with a combined surface area of 11 cm^2. The PVDF membrane used outside-in filtering and the ceramic membrane used inside-out filtering. Pure water permeability was determined for every module before it was used in the filtration experiments. ZP and PN of particles were examined through a series of jar tests under various pH and PACl doses.

Table 1: Characteristics of the two types of surface water used in the tests.

	Water A	Water B
pH	7.9	8.0
TOC (Total organic carbon) (mg C/L)	1.1	4.3
UV254 (Ultraviolet absorbance at 254 nm) (cm−1)	0.023	0.080
SUVA (UV254/TOC × 100) (L mg−1 m−1)	2.1	1.9
Turbidity (mg/L)	1.9	7.3
Alkalinity (mg CaCO3/L)	9.9	70

Jar tests

The jar tests were performed at room temperature (18 ± 3 °C) using six standard 1 L beakers and a six-paddle jar test apparatus. We used PACl (10% as Al_2O_3) as the coagulant, at five different doses (0.3, 0.6, 1.0, 2.0 and 3.0 mg Al/L) with water A, and six doses (0.3, 0.6, 1.0, 2.0, 3.0 and 5.0 mg Al/L) with water B. We adjusted pH by adding 0.1 N sodium hydroxide or hydrochloric acid prior to the addition of PACl. The pH levels after PACl addition were 5.0, 5.5, 6.0, 6.5, 7.0, and 8.0. Mixing time was set at 3 min at 160 rpm (G value = 207 s^{-1}) for all samples, after which the coagulated water was allowed to settle for 3 min. In this study, we operationally define particles larger than 0.5 μm as micro-particles. After settling, the supernatant from 2 cm below the surface, equal to 200 mL, was collected. This fraction contained both micro-particles and particles smaller than 0.5 μm. Because large particles such as micro-particles will scatter the laser light if they are contained in the sample, this study eliminated such large particles and enable the lighting of smaller ones. The 200 mL sample was then centrifuged (3000 rpm, 10 min) and filtered using 1.0 and 0.5 μm membranes. This removed the micro-particles, so that the treated solution contained only particles smaller than 0.5 μm. As there is a possibility that some particles smaller than 0.5 μm might also be removed via centrifugation process, here we operationally define the residual particles between 20 nm and 0.5 μm in size as meso-particles. The PN and ZP of the micro-particles in the supernatant and meso-particles in the filtrate after filtration through the 0.5 μm membrane were determined separately. Dissolved organic carbon (DOC) and ultraviolet absorbance at 254 nm (UV_{254}) were also determined.

Coagulation-membrane filtration experiment

A schematic diagram of the coagulation-membrane filtration apparatus is shown in Figure S1. The six experimental conditions for each of the bench-scale experiments are summarized in Table 2, based on pH and PACl dose. We also conducted membrane filtration of raw water. Water A was filtered through the PVDF membrane after coagulation, and water B through both the PVDF and ceramic membranes. The PACl and raw water (after pH adjustment) were delivered to the mixing tank by two peristaltic pumps (Master flex No.7523-90, Germany) at a constant flow rate, and were mixed at 160 rpm (GT = 37,260). The coagulation overflow ran directly into the filtration tank combined with submerged PVDF membrane module, simulating in-line coagulation (Li et al., 2011). The coagulated water from the PVDF membrane filtration tank was pumped to the ceramic membrane module by a peristaltic pump at a constant flow rate. Details of the mixing tank are summarized in Table 3.

Table 2: Pre-coagulation conditions in the bench-scale coagulation-membrane filtration experiments.

Run no.	Water A		Water B	
	pH	PACl dose (mg Al/L)	pH	PACl dose (mg Al/L)
1	5.5	1.0	5.0	1.0
2	6.5	1.0	6.0	1.0
3	7.5	1.0	7.0	1.0
4	5.5	3.0	5.0	5.0
5	6.5	3.0	6.0	5.0
6	7.5	3.0	7.0	5.0

Table 3: Coagulation conditions in the mixing tank.

Condition	Water A	Water B
Mixing volume	100 mL	50 mL
Mixing speed	160 rpm	160 rpm
GT value	37,260	37,260
Raw water flow rate	22.9 mL/min	16.2 mL/min

The PVDF membrane was operated with a flux of 1.5 $m^3/m^2/d$, which was monitored using both a digital flow meter and manual analysis with a measuring cylinder. While water A was being filtered, hydraulic backwashing was conducted at a flux of 2.25 $m^3/m^2/d$ for 30 s every 15 min. While water B was filtered, hydraulic backwashing was conducted at a flux of 2.25 $m^3/m^2/d$ for 4 min every 30 min. During the last 3 s of each backwashing, about one third of the water combined with the settled sludge in the filtration tank was drawn out (recovery rate = 97%). For water A the filtration was conducted for 13 h, and for water B it was conducted for 120 h or until the TMP reached maximum operating pressure (60 kPa). During the coagulation of water B in the ceramic membrane filtration experiment, the flux was fixed at 4 $m^3/m^2/d$, which was monitored using a digital flow meter, and hydraulic backwashing (0.5 MPa) was conducted for 1.5 s every hour. The filtration was conducted for 120 h or until the TMP reached maximum operating pressure (100 kPa). The rate of fouling was monitored by recording TMP.

Analytical methods

TOC and DOC were measured using a TOC analyzer (TOC-L; Shimadzu, Japan) and UV_{254} was measured with a UV spectrophotometer (UV-1800; Shimadzu). Residual aluminum was measured using an inductively coupled plasma-optical emission spectrometry analyzer (iCAP 7000; Thermo Fisher Scientific, USA). ZP was determined with a ZEECOM (ZC-3000 Series; Microtec, Japan). The ZEECOM was equipped with a violet laser (wave length: 405 nm, electric power: 65 mW) in addition to the microscope, and was specially manufactured for this study. In the cell channel of the ZEECOM there are two stagnant boundary layers. After determining the velocity of particles at both layers at voltages of 10–30 V, electrophoretic mobility U could be calculated using the equation $U = v/(V/L)$ (where v is particle velocity (cm/s), V is voltage (V), and L is the distance between the electrodes (cm)). ZP could then be derived from the Smoluchowski equation: $\zeta = 4\pi\eta/\varepsilon \times U \times 300 \times 300 \times 1000$ (where ζ is ZP (mV), η is viscosity of water (g cm^{-1} s^{-1}), and ε is the relative permittivity of water (dimensionless)). As the distance observed by microscope was calibrated by measuring standard particles and time was always synced with computer, the velocity calculated from the distance and time could be guaranteed. The ZP value we used was the average of the ZP values at the two stagnant boundary layers after 50–200 measurements. We used the ZEECOM's scattered light mode when measuring the ZP of the particles in the supernatant, and thus measured only the average ZP of the micro-particles, as meso-particles cannot be detected by scattered light. We used the ZEECOM's violet laser mode when measuring the ZP of the meso-particles in the filtrates following filtration through the 0.5 μm membrane. The PN of the meso-particles was counted with the software WinROOF 2013 using the pictures obtained in the ZP measurement of the meso-particles. One picture is equivalent of 0.62 mm × 0.47 mm × 0.08 mm in volume, and therefore the counts were converted into volumetric concentrations by multiplying 43,240 to the counted meso-particles. The method applied for meso-particle count was got certified by the counting of known particles determined by Coulter counter (Multisizer 4e; Beckman Coulter, USA). The PN of the micro-particles was directly counted using a Hybrid Particle Counter (MW-SK121; METAWATER).

RESULTS AND DISCUSSION

Effects of coagulation conditions on the particle number

The effects of coagulation conditions (i.e., coagulation pH and PACl dose) on the PN of micro- and meso-particles in the two types of waters are shown in Fig. 1. Without PACl addition, PN of micro-particles and meso-particles in water A were about 7.0×10^5 and 6.0×10^6 particles/mL, both of which decreased when PACl was added (Fig. 1 (a, b)). For both of water, PN of micro-particles decreased with an increase in PACl dose, but less affected by pH (Fig. 1 (a, c)). The decrease in PN was particularly dramatic for doses of 2.0 mg Al/L and below. For water A (Fig. 1 (a)), at same PACl dose, minimum PN was seen at pH 6.5–7.0. This could be explained by conventional coagulation theory that cross-linking effect would be strongest around pH 7.0, at which condition particles would aggregate into larger ones thus lead to decrease of PN. However, for water B (Fig. 1 (c)), pH seems to have little effect on PN of micro-particles. The reason could be explained by comparing water qualities in Table 1. From Table 1, the lower value of SUVA of water B indicates that water B consisted more hydrophilic substances including biopolymers than water A, whose electric charge densities are close to zero (Buffle et al., 1998), which would be removed based on the mechanism that coprecipitation with coagulants rather than charge neutralization.

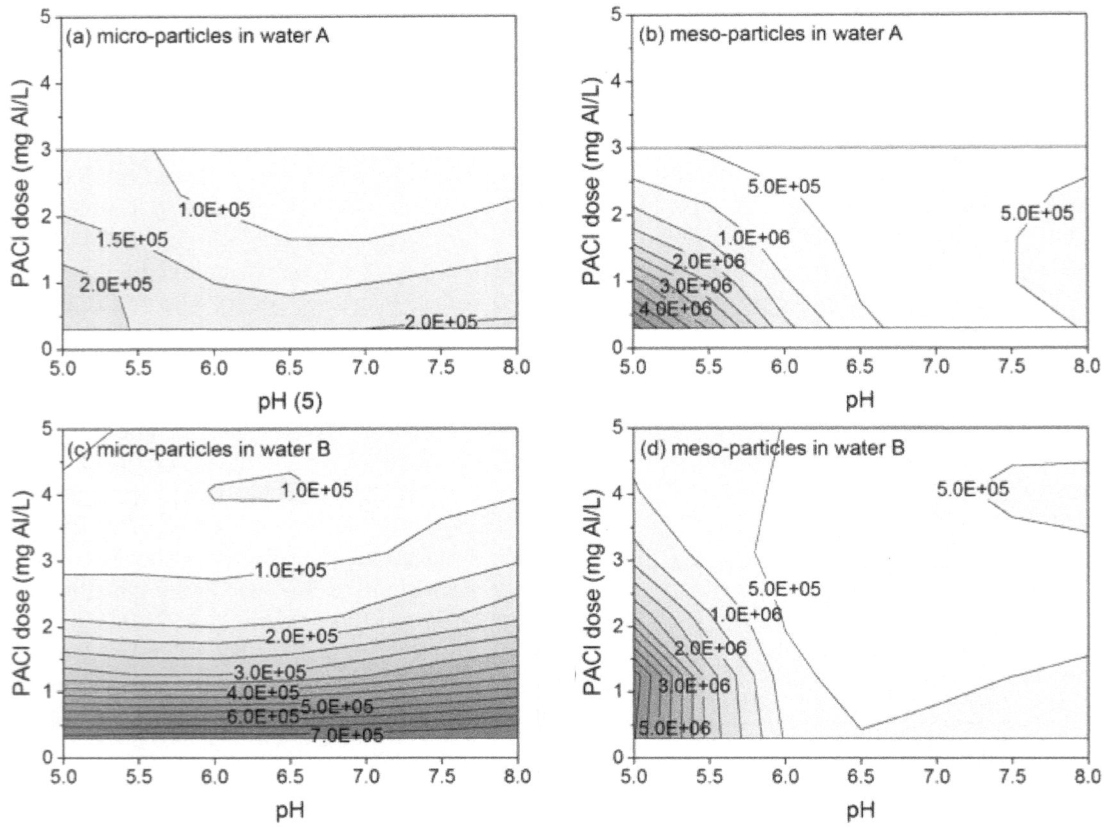

Figure 1: Effects of pH and PACl dose during coagulation on particle numbers of: (a) micro-particles in water A; (b) meso-particles in water A; (c) micro-particles in water B; and (d) meso-particles in water B.

Since meso-particles have a different size distribution than micro-particles, we predicted that their physicochemical characteristics would differ greatly. PN of meso-particles were more strongly affected by coagulation pH than by the PACl dose (Fig. 1 (b, d)). PN increased with decreasing pH, and for both source waters reached a maximum at pH 5.0. We hypothesize that this was due to elimination of the Al polymer cross-linking effect caused by the form of the aluminum changing from a polymer to dissolved Al^{3+} (Duan and Gregory, 2003). PN of micro-particles were minimized at pH 6.5–7.0, but above pH 7.0 increased once again. We hypothesize that this is because aluminum hydroxide became the main component at high pH levels, so the formation of coarse flocs was inhibited (Duan and Gregory, 2003).

Effects of coagulation conditions on zeta potential

Floc charge plays an important role in the coagulation process. We therefore suggest that the differences between meso-particles and micro-particles in terms of PN, discussed above, may have been related to differences in zeta potential (Fig. 2).

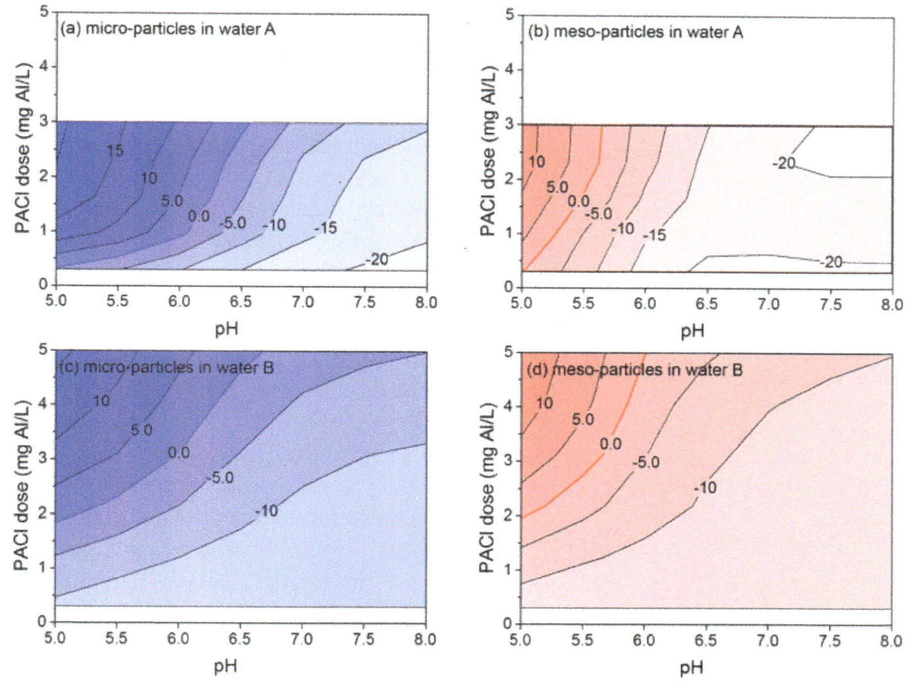

Figure 2: Effects of pH and PACl dose during coagulation on the zeta potential of: (a) micro-particles in water A; (b) meso-particles in water A; (c) micro-particles in water B; and (d) meso-particles in water B.

The particle size distribution of meso-particles have been studied occasionally (Kimura et al., 2008 and Stoller, 2009), but nothing has been reported regarding the charge of meso-particles. From Fig. 2 it is clear that the ZP characteristics of meso-particles and micro-particles differ. Under the same PACl dose, the neutralization point (ZP = 0) of meso-particles is at a lower pH than that of micro-particles. Thus while micro-particles are neutralized within a normal coagulation pH range (pH 6.0–7.0), meso-particles are not neutralized and left with a residual negative charge. In addition, the ZP of micro-particles increases linearly with PACl dose, but there is little change in the ZP of meso-particles, and even when a dose over 3.0 mg Al/L is added, the ZP cannot be neutralized at pH 6.0 and above. The similar trend was seen when using artificial water (Fig. 3).

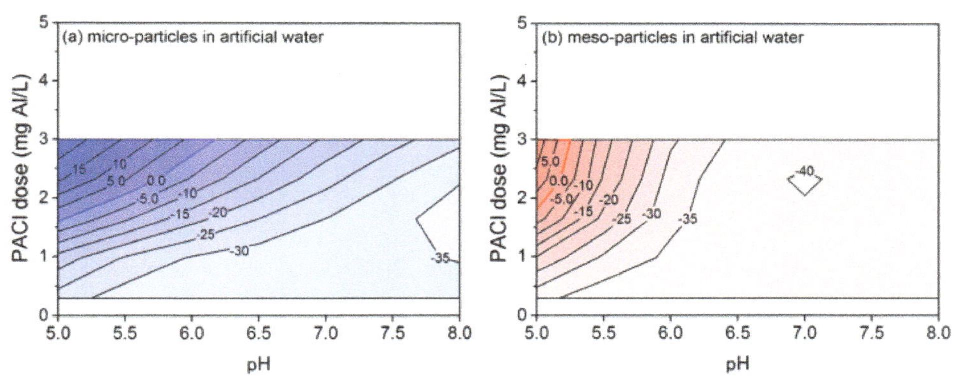

Figure 3: Effects of pH and PACl dose during coagulation on the zeta potential of (a) micro-particles and (b) meso-particles in artificial water. (The artificial water was made from humic acid sodium salt (Sigma-Aldrich, USA), and filtered through 0.5 μm filter in order to remove undissolved humic acid sodium salt, after which the particle number of meso-particles was about 8.0×10^6 particles/mL, and then used as raw water for jar tests.).

Possible reasons for the neutralization point of meso-particles occurring at a lower pH than that of micro-particles are that: (1) the positive charge of the PACl increases as pH decreases; or (2) the negative charge of the organic materials increases with decreasing pH. The charge of PACl usually increases as pH decreases (Fig. 4). The ZP of the PACl used in this study decreased by 50% from +40 mV to +20 mV as pH increased from 5.0 to 7.0. In dissolvable organic matter, dissociated protons bond with functional groups as pH decreases, so we hypothesized that ZP would increase until it reached the neutralization point. However, there was almost no change in ZP from pH 5.0 to pH 8.0 for the organic matter in water A (Figure S2). Thus, given that the ZP of the PACl was halved at pH 7.0 relative to pH 5.0, the PACl dose required to reach the neutralization point at pH 7.0 should be twice that required at pH 5.0. However, since the meso-particles were not neutralized at pH 7.0, even when 1.0 mg Al/L of PACl was added (more than twice the amount required to neutralize them at pH 5.0: 0.4 mg Al/L for water A), we assume that the charge of dissolved organic matter and coagulant does not depend solely on pH or that some other process is affecting the system. The charge neutralization mechanism of meso-particles is a topic for future investigation.

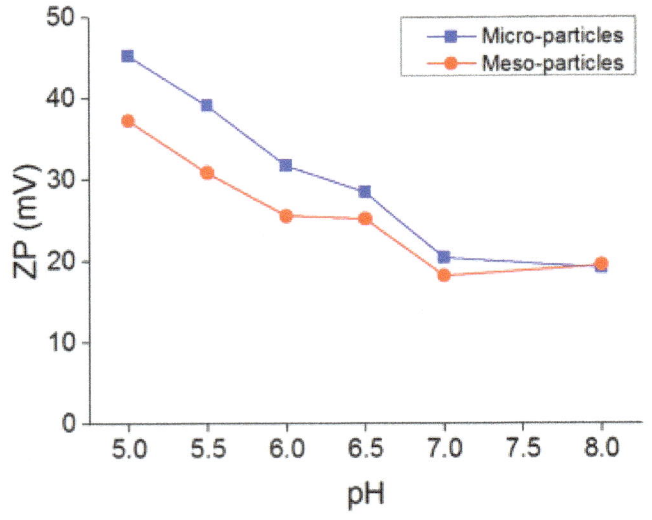

Figure 4: Effects of pH on zeta potential of pure coagulant PACl. (Micro- and meso-particles were made by the same jar tests processes mentioned in materials and methods after adding 1.0 mg Al/L of PACl into 1 L of pure water.)

Interestingly, a comparison of Fig. 2 and Fig. 3 shows that the PN of meso-particles is extremely high within the low pH range in which they are charge-neutralized. This indicates that although these particles neutralize, coarsening is slow. Since meso-particles are extremely small, Brownian motion is dominant in determining their behavior. Although van der Waals forces normally cause charge-neutralized particles to coagulate or coarsen, we believe that for meso-particles, coarsening tends not to occur under conditions where cross-linking effects are not operating, because Brownian motion causes them to move vigorously.

Extent of irreversible membrane fouling

Meso-particles are close in size to the membrane pore size, and their lodging in membrane pores is thought to contribute to the process of irreversible membrane fouling. Based on Fig. 2 and Fig. 3 we selected six coagulation conditions under which meso-particle characteristics differed substantially, and conducted bench-scale coagulation-membrane filtration experiments using these conditions. With a PVDF membrane, TMP increased as membrane fouling progressed (Fig. 5). Because we removed the cake layers that had accumulated on the membrane surface by means of physical cleaning every 30 min, the increase in TMP shown in Fig. 5 indicates the progression of irreversible membrane fouling. We defined TMP after 13 h as our indicator of the extent of irreversible membrane fouling and used this to summarize the results from the filtration experiments (Fig. 6).

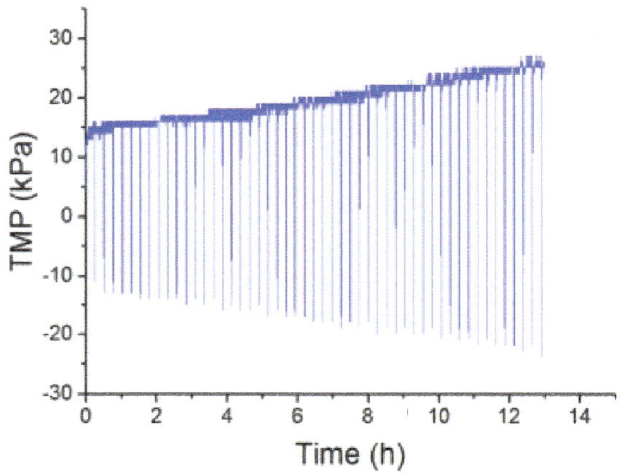

Figure 5: Change in trans-membrane pressure with time when water A was filtered using a PVDF membrane (PACl dose: 1.0 mg Al/L; pH: 6.5).

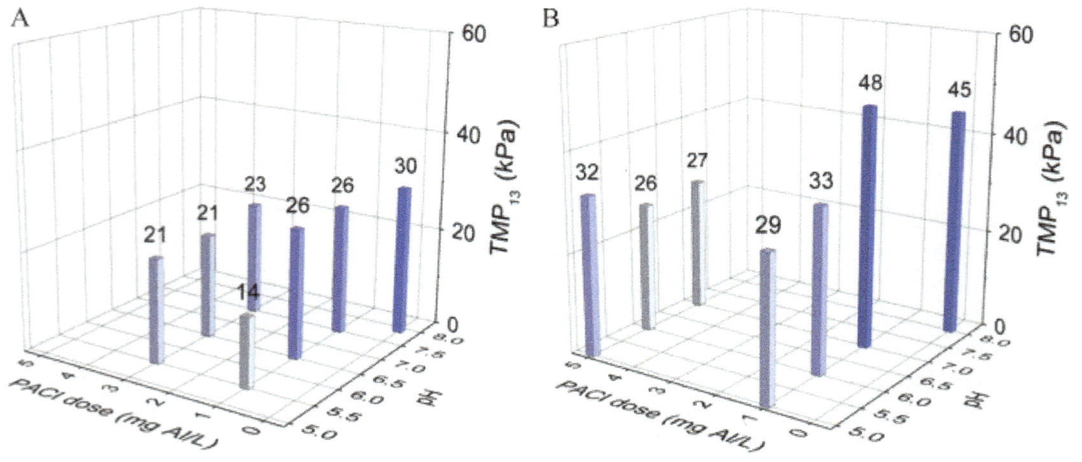

Figure 6: Effect of coagulation pH and PACl dose on irreversible membrane fouling (left: water A, right: water B). The vertical axis gives trans-membrane pressure values after 13 h (TMP_{13}), as an indicator of the extent of membrane fouling. The numbers above each column represents the TMP_{13} value at each condition.

In the tests involving water B, at pH 7.0 and PACl dose of 1.0 mg Al/L, membrane fouling was more rapid than when the source water was filtered directly (Fig. 6). This shows that setting the coagulation conditions appropriately is extremely important in controlling membrane fouling. Although for both types of water membrane fouling was controlled through coagulation under acidic conditions, conditions of best control differed for each water source. For water A these conditions were a pH of 5.5 and a PACl dose of 1.0 mg Al/L, but for water B they were a pH of 6.5–7.0 and a PACl dose of 5.0 mg Al/L. Thus the optimal coagulation conditions for controlling irreversible membrane fouling vary depending on the characteristics of the source water.

For water A, irreversible membrane fouling reduced as the PACl dose increased within the pH range 6.0–8.0, but at pH 5.5 increasing the PACl dose accelerated membrane fouling. Thus increasing the PACl dose alone cannot control irreversible membrane fouling.

Membrane fouling was controlled at low pH levels even though the PN of both meso- and micro-particles increased, for both types of water (Fig. 1 and Fig. 6). The fact that the correlation coefficient between the extent of irreversible membrane fouling and PN of both meso- and micro-particles was less than 0.5 (Fig. 7) suggests that quality may be more important than quantity in determining the rate of irreversible membrane fouling.

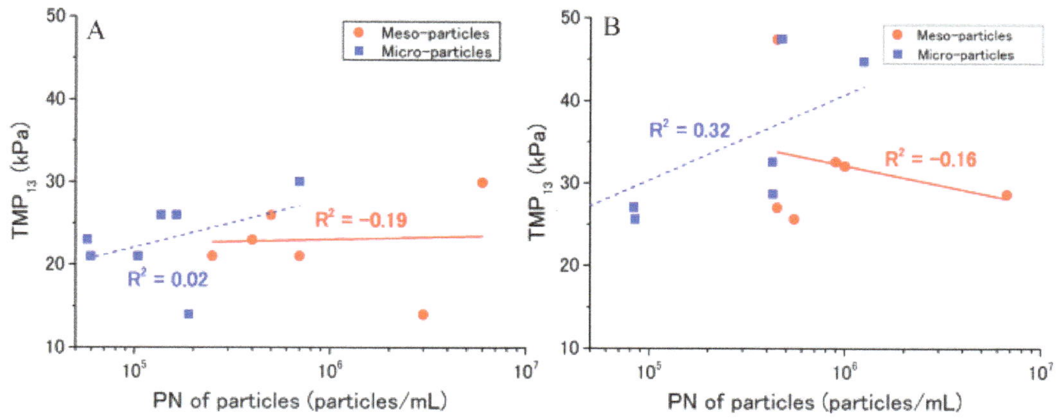

Figure 7: Relationship between particle number and the extent of irreversible membrane fouling (left: water A, right: water B). The vertical axis gives trans-membrane pressure values after 13 h.

Meso-particle characteristics contributing to the rate of irreversible membrane fouling

As the detail correlation types are still unknown, we conducted both linear and quadratic correlation analysis between particle PN or ZP and TMP (Table S2), and used the correlation types with higher corrected R^2 values. As the result, we used quadratic correlation between particle ZP and TMP, rather than linear correlation shown in Fig. 7. From Fig. 8, there was a clear correlation between particle ZP and the extent of irreversible membrane fouling for both water A and water B. For meso-particles this was a second-order correlation, with a correlation coefficient of 0.71 for water A and 0.76 for water B. Irreversible membrane fouling was best controlled, for both types of water, when meso-particle ZP approached zero. The result was similar when a ceramic membrane with a nominal pore size of 0.1 μm was used instead of the PVDF membrane (Fig. 9). This suggests that, regardless of membrane type and source water quality, irreversible membrane fouling can be best controlled by selecting coagulation conditions under which meso-particle ZP approaches zero.

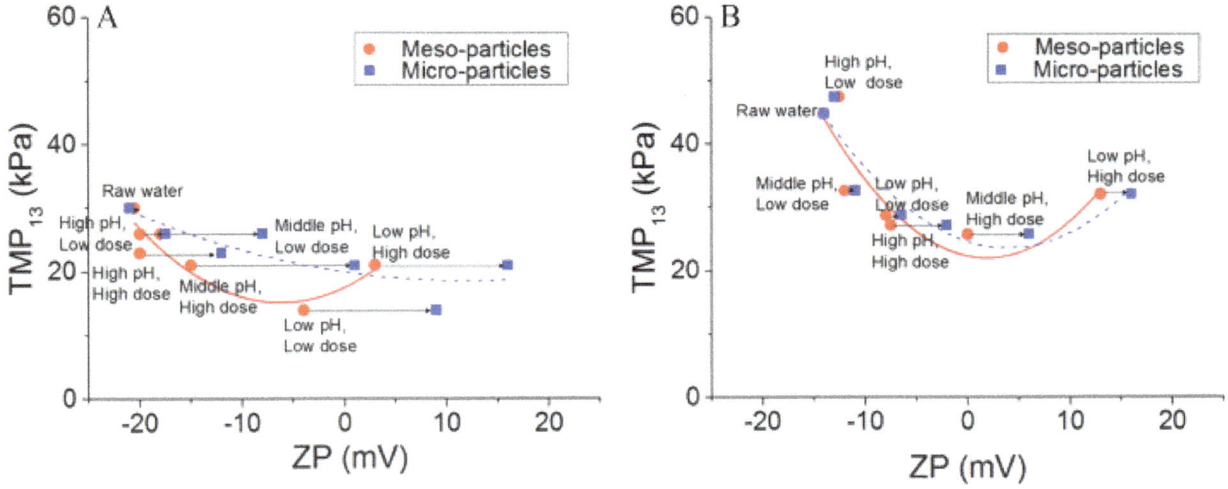

Figure 8: Relationship between particle zeta potential and the extent of irreversible PVDF membrane fouling (left: water A, right: water B). The vertical axis gives trans-membrane pressure values after 13 h.

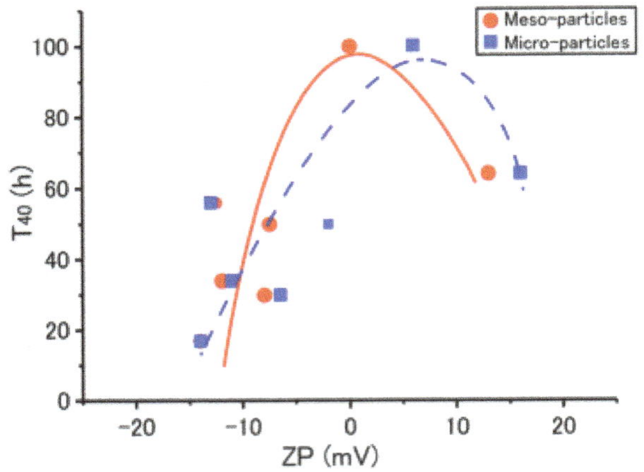

Figure 9: Relationship between particle zeta potential and extent of irreversible membrane fouling for water B with a ceramic membrane. The vertical axis gives the time at which trans-membrane pressure reached 40 kPa.

The size of meso-particles is on the same order as that of the membrane pores, and when these particles are retained inside the membrane pores they block passage and cause an increase in membrane filtration resistance. For such particles to not be removed by physical cleaning and instead remain in the membrane pores, they must be strongly adsorbed to the membrane surface. At pH 5.0–8.0, the membrane surface is negatively charged (Fig. 10), and if the meso-particles are also negatively charged, they are repelled. However, positively charged dissolved aluminum, calcium, and magnesium are present in the post-coagulation water, and the cross-linking effects due to these cations appear to cause electrostatic adsorption of the negatively charged meso-particles to the membrane surface (Jermann et al., 2007, Li and Elimelech, 2004 and Listiarini et al., 2009). In addition, when the meso-particles are positively charged they adhere electrostatically to a negatively charged membrane surface. We suggest that electrostatically adhered meso-particles are difficult to remove with physical cleaning, and thus are responsible for irreversible membrane fouling. We hypothesize therefore that when the meso-particle charge approaches zero, the strength of their electrostatic adhesion to the membrane surface is reduced and they can be removed by physical cleaning.

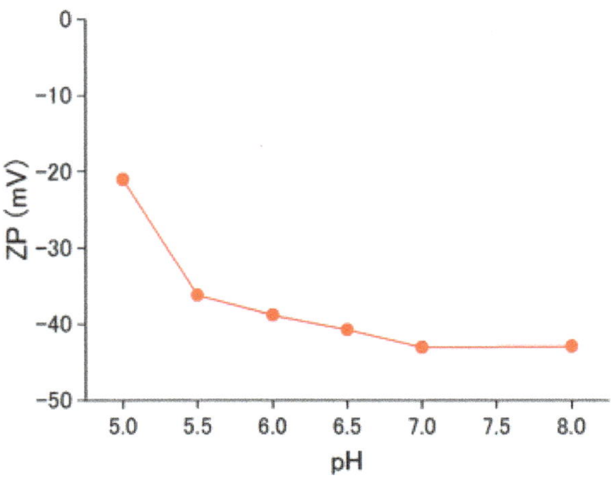

Figure 10: Effect of pH on the zeta potential of the PVDF powder used to make PVDF membranes.

This indicates that the control of ZP is more important in limiting irreversible membrane fouling than controlling PN of meso-particles and that cations play a crucial role in membrane fouling. Similarly, Jermann et al. (2007) found that humic acid adsorbed to the membrane surface due to the cross-linking effect of calcium, and on this base proposed a mechanism for the process of irreversible membrane fouling. They held that because the hydraulic resistance was greater than what the effect of humic acid could counteract when meso-particles blocked the membrane pores, even a tiny number of meso-particles led to severe fouling.

The relationship between the ZP of micro-particles and the extent of irreversible fouling was weaker than for meso-particles and varied depending on the type of source water (Fig. 8). For water B there was a second-order correlation such that membrane fouling was minimal when ZP was zero for micro-particles, just as for meso-particles. However, for water A irreversible membrane fouling continued to decrease as micro-particle ZP increased above zero. We hypothesize that this is because micro-particles that have accumulated on the membrane surface are easily removed by physical cleaning, resulting in a weaker correlation between micro-particle ZP and extent of irreversible membrane fouling.

Determining optimal coagulation conditions by focusing on meso-particle characteristics

Using the relationship between meso-particle ZP and irreversible membrane fouling, we can explore suitable coagulation conditions for coagulation-membrane filtration. Based on existing jar tests, optimal coagulation conditions have been determined based on water quality factors such as concentration of organic matter and turbidity. However, for membrane filtration we must set coagulation conditions according to other factors to control membrane fouling. Based on this study we conclude that membrane fouling can be controlled by measuring the ZP of meso-particles generated following the jar test and selecting conditions such that this value approaches zero. Both types of water used in this study had meso-particle charge neutralization points close to pH 5.5, which means coagulation must take place under more acidic conditions than for typical coagulation. When acidic coagulation took place in water B, coagulation conditions were not optimal because the residual aluminum levels were greater than that of standard tap water (0.2 mg Al/L; Fig. 11).

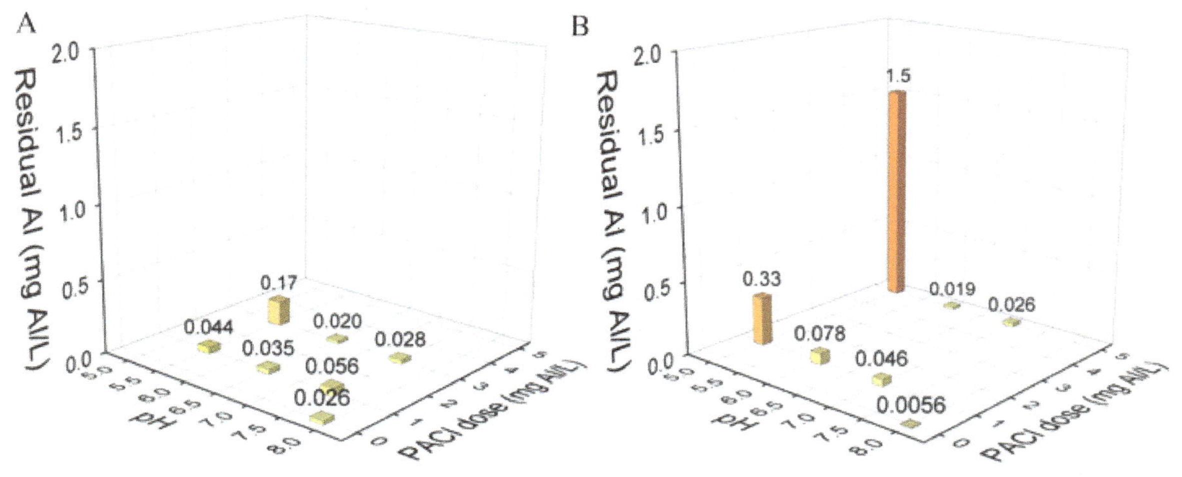

Figure 11: Effects of coagulation pH and PACl dose on residual aluminum concentration (left: water A, right: water B).

Furthermore, the acidic water would affect the efficiency of chlorination and pipe corrosion. To escape this situation, in Japan, pH must be returned to circumneutral before disinfection process. However, the cost for pH elevation after filtration would be much higher than the cost saved by prevention of the irreversible membrane fouling at acidic pH. To minimize the total cost for water purification, we believe that we have to achieve the prevention of irreversible membrane fouling under circumneutral pH condition. For this reason, we suggest that in future research it will be necessary to investigate coagulants that bring meso-particle ZP to zero at a circumneutral pH, and also to develop techniques such as pre-oxidation or pre-chlorination for the modification of meso-particle characteristics (e.g., the functional groups on meso-particles). The effects of meso-particles on ultrafiltration membrane fouling are also a topic for future research.

CONCLUSIONS

In this study, we operationally defined particles into micro- (i.e., >0.5 μm) and meso-particles (i.e., 20 nm–0.5 μm). Physicochemical characteristics including ZP and PN of them were firstly investigated with the use of a new particle tracker combined a high-output violet laser with a microscope. Then the principle factors governing the degree of irreversible membrane fouling were identified based on the repeated membrane filtration experiment using both PVDF and ceramic membranes under various coagulation conditions.

The main results obtained are shown below:

1. Identification of ZP of meso-particles remaining after coagulation was succeeded with the use of ZEECOM equipped with a violet laser.
2. The neutralization points (ZP = 0) were found to be different depending on the size of particle. Under same PACl dose, the neutralization point of meso-particles was lower in pH than that of micro-particles.
3. Under the condition where micro-particles were neutralized, meso-particles tended to remain as negatively charged particles, and it is still difficult to neutralize them even when the PACl dose was increased to 3.0 mg Al/L.
4. Irreversible membrane fouling in both PVDF and ceramic MF membrane was best controlled while meso-particle ZP approached zero.
5. PN of micro- and meso-particles had no relation with irreversible membrane fouling regardless of membrane type and source water.
6. Optimal pre-coagulation condition for the prevention of irreversible membrane fouling, or acidic pH, is not suitable for actual operation, as considering residual aluminum concentration, pipe corrosion, and chlorination efficiency. It is therefore necessary to investigate coagulants or other methods for the appropriate modification of meso-particle characteristics so as to achieve the efficient pre-coagulation from the view point of both water quality and fouling prevention.

ACKNOWLEDGMENT

The authors greatly appreciate the support of Dr. Dabide Yamaguchi from METAWATER Co., Ltd, Japan, who gave advice about the Hybrid Particle Counter (MW-SK121; METAWATER).

APPENDIX A. SUPPLEMENTARY DATA

1. Supplementary data related to this article can be found at http:// dx.doi.org/10.1016/j.watres.2016.05.076.

2. Amirtharajah, A., Mills, K.M., 1982. Rapid-mix design for mechanisms of alum coagulation. J. Am. Water Works Assoc. 74 (4), 210e216.
3. Buffle, J., Wilkinson, K.J., Stoll, S., Filella, M., Zhang, J.W., 1998. A generalized description of aquatic colloidal interactions: the three-colloidal component approach. Environ. Sci. Technol. 32 (19), 2887e2899.
4. Cho, M.H., Lee, C.H., Lee, S., 2005. Influence of floc structure on membrane permeability in the coagulation-MF process. Water Sci. Technol. 51 (6e7),143e150.
5. Choi, K.Y.J., Dempsey, B.A., 2004. In-line coagulation with low-pressure membrane filtration. Water Res. 38 (19), 4271e4281.
6. Duan, J., Gregory, J., 2003. Coagulation by hydrolysing metal salts. Adv. Colloid Interface Sci. 100e102, 475e502.
7. Gao, W., Liang, H., Ma, J., Han, M., Chen, Z.L., Han,

Z.S., Li, G.B., 2011. Membrane fouling control in ultrafiltration technology for drinking water production: a review. Desalination 272 (1e3), 1e8.

8. Huang, H., Schwab, K., Jacangelo, J.G., 2009. Pretreatment for low pressure membranes in water treatment: a review. Environ. Sci. Technol. 43 (9), 3011e3019.
9. Jermann, D., Pronk, W., Meylan, S., Boller, M., 2007. Interplay of different NOM fouling mechanisms during ultrafiltration for drinking water production. Water Res. 41 (8), 1713e1722.
10. Kimura, K., Hane, Y., Watanabe, Y., Amy, G., Ohkuma, N., 2004. Irreversible membrane fouling during ultrafiltration of surface water. Water Res. 38 (14e15), 3431e3441.
11. Kimura, K., Maeda, T., Yamamura, H., Watanabe, Y., 2008. Irreversible membrane fouling in microfiltration membranes filtering coagulated surface water. J. Memb. Sci. 320 (1e2), 356e362.
12. Kimura, K., Tanaka, K., Watanabe, Y., 2014. Microfiltration of different surface waters with/without coagulation: clear correlations between membrane fouling and hydrophilic biopolymers. Water Res. 49, 434e443.
13. Lead, J.R., De Momi, A., Goula, G., Baker, A., 2006. Fractionation of freshwater colloids and particles by SPLITT: analysis by electron microscopy and 3D excitation-emission matrix fluorescence. Anal. Chem. 78 (11), 3609e3615.
14. Lebeau, T., Lelievre, C., Buisson, H., Cleret, D., Van de Venter, L.W., Cote, P., 1998. Immersed membrane filtration for the production of drinking water: combination with PAC for NOM and SOCs removal. Desalination 117 (1e3), 219e231.
15. Lee, J.D., Lee, S.H., Jo, M.H., Park, P.K., Lee, C.H., Kwak, J.W., 2000. Effect of coagulation conditions on membrane filtration characteristics in coagulationmicrofiltration process for water treatment. Environ. Sci. Technol. 34 (17), 3780e3788.
16. Li, M., Jiang, M., Wu, C., 1997. Fluorescence and light-scattering studies on the formation of stable colloidal nanoparticles made of sodium sulfonated polystyrene ionomers. J. Polym. Sci. B Polym. Phys. 35 (10), 1593e1599.
17. Li, M., Wu, G., Guan, Y., Zhang, X., 2011. Treatment of river water by a hybrid coagulation and ceramic membrane process. Desalination 280 (1e3), 114e119.
18. Li, Q.L., Elimelech, M., 2004. Organic fouling and chemical cleaning of nanofiltration membranes: measurements and mechanisms. Environ. Sci. Technol. 38 (17), 4683e4693.
19. Listiarini, K., Chun, W., Sun, D.D., Leckie, J.O., 2009. Fouling mechanism and resistance analyses of systems containing sodium alginate, calcium, alum and their combination in dead-end fouling of nanofiltration membranes. J. Memb. Sci. 344 (1e2), 244e251.
20. Matilainen, A., Vepsalainen, M., Sillanpaa, M., 2010. Natural organic matter removal by coagulation during drinking water treatment: a review. Adv. Colloid Interface Sci. 159 (2), 189e197.
21. Perret, D., Leppard, G.G., Muller, M., Belzile, N., Devitre, R., Buffle, J., 1991. Electron microscopy of aquatic colloids: non-perturbing preparation of specimens in the field. Water Res. 25 (11), 1333e1343.
22. Singer, P.C., 1994. Control of disinfection by-products in drinking-water. J. Environ. Eng. ASCE 120 (4), 727e744.
23. Stoller, M., 2009. On the effect of flocculation as pretreatment process and particle size distribution for membrane fouling reduction. Desalination 240 (1e3), 209e217.
24. Tombacz, E., Filipcsei, G., Szekeres, M., Gingl, Z., 1999. Particle aggregation in complex aquatic systems. Colloids Surf. A Physicochem. Eng. Asp. 151 (1e2), 233e244.
25. Wiesner, M.R., Clark, M.M., Mallevialle, J., 1989. Membrane filtration of coagulated suspensions. J. Environ. Eng. ASCE 115 (1), 20e40.
26. Yamamura, H., Okimoto, K., Kimura, K., Watanabe, Y., 2014. Hydrophilic fraction of natural organic matter causing irreversible fouling

CHAPTER 8

PORTABLE LED FLUORESCENCE INSTRUMENTATION FOR THE RAPID ASSESSMENT OF POTABLE WATER QUALITY

J. Bridgeman, A. Baker, D. Brown, J.B. Boxall

Keywords: Fluorescence, LEDs, Organic matter, Microbial matter, Potable water quality

INTRODUCTION

Regulations place a duty on water companies to supply water that is wholesome at the time and point of supply, with wholesomeness defined by reference to prescribed concentrations assigned to various microbiological, chemical and physical parameters. Prescribed concentrations or values for microbiological parameters rely on proven, but dated indicator organisms, such as coliform bacteria, *E. coli* and colony counts. In addition to meeting standards, water must not contain any micro-organism or parasite at a concentration which would constitute a potential danger to human health.

The challenge to ensure microbial quality is addressed by water companies through disinfection, both during treatment processes and *via* the maintenance of a residual in most distribution systems. The most commonly used disinfectant in the developed world is chlorine. The use of chlorine, ozone, or chlorine dioxide as a disinfectant in water rich in organic matter (OM) (which also acts as a microbial food source) can lead to the occurrence of potentially carcinogenic disinfection byproducts (DBP). Consequently, water companies must manage the competing needs of biological and chemical compliance. Thus, it is imperative that companies monitor the microbial and chemical quality of drinking water at various stages of treated water supply systems, including trunk mains, service reservoirs and the distribution network.

Organic matter is ubiquitous in every water supply system and monitoring of its concentration and attributes is important for issues such as source water ecological health, treatment cost and efficacy, control of disinfection by-products and biological regrowth in distribution. Organic matter concentrations are typically assessed as total organic carbon (TOC) and/or dissolved organic carbon (DOC). Assessment requires complex and time-consuming laboratory procedures such that the data can only be used retrospectively rather than for proactive or pre-emptive management.

Another primary concern to water utilities is to ensure that the drinking water that is supplied does not pose an unacceptable health risk to consumers. As the number and type of different pathogens present in waters is extensive, varied and dependent on a range of environmental factors, it is not feasible to isolate and identify each specific pathogen on a regular basis. Hence, reliance has traditionally been placed on the measurement of total plate counts, as an overall indicator of microbial load and detection of faecal indicator bacteria and other coliform bacteria for contamination. Although these culture based tests give precise enumeration, they can take more than 30 h to perform from sample receipt to results. It is also known that, in the natural environment, often only 1% or less of microbes can be cultured in this way, leading to what has become commonly known as the "great plate count anomaly" (Staley and Konopka, 1985, Amann et al., 1995 and Allen et al., 2004).

The measurement of both organic and microbial matter thus currently relies on the collection and transportation of discrete samples that are then analysed by time consuming techniques (taking days). Such approaches are only able retrospectively to confirm (or otherwise) water quality. *In-situ* real time measurement is therefore highly desirable, as it would enable

Citation: J. Bridgeman, A. Baker, D. Brown, J.B. Boxall, Portable LED fluorescence instrumentation for the rapid assessment of potable water quality, Science of The Total Environment, Volumes 524–525, 15 August 2015, Pages 338-346, ISSN 0048-9697, http://dx.doi.org/10.1016/j.scitotenv.2015.04.050.

assurance of water quality, optimisation of process control and overall proactive and preventative operation of water supply systems. It would also be of great value to improve spatial and temporal coverage of water supply systems to fully develop and implement proactive management. Hence, there exists a pressing need for the development of novel technologies that will enable real time, *in-situ*, low powered and, ideally, continuous analysis of drinking water quality to improve the robustness of water supply system management. Brock (1987) affirmed the necessity of *in-situ* studies for describing environmental microbiology. Only then is it possible to identify the affinities, symbiosis and interactions with the surrounding environment, which are not applicable with indirect techniques (Souza et al., 2007). Similarly, the measurement of organic matter in an unaltered state makes results directly applicable only to the specific reaction conditions and water (Goslan, 2003).

Water quality is, of course, a function of inorganic pollutants as well as organics. Such inorganic pollutants may include nitrates and phosphates arising from run-off from agricultural land, heavy metals from highway run-off, ammonia from wastewater effluent discharges, arsenic occurring naturally in groundwater, or copper from household plumbing. Whilst the presence of these inorganics in water is recognised, the focus of the work reported here is on the detection of organic and microbial matter.

In the work reported in this paper, the development and deployment of a novel, LED-based instrument, capable of the *in situ* detection of fluorescence peaks T and C (surrogates for organic and microbial matter respectively) in water is described. The novelty of the work (and instrument) is demonstrated *via* the robust correlations observed between instrument results and those provided by well-established analytical methods including, for the first time, a comparison with flow cytometry data and an assessment of the fluorescence response of four individual bacteria isolated from drinking water.

BACKGROUND

Current methods of organic and microbial determination

A range of different techniques have been developed and applied to characterise and quantify the organic and microbial matrix within water. Total organic carbon (TOC) is the most comprehensive and widely adopted operational measure used to quantify the presence of organic carbon atoms in waters. TOC is often synonymous with natural organic matter (NOM) because organic contaminants in naturally sourced water supply systems generally represent an insignificant fraction of the TOC (Leenheer and Croué, 2003). More sophisticated analytical techniques which differentiate physico-chemical properties are also available, including: resin abstraction (used principally to determine hydrophobicity), high performance size exclusion chromatography (HPSEC), gas-chromatography–mass spectrometry (GC–MS) and UV–visible absorbance optical properties (Chin et al., 1994, Visco et al., 2005 and Bridgeman et al., 2011). Such techniques are predominantly laboratory-based and labour intensive, requiring extensive sample preparation and analysis, rendering them impractical in studies involving many sampling sites or frequent monitoring.

Standard methods used to test for the presence of heterotrophic bacteria (HB), total coliforms (TC), faecal coliforms (FC) and *E. coli* include, *inter alia*, the multiple tube method (presence of gas and acid at 44 °C), membrane filtration (MF), and colony-forming units (cfu) on agar media (Great Britain Department of Environment, 1983 and Cumberland et al., 2011). The genera enumerated by any of these methods are highly variable since the cultivation medium of choice, incubation temperature and time, origin (river, surface water reservoir, treated and disinfected drinking water, *etc.*), season of the year, and age of the water sample all have a significant effect on which genera will grow under these selected conditions (Allen et al., 2004). These factors, in addition to the time taken to acquire results, have stimulated research into more reliable, all-encompassing, more rapid methods of detection, such as the use of flow cytometry (*e.g.* Veal et al., 2000, Hammes et al., 2008 and Prest et al., 2013). Flow cytometric methods offer a fast, accurate and quantitative laboratory-based alternative to traditional culture based methods. However the sensitive nature of the equipment, the procedures involved, the necessity for expensive dyes (used for staining samples) and high initial capital cost of equipment, render flow cytometry inappropriate for use in frequent investigatory work in the field.

Applications of Fluorescence Spectroscopy to Water Quality Assessment

An innovative and rapid way to monitor the quality of water is to exploit the natural fluorescence of DOM. Previous research, conducted predominantly on riverine, marine and waste waters, has identified relationships between fluorescence and organic (*e.g.*Patel-Sorrentino et al., 2002, Baker et al., 2008 and Stedmon et al., 2011) and microbial (*e.g.* Determann et al., 1998 and Reynolds, 2003; Wu et al., 2003;Elliot et al., 2006) water quality. Fluorescence emitted at 400–480 nm under excitation at 300–360 nm (fulvic-like fluorescence) is indicative of the presence of organic carbon (peak C), whilst fluorescence emitted at 340–370 nm under excitation at 220–240 nm or 270–280 nm (tryptophan-like fluorescence) is indicative of microbial activity (peak T) (Coble, 1996).

Fluorescence analysis is a rapid, reagentless technique that requires only small volumes of sample and little sample preparation. An important advantage of measuring fluorescence in comparison to the more widely adopted optical measurement of UV_{254}absorbance is its sensitivity (Leenheer and Croué, 2003). Fluorescence measurements are typically 10–1000 times more sensitive than those of UV absorption spectroscopy with single molecule detection possible and discrimination of different sources of organic matter that absorb at similar wavelengths (Skoog et al., 2004, LeBlanc and Dufour, 2002 and Henderson et al., 2009). This is particularly relevant for the assessment of treated potable waters which exhibit low DOC, low UV absorbance and therefore low signal to noise ratio. Thus, the increased sensitivity of fluorescence for such analyses represents the optimum technique.

However, fluorescence analysis does have the potential to be confounded when assessing concentrated DOM samples as a result of inner-filtering, where emitted energy is absorbed and re-emitted at longer wavelengths by surrounding molecules.

Fluorescence spectroscopy has generally been reliant on expensive laboratory-based equipment, producing excitation emission matrices which are not readily quantified by the non-expert. These analysis systems typically use a high powered white-light source chosen for its high intensity and broadband spectral output, with appropriate spectral filtering to excite fluorophores in the sample under analysis. For example, a typical xenon arc lamp produces 150 W of output power in a uniform band from 200 to 2600 nm (Moe et al., 2005). Unfortunately, these light sources require high current power supplies and special handling, making them challenging to use outside of a laboratory environment.

Previous research has investigated the use of LEDs in arrays to produce a wider spectrum of light with higher spectral output (*e.g.* Hart and JiJi, 2002, Moe et al., 2005 and Dickens et al., 2011). Drawbacks of early LED-based systems include insufficient light output, instability (poor lifetime), expense and, until recently, lack of lower wavelength LEDs, so limiting their application to long wavelength ultra-violet, visible and near infrared excitable fluorescent compounds (Hart and JiJi, 2002). However, recent step change advances in manufacturing processes are now making LEDs an attractive alternative to conventional white-light sources. LEDs offer a highly energy-efficient means of producing monochromatic light. They provide a concentrated, small, cool emitter ideal for miniature analytical devices because of their reduced power of operation, size, and longer lifetime and also provide spectral control without the need for high overhead optics. Recently, the availability and reductions in cost of UV LEDs that emit to 270 nm and below, have broadened the horizons for their use in LED-based fluorescence devices for new applications. Commercially available instruments, such as the Turner Designs Cyclops™ and Chelsea Technologies Group UvLux™, reflect the popularity of LEDs in portable instruments used for targeting a range of fluorophores in a variety of applications.

The sensitivity of fluorescence spectroscopy to all microbial material (rather than the 1% which can be cultured and enumerated as colony forming units, CFU) presents an opportunity to effect a major step change in water quality assessment techniques, moving beyond the dated and limited use of indicator organisms, to enhance security of supply to customers. In this paper, the capability and efficiency of an innovative battery powered, portable, LED-based instrument in detecting peak C and peak T fluorescence at levels potentially found in potable water supply systems is demonstrated. These peaks are of interest as peak C provides a surrogate for the DOC present in the water, whilst peak T will identify any microbial growth which occurs as a result of DOC presence and insufficient residual chlorine concentration in the water. Furthermore, whilst both peaks are quenched by chlorine, a loss of chlorine residual manifests itself *via* an increase in fluorescence.

A comparison of an LED-based instrument with a traditional bench top instrument is presented to assess whether the device provides a suitable, portable, low cost and low powered alternative, before assessing the device's performance against traditional regulatory organic and microbial measures which, as commented upon earlier, are economically and operationally sub-optimal. The feasibility of applying this novel technology in the field to provide real time, *in-situ*, low power and potentially continuous analysis, hence providing an expedient indicator of changes in organic and microbial water quality to improve the robustness of water supply management is discussed.

METHODOLOGY

Instrument design

The prototype portable device has been designed to house all opto-mechanics, electronic circuitry and rechargeable battery in an enclosed toughened plastic case type structure. The case also provides space for a notepad computer *via* which the software for device control is operated. Only a small area for the sample holder is openly visible, capable of accommodating both discrete, and flow-through, standard 10 mm quartz cuvettes. When closed, the dimensions of the device are approximately 425 (l) × 300 (w) × 225 (d) mm, with a weight of approximately 3.5 kg. The battery life for routine operation of the software is dependent on the usage of the notepad computer, but is estimated at approximately 8–9 h. As this is the first prototype, further size optimization is likely. Further, it is anticipated that the notebook computer will be replaced by a data logger and simple four number display. Two LEDs, pulsed only when sampling and fired sequentially with a 2-s delay, are used to excite the water sample in the peak C and peak T regions. Emission is detected by two corresponding photomultiplier tubes (PMTs) for each excitation wavelength, measuring at the peak and the Raman emission wavelengths, respectively. The PMTs are located at the centre points of the fluorescence signal at right angles to their respective LEDs in a "T" shaped formation (Fig. 1). This formation is identical for both channels, with the peak C set being layered vertically over the peak T set, thus enclosing the cuvette when it is present. Concave collection mirrors opposite the LEDs are used to refocus the LED light into the sample, thus maximising the potential output fluorescence signal density on the active area of each PMT.

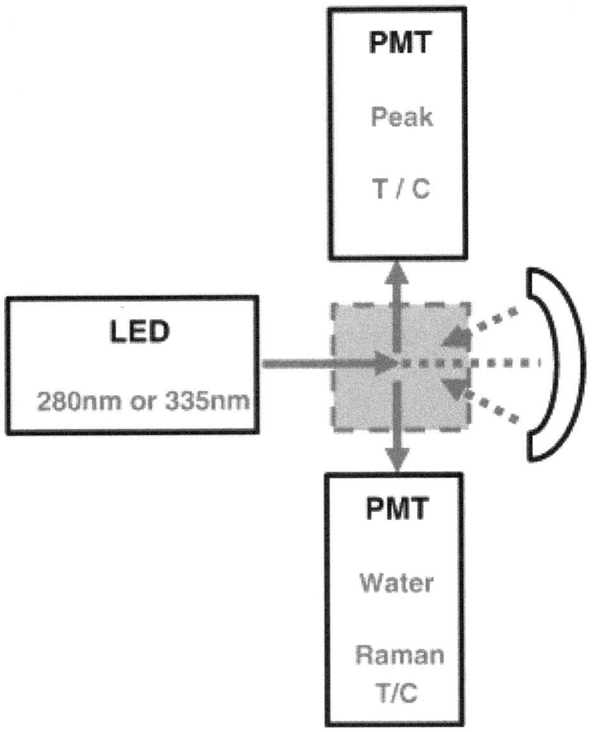

Figure 1: Plan view of arrangement of LED and PMTs in device.

The case also provides space for a notepad computer, which runs the software interface developed specifically to control the prototype device. The software user interface displays six values: *viz.* T and C fluorescence, the Raman signal of each channel (each read directly from separate PMTs), and the ratios of the T and C fluorescence signals over the corresponding Raman signal. The excitation wavelength for the T channel is 280 nm, with emission being recorded at 310 nm for the Raman signal and 350 nm for the peak T signal. For the C channel, the excitation wavelength is 335 nm and the emission is recorded at a wavelength of 380 nm for the Raman and 425 nm for the peak C signal. The emission wavelengths were selected based on findings from a previous work (Bridgeman et al., 2011). This configuration therefore permits the measurement of the ratio between the Raman signal and the corresponding emission signal and so allows the T and C peaks to be measured as a ratio of Raman intensity rather than as an absolute measurement. Normalisation of the fluorescence intensity in this way provides a useful internal standard and also means that potentially confounding effects of external variables, such as temperature fluctuations, are minimised.

Digital gain software on the instrument was incorporated to allow for the gain on the four photomultiplier tubes to be adjusted and measurement settings to be clearly defined, allowing for accurate comparison between groups of test results and to also allow for different measurement sensitivity ranges to be established, appropriate to the quality of water being inspected.

Experimental design

Laboratory investigations were conducted to (a) explore the device's sensitivity and confirm relevant settings suitable for potable water supply systems; (b) demonstrate if the LED device could match, or improve upon, a bench top device performance; and (c) compare measurement accuracy to alternative measures of organics and microbial determination.

Initial tests were undertaken on six samples of water from the Sheffield locality to optimise the instrument settings for the expected range of organic and microbial material within potable water supply systems. These six waters were selected to provide a representative spread of the types of water quality that would be expected in water supply systems from raw waters through to customer taps. Settings were adjusted to ensure that the viable measurement range was as broad as possible. (Full details are presented in the Supplementary information.) All results reported herein had consistent instrument settings. Results are presented as averages of three real repeats on each water sample which had three connected technical repeats (total of nine readings).

Validation of the sensitivity and accuracy of the portable device was demonstrated *via* a comparison of fluorescence intensities recorded on the portable device with measurements at equivalent peak C and T excitation and emission wavelengths on a bench top spectrophotometer (FluoroSENS, Gilden). For this purpose, 209 samples were collected through the three collaborative water company water supply systems, from source waters through to customer taps. The organic and bacterial content of the majority of these water samples was also assessed in the water company's (UKAS) accredited laboratories by the traditional regulatory measures of TOC, heterotrophic plate counts (HPC) at 22 °C (48 h) and 37 °C (24 h), *E. coli* and total coliforms.

Due to the previously described shortcomings of traditional bacteriological determinations, further tests were conducted on a selection of the collected water samples to compare the performance of the portable device against more selective bacteriological measures. Total bacteria counts of 48 of the samples were made using flow cytometric staining with a mixture of SYTO 9™ and propidium iodide dyes. To explore further possible bacterial specific florescence response, four distinct bacterial isolates were tested at a range of concentrations.

Laboratory methods

Fluorescence was measured in a 10 mm pathlength quartz cuvette (Starna Scientific Ltd), on both the prototype portable instrument and a bench top instrument. Emission scans were taken on the bench top device at the same peak T (280 nm) and peak C (335 nm) excitation wavelengths as the portable device, with the emitted fluorescence detected in 2 nm steps between 290–400 nm and 345–450 nm, respectively, and excitation and emission slit widths both set to 5 nm. Sets of three repeat analyses were made for each sample on both devices, with each repeat being subjected to three technical repeats. Results presented are therefore an average of nine readings. Between each test, the cuvette was thoroughly rinsed with distilled water. To maintain consistency of measurement conditions, blank scans using a sealed cell containing deionised water were run systematically to measure the intensity of the Raman line of water at 348 nm excitation wavelength. The mean Raman value during the study period was 412 intensity units. Before and after each set of tests a reading was taken on the portable and bench top device with distilled water in a sealed cuvette to check machine stability. All tests were conducted in temperature controlled laboratories at 20 °C. All fluorescence data are presented as the ratio of measured fluorescence to Raman intensity and are referred to as Raman-normalised fluorescence intensity.

Analysis of all regulatory organic and microbial measures took place at water company UKAS (United Kingdom Accreditation Service) laboratories. The majority of corresponding fluorescence measurements on the portable device were also made on-site, coupled with repeat analysis at the University of Sheffield laboratories when samples were tested again on the bench top fluorescence device. TOC was measured using an Aurora 1030 W analyser, using the non-purgeable method in which orthophosphoric acid is used to convert the total inorganic carbon to carbonic acid which is subsequently removed in a nitrogen carrier gas along with the purgeable organic carbon. The non-purgeable organic carbon (NPOC) is then oxidised using acidified sodium persulphate using heat (1030 W) as an initiator, before being sparged with nitrogen carrier gas for detection by a non-dispersive infrared detector. Results are presented as the mean of three valid measurements.

Standard methods were followed for the enumeration of both HPCs and the identification of *E. coli* and total coliforms. Heterotrophic bacteria numbers were derived from 1 ml samples cultured with 20 ml of sterile molten yeast extract agar in sterile plastic Petri dishes. Colony-forming units (CFU) were counted after incubating at either 22 °C (for 48 h) or 37 °C (for 24 h). Results are presented as CFU per 100 ml with all tests being performed in triplicate. *E. coli* and total coliforms were determined by the membrane filtration/plate count (MF/PC) method, whereby 100 ml of the sample was filtered through a sterile 0.45 µm cellulose nitrate filter (Millipore) and cultured on a sterile pad soaked in membrane lauryl sulphate broth (OXOID) and incubated for 4 h at 30 °C followed by 14 h at 37 °C. After incubation the dishes were examined for yellow and green colonies, signalling the presence of presumptive coliforms and *E. coli*, respectively. All green colonies were counted as confirmed *E. coli* and did not require any additional testing. All yellow colonies were counted as presumptive coliforms and were confirmed by sub-culturing onto two sets of TYEA confirmation plates to which an ONPG disc was added. In addition, one set of McConkey agar plates and one set of TSA agar plates were prepared. One set of TYEA plates were incubated at 44 °C, whilst all other plates are incubated at 37 °C. All tests were performed in triplicate and results presented as the mean CFU per 100 ml.

Flow cytometer analysis was performed using a BDTM LSRII Flow Cytometer System (BD Biosciences, UK). Samples were transported in sterile 50 ml falcon tubes in the dark and refrigerated at 4 °C until being analysed (within 6 h of collection). Water samples were stained with LIVE/DEAD® BacLight™ Kit (Invitrogen, UK) following the manufacturer›s instructions. BacLight staining contains two nucleic acid-binding stains: SYTO 9™ and propidium iodide (PI). SYTO 9™ can penetrate all bacterial membranes, bind to nucleic acids and stain the cells green, whereas propidium iodide (PI) only enters cells with damaged membranes and the combination of the two stains yields red fluorescent cells. 1.5 µl of component A (SYTO 9 dye, 1.67 mM/PI, 1.67 mM) and 1.5 µl of Component B (SYTO 9 dye, 1.67 mM/PI, 18.3 mM) were added to 1 ml of the sample and incubated in the dark at room temperature for 20 min prior to measurement. Both dyes were excited by a blue 488 nm laser. SYTO 9 was detected by a 505 nm long-pass and a 530/30 nm band-pass filter set and PI by a 655 nm long-pass and a 660/20 nm band-pass filter set. Data were processed and analysed using the BDFACSDiva™ software (BD Biosciences, UK), and results presented as the average of triplicate measurements.

Isolation of bacteria from tap water was undertaken following Ramalingam et al. (2013). From the pure cultures selected, a small quantity of each of the bacteria was suspended in a liquid medium and grown at 25 °C for 72 h with agitation at 150 rpm. The suspended cultures were then centrifuged at 5000 rpm for 10 min, before being washed in a small volume of distilled water and centrifuged again. This cleaning process was repeated another two times, before the pellet was removed and suspended in 30 ml of distilled water and adjusted to give an optical density of 1 (at 595 nm) for each strain. Subsequently, the bacteria suspensions were diluted to different strengths (1:2, 1:4, 1:8 and 1:12) and then tested for their fluorescence response. Additional counts of total bacteria in each dilution were also made using flow cytometry by applying the same methodology as described earlier.

RESULTS

Instrument comparisons

209 individual samples were collected at various locations through multiple water supply systems. Peak C and peak T intensities were determined for each sample using both the new portable instrument and a traditional bench top device. Fig. 2 and Fig. 3 show excellent correlation between the two instruments, with r^2 values of 0.98 and 0.77 for the full data sets for both peak C and peak T, respectively (each point displays the averages of nine readings with ± 1 s.d.). Data in Fig. 2 and Fig. 3 have been broken down into different sample types and correlations for each subset are shown in Table 1. These comparisons indicate the portable device›s ability to record the two peaks› intensities for a broad range of fluorophore concentrations typically expected in raw, partially treated and potable water samples, in a fraction of the time, cost and power used by the bench top device.

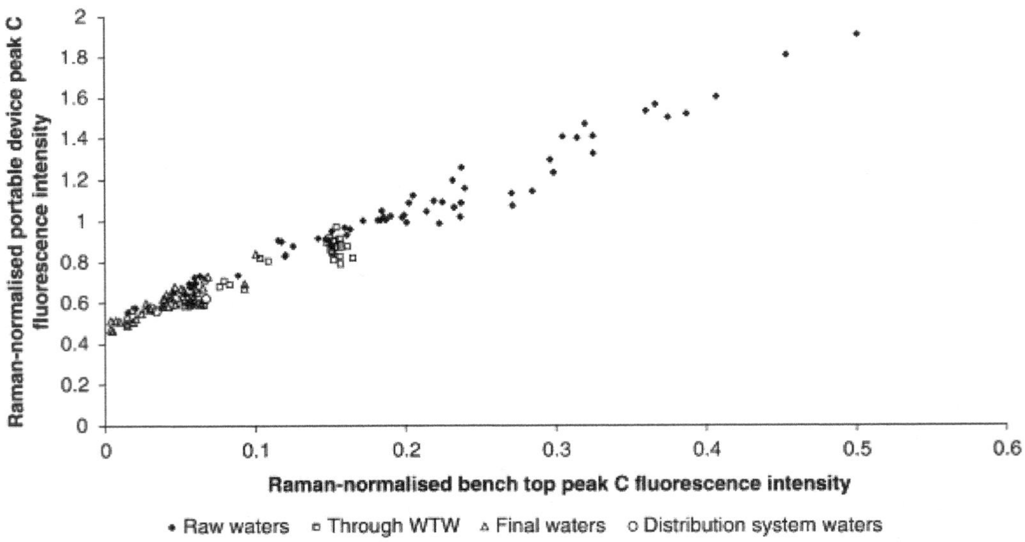

Figure 2: Peak C intensity comparison between new portable and traditional bench top devices.

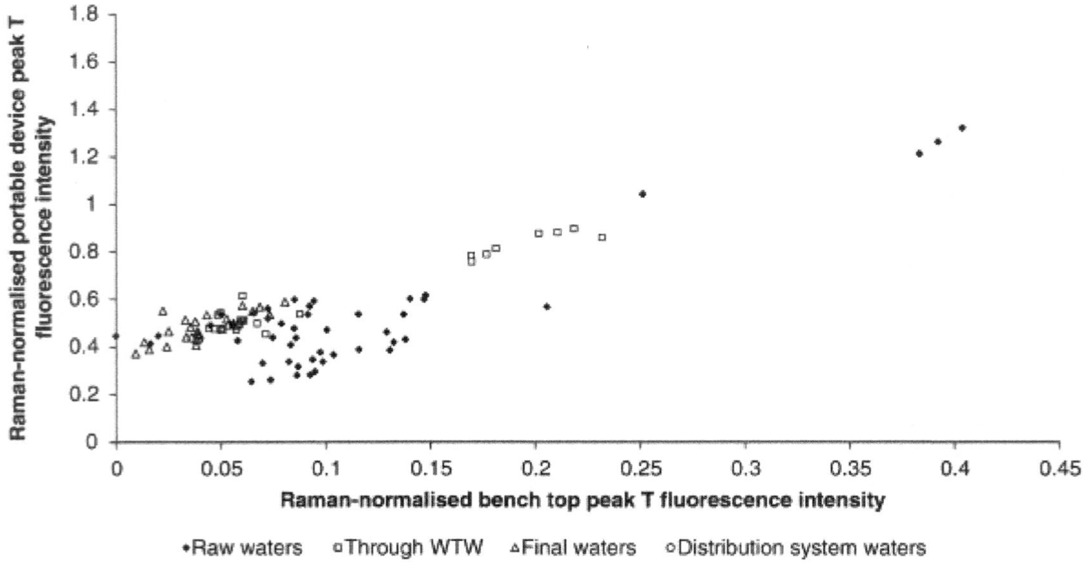

Figure 3: Peak T intensity comparison between new portable and traditional bench top devices.

Table 1: Peak C and peak T correlations between portable and bench top devices (r^2 values shown, p value < 1%).

	Peak C	Peak T
All data	0.98	0.68
Raw	0.96	0.74
Through WTW	0.90	0.94
Final water	0.79	0.58
Distribution system	0.72	nd

Comparison of peak C with total organic carbon

Peak C fluorescence intensity has been shown to correlate with TOC in a range of water types, from marine waters (Vodacek et al., 1995;Ferrari et al., 1996), a collection of surface waters (Smart et al., 1976), partially-treated waters (Bieroza et al., 2009 and Bieroza et al., 2010), sewage and industrial effluents (Hudson et al., 2008), to ground waters (Stedmon et al., 2011). As the majority of waters tested as part of this study were randomly selected samples collected as part of routine water company regulatory analyses, a variety of raw (surface and ground water), partially treated, and treated waters are represented in our comparison of TOC concentrations and portable device peak C intensities (Fig. 4). At the lower TOC concentrations (*i.e.* 2.5 mg/l and below), the relationship between peak C fluorescence and TOC approximates to a linear one. However, as the TOC concentration increases, it is noticeable that the relationship is non-linear. This is to be expected and is a result of inner-filter effects confounding the measurements at higher TOC concentrations. The relationship between fluorescence and measured TOC across the full data range can best be described using a power function, such that peak C = $260.30 \times [TOC]^{0.32}$.

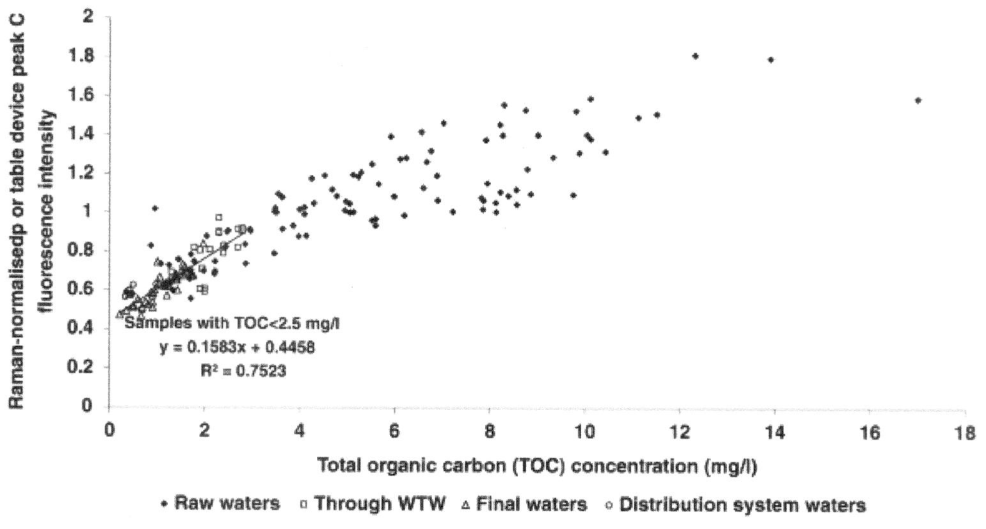

Figure 4: TOC concentration *versus* peak C intensity on portable device. (Raw water n = 107, through WTW n = 35, final waters n = 37, distribution system n = 6.)

As the majority of TOC samples were not taken in the distribution system, as regulatory requirements do not necessitate it, there is a large spread in TOC concentrations, ranging from 0.16 to 17.1 mg/l. Analysis of those TOC values typically found in potable water distribution systems (*i.e.* TOC < 2.5 mg/l, n = 107) yielded a linear correlation between fluorescence and TOC ($r^2 = 0.75$). This is encouraging considering that sensitivity settings were set for a wider spread of waters whilst performing these tests (*i.e.* the instrument was not optimised for low TOC waters).

Comparison of peak T with regulatory measures of bacteria

Work was undertaken to consider whether comparison of regulatory microbial measures with readings from the portable device would yield robust correlations between culturable microbial presence and water fluorescence, thus demonstrating the future potential use of the device as a more expedient microbial indicator than current traditional methods. However, analysis of 190 samples found no discernible relationships between peak T intensity and 2-day (37 °C) plate counts (r^2 = 0.01), between peak T intensity and 3-day (22 °C) plate counts (r^2 = 0.08), or between peak T intensity and either presumptive coliforms or *E. coli* (r^2 = 0.11 and 0.08 respectively). Whereas the waters selected for peak C comparison with TOC were taken primarily from raw, partially treated and treated samples, the majority of samples tested for regulatory microbial measures were taken in distribution systems. Hence, many of the samples analysed display zero counts, albeit still with a fluorescence response. However, it is known that living and dead cellular material, non microbial compounds, and possibly exudates can all generate a peak T fluorescence response, and it is known that 1% or less of environmental microbes can be cultured by the HPC methods, the "great plate count anomaly". Hence the lack of correlation is perhaps unsurprising. Consequently, the decision was taken to investigate peak T fluorescence in relation to flow cytometry and specific bacterial isolates.

Comparison of peak T with flow cytometer counts

Although large discrepancies between total bacterial concentrations and cultivable bacterial cell concentrations have been identified in previous studies (Staley and Konopka, 1985 and Hammes et al., 2008), the former is still not included in routine regulatory monitoring, chiefly due to the lack of appropriate analytical equipment to make it resource and economically viable. Recently, flow cytometry coupled with fluorescent staining has been demonstrated for the rapid assessment of drinking water (Hammes et al., 2008, Lautenschlager et al., 2010, Ho et al., 2012 and Prest et al., 2013), capable of providing a quantitative and accurate measure of information such as total cell concentration, bacterial viability and characteristics.

48 of the samples collected from collaborating water company distribution systems and tested using standard methods were also tested using flow cytometry. Total cell count results obtained reflect typical concentrations observed in previous studies on drinking water, in the order of 10^4 to 10^5 per ml of diverse microbial populations (Hoefel et al., 2003). Comparison of the portable device peak T intensity and these flow cytometry data,Fig. 5, reveal a reasonable correlation (r^2 = 0.56) and certainly an improvement in comparison to the regulatory microbial measures. It is interesting to note that the error bars associated with technical repeats of flow cytometry are generally greater than those associated with the new fluorescence device. The improved correlation reflects the benefits of flow cytometry identifying all bacteria in a sample rather than the culturable bacteria only. Hoefel et al. (2003) demonstrated that HPC counts in a mix of raw and potable waters were generally 2–4 log fold less than the numbers of total bacteria reported using the same stains as the BacLight™ Kit in this study. With peak T fluorescence capable of picking up an area of identifiable bacterial activity, a correlation would therefore be anticipated.

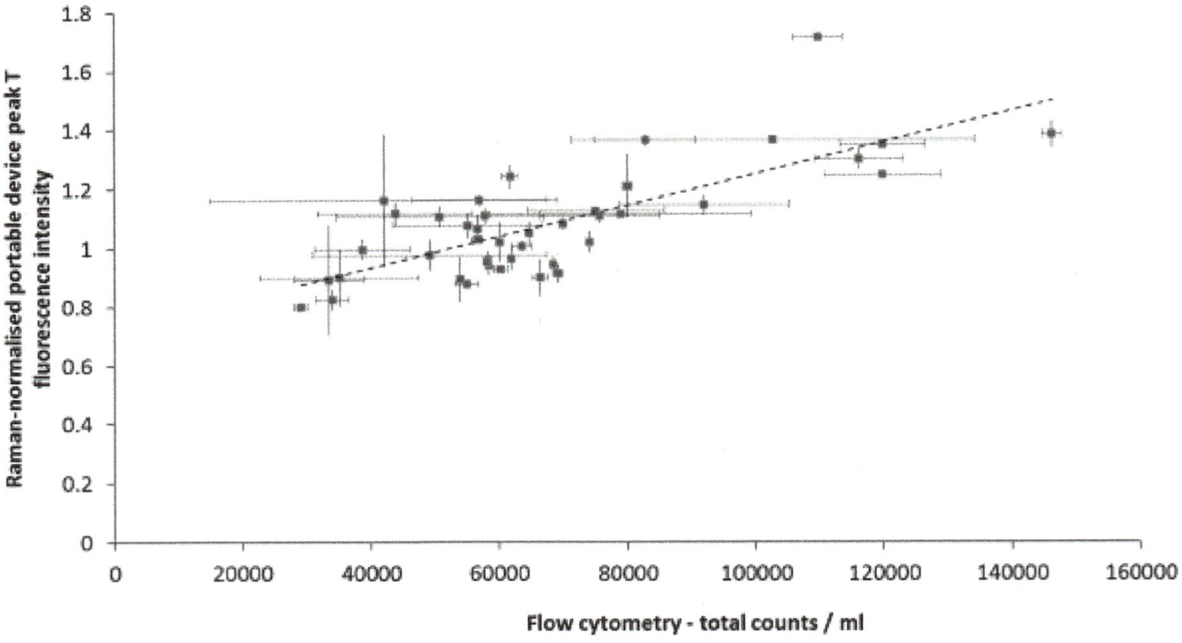

Figure 5: Comparison of flow cytometry counts of 48 samples with peak T intensity measured on the portable device.

Fluorescence response of individual bacteria

As an initial assessment of the portable device's potential for picking up the fluorescence response of individual bacteria which are found in potable water, four isolates used in a drinking water distribution related project at the University of Sheffield were assessed. Initial reasoning behind the selection of these bacterial isolates was based on their ubiquity (*Sphingobium* sp. and *Methylobacterium* sp.) in water distribution systems worldwide and also for their less common or unique presence (*Rhodococcus* sp. and*Xenophilus* sp.) in the chosen domestic drinking water (Ramalingam et al., 2013).

Each of the 4 strains of bacteria showed a strong and clear peak T fluorescence response on the portable device. A clear and different response to the dilutions (expressed at flow cytometry counts) of each of the bacteria with good correlation was observed (Fig. 6), with the two more commonly observed bacteria in distribution systems (*Sphingobium* sp.and *Methylobacterium* sp.) showing the greatest peak T intensities. It is noteworthy that the strains with the highest bacteria counts (*e.g. Rhodococcus* sp.) did not necessarily give the greatest fluorescence response, in keeping with the observations by LeBlanc and Dufour (2002) between different bacteria types. These results confirm that at least some of the peak T fluorescence which is observed in potable water can have a direct bacterial origin, but that the association is not trivial.

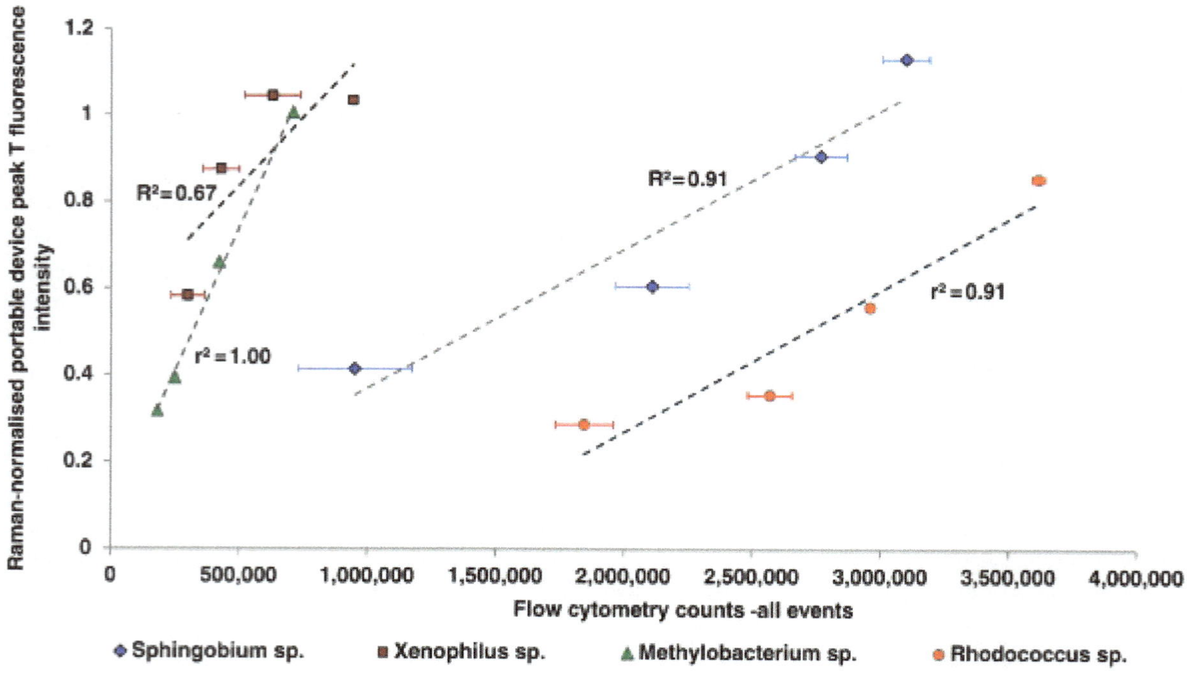

Figure 6: Portable device peak T intensities *versus* counts of dilutions of four bacteria isolates.

DISCUSSION

Common methods used to measure the microbial quality of drinking water are based on spot samples, followed by analysis *via* time-consuming laboratory tests. However, contamination events in distribution systems are typically characterised as being stochastic, short-termed and difficult to detect (Stedmon et al., 2011). Therefore the probability of spot samples, typically performed (at most) on a weekly basis on small quantities of water, picking up a contamination event before it reaches consumers is low. Of the failures that are captured by regulatory sampling, few may be attributed to known or quantified events, as associations could be hidden by the potential for bacteriological contamination to survive for long periods and even multiply if disinfection residuals are low (LeChevallier and Shaw, 1996).

Although fluorescence does not provide a direct measure of the presence or concentration of pathogens in a water supply, it does offer a potentially sensitive and online approach to detect sudden changes such as those brought about by the intrusion of contaminated water potentially bearing pathogens (Stedmon et al., 2011). Therefore, in a similar manner to how HPC tests are currently applied to understand changes in drinking water systems over time, online monitoring, using a device such as the one developed here, could be applied to alert operators to increases in general microbial activity. In contrast to HPC tests however, online monitoring provides the opportunity to build up a comprehensive picture of the range in water quality at any given location, to which any deterioration or sudden change can highlight the need for further investigation. Spot samples can only indicate that, at any given time, certain bacteria (not necessarily harmful to health) did or did not grow under the laboratory conditions and tests to which they were subjected. Further work is required to determine the device's capabilities as a continuous monitor and the potential issues which may arise when applying it in this way, such as optical fouling and the determination of appropriate base levels from which fluctuations are determined.

Although uncertainty exists as to what the peak T results in this study could be correlated with, the potential for the device to be applied as a means of assessing contamination events, such as ingress into service reservoirs or cross water connections, is reinforced by employing a combined measurement of both peak C and peak T fluorescence. The majority of bacteria associated with water supply systems, including all human pathogens, are heterotrophs. As heterotrophic bacteria require carbon, nitrogen, and phosphorus in a ratio of approximately 100:10:1 (C:N:P), organic carbon may often be a growth-limiting nutrient in drinking water (LeChevallier et al., 1991). Assessment of peak C in addition to peak T, at defined sensitivities according to water type, therefore provides a valid tool for discrimination of water sources and assessing abnormal fluctuations in water quality to give operators an indication of the need for further investigation.

In the environmental literature, peak T has been linked with measures such as biological oxygen demand (BOD) (for example; Hudson et al., 2008 and Bridgeman et al., 2013), which, acting as a measure of oxygen uptake in a water, acts itself as a surrogate for bacterial activity. It has been demonstrated with these initial findings that at least some of the peak T fluorescence which is observed in potable water can have a direct bacterial origin. Similar to a study of the bacteria *Pseudomonas aeruginosa*, isolated from an urban river (Elliot et al., 2006), a peak T response has been observed on the new device that is directly attributable to the type and quantity (albeit large) of four bacteria found in potable water. Further work is required to assess further strains of potable water bacteria to compare the peak T response of different species, to allow for a more comprehensive understanding of the root cause behind this area of fluorescence in environmental rather than laboratory cultured waters.

Thus, the foregoing demonstrates the development of a new approach for the simultaneous *in-situ*, real-time monitoring of both organic and microbial matter, with correlations with TOC ($r^2 = 0.75$) and total microbial counts ($r^2 = 0.56$). No other technology currently exists which can make both measurements simultaneously, although TOC can be measured on-line using automated TOC analysers and also determined indirectly by ultraviolet absorbance (for a review of all TOC analysis methods see Matilainen et al., 2011). The observed correlation between fluorescence and TOC shown here, although strong, is lower than that typically reported for the UV absorbance–DOC/TOC relationship, which typically has $r^2 > 0.9$ (Matilainen et al., 2002, Spencer et al., 2009 and Carter et al., 2012), but has the advantage of also providing information on total microbial counts. On-line microbial monitoring has yet to be established; the first feasibility study has recently been reported (Besmer et al., 2014) but the authors note the challenges of instrument cost and the handling of large datasets. Advantages of the fluorescence instrumentation, in addition to the ability to measure both organic and microbial matter, include low cost, small size and portability, low power requirements and a simple data output.

CONCLUSIONS

A fully operational portable LED-based device has been developed and shown to provide an accurate *in situ* assessment of the peak C and peak T response for a broad range of fluorophore concentrations typically found in potable water treatment and supply samples. The systematic validation of the results achieved by the new device and those provided by well-established analytical methods clearly demonstrate that this device offers, in a fraction of the time and cost of a traditional bench top device, the potential to promote a more proactive means of monitoring water quality. Unlike a traditional bench top device, the portability, low powered nature and ruggedness of the device allows for its use outside of laboratory conditions, and through its adjustable sensitivity, it is possible to apply the device to a range of water qualities. The device has been used in a discrete manner to measure the quality of a range of water company samples. Although correlating well with regulatory organic surrogates, the device did not provide any notable correlation with regulatory microbial measures. This highlights the difficulties in comparing the small percentage of the bacteria in a water sample which are culturable to a fluorescence response which is an area of identifiable general bacterial activity, regardless of the type or strain of bacteria.

Flow cytometer tests showed peak T intensity to correlate with an increase of total live and dead bacteria numbers in a sub-set of the samples. Tests with four bacteria isolated from a potable water tap sample showed a clear and different response between strains, although the strains with the highest bacteria counts did not necessarily give the greatest peak T fluorescence response.

The results presented here demonstrate that the novel fluorescence-based instrumentation that has been developed to assess treated water quality in real-time, *in-situ*, based on new developments in fundamental science, and innovative engineering design and deployment, represents a low cost, rugged, monitoring instrument which can be deployed extensively in supply systems to provide early indication of the need for active management of water quality. The significant scientific and engineering developments derived from this work can be used to facilitate a step change in quality assessment for the water industry with the potential for a much-needed move beyond the use of indicator organisms, and beyond measurement and reporting purely to satisfy regulatory requirements. Furthermore, the new instrument has deployment applications in the real-time determination of the microbial health and biochemical oxygen demand of surface and ground waters that will assist with enhancing understanding and improving water quality as required under the Water Framework Directive. Whilst initially focussed on the water industry, it is believed that this new capability could lead to the development and widespread adoption of a new class of fluorescence analysis tools with applications in many other academic disciplines.

ACKNOWLEDGEMENTS

The research reported here was sponsored by the Engineering and Physical Sciences Research Council under grants EP/I001379/1, EP/I001468/1 and EP/I029346/1. The authors also acknowledge the laboratory support provided by Esther Karunakaran, Isabel Douterelo and David Gaskell.

REFERENCES

1. Allen, M.J., Edberg, S.C., Reasoner, D.J., 2004. Heterotrophic plate count bacteria—what is their significance in drinking water? Int. J. Food Microbiol. 92, 265–274.
2. Amann, R.I., Ludwig, W., Schleifer, K.H., 1995. Phylogenetic identification and in situ detection of individual microbial cells without cultivation. Microbiol. Rev. 59 (1), 143–169.
3. Baker, A., Tipping, E., Thacker, S.A., Gondar, D., 2008. Relating dissolved organic matter fluorescence and functional properties. Chemosphere 73, 1765–1772.
4. Besmer, M.D., Weissbrodt, D.G., Kratochvil, B.E., Sigrist, J.A., Weyland, M.S., Hammes, F., 2014. The feasibility of automated online flow cytometry for in-situ monitoring of microbial dynamics in aquatic ecosystems. Front. Microbiol. 5, 265. http://dx.doi.org/10. 3389/fmicb.2014.00265.
5. Bieroza, M., Baker, A., Bridgeman, J., 2009. Relating freshwater organic matter fluorescence to organic carbon removal efficiency in drinking water treatment. Sci. Total Environ. 407, 1765–1774.
6. Bieroza, M., Baker, A., Bridgeman, J., 2010. Fluorescence spectroscopy as a tool for determination of organic matter removal efficiency at water treatment works. Drinking Water Eng. Sci. 3, 63–70.
7. Bridgeman, J., Bieroza, M., Baker, A., 2011. The application of fluorescence spectroscopy to organic matter characterisation in drinking water treatment. Rev. Environ. Sci. Biotechnol. 10, 277–290.
8. Bridgeman, J., Baker, A., Carliell-Marquet, C.M., Carstea, E., 2013. Determination of changes in wastewater quality through a treatment works using fluorescence spectroscopy. Environ. Technol. 34 (23), 3069–3077.
9. Brock, T.D., 1987. The study of microorganisms in situ: progress and problems. Symposium of the Society for General Microbiology, pp. 1–17.
10. Carter, H.T., Tipping, E., Koprivnjak, J.-F., Miller, M.P., Cookson, B., Hamilton-Taylor, J., 2012. Freshwater DOM quantity and quality from a two-component model of UV absorbance. Water Res. 46, 4532–4542.
11. Chin, Y.P., Aiken, G., O'Loughlin, E., 1994. Molecular weight, polydispersity, and spectroscopic properties of aquatic humic substances. Environ. Sci. Technol. 28, 1853–1858.
12. Coble, P.G., 1996. Characterization of marine and terrestrial DOM in seawater using excitation–emission matrix spectroscopy. Mar. Chem. 51 (4), 325–346.
13. Cumberland, S., Bridgeman, J., Baker, A., Sterling, M., Ward, D., 2011. Fluorescence spectroscopy as a tool for determining microbial quality in potable water applications. Environ. Technol. 33 (6), 687–693.
14. Determann, S., Lobbes, J.M., Reuter, R., Rullkotter, J., 1998. Ultraviolet fluorescence excitation and emission spectroscopy of marine algae and bacteria. Mar. Chem. 62, 137–156.
15. Dickens, J.E., Vaughn, M.S., Taylor, M., Ponstingl, M., 2011. An LED array-based light induced fluorescence sensor for real time process and field monitoring. Sensors Actuators B 158, 35–42.
16. Elliot, S., Lead, J.R., Baker, A., 2006. Characterisation of the fluorescence from freshwater, planktonic bacteria. Water Res. 40, 2075–2083.
17. Ferrari, G.M., Dowell, M.D., Grossi, S., Targa, C., 1996. Relationship between the optical properties of chromophoric dissolved organic matter and total concentration of dissolved organic carbon in the southern Baltic Sea region. Marine Chemistry 55 (3-4), 299–316.
18. Goslan, E.H., 2003. Natural organic mater character and reactivity: assessing seasonal variation in a moorland water. (EngD Thesis). University of Cranfield.
19. Great Britain Department of Environment, 1983. The Bacterial Examination of Drinking Water Supplies 1982: Methods for the Examination of Waters and Associated Materials — Reports on Public Health and Medical Subjects. HMSO, London.
20. Hammes, F., Berney, M., Yingying, W., Vital, M., Koster, O., Thomas, E., 2008. Flowcytometric total bacterial cell counts as a descriptive microbiological parameter for drinking water treatment processes. Water Res. 42, 269–277.

21. Hart, S.J., JiJi, R.D., 2002. Light emitting diode excitation emission matrix fluorescence spectroscopy. Analyst 127, 1693–1699.

22. Henderson, R.K., Baker, A., Murphy, K.R., Hambly, A., Stuetz, R.M., Khan, S.J., 2009. Fluorescence as a potential monitoring tool for recycled water systems: a review. Water Res. 43, 863–881.

23. Ho, L., Braun, K., Fabris, R., Hoefel, D., Morran, J., Monis, P., Drikas, M., 2012. Comparison of drinking water treatment process streams for optimal bacteriological water quality. Water Res. 46, 3934–3942.

24. Hoefel, D., Groobya, W.L., Monisa, P.T., Andrews, S., Sainta, C.P., 2003. Enumeration of water-borne bacteria using viability assays and flow cytometry: a comparison to culture-based techniques. J. Microbiol. Methods 55, 585–597.

25. Hudson, N., Baker, A., Ward, D., Reynolds, D.M., Brunsdon, C., Carliell-Marquet, C., Browning, S., 2008. Can fluorescence spectrometry be used as a surrogate for the biochemical oxygen demand (BOD) test in water quality assessment? An example from South West England. Sci. Total Environ. 391, 149–158.

26. Lautenschlager, K., Boon, N., Wang, Y., Egli, T., Hammes, F., 2010. Overnight stagnation of drinking water in household taps induces microbial growth and changes in community composition. Water Res. 44, 4868–4877.

27. LeBlanc, L., Dufour, E., 2002. Monitoring the identity of bacteria using their intrinsic fluorescence. FEMS Microbiol. Lett. 211, 147–153.

28. LeChevallier, M.W., Shaw, N., 1996. Factors Limiting Microbial Growth in Distribution Systems: Full Scale. American Water Works Association, Denver, Colorado.

29. LeChevallier, M.W., Schulz, W., Lee, R.G., 1991. Bacterial nutrients in drinking water. Appl. Environ. Microbiol. 1991, 857–862 (March).

30. Leenheer, J.A., Croué, J.-P., 2003. Aquatic organic matter: understanding the unknown structures is key to better treatment of drinking water. Environ. Sci. Technol. 19A (January 1, 2003).

31. Matilainen, A., Lindqvist, N., Korhonen, S., Tuhkanen, T., 2002. Removal of NOM in the different stages of the water treatment process. Environ. Int. 28, 457–465.

32. Matilainen, A., Gjessing, E.T., Lahtinen, T., Hed, L., Bhatnagar, A., Sillanpää, M., 2011. An overview of the methods used in the characterisation of natural organic matter (NOM) in relation to drinking water treatment. Chemosphere 83, 1431–1442.

33. Moe, A.E., Marx, S., Banani, N., Liu, M., Marquardt, B., Wilson, D.M., 2005. Improvements in LED-based fluorescence analysis systems. Sensors Actuators B 111–112, 230–241.

34. Patel-Sorrentino, N., Mounier, S., Benaim, J.Y., 2002. Excitation–emission fluorescence matrix to study pH influence on organic matter fluorescence in the Amazon basin rivers. Water Res. 36, 2571–2581.

35. Prest, E.I., Hammes, F., Kotzschb, S., van Loosdrecht, M.C.M., Vrowenvelder, J.S., 2013.

36. Monitoring microbiological changes in drinking water systems using a fast and reproducible flow cytometric method. Water Res. 47, 7131–7142.

37. Ramalingam, B., Sekar, R., Boxall, J.B., Biggs, C.A., 2013. Aggregation and biofilm formation of bacteria isolated from domestic drinking water. Water Sci. Technol.: Water Supply 13 (4), 1016–1023.

38. Reynolds, D., 2003. Rapid and direct determination of tryptophan in water using synchronous fluorescence spectroscopy. Water Res. 37, 3055–3060.

39. Skoog, D.A., West, D.M., Holler, F.J., Crouch, S.R., 2004. Fundamentals of Analytical Chemistry. Brooks/Cole Thomson Learning, Belmont, CA, USA.

40. Smart, P.L., Finlayson, B.L., Rylands, W.D., Ball, C.M., 1976. The relation of fluorescence to dissolved organic carbon in surface waters. Water Research 10, 805–811.

41. Souza, J.V.B., da Silva, R.M., Koshikene, D., Silva, E.S., 2007. Applications of fluorescent in situ hybridization (FISH) in environmental microbiology. J. Food Agric. Environ. 5 (3&4), 408–411.

42. Spencer, R.G.M., Aiken, G.R., Butler, K.D., Dornblaser, M.M., Striegl, R.G., Hernes, P.J., 2009. Utilizing chromophoric dissolved organic matter measurements to derive export and reactivity of dissolved organic carbon exported to the Arctic Ocean: a case study of the Yukon River, Alaska. Geophys. Res. Lett. 36, L06401. http://dx.doi.org/10.1029/ 2008GL036831.

43. Staley, J.T., Konopka, A., 1985. Measurement of in situ activities of non-photosynthetic microorganisms in aquatic and terrestrial habitats. Annu. Rev. Microbiol. 39, 321–346.

44. Stedmon, C.A., Seredynska-Sobecka, B., Boe-Hansen, R., Le Tallec, N., Waul, C.K., Arvin, E., 2011. A potential approach for monitoring drinking water quality from groundwater systems using organic matter fluorescence as an early warning for contamination events. Water Res. 45, 6030–6038.

45. Veal, D.A., Deere, D., Ferrari, B., Piper, J., Attfield, P.V., 2000. Fluorescence staining and flow cytometry for monitoring microbial cells. J. Immunol. Methods 243, 191–210.

46. Visco, G., Campanella, L., Nobili, V., 2005. Organic carbons and TOC in waters: an overview of international norm for its measurements. Microchem. J. 79, 185–191.

47. Vodacek, A., Hoge, F.E., Swift, R.N., Yungel, J.K., Peltzer, E.T., Blough, N.V., 1995. The use of in situ airborne fluorescence measurements to determine UV absorption coefficients and DOC concentrations in surface waters. Limnol. Oceanogr 40 (2), 411.

48. Wu, F.C., Evans, R.D., Dillon, P.J., 2003. Separation and characterization of NOM by highperformance liquid chromatography and on-line three-dimensional excitation emission matrix fluorescence detection. Environ Sci Technol 3716, 3687.

CHAPTER 9

THE USE OF SIMULATED WHOLE EFFLUENTS IN TOXICITY ASSESSMENTS: A REVIEW OF CASE STUDIES FROM REVERSE OSMOSIS DESALINATION PLANTS

Laura J. Falkenberg, Craig A. Styan

Keywords: Brine, Environmental impact, Ecotoxicology, Marine, Wastewater

INTRODUCTION

Global population growth and associated patterns of consumption are placing increasing pressure upon resources, particularly freshwater [1]. Water scarcity is projected to worsen across much of the globe in the coming century, with severe shortages anticipated to affect 2.7 billion people in more than 80 countries [2]. Consequently, there is growing interest in methods to enhance the availability of freshwater. In coastal regions, desalination of seawater is increasingly explored and utilised as a method to complement other sources of water supply [3]. The process of desalination has a relatively long history in the Middle East (particularly the United Arab Emirates, Kuwait and Saudi Arabia), with capabilities rapidly expanding in the United States, Europe, China and Australia [3] and [4].

Desalination of seawater can, however, have environmental impacts, with particular concern often focussed on the potential effects of the resulting wastewater discharges returned to the marine environment. These discharges can have negative effects because of their highly saline nature, and also because they contain chemicals added during processing activities including chlorination, pH adjustment, coagulation, flocculation, dechlorination, antiscaling and membrane cleaning [3] and [5]. The resulting saline and chemical-rich waste streams can adversely affect surrounding environments by degrading water and sediment quality, impairing the functioning of marine life and disrupting the intactness of ecosystems (reviewed in [3], [6] and [7]). Importantly, not only can individual contaminants (i.e., salinity or chemical additive) have a specific impact, but their effect may be exacerbated when added to the water column in combination, as occurs in the effluent from a desalination plant [8] and [9]. Thus, while some general conclusions can be drawn regarding the broad ecological effects of effluents, impacts may be relatively specific to each plant because the waste stream produced is dependent on the local seawater characteristics as well as the chemicals, processes and specific running conditions chosen for use.

The urgent need to obtain potable water in many parts of the world has meant that, historically (particularly before the 1970s), environmental issues associated with desalination had largely been overlooked or an issue of secondary concern [10] and [11]. Increasingly, however, the disposal of wastewaters is drawing attention both in terms of environmental approvals and public perception (i.e., obtaining a 'social licence to operate'). This attention is particularly strong in developed nations such as Australia where environmental issues are often high on the public agenda and desalination is becoming more common [12] and [13]. Within Australia, desalination has the potential to provide an important source of potable water to growing coastal populations (~ 85% of the population lives within 50 km of the coast) and over the last few years a number of large (~ 50–200 GL a^{-1}) reverse osmosis plants have been built across the country to augment domestic water supplies of most of the major cities [12] and [14]. Perhaps less in the public eye, but equally important economically, is the production of water by desalination for industrial applications such as mining

---

Citation: Laura J. Falkenberg, Craig A. Styan, The use of simulated whole effluents in toxicity assessments: A review of case studies from reverse osmosis desalination plants, Desalination, Volume 368, 15 July 2015, Pages 3-9, ISSN 0011-9164, http://dx.doi.org/10.1016/j.desal.2015.01.014.

— to date industry applications have mainly involved saline groundwater, but increasingly seawater is being considered as a potential intake source and receiver of waste streams [12] and [15]. Ongoing uncertainty regarding the potential impacts of wastewater from desalination plants means, however, that limits of pollutants in wastewaters set by regulators in Australia are often highly precautionary, adding significantly to the cost of establishing and operating desalination plants. Consequently, a key issue is the development of a better understanding of the salinity and toxin tolerance of marine species in the vicinity of outflows [12], as this capability would likely bring both ecological and economic benefits.

Given concern regarding the potential impacts of waste streams, it is increasingly common for assessment of their toxicity to be conducted before a desalination plant becomes operational, and often before construction even begins. By conducting assessments early in the process, Australian regulators are able to assess the likely risks to the environment and, where deemed acceptable, to set licensing conditions (e.g., permissible waste stream volumes and concentrations) before any impact occurs. Complicating such testing, however, is the potential influence of local seawater characteristics, synergies among the specific chemicals released by each plant, and likely differing responses of the complex biota in the receiving environment [1]. One approach to this kind of assessment is to simply ignore the potential influence of synergistic effects — that is, to identify the potential effect of each contaminant that may be in an effluent and then add predicted responses to forecast the combined toxicity of waste streams. While such an approach takes into account local plant designs and likely running conditions, it cannot account for any interactive effects among the chemicals considered nor facilitate an understanding of the conditions which drive an interactive response [16]. An alternative approach is to conduct whole effluent toxicity (WET) assessments using wastewater collected from similar desalination plants that are already operational elsewhere. While this method addresses concerns regarding potential synergistic effects of multiple chemicals, the local context may be overlooked. That is, identifying similar plants can be difficult, with potential that the closest similar plant has differing seawater characteristics. Further, even where similar plants do exist within a suitable distance and samples can be obtained, information about associated running conditions and procedures may not be something the plant operators are willing to provide if this is outside of their normal licence conditions. Without such information, just how similar the actual (sampled) effluent under real running conditions is to that expected from the theoretical plants is difficult to gauge. In turn, this means that the results from WET testing may be hard to relate to the potential toxicological impacts of planned plant [8].

A need exists, therefore, for a way to consider the potential impacts of the whole (combined) effluents expected to be produced by proposed desalination plants which takes into account the specific characteristics of local environment, plant design and likely running conditions. An emerging way to do this is by WET testing of simulated whole effluent(s) manufactured to represent the outflow which will likely be produced by a specific plant under predicted running conditions. While this method holds promise and is increasingly utilised and advocated in literature resulting from applied studies (such as consultants reports and government documents, e.g., [17], [18] and [19]), it has received relatively little attention in the academic literature and not, to our knowledge, been used widely outside of Australia. Here, our overall objective was to consider the potential utility of simulated whole effluents in toxicity assessments by reviewing three Australian case studies. Specifically, we consider the Cape Riche, Victoria and Gorgon reverse osmosis desalination plants. We have chosen this subset of Australian desalination plants as they provide a good indication of the various methods and approaches used in testing of simulated whole effluents (described below). In this review we summarise the background to each project, effluents simulated, models and hypotheses of interest, WET testing done, results (largely in terms of management applications), and assess any features of the assessments which constrained the efficacy of this approach.

CASE STUDIES

Simulated whole effluents are increasingly used in toxicity assessments conducted prior to and during the development of desalination plants. In a growing number of cases, reports summarising these assessments in terms of the specific approach used and results obtained are made freely available, which enables their comparison. Here, we review three such projects and their reports, specifically; the Cape Riche seawater desalination plant summarised in [5], the Victorian Desalination Project constructed in Wonthaggi as detailed in [20], and the Gorgon Development at Barrow Island documented in [8]. These reports form the basis of the following comparison(s), with detail included in this section obtained from them unless otherwise indicated.

Cape Riche Seawater Desalination Plant, South Western Australia (Cape Riche) [5]

The Cape Riche Desalination Plant was proposed to be located in the South Coast region of Western Australia, approximately 90 km east north-east of Albany and 5 km west of Cape Riche [5] and [21]. The plant was to have the capacity to supply of 12 GL of treated water per year via pipeline to the Southdown Magnetite Project (an open pit magnetite mine) [5] and [21]. The desalinated water was to be primarily used as process water at the mine site, meeting 85% of make-up water requirements, with the potential for some to be further treated at the site and used as potable water if required. Following final environmental approvals for the desalination plant [21], however, plans for the mine were put on hold in November 2012.

For ecotoxicological assessment of the Cape Riche plant, a simulated effluent was created in the laboratory with chemical specifications matching those expected to be found in the wastewater from the proposed plant. The simulated effluent was produced by collecting seawater from near the expected site at Cape Riche and transporting it to a commercial ecotoxicology laboratory several hundred kilometres away in Perth. Once in the laboratory, the seawater was passed through a laboratory-scale reverse osmosis membrane to simulate the desalination process and then dosed with an antiscalant considered likely to be used in the planned plant; specifically, 'Nalco Permatreat PC-1020' [5]. Given that studies previously performed in Western Australia had identified elevated salinity and residual levels of antiscalant were of most significance when assessing the toxicity of desalination plant effluent discharges, it was considered important to assess a high salinity brine effluent including an anticipated residual concentration of antiscalant for testing [5] (Table 1). In Australia, environmental regulatory authorities usually adopt recommendations from ANZECC and ARMCANZ[22] and, where possible, assess potential environmental effects of chemical releases on ecosystems via WET testing on the widest range of (local) species possible, including representatives of different trophic levels [23]. Consequently, the toxicity of the Cape Riche simulated effluent was assessed for a range of marine organisms from six species considered representative of those found in the temperate waters of southern Western Australia where Cape Riche is located [5] (Table 2).

Table 1: Summary of the simulated whole effluent treatments used in the Cape Riche, Victoria and Gorgon case studies.

Number	Treatment composition
Cape Riche	
1	Brine+antiscalant
Victoria	
1	Saline concentrate (brine)+antiscalant
2	Saline concentrate (brine)+antiscalant+membrane cleaning surfactant (worst case scenario)
3	Saline concentrate (brine)+antiscalant+membrane cleaning surfactant (worst case scenario)+biocide+prechlorinated
4	Saline concentrate (brine)+antiscalant+membrane cleaning surfactant (long term average)+prechlorinated
Gorgona	
1	Brine only (no other additives)
2	Veolia brine effluent+backwash
3	Veolia brine effluent+backwash+CIP (variant i)
4	Veolia brine effluent+backwash+CIP (variant iv)
5	ITT brine effluent+backwash
6	ITT brine effluent+backwash+CIP (variant ii)
7	Osmoflo brine effluent+backwash
8	Osmoflo brine effluent+backwash+CIP (variant iii)
9	Osmoflo brine effluent+backwash+CIP (variant iv)
10	GE brine effluent+backwash
11	GE brine effluent+backwash+CIP (variant v)
12	Backwash only

Note that for the Gorgon case study, each 'brine effluent' contains the relevant biocides, flocculants and antiscalants.

Table 2: Summary of the test species, and taxonomic groups they belong to, used in the Cape Riche, Victoria and Gorgon case studies.

Taxonomic group	Species	Cape Riche	Victoria	Gorgon
Microalgae	Isochrysis galbana	×		×
	Nitzschia colsterium		×	
Macroalgae	Ecklonia radiata	×		
	Hormosira banksii		×	
Annelid	Diopatra dentata			×
Echinoderm	Heliocidaris erythrogramma	×		
	Heliocidaris tuberculata		×	×
Mollusc	Mytilis edulis	×		
	Mimachlamys asperrima		×	
	Saccostrea glomerata			×

Crustacean	Gladioferens imparipes	×		
	Allorchestes compressa		×	
	Penaeus monodon			×
Fish	Pagrus auratus	×		
	Seriola lalandi	×	×	
	Lates calcarifer			×

The testing of six species allowed the calculation of a Species Sensitivity Distribution (SSD) for the simulated effluent, enabling predictions of the dilution factors necessary to achieve various levels of species protection if this effluent was to be released back into the waters near Cape Riche. Specifically, the chronic dilution factor calculated for the 99% level of species protection was 53 (Table 3); that is, testing predicted that effluent would need to be diluted 53-fold to result in a concentration such that 99% of species would not be expected to be affected by long term exposure. The dilution factors required to achieve (lower) protection levels of 95%, 90% and 80% of species were also reported and were 40, 35 and 28 respectively (Table 3). These dilution factors can be of use in two key ways. Firstly, from an engineering perspective, such information is important for designing adequate dispersion and mixing of effluents with receiving waters; e.g., ensuring systems have enough pressure, flow, and outfall nozzle jet dimensions to provide sufficient energy to get effective mixing that achieves the suggested rates of dilution quickly. Secondly, from the perspective of an environmental regulator, such information can be combined with modelled forecasts of the location, likely rates of mixing and areas of increased concentrations to assess the potential for impacts on local ecosystems in terms of both severity and spatial extent.

Table 3: Summary of the reported level of species protection (%), effluent (%) and safe dilution factor for the Cape Riche, Victoria and Gorgon case studies.

Sample code	Level of species protection (%)	Effluent (%)	Safe dilution factor
Cape Riche[a]		Chronic	
1	99	1.9	53
1	95	2.5	40
1	90	2.9	35
1	80	3.6	28
Victoria[b]		Acute and chronic	
1	99	3.2	31
2	99	6.9	14
3	99	5.2	19
4	99	7.0	14
Gorgon[c]		Acute and chronic	
All reported together[d]	99	10.3	9.7
	95	12.9	7.8
	90	14.1	7.1

a Reported here for Cape Riche are effluent (%) and safe dilution factors calculated using the EC10 values.

b Victoria case study combined EC10 and LC10 data using an acute to chronic ratio (ACR) of 2.5.

c Gorgon case study combined the EC10 and LC10 data using an ACR correction value of 2.

d 12 effluents were tested in the Gorgon case study, however, given that no effluent was found to be more toxic than the others individual effluent (%) and safe dilution factors were not reported.

The ecotoxicological assessment conducted for the Cape Riche desalination plant is a good example of how a simple approach of simulating effluents and using them in WET testing can provide information before plants are constructed, particularly where focus is placed on locally relevant species and receiving waters [24]. We note that while only one simulated effluent was tested, the number of species assessed (six) and logistics of using locally specific source water for creating test effluents still represents a substantial amount of work. We do highlight, however, that by using this simple approach the derived dilution factors are really only representative of a limited set of (idealised) running conditions—essentially those simulated. These simulated conditions may or may not produce the range of effluents resulting from the actual final plant, which would probably operate in varying and possibly unexpected conditions. For example, it is impossible to conclude the toxic effects that would result if a different antiscalant was used in the eventual plant and/or chemicals resulting from desalination processes were also included in the effluent streams (e.g., chemicals involved in chlorination, pH adjustment, coagulation, flocculation, dechlorination and membrane cleaning). Given this, the dilution factor may need to be modified under different running conditions not considered in this study. An additional limitation of testing just one simulated effluent is that the cause(s) of toxicity in the simulated effluent stream cannot be unravelled (i.e., a hypothesis considering the relative influence of brine vs. antiscalant vs. brine + antiscalant could not be tested). Accordingly, both of our other case studies involved more complex sets of treatments and tests.

Victorian Desalination Plant, Bass Coast (Victoria) [20]

The Victorian Desalination Plant was constructed in the south east of Australia, on the north coast of Bass Strait. The plant was designed to supply potable water to the supply system for the state capital of Melbourne (population ~ 4 million people) and possibly other regional supply systems [25]. Although the completed plant is now capable of supplying up to 150 GL of high quality drinking water each year, it has been placed in standby mode and will not be producing water for the short-term foreseeable future [26].

Before construction, WET testing was undertaken on potential effluents as part of the approval process for this plant, specifically EPA Works Approval Condition 2.3 and Performance Requirement PR217 [20] and [27]. Testing was done on four effluents simulated to encompass a range of operating conditions, specifically the long-term average outflow, worst case scenario within normal operating conditions, absolute worst case, and long-term average outflow composition together with cleaning in place (CIP) chemicals (Table 1). These effluents enabled consideration of models and hypotheses regarding different levels of production and some of the different processes involved in desalination (summarised in Table 4).

Table 4: The alternative models and hypotheses that the Victoria and Gorgon case studies tested.

Case study[a] and option	Model	Hypothesis[b]
Victoria		
A	Additives exert acute toxicity	2 or 3 or 4>1
B	Additives do not exert any acute toxicity	2=3=4=1
C	Additives interact to decrease toxicity	2 or 3 or 4<1
Gorgon		
A	Brine per se is the most toxic component of the effluent and the presence or absence of flocculants, biocides and/or other backwashing or CIP components does not affect toxicity	1=2=3=4=5=6=7=8=9=10=11>12
B	Backwash is the most toxic component of effluent and the presence or absence of other components and/or brine does not affect toxicity	2=3=4=5=6=7=8=9=10=11=12>1
C	Backwash and brine is more toxic in combination than either alone and other components do not affect toxicity	2=3=4=5=6=7=8=9=10=11>1, 12
D	Certain supplier's effluents are more toxic than others, regardless of CIP chemicals used	2=3=4And 5=6And 7=8=9And 10=11And 2>1, 12 Or 5>1, 12 Or 7>1, 12 Or 10>1, 12And not 2=3=4=5=6=7=8=9=10=11
E	CIP chemicals are the most toxic component, or make whole effluents more toxic, regardless of which CIP variant is used or supplier	3=4=6=8=9=11>1, 2, 5, 7, 10, 12
F	Some specific CIP variants are the most toxic or make a whole effluent more toxic, regardless of which supplier's treatment they are used with	3>1, 2, 4, 5, 6, 7, 8, 9, 10, 11, 12Or 4=9>1, 2, 3, 5, 6, 7, 8, 10, 11, 12Or 6>1, 2, 3, 4, 5, 7, 8, 9, 10, 11, 12Or 8>1, 2, 3, 4, 5, 6, 7, 9, 10, 11, 12Or 11>1, 2, 3, 4, 5, 6, 7, 8, 9, 10, 12
G	CIP chemicals are the most toxic component or make whole effluents more toxic, regardless of which CIP variant is used, but depending on which supplier's treatment they are used with	3=4>1, 2, 12And 6>1, 5, 12And 8=9>1, 7, 12And 11>1, 10, 12And not 3=4=5=8=9=11

H	Some CIP variants make a whole effluent more toxic, depending on which supplier's treatment they are combined with	3>1, 2, 4, 5, 6, 7, 8, 9, 10, 11, 12Or 4>1, 2, 3, 5, 6, 7, 8, 9, 10, 11, 12Or 6>1, 2, 3, 4, 5, 7, 8, 9, 10, 11, 12Or 8>1, 2, 3, 4, 5, 6, 7, 9, 10, 11, 12Or 10>1, 2, 3, 4, 5, 6, 7, 8, 9, 11, 12Or 11>1, 2, 3, 4, 5, 6, 7, 8, 9, 10, 12

a No alternative models were assessed for Cape Riche given that only one simulated whole effluent was tested so this case study is excluded from the table.

b The effluents in the 'Hypothesis' are defined in Table 1 . In these hypotheses '1 > 2, 3' indicates that the toxicity of the whole effluent 1 will be greater than for whole effluent 2 and whole effluent 3 while '1 = 2' suggests that there will be no detectable difference in toxicity between whole effluents 1 and 2. Note that in bioassays the parameter chosen to measure 'toxicity' (e.g., LC10 or EC10) is measured as less when toxicity is greater.

The four simulated effluents were produced using seawater obtained from near the location of the intake/outfall streams in Bass Straight (Wonthaggi) which was sent to a university based research centre in a different state (~ 1000 km away in Sydney, New South Wales) that specialises in reverse osmosis membrane desalination techniques. The water was pre-treated according to the requirements for each simulated effluent before being passed through a laboratory-scale reverse osmosis unit and processed until the required salinity was achieved. The processed brine was then used in testing conducted at a specialist ecotoxicology laboratory (also in Sydney, New South Wales). The tests were undertaken on six species found in the waters of Victoria or New South Wales with those from New South Wales closely related to species found in Victoria [27].

Similarly to the work done for Cape Riche, WET testing for the Victorian Desalination Plant was used to identify the safe dilution factors necessary for 99% (and other) level(s) of species protection. Here, this was determined for each of the four effluents separately, with the safe dilution factors ranging from 14 to 31 (Table 3). The ranking of the different effluents (as defined in Table 1) from most to least toxic based on safe dilution factors was 1, 3 and then equally 2 and 4. Interestingly, this result suggests that effluent 1 (which represents normal operating conditions) may be more toxic than those which represent the worst case scenarios or streams with additional chemicals Zeffluents 2, 3 and 4). Perhaps more importantly, the comparison of effluents did not suggest that the presence of biocide and/or surfactant in waste streams exacerbates the toxicity of the effluent, at least at the concentrations predicted. Although this approach required a lot (four times) more ecotoxicology work than an approach considering a single running condition, the obvious advantage is that it accounts for a wider range of operational conditions. That is, it was possible to consider effluents with different chemical constituents as could be produced as a result of varying running conditions and production schedules used over time, a reality in any operational desalination plant. Of course, as in the Cape Riche case study, the relevance of the results of WET testing is still related to how well the predicted (and tested) simulated effluents match the range of eventual effluents produced by the constructed plant.

Gorgon desalination plant, Barrow Island, Western Australia (Gorgon) [8]

The Gorgon desalination plant was constructed on Barrow Island, around 60 km off the northwest coast of Western Australia [28]. This desalination plant was required to provide water for the construction and maintenance of the Gorgon Liquid Natural Gas Development, a facility for processing and exporting liquefied natural gas (LNG) to international and domestic markets [28]. Given its offshore location, the continuous provision of locally produced freshwater was critical for the plant and exporting facility, during both the construction phase and ongoing operations. The remote nature of Barrow Island also meant that for logistic reasons all desalination waste would need to be released into the marine environment as effluent and as such needed to be assessed by local environmental regulatory authorities [8].

As with the other case studies, consultation with the relevant environmental regulatory authorities regarding marine waste discharges led to a recommendation for WET testing on simulated effluents to assess the potential toxicity of wastewater. Additionally, as for the Victorian plant, regulators were concerned with the potential for synergistic effects of combined waste streams incorporating effluent from brine production as well as periodic inputs from processes such as membrane cleaning and backwashing [8]. The Gorgon testing was, however, further complicated as at the time approvals were being sought a final decision had not yet been made about which of several alternative reverse osmosis systems might be employed. Each potential system used slightly different processing schedules and at least one unique chemical (i.e., proprietary to the supplier), complicating decisions regarding which simulated effluents should be considered in testing. In contrast to previous case studies outlined above, the effluents simulated and used in Gorgon WET testing were, therefore, selected not only to cover a number of operating processes (n = 6), but also a range of potential suppliers with unique chemicals/application rates (n = 4). This complexity led to the identification of 24 unique potential whole effluents that could be released and, ideally, needed to be assessed via WET testing. The approach taken to rationalise this large number of potential effluents was to test only a subset (n = 10, plus two control treatments) and look for evidence of differences in toxicity among these using relatively complex statistical approaches [8]. That is, 12 simulated effluents were considered which covered a range of suppliers and processes (summarised in Table 1) with the relative effects then used to examine a range of different hypotheses and models regarding relative toxicities and interactive effects (described in Table 4). If any evidence of synergistic toxicity was found further WET testing would then have been done on other (initially not-tested) combinations to resolve patterns of toxicity further. Ultimately though, this was not required because there was no evidence of any waste streams actually being more toxic than just the brine component[8].

Given the added complexity in experimental design associated with considering a greater number of treatments in this monitoring programme, it was necessary to simplify the methods used to simulate effluents and conduct testing in an established ecotoxicology laboratory capable of dealing with the large volume. As in the other case studies, effluents for the potential Gorgon plant were simulated taking into account local seawater composition at Barrow Island and calculations about likely dosing rates and frequencies of chemicals used during different processes; for Gorgon, however, this was repeated for each of the four suppliers. The simulated effluents were created using sea water collected from near Barrow Island which was shipped to the commercial ecotoxicology laboratory (based in Sydney, New South Wales). In the laboratory, brine was simulated by concentrating the sea water via evaporation rather than reverse osmosis, with the appropriate chemicals added to each effluent at modelled concentrations. Although the number of marine species (six) tested was similar to that used in other case studies, the species selected for the Gorgon study were those that could be tested reliably in the ecotoxicology laboratory, rather than organisms sourced specifically from Barrow Island. While the species tested were all relevant to Barrow Island (Barrow Island was within the distributional range of most of the species), two were close relatives of species found at Barrow Island rather than actually being found there (Table 2). Moreover, all of the test organisms were sourced from the wild or aquaculture near the (temperate) testing location in Sydney (1000 s of kilometres away from Barrow Island) and simulated effluent was diluted using water collected from the Sydney region.

Testing of the 12 simulated effluents failed to detect evidence that any were more toxic than others, and none of the eight proposed hypotheses regarding differences in toxicity of the effluents were supported. Rather, this testing revealed that the toxicity of simulated whole effluents composed of just brine or sodium hypochlorite in seawater was not detectable as different from those which contained additional chemicals that would be used during the pre-treatment, post-treatment or cleaning in place processes; i.e., there was no evidence of synergistic toxic effects. Given the similarity among effluents, a single PC99 and safe dilution factor were calculated, which indicates that a dilution factor of 9.7 would be appropriate (Table 3). In addition, the safe dilution factor was also calculated using an alternative and more conservative approach based on the minimum concentration of whole effluent where no biological effect was detected (NOEC) and the application of a safety factor of 10. As anticipated, this method indicated that a more conservative dilution factor of 40 should be sufficient to protect 99% of species.

DISCUSSION

Desalination of seawater can alleviate water scarcity in coastal areas and provide benefits to human populations, yet this process may also negatively affect biota of receiving waters as it creates wastewater containing high levels of brine and chemical contaminants (reviewed in [3] and [6]). Current techniques used to forecast the potential effects of wastewater discharged from proposed desalination plants are imperfect [29], but the use of simulated whole effluents in WET testing may allow improved predictions of toxicity. Here, we have reviewed three case studies (Cape Riche, Victoria, Gorgon) in which simulated whole effluents were used to estimate safe dilution levels of outflows before the plants were constructed (summarised in Table 3). In two of these case studies (i.e., Victoria, Gorgon), the testing of multiple simulated whole effluents also provided a way to assess the potential (relative) toxicity of different contaminants and waste streams and, in one case (i.e., Gorgon), this information was obtained even before the supplier of the desalination plant was selected [8] and [20]. Interestingly, where the relative toxicity of different effluent components was considered, salinity was identified as having the strongest influence, with a lack of interaction among contaminants (synergistic or otherwise) [8] and [20].

Central to the use of simulated whole effluents in testing for toxicity is an assumption that the effluent tested is similar to that which will be produced by an operational plant. As in the case studies considered here, efforts are made to maximise the similarities between the effluent simulated and actually produced. Given that the plants are not, however, producing or releasing effluents at the time of testing, approximations need to be made. Consequently, differences may result where desalination plant management is modified relative to that proposed before construction and/or operation (i.e., altered addition of cleaners, anti-scalants or other products) [30]. Additionally, although the effluents simulated, and tested, are often developed to mimic the long-term average outflow (i.e., the Cape Riche case study), it may be the pulsed addition of particular effluent combinations produced during shorter-term worst case scenario events that affect the biota more strongly. This potential was recognised in one of the case studies considered here, the Victorian Desalination Plant, where simulated effluents tested encompassed a range of operating conditions, specifically the long-term average outflow, worst case scenario within normal operating conditions, absolute worst case, and long-term average outflow composition together with cleaning in place (CIP) chemicals [20].

As the effluents used in testing are simulated, potential exists that methods used in their production may not necessarily produce effluent of the same condition as an actual, operational plant. Again, although efforts were made to mimic production methods, often these could not be identical to those used in a full-scale desalination plant. For example, to simulate saline brine in these case studies a variety of methods were used (i.e., passed through a laboratory-scale reverse osmosis membrane in Cape Riche and Victoria and evaporated in Gorgon; [20], [8] and [5] respectively), which may not modify the water column in the same way as reverse osmosis in an operational plant (i.e., RO can affect concentrations of some trace elements like Boron; [8]). Further, as testing was typically conducted at a relatively large distance from the location where the desalination plant would be located, further differences may result if the substances, and their concentrations within the wastewater, differ from that anticipated due to the natural composition of the treated water (i.e., phosphates, nitrates and ammonium) [30]. That is, while protocols typically specify that local receiving water is used as dilution water [24], this was not achieved in the Gorgon case study [8]. Whether this would make a difference to the toxicity of whole effluents is unclear, but at least some differences are known in brine that gets produced between the desalination processes and at different locations. These issues can, however, only be addressed when a real plant is running and water quality can be directly measured.

While there are some drawbacks associated with the use of simulated whole effluents in toxicity assessments, a key advantage is that they facilitate the consideration of various waste streams, potentially before the plant is constructed, enabling a better-informed decision making process. The complexity of alternative hypotheses and models that can be assessed using this method is only limited by the number and diversity of waste effluent streams that are simulated. Here, the case studies considered incorporated a range of simulated whole effluents (i.e., n = 1, 4 or 12) which enabled different complexities of models to be assessed, from the very simple (i.e., in Cape Riche the toxicity of a single effluent) to the complex (i.e., in Victoria and Gorgon the toxicity of specific combinations of chemicals likely in different processes and running conditions) (summarised in Table 4). Where the range of simulated whole effluents used to test hypotheses is narrow, this can result in greater uncertainty regarding which component of the effluent is driving the observed toxicity, leading to the establishment of conservative management targets. That is, in the case of Cape Riche the dilution factor was set to be that which reduced the level of all contaminants (including salinity) to levels below the recommended marine water quality guideline trigger values [5]. If testing had identified the influential component then management might have been subsequently developed to target that specific component, potentially resulting in more moderate targets.

While rigorous testing using multiple simulated effluents will likely be of benefit when it comes to establishing management targets, there are practical issues which may limit its implementation. That is, as the number of processes and potential suppliers considered in tested hypotheses increase, the possible combinations of chemicals expand drastically. The resources, particularly time, required to run assays using standard WET approaches mean that in many instances all of the identified combinations cannot be explicitly tested. This limitation was influential in the Gorgon case study in which, given the range of processes and potential suppliers considered, twenty-four whole effluents were identified that could have been tested [8]. In recognition of the difficulties associated with considering this number of effluents, the testing programme was designed to initially consider a representative subset of effluents (ten of the potential effluents plus two control treatments). If a result was found that supported multiple models or indicated the existence of interactive effects, additional assessments would have then been subsequently conducted [8]. Although this approach enabled a broad range of simulated effluents to be considered, potential remains that a particular effluent may, for some unanticipated reason, be more toxic than those that were tested, highlighting the need for development of a method that facilitates the consideration of more effluents.

A key aspect of effective WET testing is that it considers organisms living in the receiving environment likely to be impacted. For example, the use of appropriate (local) species is of particular interest and importance in Australian studies given its geologically long separation from most Northern Hemisphere continents [23]. Thus, it is interesting to note that the case studies considered here undertook testing on a similar suite of taxa (in some cases the same species), despite the fact that they were considering three geographically distant areas, on opposite sides of the continent, in both tropical and temperate zones (summarised in Table 2). Although recognised as important, ensuring testing uses organisms found in the receiving environment or are closely related to organisms that are is currently difficult given that most assays are highly technical, and so are often conducted offsite (described in [24]). The most striking example of this in our case studies was from Gorgon; while the anticipated impact site was located in north-west Western Australia, testing was conducted in Sydney, approximately 3750 km away. While it was recognised that sourcing test organisms from local populations near Barrow Island would have been the best option, this was not possible due to potential quality control issues associated with transportation of marine organisms from such a remote location to an ecotoxicological laboratory in a capital city. The organisms chosen, therefore, were those which could be sourced near the testing location in Sydney, where the ecotoxicology laboratory contracted to do the work was based [8]. Further, given the complexity of tests conducted, the organisms ultimately selected were those that the ecotoxicology company had a large amount of experience working with, both in terms of husbandry and quality assurance tests [8]. This is particularly important given that key biota in particular areas, potentially including those for which ecotoxicological tests have not been developed, could be more sensitive to contaminants than those species typically considered [24] and [29]. Consequently, estimates may be improved if methods were used which enabled consideration of a greater diversity of species, including those of the environment likely to be impacted, as may occur if the method could be implemented outside of traditional WET testing facilities.

CONCLUSIONS

As desalination is increasingly used to alleviate water scarcity, potential exists that this process may impact marine environments. Consequently, new approaches for identifying potential effects and enabling their mitigation are likely to be increasingly desirable. The use of simulated whole effluents allows a variety of hypotheses and models specific to the contaminants and area of interest to be directly assessed, an outcome not possible if using the more traditional methods of forecasting likely impacts of desalination effluent disposal (e.g., considering the impacts of currently active plants located elsewhere or identifying the effects of multiple effluent contaminants separately). While representing an improvement in impact assessments of plants before their construction, our review highlights that these tests remain imperfect with issues largely linked to the WET methods used, specifically the time these tests take to conduct and limited species which can be considered. If the methods applied in such testing continue to improve we anticipate that over time the use of simulated effluents in WET testing will increase. This method could facilitate site-specific investigations and the derivation of relevant local management targets across a wide range of geographic areas for different desalination plants and processes, potentially well before plants are constructed. Given these potential benefits, testing using simulated whole effluents could become an integral component of regulatory approval mechanisms.

ACKNOWLEDGEMENTS

This paper was possible because the companies responsible for the monitoring programmes outlined have made detailed information freely available on the internet, which we gratefully acknowledge. We also acknowledge the comments from an anonymous reviewer. Funds provided to UCL by BHP Billiton Sustainable Communitiessupported LJF.

REFERENCES

1. D.A. Roberts, E.L. Johnston, N.A. Knott, Impacts of desalination plant discharges on the marine environment: a critical review of published studies, Water Res. 44 (2010) 5117–5128.
2. R. Barker, B. van Koppen, T. Shah, A Global Perspective on Water Scarcity and Poverty: Achievements and Challenges for Water Resources Management, International Water Management Institute (IWMI), Colombo, Sri Lanka, 2000.
3. S. Lattemann, T. Höpner, Environmental impact and impact assessment of seawater desalination, Desalination 220 (2008) 1–15.
4. G.A. Tularam, M. Ilahee, Environmental concerns of desalinating seawater using reverse osmosis, J. Environ. Monit. 9 (2007) 805–813.
5. Intertek-Geotechnical, Toxicity assessment for a simulated desalination brine effluent, Report Number ECX11-0704, Prepared for GHD Pty Ltd, June 2011, IntertekGeotechnical Services Pty Ltd., Welshpool, Western Australia, 2011.
6. J.J. Sadhwani, J.M. Veza, C. Santana, Case studies on environmental impact of seawater desalination, Desalination 185 (2005) 1–8.
7. R. Miri, A. Chouikhi, Ecotoxicological marine impacts from seawater desalination plants, Desalination 182 (2005) 403–410.
8. RPS, Whole effluent toxicity testing of simulated reverse osmosis brine effluents, Gorgon Development Barrow Island, Volume 1 of 2, Report prepared for Chevron Australia Pty Ltd, December 2008, Subiaco, Western Australia, 2008.
9. E.S. Darling, I.M. Côté, Quantifying the evidence for ecological synergies, Ecol. Lett. 11 (2008) 1278–1286.
10. I. Safrai, A. Zask, Reverse osmosis desalination plants—marine environmentalist regulator point of view, Desalination 220 (2008) 72–84.
11. A. Hashim, M. Hajjaj, Impact of desalination plants fluid effluents on the integrity of seawater, with the Arabian Gulf in perspective, Desalination 182 (2005) 373–393.
12. NCEDA, Australian Desalination Research Roadmap, Murdoch University, Rockingham, Western Australia, 2010. 98.
13. N. Cannesson, P.M. Johnstone, M.A. Mitchell, S.F. Boerlage, Minimizing community, environmental, and marine impacts at the Gold Coast Desalination Plant, IDA Journal of Desalination and Water Reuse, 1, 2009, pp. 74–87.
14. Arup, Olympic Dam Expansion, Supplementary Environmental Impact Statement, Appendix H3 Comparisons with Other Desalination Plants Report Prepared for BHP Billiton Olympic Dam Corporation Pty Ltd, Australia, 2011.
15. S. Trousdale, E. Henderson, Innovation addresses environmental issues at Sydney desalination plant, IDA Journal of Desalination and Water Reuse, 1, 2009, pp. 52–63.
16. E.J. Mitchell, J.E. Burgess, R.M. Stuetz, Developments in ecotoxicity testing, Rev. Environ. Sci. Biotechnol. 1 (2002) 169–198.
17. EPA, Ammonia and urea plants, Burrup Peninsula; change to split statement 614 into two separate statements, Section 46 Report and Recommendations of the Environmental Protection Agency, October 2005, Perth, Western Australia, 2005.
18. Arup/ENSR, Olympic Dam Expansion, Draft Environmental Impact Statement 2009, Main Report Volume 1, Report Prepared for BHP Billiton Olympic Dam Corporation Pty Ltd, December 2008, Australia, 2009.
19. GHD, Iron Ore Holdings Limited, Cape Preston East Environmental Studies, Brine Dispersion Study, January 2013, Perth, Western Australia, 2013.
20. TDJV, Victorian Desalination Project, Ecotoxicology Assessment, RP-TDV-EN-0-X- 000-0004-F-02, Melbourne, Victoria, 2010.
21. EPA, Cape Riche Desalination Plant Given Conditional Approval: EPA Report 1431, Release Date 13 February 2012, Government of Western Australia, Western Australia, 2012.
22. ANZECC, ARMCANZ, Australian and New Zealand Guidelines for Fresh Water and Marine Water Quality, Australian and New Zealand Environment and Conservation Council/Agriculture and Resource Management Council of Australia and New Zealand, Canberra, 2000.
23. M.S.J. Warne, G. Batley, O. Braga, J. Chapman, D. Fox, C. Hickey, J. Stauber, R. Van Dam, Revisions to the derivation of the Australian and New Zealand guidelines for toxicants in fresh and marine waters, Environ. Sci. Pollut. Res. 21 (2014) 51–60.
24. R. Van Dam, J. Chapman, Direct toxicity assessment (DTA) for water quality guidelines in Australia and New Zealand, Australas. J. Ecotoxicol. 7 (2002) 175–198.
25. M.S.J. Warne, Derivation of the Australian and New Zealand water quality guidelines for toxicants, Australas. J. Ecotoxicol. 7 (2002) 123–136.
26. Melbourne Water, Desalination, Melbourne Water, Melbourne, Victoria, 2014.
27. M.S.J. Warne, Ecotoxicity of Effluents From the Proposed Victorian Desalination Plant: Final Report, CSIRO: Water for a Healthy Country National Research Flagship, Glen Osmond, South Australia, 2010.
28. Chevron, Gorgon Project Overview, Gorgon Project

Factsheet January 2014, Chevron Australia, Perth, Western Australia, 2014.

29. P.M. Chapman, Whole effluent toxicity testing—usefulness, level of protection, and risk assessment, Environ. Toxicol. Chem. 19 (2000) 3–13.

30. J.L. Sánchez-Lizaso, J. Romero, J. Ruiz, E. Gacia, J.L. Buceta, O. Invers, Y. Fernández Torquemada, J. Mas, A. Ruiz-Mateo, M. Manzanera, Salinity tolerance of the Mediterranean seagrass Posidonia oceanica: recommendations to minimize the impact of brine discharges from desalination plants, Desalination 221 (2008) 602–607.

CHAPTER **10**

ULTRAVIOLET RADIATION AS A BALLAST WATER TREATMENT STRATEGY

Ranveig Ottoey Olsen, Friederike Hoffmann, Ole-Kristian Hess-Erga, Aud Larsen, Gunnar Thuestad, Ingunn Alne Hoell

Keywords: Tetraselmis suecica, Ultraviolet irradiation, Esterase substrate, Flow cytometry, Inactivation, Dark incubation

INTRODUCTION

Ships use water as ballast to ensure stability and trim during the voyage, and ambient water is pumped into ballast tanks in the hull of the ships. It is traditionally discharged without any treatment and represents a global vector for aquatic invasion. A multitude of organisms like virus, bacteria, algae and zooplankton are carried around the world in ship's ballast tanks (David et al., 2007, Drake et al., 2007 and Hallegraeff and Bolch, 1991). Some organisms survive in ballast tanks and are released into new environments. If nonindigenous species adapt and establish in a new environment, they might have an impact on the native species and cause ecological change in the ocean (Gollasch et al., 2015 and Ruiz et al., 1997). It is of importance to minimize and prevent dispersal of species by ballast water discharge to hinder potential harm to ecosystems, the economy, or human health (Ruiz et al., 2000).

In 2004 the International Maritime Organization (IMO) established standards for ballast water treatment through the International Convention for the Control and Management of Ship's Ballast Water and Sediments (International Maritime Organization, 2004). Regulation D-2 of the Convention sets the standard regarding category and concentration of organisms at discharge. The Convention will enter into force 12 months after being ratified by 30 States representing 35% of the merchant shipping tonnage. In August 2015 44 States, representing 32.86% of the world tonnage, have ratified the Convention. The upcoming IMO regulations have led to development of various ballast water treatment systems (BWTS) that facilitate disinfection of ballast water (David and Gollasch, 2015, Delacroix et al., 2013, Lloyd's Register Marine, 2015a, Lloyd's Register Marine, 2015b, Stehouwer et al., 2015, Werschkun et al., 2012 and Werschkun et al., 2014). All BWTS have to be approved by national authorities according to IMO regulations and/or the regulations of other national bodies (e.g. U.S. Coast Guard (USCG)).

When selecting and installing a BWTS, the shipping companies have to consider different technical and operational aspects (Lloyd's Register Marine, 2015a and Lloyd's Register Marine, 2015b). The BWTS use a range of different treatment technologies, from processing the water with solid–liquid separation to chemical- (active substances) and/or physical disinfection (e.g. UV). The main operational cost for UV based BWTS is related to power consumption (Werschkun et al., 2014). Ship owners can reduce such costs by lowering the UV intensity, providing that the ship's discharged ballast water still complies to Regulation D-2 (International Maritime Organization, 2008a). It is therefore of interest to determine the lowest lethal UV dose and to estimate the time required for inactivation when stored in ballast tanks after irradiation.

UV irradiation is performed either by low pressure (LP) or medium pressure (MP) UV lamps (Oguma et al., 2002, Werschkun et al., 2012 and Zimmer and Slawson, 2002). LP lamps emit UV-C radiation, primarily at 254 nm, which is most efficiently absorbed by nucleic acids and causes DNA damages (Sinha and Häder, 2002). UV induced DNA damages can be reversed by DNA repair mechanisms, referred to as photoreactivation and dark repair (Sancar and Sancar, 1988

Citation: Ranveig Ottoey Olsen, Friederike Hoffmann, Ole-Kristian Hess-Erga, Aud Larsen, Gunnar Thuestad, Ingunn Alne Hoell, Ultraviolet radiation as a ballast water treatment strategy, Volume 103, Issues 1–2, 15 February 2016, Pages 270-275, ISSN 0025-326X, http://dx.doi.org/10.1016/j.marpolbul.2015.12.008.

and Sinha and Häder, 2002). MP UV lamps emit radiation spanning the UV-A, -B and -C bands causing additional damage to proteins and enzymes. For instance, UV-B radiation can affect key components in photosynthesis (Fiscus and Booker, 1995, Holzinger and Lütz, 2006 and Kottuparambil et al., 2012), causing energy deprivation in phytoplankton cells. Thus, it has been argued that MP UV lamps can cause a higher degree of inactivation compared to LP UV lamps (Kalisvaart, 2001 and Oguma et al., 2002).

UV irradiation can leave cells in different conditions (live, dead or damaged), whereof the viability of damaged cells at discharge is uncertain (Olsen et al., 2015). Damaged cells can be unculturable, though they can be metabolic active and may pose a health risk (Oliver, 2010). Further, cellular DNA repair mechanisms can restore the genetic information (Sancar and Sancar, 1988, Sinha and Häder, 2002 and Zimmer and Slawson, 2002) causing the cell to grow and replicate after discharge (Liebich et al., 2012, Martínez et al., 2012 and Martínez et al., 2013). Additionally, the terminology describing the organisms at discharge can be confusing or unclear. The IMO Convention refers to "viable" organisms (International Maritime Organization, 2004), and the Guidelines for approval of ballast water management systems (G8) define "viable organisms" as "organisms and any life stages thereof that are living" (International Maritime Organization, 2008a). USCG also uses the term "living" (United States Coast Guard, 2012).

Determining the condition of UV irradiated cells is a complex task. On the other hand, cheap, fast and reliable methods to analyze ballast water are necessary for approval of BWTS technologies and for compliance testing of ballast water discharge (International Maritime Organization, 2013). Testing for compliance can be performed in two steps; an indicative and a detailed analyses. An indicative analysis is a relatively simple and quick measurement that gives a rough estimate of the number of viable organisms in the ballast water at discharge. Examples of indicative analysis methods are e.g. BallastCAM and various fluorescence or ATP detections (Drake et al., 2014, First and Drake, 2013,First and Drake, 2014, Gollasch and David, 2012, Gollasch and David, 2015 and van Slooten et al., 2015). If an indicative analysis shows compliance to Regulation D-2, there is no need for a detailed analysis. Should the indicative analyses be non-compliant, however, a detailed analysis must be undertaken to give robust and direct measurements determining the concentration of viable organism in ballast water discharge according to Regulation D-2. Quantification of live bacteria traditionally relies on cultivation methods, which is time-consuming and may give false negatives as several species are uncultivable although viable (Roszak and Colwell, 1987 and Staley and Konopka, 1985). Flow cytometry (FCM) has been suggested as a promising method for detailed analysis (International Maritime Organization, 2013 and Peperzak and Gollasch, 2013). FCM facilitates rapid detection, enumeration and characterization of organisms in combination with fluorescent dyes, and enables to study populations and communities indirectly (Peperzak and Brussaard, 2011 and Shapiro, 2000).

Previously a FCM protocol was developed to distinguish between live and dead*Tetraselmis suecica* cells (Olsen et al., 2015). For UV irradiated samples the FCM protocol could not distinguish between live and damaged cells, as the latter contain both dying and repairable cells. The current study uses the FCM protocol to elaborate on different UV doses and the effect of dark incubation on inactivation of the algae *T. suecica*, to simulate a ballast water treatment and subsequent transport. Our specific objectives were to:

1) Determine the minimum UV dose that permanently inactivates the algae.
2) Quantify effects of different UV doses on *T. suecica*.
3) Estimate the time of dark incubation required to permanently inactivate the algae treated with UV doses lower than minimum permanently inactivation dose.
4) Provide recommendations for ballast water management.

MATERIAL AND METHODS

The phytoplankter specie *T. suecica* (Strain K-0297, Scandinavian Culture Collection of Alga and Protozoa, University of Copenhagen, Denmark) was selected as a test organism. It was cultured in 24 PPT artificial sea water (Marine SeaSalt, Tetra, Melle, Germany) added 0.12% Substral (The Scotts Company (Nordic) A/S, Naverland, Glostrup, Denmark), at 15 °C, 100 rpm, 14:10 light:dark cycle and 90 lx light intensity (Flora-Glo, T8, 20 W). The culture was diluted in growth medium to a density of 10^4 live cells ml^{-1} prior to irradiation, monitored by FCM.

Irradiation was performed using a collimated beam MP UV lamp (800 W) (BestUV, Hazerswoude, The Netherlands) (Olsen et al., 2015). For each experiment three samples of 15 ml diluted *T. suecica* culture were irradiated with the same UV dose in a petri dish (inner diameter 6 cm, culture depth 7 mm) while mixed with a 1 × 0.4 cm magnetic stir bar at 200 rpm in room temperature (RT). The intensity (mW/cm^2) of the UV lamp was fixed and the exposure times used were 155, 233, 311, 622 and 1244 s for UV doses 100, 150, 200, 400 and 800 mJ/cm^2, respectively. The irradiated samples were transferred to sterile 50 ml polypropylene tubes (Fisher Scientific), so was the control samples, including 2 × 15 ml non-irradiated cells and 10 ml dead cells. The dead cells were killed by fixation with formaldehyde at 5% final concentration (36.5–38% formaldehyde, Sigma-Aldrich). All tubes were wrapped in aluminum foil and incubated in the dark with loosened lids at 15 °C.

First, a pre-study over 5 days was performed to observe the inactivation effect of different UV doses and dark incubation, and to test whether this effect was interpretable with FCM. This was followed by two complete experiments, denoted as exp-I and exp-II, and an overview of the set-up for these experiments is given in Fig. 1.

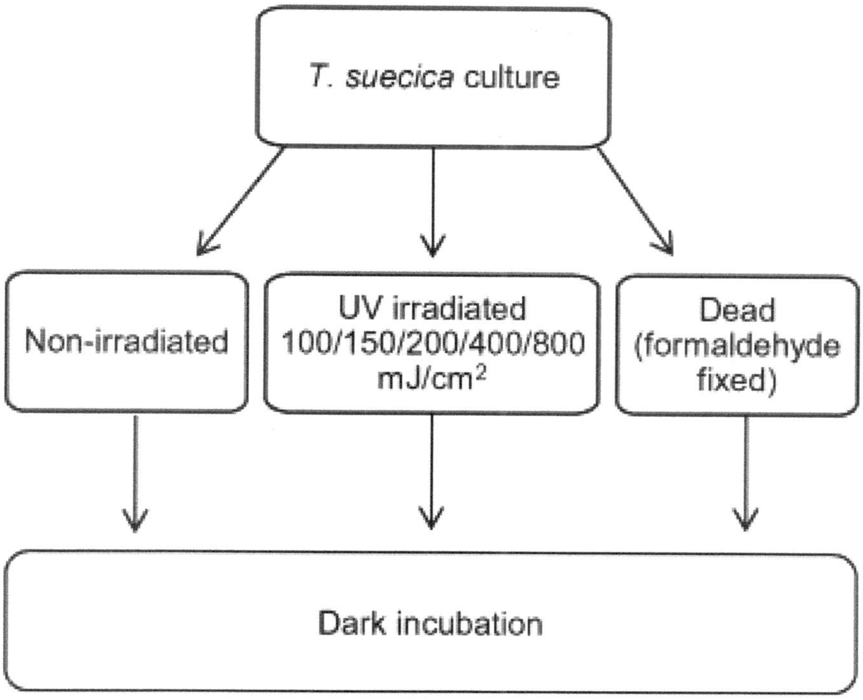

Figure 1: Experimental set-up showed by a flow diagram. This set-up was followed in the pre-study, exp-I and exp-II.

For FCM analysis, the samples were stained with 5-carboxyfluorescein diacetate acetoxymethyl ester (CFDA-AM) and analyzed with an Attune Acoustic Focusing Cytometer (Olsen et al., 2015). The samples in the pre-study were analyzed at days 1, 3 and 5 after treatment. In exp-I samples were analyzed at days 1, 3, 6, 9, 13 and 22, and in exp-II samples were analyzed at days 1, 3, 6, 10, 15 and 22 after treatment (Fig. 1). The samples in exp-I and -II were analyzed at different intervals due to logistics. A previously defined gate (i.e. a collection of single cell FCM-signals) in the FCM dot plots was used for analysis. The gate was defined based on plate count results from non-irradiated cells, and regression analysis was used for validation of the gate (Olsen et al., 2015).

To determine the number of culturable cells, plate count analysis was performed 1 day after UV treatment (Olsen et al., 2015). Also at day 1, a most probable number (MPN) analysis was performed. The samples were diluted in growth medium (24 PPT artificial sea water added 0.12% Substral) in 10-fold series up to 10^{-4} dilution to a total volume of 1 ml pr. well in 48 well plates (Greiner Bio-One, Austria) and incubated at 15 °C, 14:10 light:dark cycle and 90 lx light intensity. Positive growth was determined by a change in color into green as detected by the eye, and scored against the MPN table for a three-replicate design from FDA›s Bacterial Analytical Manual (U. S. Food and Drug Administration (FDA), 2010), which gives rough results in intervals.

When the numbers of live/damaged cell signals in FCM dot plots were approximately 10% of the total number of cells, a regrowth check was performed for verification. For exp-I this procedure was carried out at day 22 for samples treated with 100–200 mJ/cm^{2}and at day 2 for samples treated with 400 and 800 mJ/cm^2, and for exp-II at day 20 for samples treated with 100–200 mJ/cm^2 and at day 3 for samples treated with 400 and 800 mJ/cm^2. 1 ml of the sample was added to 9 ml growth medium in 50 ml Erlenmeyer flasks. No visible change to green color indicated that there were no reproductive cells in the sample. The flasks, trays and plates were incubated at 15 °C in the dark for 3 weeks.

RESULTS

FCM analysis in the pre-study showed that inactivation increased with higher UV doses and during the dark incubation period (data not shown). Based on these results, two complete experiments were carried out (exp-I and -II), and these results are presented below.

UV irradiation with doses $\geq$ 400 mJ/cm^2 inactivated the algae permanently as demonstrated by all analysis methods. FCM analyses (Fig. 2b, d) of samples irradiated with 400 and 800 mJ/cm^2 displayed < 4% and $\leq$ 0.1% live/damaged cell signals after 1 day, respectively. The numbers of live and damaged cells remained at this level or were further reduced during the incubation period. Plate count and MPN did not show any growth at UV doses $\geq$ 400 mJ/cm^2 and neither did regrowth check performed at days 2 and 3 for exp-I and exp-II, respectively.

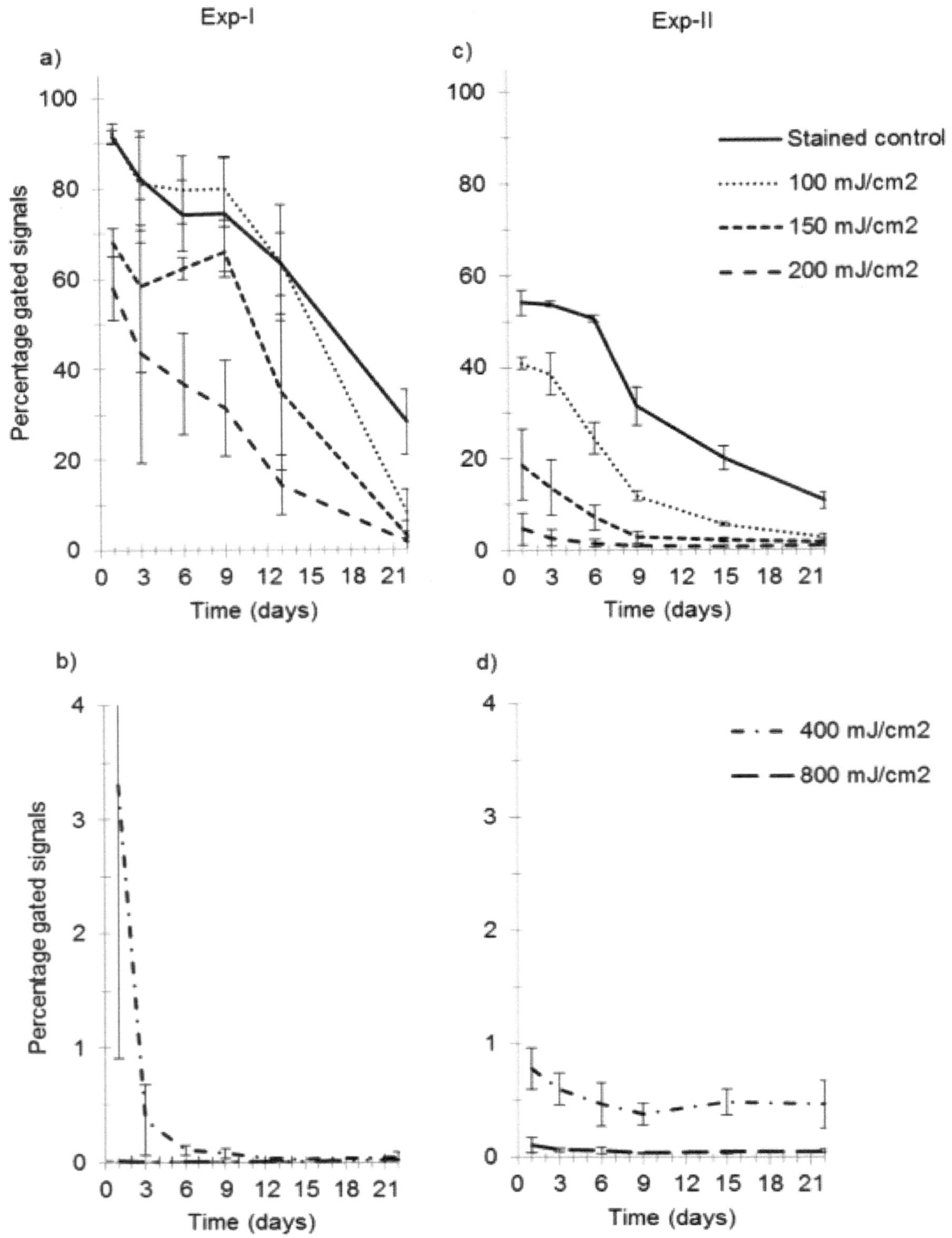

Figure 2: Line graphs showing % gated signals (= live and damaged cells) of the total number of cells (= live, damaged and dead cells) from exp-I (a, b) and exp-II (c, d). Error bars indicate 1 standard deviation of 3 replicates.

As observed in the pre-study, FCM analysis of UV irradiated samples showed a relationship between inactivation and UV doses. This was examined at day 1, when an immediate effect of UV treatment was observed, and again the number of live/damaged cells decreased with higher UV doses. Table 1 shows a comparison of the results from FCM, MPN and plate count from exp-I and -II. The control samples were all in the same range, but MPN results showed > 11,000 cells ml^{-1}, indicating that the samples should have been diluted further. The results for the UV irradiated samples, varied when using different analysis methods. In exp-I there were good agreements between live/damaged cell numbers obtained by FCM and MPN analysis, but the plate count analysis resulted in considerably lower numbers at day 1. For exp-II, comparable results were obtained using plate count and MPN analysis whereas the results using FCM gave considerable higher live/damaged cell numbers.

Table 1: FCM, plate count and MPN results from exp-I and exp-II. Results are all in cells ml^{-1}. When 0 cfu was detected by the plate count method, the values show "< 10" as the results are obtained by 100 µl being spread on the agar plates. The MPN values show "< 3" when no visible green color was observed, according to the MPN table (U. S. Food and Drug Administration (FDA), 2010). n.d. = no data. 1 standard deviation (±) for FCM and plate counts, as well as 95% confidence intervals for MPN, are in brackets.

Experiment	Analysis	Time (days)	Control	100 m J/cm²	150 m J/cm²	200 mJ/cm²	400 m J/cm²	800 m J/cm²
Exp-I	FCM	1	8231 (1222)	13,902 (572)	10,003 (318)	8531 (1437)	488 (348)	3 (1)
		3	5248 (183)	12,202 (1012)	8896 (2926)	6402 (3494)	55 (45)	2 (1)
		6	4373 (1185)	12,221 (1601)	9453 (343)	5468 (1959)	16 (6)	2 (2)
		9	8485 (2377)	12,218 (1549)	10,195 (1183)	4812 (1386)	12 (6)	1 (1)
		13	8258 (714)	9949 (2110)	5533 (2666)	2239 (958)	5 (2)	2 (2)
		22	4160 (968)	1354 (970)	512 (627)	304 (321)	8 (7)	5 (4)
	MPN	1	>11,000 (4200–40,000)	11,000(1800–41,000)	11,000(1800–41,000)	11,000(1800–41,000)	<3 (0–9.5)	<3 (0–9.5)
	Rate count	1	8667 (577)	7467 (503)	8400 (1249)	5040 (170)	<10	<10
Exp-II	FCM	1	10,419 (883)	9016 (332)	3946 (1464)	973 (682)	169 (37)	21 (12)
		3	9915 (198)	8429 (1518)	2893 (1203)	617 (382)	126 (23)	13 (2)
		6	10,688 (954)	5320 (966)	1587 (607)	344 (197)	100 (40)	11 (6)
		10	6366 (577)	2663 (363)	597 (275)	180 (45)	73 (11)	7 (1)
		15	4327 (476)	1190 (167)	427 (74)	172 (53)	105 (22)	9 (3)
		22	2441 (481)	642 (141)	406 (223)	214 (57)	111 (50)	11 (3)
	MPN	1	n.d.	4600(900–20,000)	930(180–4200)	<3 (0–9.5)	<3 (0–9.5)	<3 (0–9.5)
	Rate count	1	10,187 (2510)	4107 (508)	950 (416)	48 (50)	<10	<10

For samples treated with UV doses 100, 150 and 200 mJ/cm² inactivation of cells was dependent on time of dark incubation as demonstrated by FCM (Fig. 2a, c). Generally, the numbers of live/damaged cells decreased during incubation, and were fewer in UV irradiated samples than in the stained controls. In exp-I the percentage of live/damaged cells in the UV irradiated samples (Fig. 2a) decreased throughout the incubation period and amounted to ≤ 3% at day 22. The samples treated with 100 mJ/cm² behaved similar to the stained controls during incubation, but at day 22 the percentage of live/damaged cells was lower than the stained controls also for samples treated with this UV dose. In exp-II (Fig. 2c) inactivation increased during incubation, and < 3% live/damaged cells were observed at days 22, 10 and 3 in the samples treated with 100, 150 and 200 mJ/cm², respectively. Regrowth checks performed at days 22 and 20 for exp-I and exp-II, respectively, were negative for all UV irradiated samples and positive for the control samples (data not shown).

Considerable variations were observed between the results from the two experiments when looking at a detailed level. Firstly, the percentage of live/damaged cells in the stained controls varied at day 1 (Fig. 2a, c), being 92% and 54% in exp-I and exp-II, respectively. Secondly, for the samples UV irradiated with ≤ 200 mJ/cm², the inactivation rate varied between the experiments (Fig. 2a, c) and the number of FCM live/damaged cells fluctuated (Table 1). Thirdly, some replicates showed large standard deviations; most evident in exp-I (Fig. 2a). Fourthly, at the first days after UV irradiation, it was observed more cells in the treated samples than in the controls (Table 1).

DISCUSSION

The aim of this study was to evaluate inactivation by different UV doses and dark incubation on the algae *T. suecica*, to give recommendations for ballast water management regarding treatment and transport.

UV doses 400 and 800 mJ/cm^2 rendered *T. suecica* cells unculturable and without esterase activity 1 day after irradiation whereas doses ≤ 200 mJ/cm^2 did not necessarily inactivate the cells. This indicates that the minimum UV dose that permanently inactivates this algal specie is somewhere between 200 and 400 mJ/cm^2 which is similar to the dose Ou et al. (2012) found to be lethal after UV-C radiating the cyanobacteria*Microcystis aeruginosa* (Ou et al., 2012). The samples irradiated with UV doses 100, 150 and 200 mJ/cm^2 contained culturable and esterase active *T. suecica* cells 1 day after irradiation, but the inactivation increased with higher UV doses. Our results are in line with previous studies of freshwater green algae *Chlorella ellipsoidea*, *Chlorella vulgaris*, and *Scenedesmus quadricanda*, and the cyanobacteria *M. aeruginosa* which showed limited sensitivity to UV-C irradiation with doses ≤ 200 mJ/cm^2 (Ou et al., 2012 and Tao et al., 2010).

Comparing the different analysis methods at day 1 for the samples UV irradiated with doses ≤ 200 mJ/cm^2 revealed that the numbers of FCM gated cells were higher than the numbers of cfu detected by plate count. Such discrepancy between plate count and FCM results has also been observed in other studies of UV irradiated bacteria and alga (Kramer and Muranyi, 2014, Olsen et al., 2015 and Schenk et al., 2011). UV induced DNA damage can block transcription and replication, inhibiting growth and reproduction (Oguma et al., 2002 and Sinha and Häder, 2002). DNA damaged cells are not detected as live by growth assays, though they can express activity (Davey, 2011, Hammes et al., 2011 and Villarino et al., 2003); explaining the contradicting results from FCM and plate count analysis.

Plate count and MPN analysis are based on reproductive capacity and one could expect these growth assays to give comparable results. However, the results obtained by the MPN were similar to the plate count results in exp-II and to FCM in exp-I. This illustrates the challenge of getting reproducible results when analyzing UV irradiated organisms with methods that analyze different cellular characteristics. Further, growth assays may introduce errors as a majority of the microbes are uncultivable (Roszak and Colwell, 1987 and Staley and Konopka, 1985). In addition, the MPN positive growth was determined by the eye, though fluorometry determination would have been a more objective analysis method and could potentially have given a slightly different outcome. However, the MPN method was validated by another researcher and by microscopy.

The FCM results showed that inactivation increased with the time of dark incubation. Phytoplankton are photosynthetic organisms requiring light as an energy source (McMinn and Martin, 2013) and dark incubation will naturally affect their viability over time (Gollasch and David, 2010), especially if the cells continue with unchanged activity causing them to run out of energy (Jochem, 1999). Additionally, incubation in darkness will limit photoreactivation; the repair mechanism which removes UV induced DNA lesions and reverses damages by using the energy of light (Sinha and Häder, 2002), leaving the cells unrepaired and dying. Other studies have reported limited or no photoreactivation in UV irradiated *Escherichia coli* during incubation at darkness (Oguma et al., 2001, Yin et al., 2015 and Zimmer and Slawson, 2002). In agreement with other studies the control samples contained reproducible cells even after 3 weeks of dark incubation (Gollasch and David, 2010 and Olsen et al., 2015). *T. suecica*'s ability to survive in darkness over time implies it does not overrate effects of UV treatments and makes it an appropriate indicator organism for ballast water monitoring.

Some of the variation on results between exp-I and -II, such as variation in the stained controls day 1, difference in inactivation rate when irradiated with doses ≤ 200 mJ/cm^2, large standard deviations and more cells in treated samples than in untreated ones, indicates that the algal culture used in exp-I contained a greater portion of fresh and healthy cells than the culture used in exp-II. Natural sea water shows great variability and contains a diverse community of algae species in different cellular phases varying with time and space and may have different tolerance and response to the same UV doses applied (Rastogi et al., 2010, Sinha et al., 1998 and Xiong et al., 1997). Although not intentionally, our study indeed demonstrated that the UV treatment/dark incubation required for inactivation is dependent on the status of the ballast water inhabiting organisms. Therefore, our results imply that organisms in ballast water treated with UV doses 100–200 mJ/cm^2 are inactivated when left in dark ballast tanks over a period of time. Our results also demonstrate the challenge of giving recommendations regarding duration of dark incubation needed for inactivation when using lower UV doses. It is, however, important to keep in mind that controllable experiments in a laboratory differ from a flow-through chamber in a commercial BWTS and that a BWTS comprise two or more treatment stages, enhancing the inactivation efficiency. Other factors can, however, also influence the inactivation efficiency in a BWTS, like biotic and abiotic particles in the sea water protecting the microbes during UV irradiation (Hess-Erga et al., 2008 and Tang et al., 2011) and a BWTS has to be optimal under the prevailing circumstances whatever factors exist that may prevent inactivation.

UV irradiation as a treatment technology has been criticized due to the uncertainty regarding inactivation at discharge but is more environmental friendly than chemical disinfection and creates no harmful by-products (Jung et al., 2012). In addition, UV irradiation represents little risk for the operators and less training is required to run the systems (International Maritime Organization, 2008b).

This study was performed in a laboratory with cultures of *T. suecica* and with the MP UV lamp as the sole treatment source. For future studies it would be of interest to do experiments with *T. suecica* in real BWTS with MP UV technology. Such studies would improve our ability to give recommendations regarding UV doses and duration of dark incubation of *T. suecica*. To evaluate whether our FCM protocol is applicable to other microbes, laboratory studies should be performed with natural sea water irradiated with different UV doses. If analysis of natural sea water is feasible with our FCM protocol, the experiment needs to be repeated in a BWTS to evaluate inactivation in a system where it is intended to be used. The microbial community in ballast water is very diverse and organisms will vary with season, location and environmental conditions. The level of metabolic activity can vary between various algae species, and the response to environmental changes and inactivation treatments can differ (Jochem, 1999, Jochem, 2000 and Olsen et al., 2015). Although there are indications that the majority of phytoplankton species can be detected by the esterase substrates fluorescein diacetate (FDA) and 5-chloromethylfluoorescein diacetate (CMFDA) (Peperzak and Brussaard, 2011), we are therefore aware that fluorescing signal from esterase substrates can vary over a large range of intensities (Dorsey et al., 1989). Our recommendation at this point is thus to treat the ballast water with UV doses close to 400 mJ/cm^2 in order to permanently inactivate the organisms. The variable results for UV doses 100–200 mJ/cm^{2}demonstrate the challenge of giving unambiguous recommendations on duration of dark incubation needed for inactivation.

ACKNOWLEDGMENTS

This research was founded by the Norwegian Research Council (project BallastFlow, project no. 208653) and Knutsen OAS Shipping AS, and supported by Solstad Shipping, Stord/Haugesund University College, VRI Rogaland, UH-nett Vest and TeknoVest. We thank Sandra Schöttner (UiB, Bergen, Norway), Stephanie Delacroix, August Tobiesen (Norwegian Institute for Water Research, Oslo, Norway) and Per Lothe (Knutsen OAS Shipping AS, Haugesund, Norway) for helpful discussions. Special thanks to Sandra Schöttner for assistance with the UV lamp and experimental set-up.

REFERENCES

1. Davey, H.M., 2011. Life, death, and in-between: meanings and methods in microbiology. Appl. Environ. Microbiol. 77 (16), 5571–5576. http://dx.doi.org/10.1128/aem. 00744-11.
2. David, M., Gollasch, S., 2015. Ballast water management systems for vessels. In: David, M., Gollasch, S. (Eds.), Global Maritime Transport and Ballast Water Management vol. 8. Springer, Netherlands, pp. 109–132.
3. David, M., Gollasch, S., Cabrini, M., Perkovic, M., Bosnjak, D., Virgilio, D., 2007. Results from the first ballast water sampling study in the Mediterranean Sea — the Port of Koper study. Mar. Pollut. Bull. 54 (1), 53–65. http://dx.doi.org/10.1016/j.marpolbul.2006. 08.041.
4. Delacroix, S., Vogelsang, C., Tobiesen, A., Liltved, H., 2013. Disinfection by-products and ecotoxicity of ballast water after oxidative treatment — results and experiences from seven years of full-scale testing of ballast water management systems. Mar. Pollut. Bull. 73 (1), 24–36. http://dx.doi.org/10.1016/j.marpolbul.2013.06.014.
5. Dorsey, J., Yentsch, C.M., Mayo, S., McKenna, C., 1989. Rapid analytical technique for the assessment of cell metabolic activity in marine microalgae. Cytometry 10 (5), 622–628. http://dx.doi.org/10.1002/cyto.990100518.
6. Drake, L.A., Doblin, M.A., Dobbs, F.C., 2007. Potential microbial bioinvasions via ships' ballast water, sediment, and biofilm. Mar. Pollut. Bull. 55 (7–9), 333–341. http://dx.doi. org/10.1016/j.marpolbul.2006.11.007.
7. Drake, L.A., Tamburri, M.N., First, M.R., Smith, G.J., Johengen, T.H., 2014. How many organisms are in ballast water discharge? A framework for validating and selecting compliance monitoring tools. Mar. Pollut. Bull. 86 (1–2), 122–128. http://dx.doi.org/10. 1016/j.marpolbul.2014.07.034.
8. First, M.R., Drake, L.A., 2013. Approaches for determining the effects of UV radiation on microorganisms in ballast water. Manag. Biol. Invasion. 4 (2), 87–99. http://dx.doi. org/10.3391/mbi.2013.4.2.01.
9. First, M.R., Drake, L.A., 2014. Life after treatment: detecting living microorganisms following exposure to UV light and chlorine dioxide. J. Appl. Phycol. 26 (1), 227–235. http:// dx.doi.org/10.1007/s10811-013-0049-9.
10. Fiscus, E.L., Booker, F.L., 1995. Is increased UV-B a threat to crop photosynthesis and productivity? Photosynth. Res. 43 (2), 81–92. http://dx.doi.org/10.1007/bf00042965.
11. Gollasch, S., David, M., 2010. Algae viability measurements over time. Final report. Prepared for the Interreg IVB North Sea Ballast Water Opportunity project, p. 6.
12. Gollasch, S., David, M., 2012. On board tests of the organism detection tools BallastCAM, FluidImaging, USA, Hach-PAM-fluorometer, USA, and Walz-Water-PAM-fluorometer. Results and findings. Prepared for the Interreg IVB North Sea Ballast Water Opportunity project, p. 10.
13. Gollasch, S., David, M., 2015. Ballast water sampling and sample analysis for compliance control. In: David, M., Gollasch, S. (Eds.), Global Maritime Transport and Ballast Water Management vol. 8. Springer, Netherlands, pp. 171–223.
14. Gollasch, S., Minchin, D., & David, M. (2015). The transfer of harmful aquatic organisms and pathogens with ballast water and their impacts. In M. David & S. Gollasch (Eds.), Global Maritime Transport and Ballast Water Management (vol. 8, pp. 35- 58): Springer Netherlands.
15. Hallegraeff, G.M., Bolch, C.J., 1991. Transport of toxic dinoflagellate cysts via ships ballast water. Mar. Pollut. Bull. 22 (1), 27–30. http://dx.doi.org/10.1016/0025-326×(91)90441-t.
16. Hammes, F., Berney, M., Egli, T., 2011. Cultivation-independent assessment of bacterial viability. In: Müller, S., Bley, T. (Eds.), High Resolution Microbial Single Cell Analytics vol. 124. Springer, Berlin Heidelberg, pp. 123–150.
17. Hess-Erga, O.K., Attramadal, K.J.K., Vadstein, O., 2008. Biotic and abiotic particles protect marine heterotrophic bacteria during UV and ozone disinfection. Aquat. Biol. 4 (2), 147–154. http://dx.doi.org/10.3354/ab00105.
18. Holzinger, A., Lütz, C., 2006. Algae and UV irradiation: effects on ultrastructure and related metabolic functions. Micron 37 (3), 190–207. http://dx.doi.org/10.1016/j. micron. 2005.10.015.
19. International Maritime Organization, 2004. International Convention for the Control and Management of Ships' Ballast Water and Sediments (BWM/CONF/36).
20. International Maritime Organization. (2008a). Guidelines for Approval of Ballast Water Management Systems (G8). Resolution MEPC.174(58).
21. International Maritime Organization, 2008b. Marine

environment protection committee. Resolution MEPC.168(57). Procedure for Approval of Ballast Water Management Systems that Make Use of Active Substances (G9).

22. International Maritime Organization, 2013. Sub-committee on bulk liquids and gases. BLG 17/18. Report to the Maritime Safety Committee and the Marine Environment Protection Committee.

23. Jochem, F.J., 1999. Dark survival strategies in marine phytoplankton assessed by cytometric measurement of metabolic activity with fluorescein diacetate. Mar. Biol. 135 (4), 721–728. http://dx.doi.org/10.1007/s002270050673.

24. Jochem, F.J., 2000. Probing the physiological state of phytoplankton at the single-cell level. Sci. Mar. 64 (2), 183–195.

25. Jung, Y.J., Yoon, Y., Pyo, T.S., Lee, S.-T., Shin, K., Kang, J.-W., 2012. Evaluation of disinfection efficacy and chemical formation using MPUV ballast water treatment system (GloEnPatrol (TM)). Environ. Technol. 33 (17), 1953–1961. http://dx.doi.org/10.1080/09593330.2012.655315.

26. Kalisvaart, B.F., 2001. Photobiological effects of polychromatic medium pressure UV lamps. Water Sci. Technol. 43 (4), 191–197.

27. Kottuparambil, S., Shin, W., Brown, M.T., Han, T., 2012. UV-B affects photosynthesis, ROS production and motility of the freshwater flagellate, Euglena agilis Carter. Aquat. Toxicol. 122–123, 206–213. http://dx.doi.org/10.1016/j.aquatox.2012.06.002.

28. Kramer, B., Muranyi, P., 2014. Effect of pulsed light on structural and physiological properties of Listeria innocua and Escherichia coli. J. Appl. Microbiol. 116 (3), 596–611. http://dx.doi.org/10.1111/jam.12394.

29. Liebich, V., Stehouwer, P.P., Veldhuis, M., 2012. Re-growth of potential invasive phytoplankton following UV-based ballast water treatment. Aquat. Invasions 7 (1), 29–36. http://dx.doi.org/10.3391/ai.2012.7.1.004.

30. Lloyd's Register Marine, 2015a. Understanding ballast water management. Guidance for shipowners and operators. Retrieved 21. August 2015, from http://www.lr.org/en/_ images/213-35824_Understanding_Ballast_Water_Management_0314_tcm155- 248816.pdf.

31. Lloyd's Register Marine, 2015b. Available ballast water treatment systems. Retrieved 5. October 2015, from http://www.lr.org/en/marine/consulting/environmentalservices/ballastwatermanagement.aspx.

32. Martínez, L.F., Mahamud, M.M., Lavin, A.G., Bueno, J.L., 2013. The regrowth of phytoplankton cultures after UV disinfection. Mar. Pollut. Bull. 67 (1–2), 152–157. http://dx.doi. org/10.1016/j.marpolbul.2012.11.019.

33. Martínez, L.F., Mahamud, M.M., Lavín, A.G., Bueno, J.L., 2012. Evolution of phytoplankton cultures after ultraviolet light treatment. Mar. Pollut. Bull. 64 (3), 556–562. http://dx. doi.org/10.1016/j.marpolbul.2011.12.021.

34. McMinn, A., Martin, A., 2013. Dark survival in a warming world. Proc. R. Soc. B Biol. Sci. 280 (1755), 7. http://dx.doi.org/10.1098/rspb.2012.2909.

35. Oguma, K., Katayama, H., Mitani, H., Morita, S., Hirata, T., Ohgaki, S., 2001. Determination of pyrimidine dimers in Escherichia coli and Cryptosporidium parvum during UV light inactivation, photoreactivation, and dark repair. Appl. Environ. Microbiol. 67 (10), 4630–4637. http://dx.doi.org/10.1128/aem.67.10.4630-4637.2001.

36. Oguma, K., Katayama, H., Ohgaki, S., 2002. Photoreactivation of Escherichia coli after low or medium-pressure UV disinfection determined by an endonuclease sensitive site assay. Appl. Environ. Microbiol. 68 (12), 6029–6035. http://dx.doi.org/10.1128/aem. 68.12.6029–6035.2002.

37. Oliver, J.D., 2010. Recent findings on the viable but nonculturable state in pathogenic bacteria. FEMS Microbiol. Rev. 34 (4), 415–425. http://dx.doi. org/10.1111/j.1574- 6976.2009.00200.x.

38. Olsen, R.O., Hess-Erga, O.K., Larsen, A., Thuestad, G., Tobiesen, A., Hoell, I.A., 2015. Flow cytometric applicability to evaluate UV inactivation of phytoplankton in marine water samples. Mar. Pollut. Bull. 96 (1–2), 279–285. http://dx.doi.org/10.1016/j. marpolbul. 2015.05.012.

39. Ou, H., Gao, N., Deng, Y., Qiao, J., Wang, H., 2012. Immediate and long-term impacts of UVC irradiation on photosynthetic capacity, survival and microcystin-LR release risk of Microcystis aeruginosa. Water Res. 46 (4), 1241–1250. http://dx.doi.org/10.1016/j. watres.2011.12.025.

40. Peperzak, L., Brussaard, C.P.D., 2011. Flow cytometric applicability of fluorescent vitality probes on phytoplankton. J. Phycol. 47 (3), 692–702. http://dx.doi. org/10.1111/j. 1529-8817.2011.00991.x.

41. Peperzak, L., Gollasch, S., 2013. NIOZ flow cytometer workshop, comparing organism detection instruments in measuring 2–10 μm and 10–50 μm plankton cells. Final Report, Prepared for Interreg IVB Project Ballast Water Opportunity, p. 66.

42. Rastogi, R.P., Richa, Sinha, R.P., Singh, S.P., Haeder, D.-P., 2010. Photoprotective compounds from marine organisms. J. Ind. Microbiol. Biotechnol. 37 (6), 537–558. http://dx.doi.org/10.1007/s10295-010-0718-5.

43. Roszak, D.B., Colwell, R.R., 1987. Survival strategies of bacteria in the natural environment. Microbiol. Rev. 51 (3), 365–379.

44. Ruiz, G.M., Carlton, J.T., Grosholz, E.D., Hines, A.H., 1997. Global invasions of marine and estuarine habitats by non-indigenous species: mechanisms, extent, and consequences. Am. Zool. 37 (6), 621–632.

45. Ruiz, G.M., Rawlings, T.K., Dobbs, F.C., Drake, L.A., Mullady, T., Huq, A., Colwell, R.R., 2000. Global spread of microorganisms by ships. Nature 408 (6808), 49–50. http://dx.doi. org/10.1038/35040695.

46. Sancar, A., Sancar, G.B., 1988. DNA-repair enzymes. Annu. Rev. Biochem. 57, 29–67.

47. Schenk, M., Raffellini, S., Guerrero, S., Blanco, G.A., Alzamora, S.M., 2011. Inactivation of Escherichia coli, Listeria innocua and Saccharomyces cerevisiae by UV-C light: study of cell injury by flow cytometry.

LWT Food Sci. Technol. 44 (1), 191–198. http://dx. doi. org/10.1016/j.lwt.2010.05.012.

48. Shapiro, H.M., 2000. Microbial analysis at the single-cell level: tasks and techniques. J. Microbiol. Methods 42 (1), 3–16. http://dx.doi.org/10.1016/S0167- 7012(00)00167-6.

49. Sinha, R.P., Häder, D.P., 2002. UV-induced DNA damage and repair: a review. Photochem. Photobiol. Sci. 1 (4), 225–236. http://dx.doi.org/10.1039/b201230h.

50. Sinha, R.P., Klisch, M., Gröniger, A., Häder, D.P., 1998. Ultraviolet-absorbing/screening substances in cyanobacteria, phytoplankton and macroalgae. J. Photochem. Photobiol. B Biol. 47 (2–3), 83–94. http://dx.doi.org/10.1016/S1011-1344(98)00198-5.

51. Staley, J.T., Konopka, A., 1985. Measurement of in situ activities of nonphotosynthetic microorganisms in aquatic and terrestrial habitats. Annu. Rev. Microbiol. 39, 321–346. http://dx.doi.org/10.1146/annurev.mi.39.100185.001541.

52. Stehouwer, P.P., Buma, A., Peperzak, L., 2015. A comparison of six different ballast water treatment systems based on UV radiation, electrochlorination and chlorine dioxide. Environ. Technol. 36 (15), 2094–2104. http://dx.doi.org/10.1080/09593330.2015. 1021858.

53. Tang, K.W., Dziallas, C., Grossart, H.P., 2011. Zooplankton and aggregates as refuge for aquatic bacteria: protection from UV, heat and ozone stresses used for water treatment. Environ. Microbiol. 13 (2), 378–390. http://dx.doi.org/10.1111/j.1462-2920. 2010.02335.x.

54. Tao, Y., Zhang, X., Au, D.W.T., Mao, X., Yuan, K., 2010. The effects of sub-lethal UV-C irradiation on growth and cell integrity of cyanobacteria and green algae. Chemosphere 78 (5), 541–547. http://dx.doi. org/10.1016/j.chemosphere.2009.11.016.

55. U. S. Food and Drug Administration (FDA), 2010. Bacterial analytical manual, appendix 2 most probable number from serial dilutions. Retrieved 21. August 2015, from http:// www.fda.gov/Food/FoodScienceResearch/LaboratoryMethods/ucm109656.htm.

56. United States Coast Guard, 2012. Standards for Living Organisms in Ships' Ballast Water Discharged in U.S. Waters. Federal Register (Retrieved 21. August 2015, from http://www.gpo.gov/fdsys/pkg/FR-2012-03-23/pdf/2012-6579.pdf).

57. van Slooten, C., Wijers, T., Buma, A.G.J., Peperzak, L., 2015. Development and testing of a rapid, sensitive ATP assay to detect living organisms in ballast water. J. Appl. Phycol. 27 (6), 2299–2312. http://dx.doi.org/10.1007/s10811-014-0518-9.

58. Villarino, A., Rager, M.N., Grimont, P.A.D., Bouvet, O.M.M., 2003. Are UV-induced nonculturable Escherichia coli K-12 cells alive or dead? Eur. J. Biochem. 270 (12), 2689–2695. http://dx.doi.org/10.1046/j.1432-1033.2003.03652.x.

59. Werschkun, B., Banerji, S., Basurko, O.C., David, M., Fuhr, F., Gollasch, S., ... Höfer, T., 2014. Emerging risks from ballast water treatment: the run-up to the international ballast water management convention. Chemosphere 112, 256–266. http://dx.doi.org/10. 1016/j.chemosphere.2014.03.135.

60. Werschkun, B., Sommer, Y., Banerji, S., 2012. Disinfection by-products in ballast water treatment: an evaluation of regulatory data. Water Res. 46 (16), 4884–4901. http:// dx.doi.org/10.1016/j.watres.2012.05.034.

61. Xiong, F.S., Komenda, J., Kopecky, J., Nedbal, L., 1997. Strategies of ultraviolet-B protection in microscopic algae. Physiol. Plant. 100 (2), 378–388.

62. Yin, F., Zhu, Y., Koutchma, T., Gong, J., 2015. Inactivation and potential reactivation of pathogenic Escherichia coli O157:H7 in apple juice following ultraviolet light exposure at three monochromatic wavelengths. Food Microbiol. 46, 329–335. http://dx. doi.org/10.1016/j.fm.2014.08.015.

63. Zimmer, J.L., Slawson, R.M., 2002. Potential repair of Escherichia coli DNA following exposure to UV radiation from both medium- and low-pressure UV sources used in drinking water treatment. Appl. Environ. Microbiol. 68 (7), 3293–3299. http://dx.doi.org/ 10.1128/aem.68.7.3293-3299.2002.

CHAPTER 11

SIGNIFICANCE OF COAGULATION KINETICS ON MAINTAINING MEMBRANE PERMEABILITY IN AN MBR FOLLOWING REACTIVE COAGULANT DOSING

O. Autin, F. Hai, S. Judd, E.J. McAdam

Keywords: Membrane performance enhancer, Membrane bioreactor, Direct visual observation

INTRODUCTION

Membrane bioreactors (MBRs) have become a standard technology for wastewater treatment. However, understanding how to ensure sustained membrane permeability remains a key operational consideration [15] and [4]. Several approaches have been introduced in order to reduce fouling including intermittent suction [16], backflushing [17], module design improvement [21] or the optimisation of aeration such as with the LEAPmbr aeration technology which has enabled dramatic reductions in operational energy cost to be realised [8]. Nevertheless membrane fouling remains important, particularly understanding how to respond to system perturbation, and as such it is suggested that MBR cost could be further reduced by better control of membrane fouling[20].

Membrane fouling is a consequence of the interaction between the membrane and a complex mixture comprising colloids, bacterial flocs and dissolved macromolecules [21]. It has been widely reported that soluble microbial products (SMP) which comprise mainly of proteins and polysaccharides are primarily responsible for the clogging and blocking of membrane pores [22] and [23]. As these compounds are generally colloidal in nature, their transport is primarily controlled by Brownian motion and as such the shear forces applied either through pumping (in side stream) or aeration (in immersed) are insufficient to provide back-transport toward the bulk which results in preferential colloidal deposition at the membrane [29]. Deposition of these high molecular weight compounds result in formation of a highly hydrated gel matrix into which microorganisms are embedded, resulting in a significant resistance to permeate flow during membrane operation [25].

To ameliorate the potential impact of SMP on membrane fouling, several authors have introduced the use of coagulants [20], [24], [4] and [5] which enable the assimilation of SMP compounds into aggregates (or flocs), that are then more strongly influenced by inertial lift and shear induced diffusion, thereby promoting back-transport away from the membrane surface into the bulk [20] and [24]. Numerous chemical compounds have been trialled including metal salts, biopolymers, starch [4] and [5] and organic polymers and have been considered particularly pertinent for reactive dosing of the coagulant to limit the impact of sudden SMP release in response to process perturbation, such as saline shock [9], [10] and [6]. Woznaik [24] described the use of MPE50 which is a modified biopolymer with a net cationic charge that has been demonstrated to effectively reduce membrane fouling both at laboratory and full scale [24], [4] and [5]. Hwang et al.[14] suggested that the cationic structure of MPE50 enabled the neutralisation of the negatively charged colloidal biopolymers, thereby encouraging floc growth, which can be considered analogous to conventional coagulation-flocculation. However, in conventional application, coagulation and flocculation processes are generally configured in series. In the coagulation step, coagulant is added within a high shear zone with an average velocity gradient (G) exceeding 1500 s^{-1} to ensure the coagulant is distributed homogeneously in order to achieve particle destabilisation and charge neutralisation. Flocculation

Citation: O. Autin, F. Hai, S. Judd, E.J. McAdam, Significance of coagulation kinetics on maintaining membrane permeability in an MBR following reactive coagulant dosing, Journal of Membrane Science, Volume 516, 15 October 2016, Pages 64-73, ISSN 0376-7388, http://dx.doi.org/10.1016/j.memsci.2016.06.008.

then proceeds in a secondary shear zone with a gentler average velocity gradient of around 100–200 s^{-1} to encourage successful collision of destabilised particles and is provided with a retention time of up to 60 min to ensure subsequent floc growth is complete [3].

For comparison, there is no high shear zone specified within present MBR design to incorporate coagulant dosing. Furthermore, the average velocity gradient imposed by coarse bubble aeration in MBR is around 150–200 s^{-1}[11], [12] and [13]. Consequently, without a clearly delineated high shear zone to provide homogeneous distribution of the coagulant, and with gas mixing in MBR providing a velocity gradient which is equivalent to flocculation, it is suggested that coagulation-flocculation will be rate limited within MBR which will cause a delay in the time from which coagulant is reactively dosed to when colloids destabilise and form aggregates. As flocculators are generally designed with residence times of up to 60 min to enable sufficient time for aggregate growth, it is posited that with direct dosing of MPE50 into the MBR, the time at which coagulant is added during the filtration cycle could therefore be of significance to sustaining membrane permeability during filtration. The aim of this study is therefore to establish the significance of coagulation-flocculation kinetics on membrane permeability following the reactive dosing of coagulant. The widely used commercially available cationic polymer MPE50 was employed as the coagulant and a model particle suspension was employed to provide a controlled testing environment, comprising of the biopolymers bovine serum albumin (BSA) and sodium alginate (Na-alginate), as surrogates for protein and polysaccharide, and 3 μm Firefli™ microspheres were included to provide a bimodal size distribution characteristic of an MBR sludge .

MATERIAL AND METHODS

Cross-flow cell linked with direct visual observation

The crossflow cell was constructed of Perspex and comprised a flow channel with a width of 0.012 m, length 0.21 m and height 0.006 m. A viewing window was engineered within a recess of the upper cell wall (Fig. 1). The opening was lined with an o-ring to provide a watertight seal to a coverslip (n=1.5, or 0.17 mm thick) which was held in place by an annular clamp. The microscope (Leica DM5500B Microsystems, Milton Keynes, UK) was fitted with a HC PL FL 10X/0.30 lens which provided a wide range of free working distance (around 11 mm) which is sufficient to incorporate the depth of field below the viewing window to the membrane surface. A Polyvinylidene fluoride (PVDF) hollow fibre membrane with a nominal pore size of 0.04 μm (Zeeweed, GE Power and Water, Ontario, USA) was used in this study and had an outside diameter of 0.0019 m, yielding an active membrane surface area of 0.00125 m^2. The depth from the outer diameter of the hollow fibre membrane to the underside of the viewing window was around 2.5 mm which was sufficient to allow active fluid transport around the fibre.

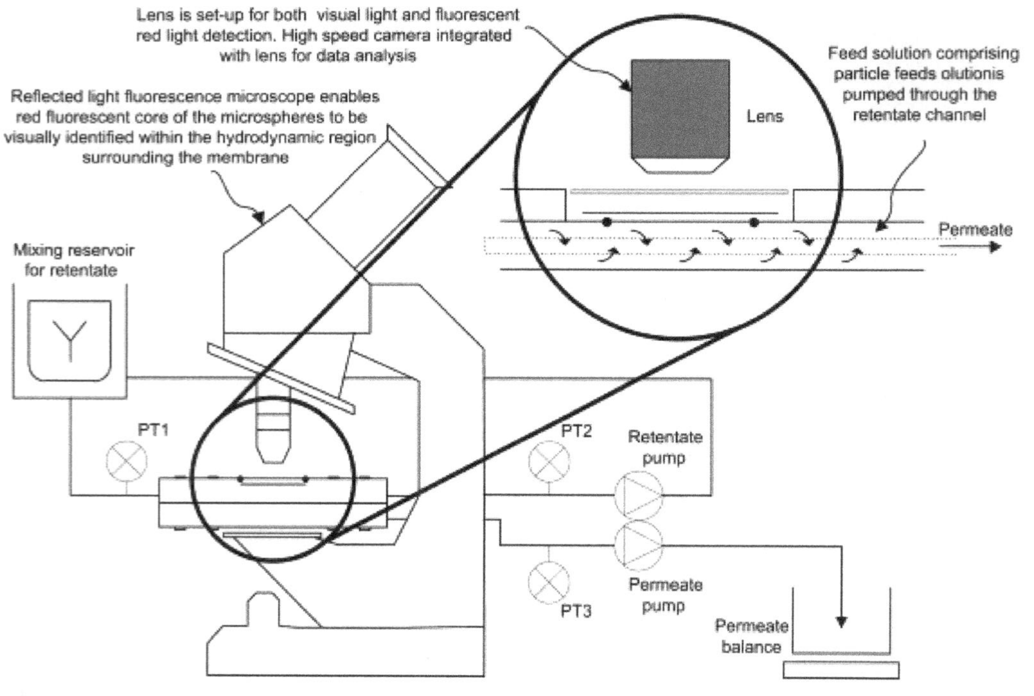

Figure 1: Experimental set-up comprising direct visual observation cell equipped with reflected light fluorescence microscope.

The microscope used for direct observation was fitted with a high speed, high-sensitivity digital camera (DFC365 FX, Leica Microsystems, Milton Keynes, UK) for image and video recording. Analysis of cake height using static images was undertaken using Leica Application Suite software. To quantify particle back transport velocity from video footage, VideoStudio software was used (Ulead, Malavida). To enhance particle tracking, a fluorescence enabled microscope was used in conjunction with Firefli™ fluorescent red 3 μm polystyrene microspheres which exhibited excitation/ emission at 542/612 nm. At the specified VL (0.011 m s^{-1}) and particle concentration (25 mg l^{-1}) visual determination of individual particles with diameters exceeding 0.5 μm can be achieved using this reflected light fluorescence direct visual observation method [19].

Feed solution was recirculated at a velocity (*VL*) of 0.011 m s^{-1} through the retentate channel (520S, Watson Marlow, Falmouth, UK). This represents a Reynold's number (*Re*) of 88 which is indicative of laminar flow. Since the relative flux is low with respect to*VL*, and making the assumption of a rectangular channel, shear rate (*θ*) was fixed to 11 s^{-1}. Both permeate and retentate were fed back to the feed reservoir, which was only slowly mixed to avoid sedimentation. Each permeability assessment was undertaken over 120 min at a fixed flux of 100 l m^{-2} h^{-1}. To determine trans-membrane pressure (TMP), pressure transducers (PT) were sited on the permeate line (−0.5 to +0.5 bar(g)), upstream and downstream of the retentate channel (0–1 bar(g)). Each PT had a reported sensitivity of less than 0.25% of range. The analogue signal arising from each PT was collected with an ADC-24 data logger and analysed with the Picolog data acquisition software (Pico technology, St Neots, UK). Fluid temperature was controlled at 20 °C. Permeate was collected on a balance (Symmetry PR-precision toploading balance, Cole-Parmer, London, UK) and the mass recorded directly onto a PC. Residual resistance from caking, pore plugging and irreversible adsorption (*Rres*) was calculated from [18] and [29]:

$$R_m = \frac{\Delta TMP}{\mu J_W} \quad (1)$$

$$R_{res} = \frac{\Delta TMP}{\mu J_{ps}} - R_m \quad (2)$$

where *Rm* is the membrane resistance determined during the filtration of deionised water at flux, *Jw* (100 l m^{-2} h^{-1}), and *Jps* is the flux used for filtration of the particle solution. Reproducibility of TMP data produced from the experimental set-up was assessed using the full particle matrix (sodium alginate, bovine serum albumin and 3 μm microspheres) and an average error of 16% recorded between replicates (Fig. 2).

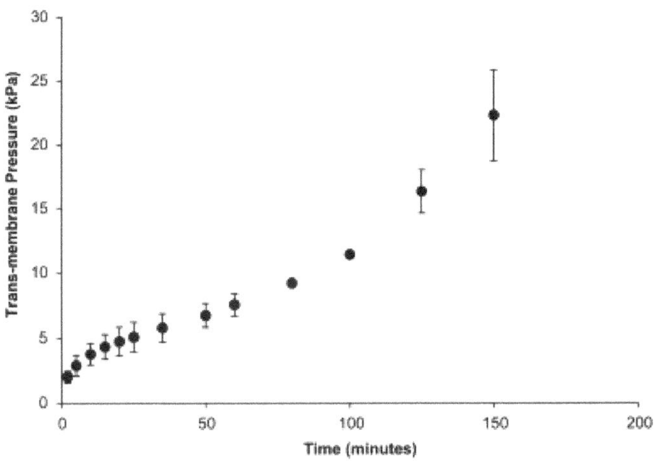

Figure 2: TMP profile during filtration of 50 mgBSA l^{-1}, 50 mgNa-Alg l^{-1} and 25 mg l^{-1} of 3 μm particles for several experiments (average error from mean, 16%). Cross-flow velocity 11 mm s^{-1} (wall shear rate 11 s^{-1}); flux 100 l m^{-2} h^{-1}.

Chemicals and analytical methods used

Bovine serum albumin (BSA, 200 mg mL^{-1}) and sodium alginate (Na-alg,) were used as model biopolymers (Sigma-Aldrich, Poole, UK). Fluorescent enabled 3 μm latex microspheres were used to provide a bi-modal particle size distribution, providing a strictly controlled particle diameter (uniformity <5%) and shape that are also easily traceable within a complex matrix owing to their fluorescent core (Fisher Scientific, Loughborough, UK). Model solutions were prepared using initial concentrations of fluorescent microspheres, sodium alginate and BSA of 25 mg l^{-1}, 50 mg l^{-1} and 50 mg l^{-1}respectively, diluted in ultrapure water (Purelab Option – S7/15, 18.2 MΩ cm^{-1} and TOC <3 ppb) to which 20 mM sodium chloride (NaCl) and 1 mM sodium hydrogen carbonate ($NaHCO_3$) were added to increase ionic strength and buffering capacity of the water (Fisher Scientific, Loughborough, UK). After each experiment, the membrane was rinsed with ultrapure water for 15 min, followed by a backwash operation at 40 l m^{-2} h^{-1} for 15 min and then chemically cleaned overnight with 500 mg L^{-1} of sodium hypochlorite solution (Sigma-Aldrich, Poole,). The cationic polymer MPE50 was selected for use in this study, the efficiency of which has been previously demonstrated in MBR matrices following screening a broad range of potential additives [4]. As such, a comparison with other potential polymers was not made. The chemical MPE50 was kindly supplied by Nalco (Leiden, The Netherlands).

Zeta potential was determined using laser Doppler electrophoresis (Malvern Zetasizer 2000, Malvern Instruments Ltd., Malvern, UK). Particle size was measured using laser diffraction (Malvern Mastersizer 2000, Malvern Instruments, UK). Protein and polysaccharide concentrations were quantified using the modified Lowry method [27]and phenol-sulfuric acid method [28] respectively. Kinetic experiments were undertaken on a jar tester to determine the rate of aggregation initiated following inclusion of MPE50. Jar tester samples were prepared in 1 L samples and were subject to a stirring at 50 rpm, equivalent to a mean velocity gradient (*G*) of 44 s^{-1}.

RESULTS

Determination of an appropriate MPE50 dose for the particle matrix

The efficacy of MPE50 for removing biopolymers from bulk solution into aggregates was assessed within water comprising a particle complex of 50 mg l^{-1} of the protein Bovine Serum Albumin (BSA), 50 mg l^{-1} of the polysaccharide sodium alginate (Na-Alg) and 25 mg l^{-1} of 3 μm particles (Fig. 3). An increase in the removal of BSA from 3% to 18% was demonstrated following an increase in MPE50 dose from 1 to 200 mg l^{-1} (Fig. 3(a)). However, a plateau was reached once the MPE50 dose had increased above 200 mg l^{-1}, which suggested that further increase in protein removal efficiency was not attainable. Sodium alginate removal efficiency increased from 5% to 65% once MPE50 dose increased from 100 mg l^{-1} to 200 mg l^{-1} and then markedly declined to 23% upon increasing the MPE50 dose to 500 mg l^{-1} (Fig. 3(b)). This is analogous to behaviour observed by [4] when dosing MPE50 within a similar concentration range to promote coagulation of protein and polysaccharides derived from soluble microbial products (SMP) produced from an MBR.

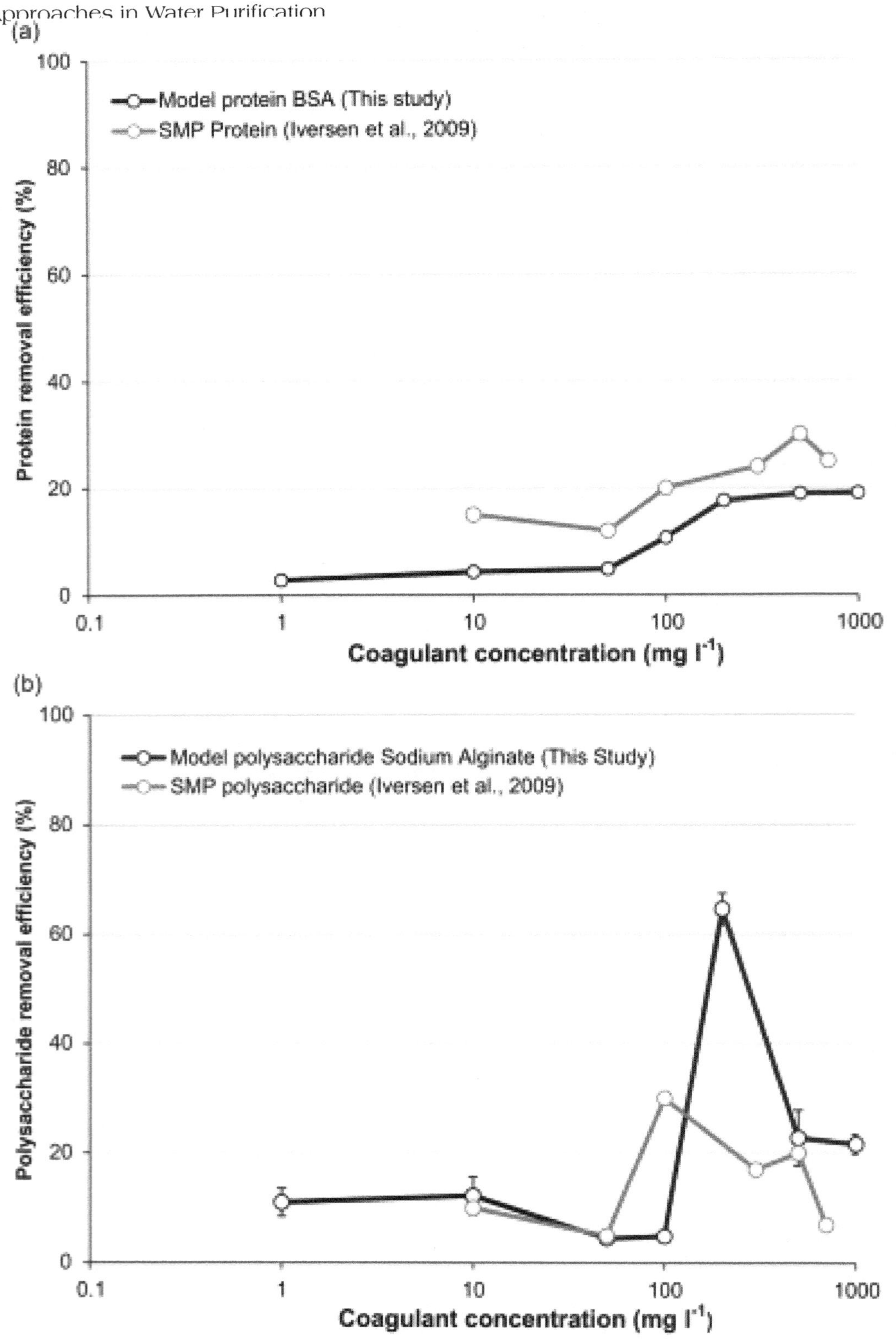

Figure 3: Characterisation of coagulant dose in the model system comprising 50 mgBSA l^{-1}, 50 mgNa-Alg l^{-1} and 25 mg l^{-1} of 3 µm particles. Reference data for MPE50 dosing in MBR sludge extrapolated from [4].

Aggregation and floc growth for the particle matrix following MPE50 dosing

Zeta potential of the individual biolpolymers and of the combined particle matrix was determined (Fig. 4). Isoelectric points (IEP) of pH 1.7 and pH 4.7 were determined for Na-Alg and BSA respectively which are coincident with the literature [7]. The zeta potential of MPE50 varied between 38.9 mV and 24.7 mV across the pH range evaluated (pH 3.6–12.3). When each of the particle types were combined within one solution (BSA, Na-alg and 3 µm particles), the zeta potential curve was between those recorded for BSA and Na-Alg and subsequently shifted to the right following a 200 mg l^{-1} addition of MPE50.

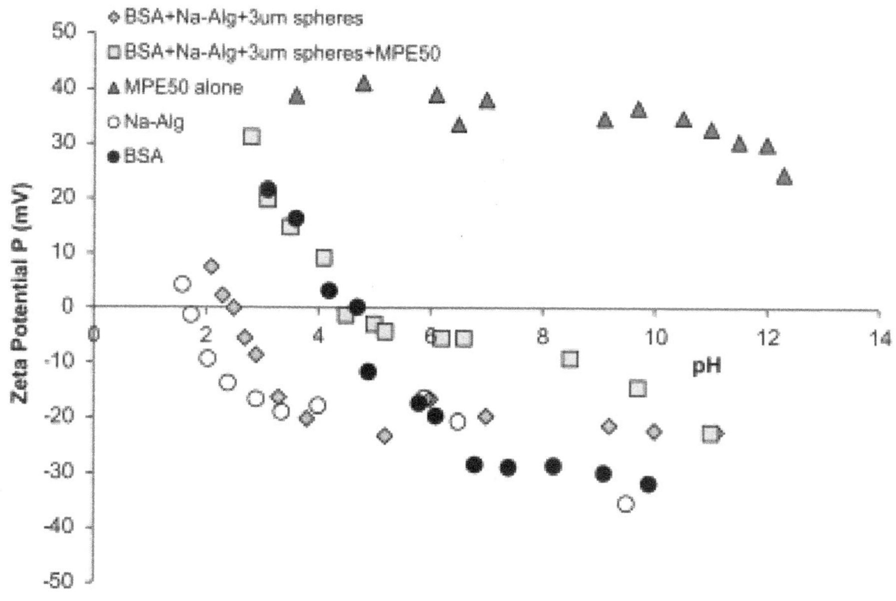

Figure 4: Zeta potential of single and complex systems.

The effect of both MPE50 and particle type on final aggregate size was determined (Fig. 5). MPE50 was dosed 10 min after the trial commenced. When either BSA or Na-alg was combined with 3 μm spheres, the impact of MPE50 addition on floc growth was negligible. When MPE50 was dosed into a mixture of Na-alg and BSA, particle size increased from less than 1 μm to over 9 μm. A notable increase in particle size was observed through addition of 3 μm spheres to the Na-alg/BSA mixture, reaching a maximum d_{50} of 230 μm around 20 min after MPE50 addition.

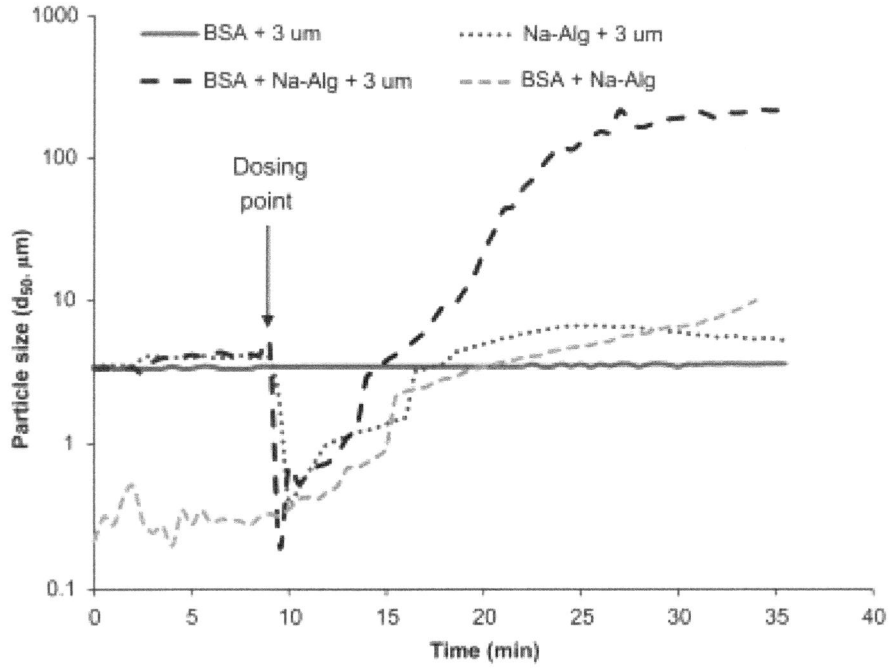

Figure 5: Particle size analysis undertaken for systems of increasing particle complexity (Malvern Mastersizer, 2000, Malvern Instruments). A 200 mg l^{-1} MPE50 dose added after around 10 min.

Establishing uptake rate of biopolymers into the floc following MPE50 dosing

The kinetics of aggregation was studied in batch using the complete particle matrix comprising 50 mgBSA l^{-1}, 50 mgNa-alg l^{-1} and 25 mg l^{-1} of 3 μm particles. Following application of a 200 mg l^{-1} dose of MPE50 to the particle matrix, water samples were subject to filtration through 0.22 μm, 0.45 μm, 2.7 μm and 8 μm membranes. The remaining sodium alginate and BSA concentrations resident in each permeate fraction then provided indication of the minimum size of aggregate that the biopolymers had assimilated into following dosing (Fig. 6 and Fig. 7).

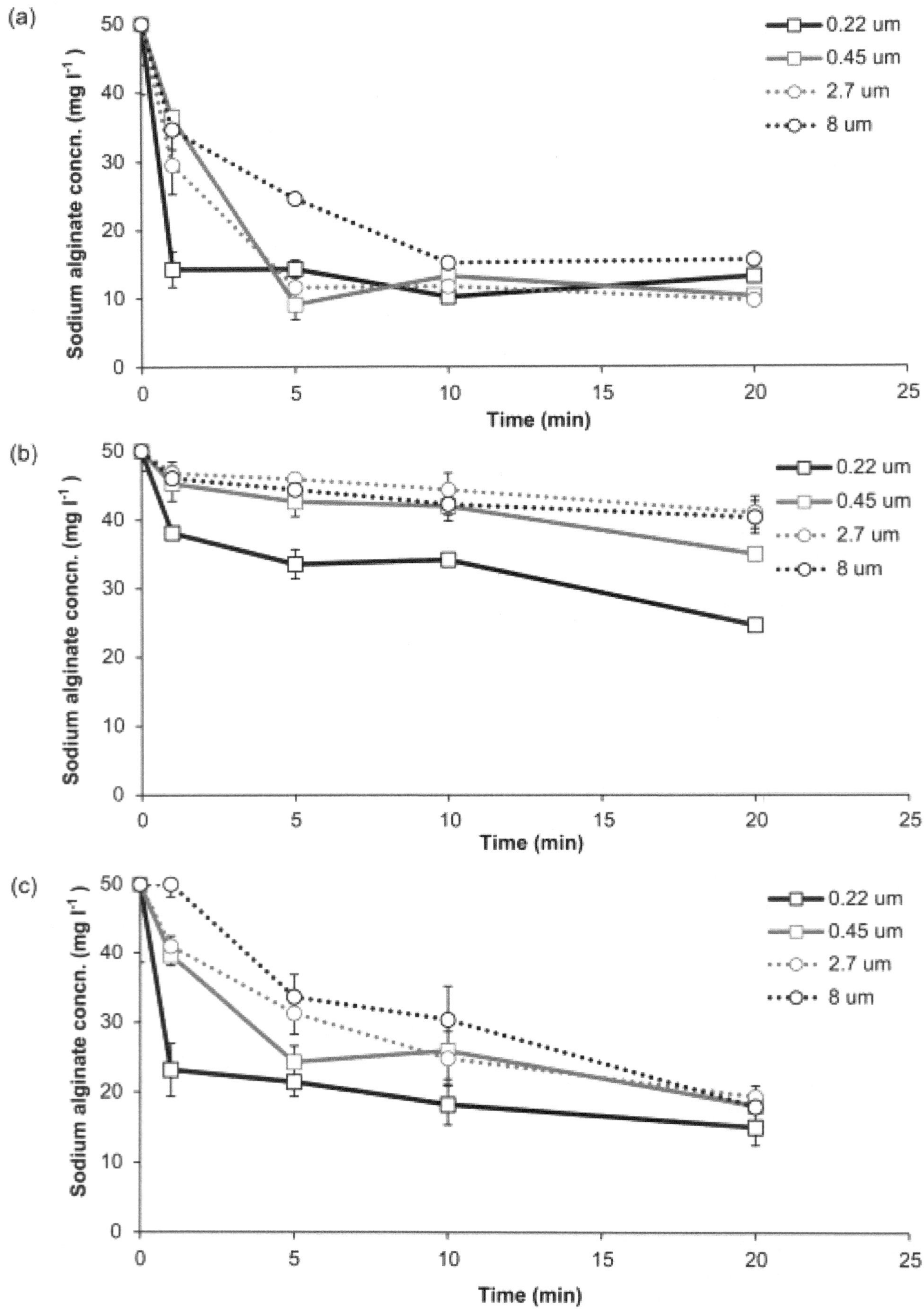

Figure 6: Determination of the kinetics of aggregation amongst the individual particulate constituents: (a) BSA (50 mg l^{-1}), Sodium alginate (50 mg l^{-1}), MPE50 (200 mg l^{-1}); (b) Sodium alginate (50 mg l^{-1}), 3 µm particles (25 mg l^{-1}), MPE50 (200 mg l^{-1}); (c) BSA (50 mg l^{-1}), Sodium alginate (50 mg l^{-1}), 3 µm particles (25 mg l^{-1}), MPE50 (200 mg l^{-1}).

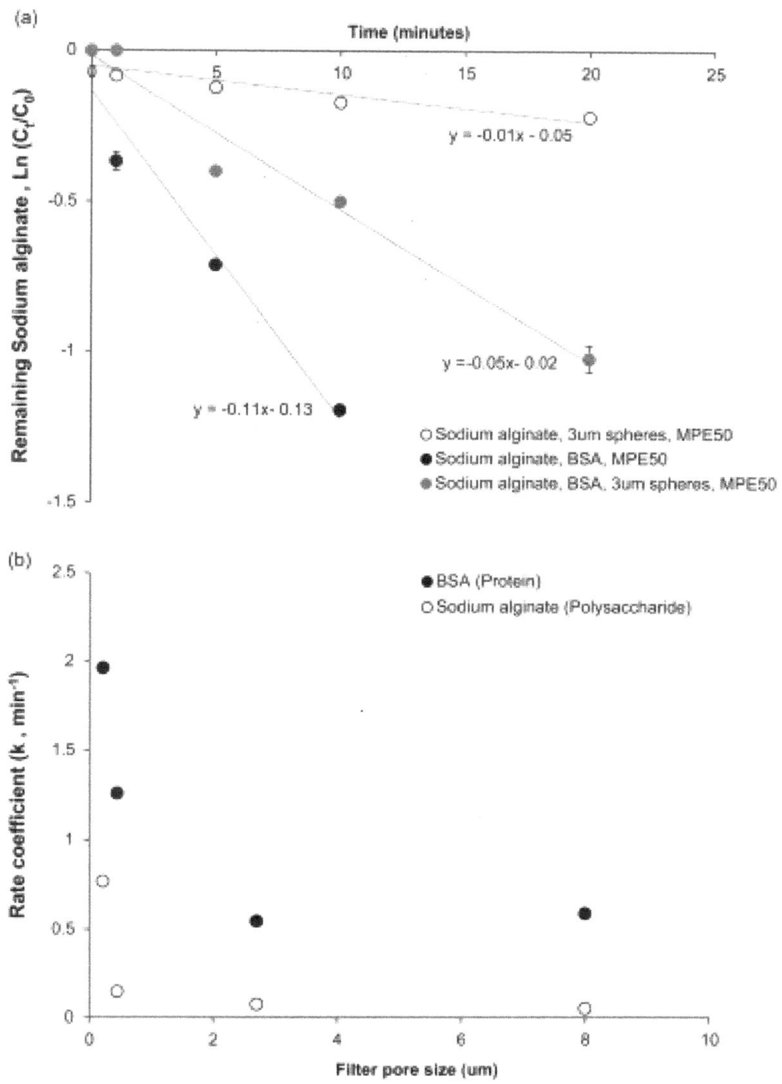

Figure 7: (a) Derivation of the rate coefficient from the aggregation experiments undertaken with different particle compositions in the presence of MPE50. Analysis of the <8 µm fraction shown; (b) The final rate coefficient determined for the full particle matrix (BSA, Sodium alginate, 3 µm spheres, MPE50) within each filtered fraction.

An initial assessment of the impact of the various fractions of the particle matrix was undertaken using (a) BSA/ sodium-alginate, (b) sodium alginate/3 µm particles, and finally (c) the complete particle matrix comprising BSA/sodium-alginate/3 µm particles (Fig. 6). During batch testing, the highest uptake of sodium alginate into larger aggregates following MPE50 addition was determined for the BSA/sodium alginate and BSA/sodium alginate/3 µm particle mixtures (around 30 mg l^{-1} consumed). However, the rate coefficient (k) which described the rate of uptake of sodium-alginate into aggregates greater than 8 µm was markedly higher for the BSA/sodium alginate mixture (k 0.11 min^{-1}) than for the sodium alginate/3 µm and BSA/sodium alginate/3 µm mixtures which were 0.01 min^{-1} and 0.05 min^{-1} respectively (Fig. 7(a)). Analysis of the rate of uptake of both sodium-alginate and BSA within the complete particle mixture (sodium-alginate/ BSA/3 µm spheres) demonstrated uptake into smaller aggregates was considerably higher than larger aggregates and furthermore that protein followed a similar trend to sodium alginate but the established rate of uptake into aggregates was considerably faster (Fig. 7(b)).

Impact of reactive MPE50 dosing on membrane fouling

The control test used a complex comprising of BSA, Na-alg and 3 µm Firefli™ microspheres but without the addition of MPE50 (Fig. 8). The TMP profile comprised of a classical two stage shape where an initial filtration stage characterised by a dP/dt_{phase1} 0.08 kPa min^{-1} was followed by a sharp rise in TMP of dP/dt_{phase2} 0.87 kPa min^{-1} at around 55 min leading to a final residual resistance (R_{res}) of 56×10^{10} m^{-1} recorded after 120 min of filtration. Dosing of MPE50 was undertaken at 0 and then 30 min after filtration had commenced to ascertain the importance of dosing time for reactive control of membrane fouling (Fig. 8). For the first 30 min of filtration, the growth in cake resistance was ostensibly similar to that without MPE50 addition. However, following further filtration, cake resistance declined and was coincident with a reduction in cake height from a maximum of 10 µm, measured after 15 min of filtration, to below 3 µm when measured after 120 min of filtration. When MPE50 dosing was applied 30 min after filtration, a marked increase in residual resistance was immediately observed (Fig. 8).

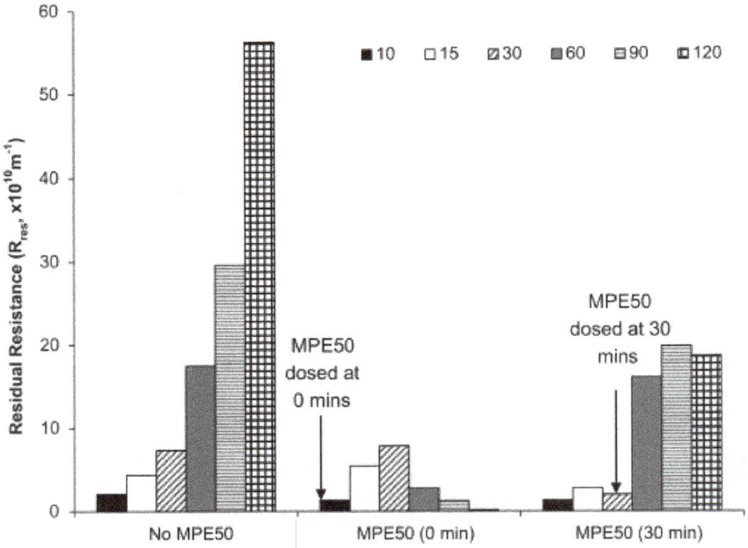

Figure 8: Residual resistance determined during the course of filtration for a particle complex comprising 50 mg l^{-1}BSA, 50 mg l^{-1} Sodium alginate and 25 mg l^{-1} 3 μm spheres. For tests conducted with MPE50 (dosing times of 0 mins and 30 mins after filtration is initiated), a 200 mg l^{-1} dose was applied. Cross-flow velocity, 11 mm s^{-1}; flux 100 l m^{-2} h^{-1}.

Before dosing of MPE50, direct observation evidenced axial migration of 3 μm Firefli™ particles in the bulk flow (circled in red) above a small static cake (<5 μm) which had formed at the membrane surface (Fig. 9(a)). Within one minute of dosing MPE50 at 30 min, a fluidised cake of around 25 μm in depth was noted to have formed above the static cake (Fig. 9(b)). Several individual 3 μm Firefli™ particles were observed within this fluidised layer. However, due to the low number density of 3 μm particles present within this fluidised polarised region above the stagnant cake, it was postulated that this was predominantly comprised of the biopolymers BSA and sodium alginate. Direct observation of the membrane wall 15 min after dosing evidenced that the fluidised layer had dissipated and growth of large aggregates had initiated which were visibly observed to undergo axial migration in the bulk flow away from the membrane wall, as well as floc rolling along the membrane surface in the direction of flow (Fig. 9(c)). The formation of the fluidised region immediately after dosing MPE50 at 30 min coincided with a rapid increase in *Rres* of over 800% to 16.1×10^{10} m^{-1}. Residual resistance did not markedly increase following continued filtration and the final *Rres* measured at 120 min was lower than in the absence of coagulant (Fig. 8). A fluidised region was not observed when MPE50 was dosed at the commencement of filtration (0 min), although a final *Rres* of 1.2×10^{10} m^{-1} was recorded which is an order of magnitude below the cake resistance recorded following MPE50 dosing at 30 min after filtration had commenced (*Rcake*19×10^{10} m^{-1}).

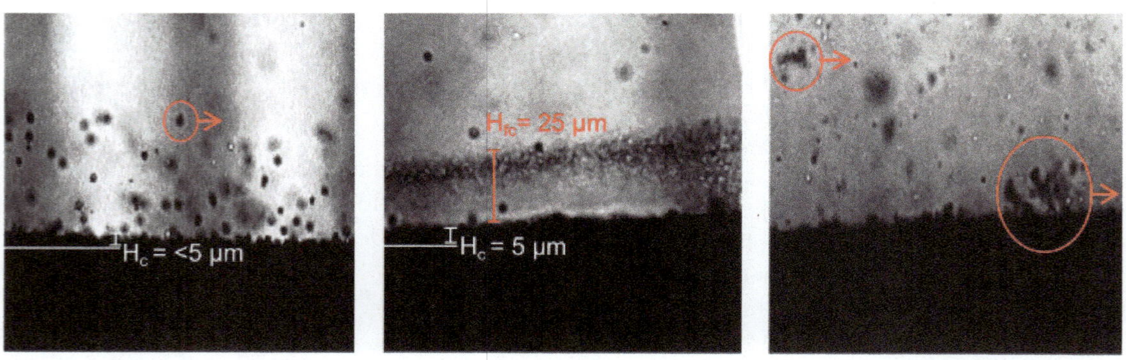

Figure 9: Direct observation of fouling before and after the addition of MPE50 at the 30 min dosing point: a) 2 min before MPE50 addition; b) 1 min after MPE50 addition); and, C) 15 min after MPE50 addition.

Fluorescence microscopy was used to evidence the presence of 3 μm Firefli™ particles within the aggregates formed following MPE 50 dosing (Fig. 10). Volume and concentration constrained the capability to use standard methods (e.g. laser diffraction) for particle size distributions. However, the comparative images qualitatively evidence the formation of considerably larger aggregates of around 80 μm following MPE50 dosing at 30 min. Interestingly, whilst the addition of MPE50 at 0 min of filtration resulted in the lowest final cake resistance, it is also the condition in which the aggregates were the smallest, stabilising at around 10 μm in size after only 5 min following MPE50 dosing. Particle velocity of the 3 μm Firefli™ particles in the polarised region of the membrane was also ascertained (Fig. 11). At the onset of filtration, particle velocity was approximately 800 μm s^{-1} at a distance of 60 μm from the membrane surface. This value is much lower than the superficial liquid velocity of 11 mm s^{-1} (*VSL* 11,000 μm s^{-1}) indicating the heterogeneous distribution of fluid velocity within the channel as previously demonstrated by [26]. When MPE50 dosing was not applied

(Fig. 11(a)), the apparent particle velocity was greatly reduced. To illustrate, after only 60 min of filtration, apparent particle velocity at 60 μm from the membrane surface was less than 200 μm s^{-1} and the 3 μm Firefli™ particles were ostensibly increasingly captured by the development of a viscous gel. In contrast, when using MPE50, the apparent particle velocity was maintained reasonably constant for the duration of the filtration cycle, measuring over 400 μm s^{-1} after 120 min of filtration (Fig. 11(b) and (c)).

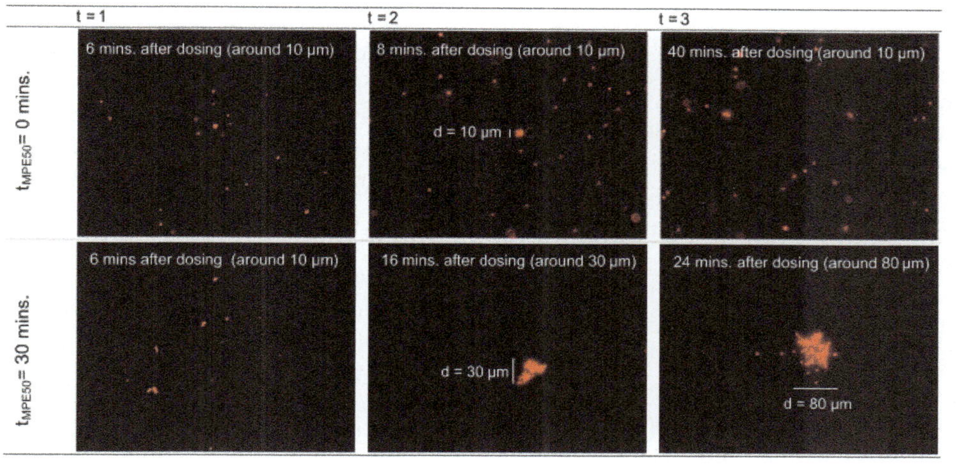

Figure 10: Aggregates visualised in the retentate channel following injection of MPE50.

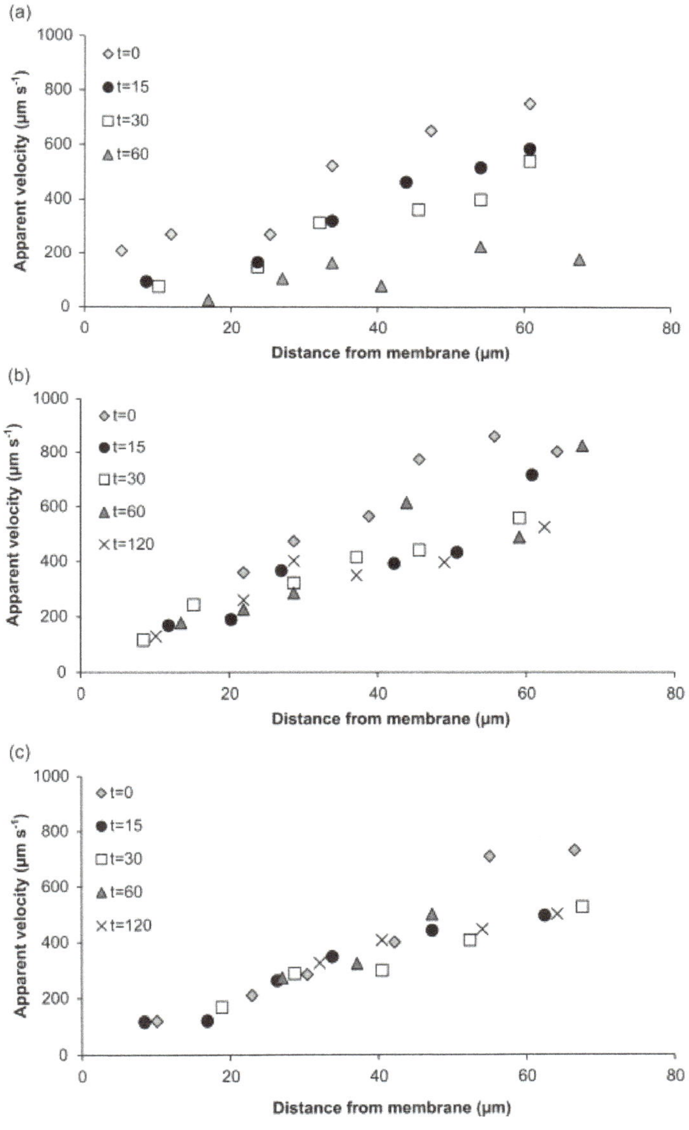

Figure 11: Particle transport velocity in fluidised region at differing time intervals during filtration. Experimental conditions as follows: (a) without addition of MPE50; (b) addition of MPE50 at commencement of filtration (0 min); and, (c) addition of MPE50 following a period of filtration.

DISCUSSION

In this study, the significance of coagulation kinetics on the stabilisation of membrane permeability following the reactive dosing of coagulant has been investigated within a model particle system comprising of a bimodal particle size distribution. An optimum MPE50 dose of 200 mg l^{-1} was identified by a 'peak' in polysaccharide removal efficiency and is analogous to previous observation of dosing MPE50 into real SMP where introducing an excess of chemical resulted in deflocculation [4] and [5]. Koseoglu et al. [5]explained this phenomenon by the adsorption of excess cationic polymer which then reversed the surface charge to net positive thereby promoting deflocculation via electrostatic repulsion. Importantly, the MPE50 dose identified in this study is within the range typically used in industry and resulted in elimination of between 10 and 30 mg l^{-1} of biopolymer which is coincident with the literature (Fig. 3). However, substantive flocculation was noted only to occur within bimodal particle systems when both protein and polysaccharide were present together (3 μm: Na-alg:BSA, d_{50} over 200 μm, Fig. 5). Hwang et al. [14] determined a d_{50} floc size of 179 μm when maintenance dosing MPE50 into the activated sludge of an MBR which compared to a d_{50} of only 101 μm without dosing. In this study, it is asserted that the first stage of flocculation was the initial formation of aggregates comprised of BSA and Na-alginate. This is analogous to behaviour observed by [7] in which formation of BSA: Na-alg soluble complexes was promoted by lowering zeta potential sufficient to permit the positively charged micro-regions on the protein to bind onto the negatively charged carboxyl groups on the alginate. Measurement of the aggregation kinetics confirmed that the assimilation of Na-alginate into aggregates is augmented when in the presence of BSA (Fig. 7(a)). Wilen et al. [30] positively correlated protein contained within the extracellular polymeric substances (EPS) to flocculating ability of activated sludge. This can be ascribed to the numerous multifunctional charged groups resident within the polypeptide structure of proteins that can promote aggregation [7] and was evidenced in this study by the high rate coefficient determined for the uptake of protein within the flocs formed when present in the complete particle mixture (Fig. 7(b)). It is therefore suggested that flocculation within this model particle system proceeds by the initial aggregation of BSA: Na-alginate into a polymer network, which then traps and embeds the 3 μm particles within their structure.

During filtration experiments, this initial phase of aggregation between BSA and Na-alginate was identified by the formation of a large fluidised cake (25 μm, Fig. 9(b)) which formed in the region local to the membrane surface after MPE50 was added 30 min into filtration (Fig. 9). Jiang et al. [1] undertook modelling of the backtransport velocity for a particle size range symptomatic of the bimodal distribution ordinarily noted in MBR studies and showed that submicron particles (such as biopolymers) had a high probability for deposition at the membrane surface as the convective forces employed during filtration often exceeded the shear forces applied by CFV, even within side-stream systems where CFV exceeding 0.5 m s^{-1} are often employed. In this study, it is asserted that the fluidised cake was formed following diffusion of MPE50 into a concentrated region of BSA and Na-alginate which had polarised local to the membrane surface as the shear conditions promoted by the CFV employed were insufficient to provide back transport for this particle size group (θ, 11 s^{-1}). A considerable rise in TMP was coincident with the development of the fluidised cake (Fig. 8). Thus whilst the addition of MPE50 reduced zeta potential, it is postulated that fouling was dominated by deposition within the polarised layer rather than specifically through promotion of particle-membrane interaction [31] and [7]. Neemann et al. [7] identified that it was at the initiation of the non-covalent protein-polysaccharide network where the highest fouling rate was determined. It is therefore suggested that this initial stage of protein-polysaccharide aggregation caused the formation of small particles (of around 10 μm or less, Fig. 5) which preferentially migrated toward the membrane forming a highly resistive cake layer. Around fifteen minutes after MPE50 was added, the fluidised cake had dissipated and large aggregates were seen to have developed (Fig. 9(c)) which were evidenced to comprise of 3 μm Firefli™ microspheres (Fig. 10). The formed aggregates were seen to be transported within the bulk flow in addition to undergoing migration through 'floc rolling' along the membrane wall. This migratory behaviour can be accounted for by the increased particle size which enhances back transport. To illustrate, using the shear induced diffusivity (SID) model set out in [2] (Eq. (3)), particles with diameters exceeding 80 μm could be subject to SID forces that would balance the convective force of permeation and hence favour particle migration away from the membrane surface:

$$J = 0.0595\gamma_0\left(\frac{a^4}{L}\right)^{1/3}\ln\left(\frac{\varphi_w}{\varphi_b}\right) \quad (3)$$

When dosing MPE50 at the onset of filtration (0 min), the same increase in aggregate size was not observed (Fig. 10) which suggests that the process of aggregation is favoured when a concentrated layer of biopolymers/microspheres has developed at the membrane surface. Polymeric coagulants ordinarily exhibit low diffusivities and high viscosities thus mixing is generally recommended to ensure enhanced dispersion of the polymer through the particle matrix using velocity gradients exceeding 1500 s^{-1} for up to 600 s [3]. Whilst a comparatively low shear profile was adopted in this study, it is asserted that the presence of the fluidised cake enhanced the probability for collision of destabilised particles through an applied concentration gradient which encouraged growth of large flocs. Importantly, dosing MPE50 was observed to reduce fouling whether added at the initiation of filtration or during the filtration cycle, which conforms with earlier observations [4] and [5]. However, residual resistance was lowest by dosing MPE50 at the onset of filtration (Fig. 8), which is ascribed to the early initiation of biopolymer aggregation which limited the development of a polarised layer of biopolymers at the membrane surface. Consequently, it is suggested that the low shear conditions initiated in MBR are not preferable to direct dosing of high concentration coagulants during filtration due to the complexity of the coagulation kinetics involved. Instead high concentration coagulant doses should be applied before or at the onset of filtration. These findings are corroborated at full scale where MPE50 dosing is recommended after chemical cleaning and before the initiation of filtration which will promote biopolymer assimilation within floc structures before permeation begins [24].

CONCLUSIONS

In this study, the impact of kinetically controlled floc growth on sustaining membrane permeability following reactive coagulant dosing was determined. Floc formation was indicated to comprise of two stages: (i) an initial destabilisation phase in which the charge on the protein and polysaccharide was sufficiently neutralised to encourage complexation; and (ii) entrapment of the coarse model particles (3 µm Firefli™ microspheres) in the polymeric complex during the floc growth phase. Floc growth was characterised by an expected time lag as with conventional flocculation systems and as the kinetics immediately favoured biopolymer aggregation, when coagulant was dosed during the filtration cycle, the intermediate biopolymer complexes (comprised of protein and polysaccharide) were preferentially transported toward the membrane causing fouling. In comparison, when coagulant was dosed at the onset of filtration, membrane fouling was constrained. It is asserted that by dosing at the onset of filtration: (i) early development of biopolymer aggregation is initiated which inhibits transport of the individual biopolymers to the membrane; and (ii) by dosing coagulant in the absence of a developed polarised layer, formation of biopolymer complexes local to the membrane is obviated. However, when dosing coagulant at the onset of filtration, only limited floc growth occurred which can be explained by the low applied wall shear rate and the absence of a 'polarised' region ostensibly promoting floc growth, which in turn enhances the probability for particle back-transport from the membrane. Based on results from the model particle system studied, it is proposed that reactive coagulant dosing is best undertaken in the absence of filtration and within a moderate shear environment for sufficient time to promote floc growth before filtration is initiated.

ACKNOWLEDGEMENTS

The authors would like to thank the Engineering and Physical Sciences Research Council for funding provided through the First Grant Scheme (EPSRC, EP/K010360/1) and Impact Acceleration Account. We would like to thank employees at Nalco for their support during the experimental phase of this work. The findings and conclusions contained within are those of the authors and do not necessarily reflect positions or policies of the funders, or Nalco. Enquiries for access to the data referred to in this article should be directed to: researchdata@cranfield.ac.uk.

REFERENCES

1. T. Jiang, M.D. Kennedy, C. Yoo, I. Nopens, W. van der Meer, H. Futselaar, J. C. Schippers, P.A. Vanrolleghem, Controlling submicron particle deposition in a side-stream membrane bioreactor: A theoretical hydrodynamic modelling approach incorporating energy consumption, J. Membr. Sci. 297 (2007) 141–151.
2. H. Li, A.G. Fane, H.G.L. Coster, S. Vigneswaran, An assessment of depolarisation models of crossflow microfiltration by direct observation through the membrane, J. Membr. Sci. 172 (2000) 135–147.
3. O.P. Sahu, P.K. Chaudhari, Review on chemical treatment of industrial wastewater, J. Appl. Sci. Environ. Manag. 17 (2013) 241–257.
4. V. Iversen, R. Mehrez, R.Y. Horng, C.H. Chen, F. Meng, A. Drews, B. Lesjean, M. Ernst, M. Jekel, M. Kraume, Fouling mitigation through flocculants and adsorbents addition in membrane bioreactors: Comparing lab and pilot studies, J. Membr. Sci. 345 (2009) 21–30.
5. H. Koseoglu, N.O. Yigit, V. Iversen, A. Drews, M. Kitis, B. Lesjean, M. Kraume, Effects of several different flux enhancing chemicals on filterability and fouling reduction of membrane bioreactor (MBR) mixed liquors, J. Membr. Sci. 320 (2008) 57–64.
6. T. Wozniak, Comparison of a conventional municipal plant and an MBR plant with and without MPE, Desalin. Water Treat. 47 (2012) 341–352.
7. F. Neemann, S. Rosenberger, B. Jefferson, E.J. McAdam, Non-covalent proteinpolysaccharide interactions and their influence on membrane fouling, J. Membr. Sci. 446 (2013) 310–317.
8. P. Côté, T. Young, S. Smoot, J. Peeters, Energy consumption of MBR for municipal wastewater treatment: current situation and potential, Proc. Water Environ. Fed. 14 (2013) 479–492.
9. E.J. McAdam, M. Pawlett, S.J. Judd, Fate and impact of organics in an immersed membrane bioreactor applied to brine denitrification and ion exchange regeneration, Water Res. 44 (2010) 69–76.
10. E. Reid, X. Liu, S.J. Judd, Effect of high salinity on activated sludge characteristics and membrane permeability in an immersed membrane bioreactor, Water Res. 283 (2006) 164–171.
11. S. Delgado, R. Villarroel, E. González, Effect of the shear intensity on fouling in submerged membrane bioreactor in wastewater treatment, J. Membr. Sci. 311 (2008) 173–181.
12. S. Jamal Khan, C. Visvanathan, V. Jegatheesan, R. Ben Aim, Influence of mechanical mixing rates on sludge characteristics and membrane fouling in MBRs, Sep. Sci. Technol. 43 (2008) 1826–1838.
13. E.J. McAdam, S.J. Judd, E. Cartmell, B. Jefferson, Influence of substrate on fouling in anoxic immersed membrane bioreactors, J. Membr. Sci. 41 (2007) 3859–3867.
14. B.-K. Hwang, W.-N. Lee, P.-K. Park, C.-H. Lee, I.-S. Chang, Effect of membrane fouling reducer on cake structure and membrane permeability in membrane bioreactor, J. Membr. Sci. 288 (2007) 149–156.
15. A. Drews, membrane fouling in membrane bioreactors – Characterisation, contradictions, cause and cures, J. Membr. Sci. 363 (2010) 1–28.
16. K. Yamamoto, M. Hiasa, T. Mahmood, T. Matsuo, Direct solid-liquid separation using hollow fibre membrane in an activated sludge aeration tank, Water Sci. Technol. 21 (1989) 43–54.
17. P. Cote, H. Buisson, C. Pound, G. Arakaki, Immersed membrane activated sludge for the reuse of municipal wastewater, Desalination 113 (1997) 189–196.
18. E.J. McAdam, E. Cartmell, S.J. Judd, Comparison of dead-end and continuous filtration conditions in a

denitrification membrane bioreactor, J. Membr. Sci. 369 (2011) 167–173.

19. O. Autin, E.J. McAdam, Method and apparatus for monitoring particles in a liquid, GB Patent Application No. 1516481.7 (2015).

20. S.-H. Yoon, J.H. Collins, D. Musale, S. Sundararajan, S.-P. tsai, G.A. Hallsby, J. F. Kong, J. Koppes, P. Cachia, Effects of flux enhancing polymer on the characteristics of sludge in membrane bioreactor process, Water Sci. Technol. 51 (2005) 151–157.

21. I.S. Chang, P. Le Clech, B. Jefferson, S. Judd, membrane fouling in membrane bioreactors for wastewater treatment, J. Environ. Eng. (2002) 1018–1029.

22. S. Rosenberger, C. Laabs, B. Lesjean, R. Gnirss, G. Amy, M. Jekel, J.-C. Schrotter, Impact of colloidal and soluble organic material on membrane performance in membrane bioreactors for municipal wastewater treatment, Water Res. 40 (2006) 710–720.

23. A. Drews, J. Mante, V. Iversen, M. Vocks, B. Lesjean, M. Kraume, Impact of ambient conditions on SMP elimination and rejection in MBRs, Water Res. 41 (2007) 3850–3858.

24. T. Wozniak, MBR design and operation using MPE-technology (Membrane Performance Enhancer), Desalination 250 (2010) 723–728.

25. S. Nataraj, R. Schomacker, M. Kraume, I.M. Mishra, A. Drews, Analyses of polysaccharide fouling mechanisms during crossflow membrane filtration, J. Membr. Sci. 308 (2008) 152–161.

26. Y. Marselina, Lifia, P. Le-Clech, R.M. Stuetz, V. Chen, Characterisation of membrane fouling deposition and removal by direct observation technique, J. Membr. Sci. 341 (2009) 163–171.

27. B. Frolund, R. Palmgren, K. Keiding, P.H. Nielsen, Extraction of extracellular polymers from activated sludge using a cation exchange resin, Water Res. 30 (1996) 1749–1758.

28. M. Dubois, K.A. Gilles, J.K. Hamiltin, P.A. Rebers, F. Smith, Colorimetric method for determination of sugars and related substances, Anal. Chem. 28 (1956) 350–356.

29. E.J. McAdam, S.J. Judd, Immersed membrane bioreactors for nitrate removal from drinking water: Cost and feasibility, Desalination 231 (2008) 52–60.

30. B.-M. Wilén, B. Jin, P. Lant, The influence of key chemical constituents in activated sludge on surface and flocculating properties, Water Res. 37 (2003) 2127–2139.

31. O. Ojajuni, S. Holder, G. Cavalli, J. Lee, D.P. Saroj, Rejection of caffeine and carbamazepine by surface-coated PVDF hollow-fiber membrane system, Ind. Eng. Chem. Res. 55 (2016) 2417–2425.

CHAPTER 12

CONTROLLING MOLECULAR WEIGHT CUT-OFF OF PEEK NANOFILTRATION MEMBRANES USING A DRYING METHOD

João da Silva Burgal, Ludmila Peeva, Patrizia Marchetti, Andrew Livingston

Keywords: Poly(ether ether ketone), Organic solvent nanofiltration, Drying method, Control of molecular weight cut-off, Statistical analysis

INTRODUCTION

Organic Solvent Nanofiltration (OSN) membranes can be used for separation in the chemical and pharmaceutical industry to perform concentration and purification and solvent recovery. Recently, it was shown how OSN could be used for catalytic reactions with reaction and separation occurring in situ under high temperature and basic conditions [1]. One of the main challenges of fabricating suitable OSN membranes is to have the right MWCO to perform the separation of interest. The most widely used method for manufacturing polymeric membranes is the phase inversion method. This method involves four mains steps: dissolving a polymer in an appropriate solvent (dope solution); membrane casting; phase inversion (wet or dry); and membrane post-treatment [2].

It is known that it is possible to manipulate the membrane performance by varying the composition of the dope solution, varying the conditions during the phase inversion step or via a post-treatment step (drying, conditioning or crosslinking) [3] and [4].

In the dope preparation step it is necessary to take into account the polymer concentration, the addition of volatile solvents, non-solvents (or 'bad' solvents) and pore forming additives [4]. It has been observed that polymer concentration has a significant effect on the viscosity which in turn affects the performance of the final membrane (higher concentration leads to higher selectivity but lower permeance) [5]. Volatile solvents such as ethylether (EE), tetrahydrofuran (THF) or dioxane, could be added to the dope solution in order to produce integrally skinned asymmetric membranes via the dry/wet method (where the evaporation step is essential) [6] and [7]. By allowing partial evaporation of the volatile solvent between the casting and immersion step, a skin-layer with elevated polymer concentration can be formed. Pore forming additives can be used to increase permeability and porosity with or without compromising the selectivity. For example addition of LiCl or $LiNO_3$ to poly(amide-hidrazide) (PAH) casting solutions result in a higher permeability without lowering selectivity [5]. Besides inorganic additives it is also common to use organic additives such as glycerol, polyethylene glycol (PEG) or polyvinylpyrrolidone (PVP); for example, adding maleic acid (MA) to cellulose triacetate (CTA) increases the porosity and permeability [5] and [8].

The choice of the non-solvent (coagulation bath) affects the membrane morphology as well. The higher the rate of exchange between solvent and non-solvent the higher the porosity of the membrane; this is the case, for example, of the N-Methyl-2-pyrrolidone (N-NMP) /water (solvent/non-solvent) pair for polysulfone (PSf) membranes. It is also common to use additives like alcohols or DMF to vary the exchange rate of solvent/non-solvent. Another factor to take into account in the coagulation bath is the temperature. In general, an increase in the temperature of the coagulation bath leads to a higher exchange rate and consequently to a more porous structure [9].

In the post-casting treatment several factors such as temperature and time of evaporation, relative humidity of the

Citation: João da Silva Burgal, Ludmila Peeva, Patrizia Marchetti, Andrew Livingston, Controlling molecular weight cut-off of PEEK nanofiltration membranes using a drying method, Journal of Membrane Science, Volume 493, 1 November 2015, Pages 524-538, ISSN 0376-7388, http://dx.doi.org/10.1016/j.memsci.2015.07.012.

air and air velocity (if a convective flow is applied) can affect the membrane performance. In terms of evaporation time there seems to be two contradictory effects. For membranes prepared from PA (polyamide), PAH (poly(amide hydrazide) and CA (cellulose acetate) the flux decreases and the rejection increases with increasing evaporation time. However, Soroko et al. [10] and See-Toh et al. [7] have concluded that increasing evaporation time reduces the flux but has no effect on rejection for polyimide (PI) membranes. In terms of temperature of evaporation, Young et al. [11] found that for poly(ethylene-co-vinyl alcohol) (EVAL)-although not used for OSN – a membrane structure with a particulate morphology was obtained at low temperatures after all the casting solution evaporated, while the rise in the evaporation temperature changed EVAL membrane structure from a particulate to a dense morphology. In addition, for PAH it was observed that with increasing temperature lower fluxes and higher rejections were obtained reaching a plateau at 100 °C; above this temperature inverse behaviour occurred in terms of solute rejection (probably due to polymer degradation) [9] and [12].

In order to stabilise and improve membrane performance there are several post-treatments that can be used such as annealing the membrane in water or under dry conditions, exposure to concentrated mineral acids, plasma treatment, drying with the solvent exchange technique and treatment with conditioning agents [5], [13], [14],[15] and [16].

For most membranes prepared by wet phase inversion it is common for membranes to be stored under wet conditions because the structure of the membrane changes ("collapses") when the membrane is subjected to a drying process. In the case of ultrafiltration membranes (and nanofiltration as well) drying almost without exceptions induces irreversible loss of solvent permeance which is thought to be related to the collapse of the nodular structure [14]. In fact, using a multiple solvent exchange procedure can minimise the risk of nodule collapse upon drying. In this procedure, the residual non-solvent present in the membrane after immersion is replaced by a first solvent, which is miscible with the non-solvent; this solvent is then replaced by a more volatile solvent, which can be removed easily by evaporation to obtain a dry membrane[17], [18] and [19]. One way to describe this nodular collapse is by using the theory introduced by Brown [20] for polymer latex particles during film formation. Beerlage [14]used this theory for PI ultrafiltration membranes and related the capillary forces (F_c) with the resistance of the matrix to deformation (F_r) developed by Brown. F_c is given by Eq.(1) where γ ($N\ m^{-2}$) is the surface tension of the gas/liquid interface inside the pores, r_p(m) is the pore radius, θ (deg) is the contact angle between the liquid and the membrane material and A (m^2) is the pore cross sectional area. F_r is given by Eq. (2) where E ($N\ m^{-2}$) is the tensile modulus of the polymer material and is a measure of the pore wall elasticity. According to this approach if $F_c > F_r$ then collapse of the nodular structure will occur (Eq. (3)).

$$F_c = \frac{2\gamma}{r_p} \cdot \cos\theta \cdot A \tag{1}$$

$$F_r = 0.37E \cdot A \tag{2}$$

Thus for any given pore size the pore will collapse if

$$\frac{2\gamma \cdot \cos\theta}{r_p} > 0.37 \times E \tag{3}$$

Based on the decrease of surface tension (for example via solvent exchange) it is possible to maintain the pore structure of a membrane (i.e. to minimise the capillary force) if the strength of the matrix is high enough [14], [18], [19], [20] and [21].

In a previous work we introduced the excellent stability and performance of native PEEK nanofiltration membranes prepared in our laboratory [22]. This research work communicates a detailed investigation of the production of non-sulphonated nanofiltration PEEK membrane with a MWCO around 350–500 g mol^{-1}, resistant to polar aprotic solvents (such as DMF), high temperature, acids and bases [1]. Different factors affecting membrane separation performance are studied including polymer concentration in the dope solution and membrane post-treatment procedures. It has been shown that the post-manufacturing membrane drying step is of vital importance for the membrane nanofiltration performance.

METHODS

Materials

2,4-Diphenyl-4-methyl-1-pentene (α-methyl styrene dimer) and methane sulphonic acid (MSA) were obtained from Sigma-Aldrich. N,N-Dimethylformamide (DMF), tetrahydrofuran (THF), propanone (acetone), 2-propanol (IPA), methanol (MeOH), ethanol (EtOH), n-hexane, heptane, acetonitrile and sulphuric acid (SA) 95 vol% were obtained from VWR UK. VESTAKEEP® 2000P and 4000P were kindly obtained from Evonik Industries; VICTREX® 150P and 450P were kindly donated by VICTREX®. The styrene oligomers standards with a molecular weight distribution of 580 (PS580) and 1300 (PS1300) were obtained from Agilent Technologies Deutschland GmbH, Germany. All reagents were used as received without any further purification.

Membrane Preparation

PEEK powder VESTAKEEP® 4000P was dissolved at a concentration of 8 wt%, 10 wt% and 12 wt% in a mixture of 3:1 wt% methanesulfonic acid (MSA) and sulphuric acid (SA) by mechanical stirring (IKA RW 20 digital) at 20 °C until complete homogenisation of polymer solution. Prior to casting the polymer solution was left 72–96 h at 20 °C until complete removal of air bubbles. The membranes were cast using a bench top laboratory casting machine (Elcometer 4340 Automatic Film Applicator) with a blade film applicator (Elcometer 3700) set at 250 μm thickness. The polymer dope solution obtained was poured into the blade and cast on a non-woven polypropylene support (Novatex 2471, Freudenberg Filtration Technologies Germany) with a transverse speed of 0.5 cm s^{-1}. Following this, the membranes were immersed in a deionised (DI) water precipitation bath at 20 °C; the water in the bath was changed several times until pH was 6–7. For the membranes produced with a 12 wt% polymer concentration a solvent exchange from water to IPA, MeOH, n-hexane, EtOH, acetone, THF, heptane or acetonitrile was performed. Finally, the membranes were left to dry at either 20 °C, 40 °C, 80 °C, 120 °C or 140 °C[1] (reported as the glass transition temperature for PEEK) [12,23]. Membrane preparation steps are represented in Fig. 1. The viscosity of the dope solution was measured immediately after casting using a rotary viscometer (LV-2020

Rotary Viscometer Cannon instruments, S16 spindle) and all values were recorded at 1 rpm spindle speed and 20 °C.

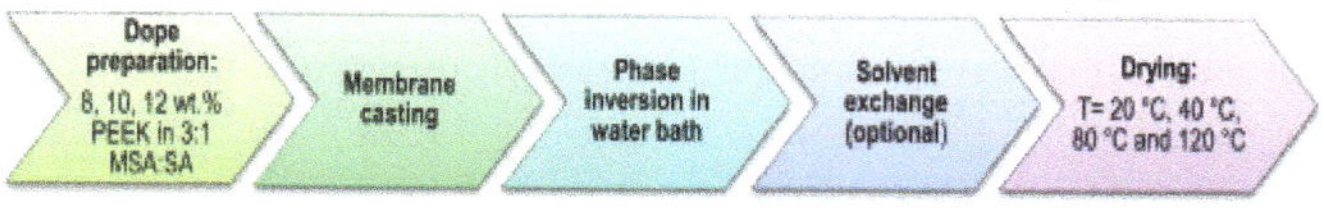

Figure 1: Schematic representation of the steps involved in the PEEK membrane preparation.

All the membrane formation steps were performed in an air conditioned room set at 20 °C and with a relative humidity (RH) in the range of 30–40%.

Membrane characterisation

Thermal gravimetric analysis (TGA)

TGA measurements of PEEK samples were performed using a TGA Q500 (TA Instruments) and 100 µL platinum pans. The measurements were done under nitrogen and oxygen atmosphere and a gas flow of 40 mL min^{-1} for nitrogen and 60 mL min^{-1} for air. The heating rates varied between 10 K min^{-1} and 40 K min^{-1} and each sample was maintained at the target temperatures – 20 °C, 40 °C, 80 °C, 100 °C and 120 ° – for 400 min (isothermal step).

Dynamic mechanical analysis (DMA)

Rectangular specimens of water "wet" membranes having a size of 35 mm×6 mm×0.2 mm (*L*×*W*×*H*) were used for the dynamic mechanical experiments. Dynamic mechanical thermal analyser (Tritec 2000DMA, TA Instruments) was used for the evaluation of the dynamic modulus (stiffness) and mechanical damping (tan). Membrane properties were measured over the temperature range from 25 to 120 °C at a heating rate of 2 K min^{-1}. The tests were carried out at 1 Hz with a displacement of 0.05 mm.

Dynamic scanning calorimetry (DSC)

Changes in the degree of crystallinity of the samples during drying were observed by differential scanning calorimetry (DSC) (DSC Q200, TA Instruments). Samples were heated from 20 to 400 °C at a constant ramp rate of 10 °C min^{-1} in DSC aluminium pans (heating cycle 1). After cooling down at a rate of 10 °C min^{-1} to 20 °C the samples were heated using the same method as the one used in heating cycle 1 (heating cycle 2). A sharp peak at about 330–340 °C is characteristic of PEEK crystal melting. The area under the melting curve was used to calculate the heat required for the melting process. The heat of melting for a 100% crystalline PEEK sample is 130 J g^{-1}[23]. Thus, the ratio of the two heats of melting was calculated to obtain the degree of crystallinity of the sample.

Scanning electron microscopy (SEM)

For cross-section imaging a membrane sample was broken in liquid nitrogen and pasted vertically onto SEM stubs covered with carbon tape. For surface imaging a membrane sample was cut and pasted horizontally onto SEM stubs covered with carbon tape. The samples were then coated with a chromium-layer in an Emitech K575X peltier under an argon atmosphere to reduce sample charging under the electron beam. SEM pictures of the surface and cross section of membrane samples were recorded using a Scanning electron microscope of low resolution (JEOL 6400) at 20 KV and under dry conditions at room temperature.

Membrane performance and analysis

ZIn order to test the membranes a rig with 8 membrane cross-flow cells was used (effective membrane area=14 cm^2 at each cell, see Fig. 2). PEEK membranes were initially conditioned by passing pure solvent through at 30 °C and 30 bar (for 1 h). Polystyrene standard solution was then poured into the feed reservoir and the system was pressurised again up to 30 bar and the temperature set at 30 °C.

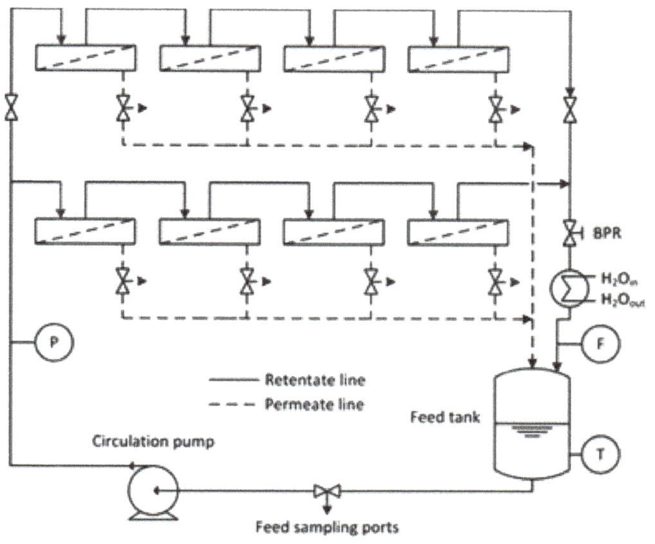

Figure 2: Schematic representation of the 8 cells cross-flow rig used in this study. P – pressure gauge; T – thermocouple; F – flow metre; BPR – back pressure regulator.

The polystyrene standard solution was prepared by dissolving 2,4-Diphenyl-4-methyl-1-pentene (dimer, M_W=236 g mol^{-1}) and Polystyrene Standards with a M_W ranging from 295 to 1995 g mol^{-1} (homologous series of styrene oligomers (PS)) in DMF or THF at a concentration of 1 g L^{-1} each M_W. Permeate and retentate samples were collected at different time intervals for rejection determination. Concentrations of PS in permeate and retentate samples were analysed using an Agilent HPLC system with a UV/Vis detector set at a wavelength of 264 nm. Separation was accomplished using an ACE 5-C18-300 column (Advanced Chromatography Technologies, ACT, UK). A mobile phase comprising 35 vol% analytical grade water and 65 vol% tetrahydrofuran (THF) both containing 0.1 vol % trifluoroacetic acid was used [24].

The flux (J) and permeance (L_p) were determined using Eqs. (4) and (5) and the rejection (R_i) of PS was evaluated applying Eq. (6). The corresponding MWCO curves were obtained from a plot of the rejection of PS versus their molecular weight. To eliminate the effect of compaction typically observed over the first 2–5 h of experiment, for membrane performance comparison purposes only the steady state flux (after 24 h) was considered (steady state flux was considered achieved when two flux (permeance) measurements within a 1 h interval showed the same value within ±0.02 L m^{-2} h^{-1} bar^{-1}).

$$J\left[\mathrm{L\,h^{-1}\,m^{-2}}\right] = \frac{\text{Flow rate}\left[\mathrm{L\,h^{-1}}\right]}{\text{Membrane area}\left[\mathrm{m^2}\right]} \quad (4)$$

$$L_p\left[\mathrm{Lh^{-1}m^{-2}bar^{-1}}\right] = \frac{J\left[\mathrm{Lh^{-1}m^{-2}}\right]}{\Delta p\ [\mathrm{bar}]} \quad (5)$$

$$R_i\ [\%] = \left(1 - \frac{C_{P,i}}{C_{F,i}}\right) \times 100 \quad (6)$$

Experimental Design

The methodology used in this study was based on the comparison of PEEK membranes dried at 120 °C and produced either with different polymer concentrations (8 wt%, 10 wt% and 12 wt%) or dried from different solvents (solvent exchange) in terms of performance (permeance and rejection). For each of the different membranes four replications were performed in order to have a statistically robust sample. All the results were analysed using *F*-test. For the permeance data the *F*-test was used for permeance values obtained after 24 h. For rejection data the *F*-test was applied to each individual polystyrene (PS), i.e. for each solute size (different MW) the four different membranes were compared with each other. Statistical significance was considered at $p<0.05$. Data are presented as means ± standard deviation of the mean (SDM).

RESULTS AND DISCUSSION

Control of pore collapsing for tuning MWCO

The effect of polymer concentration and drying temperature

As reported previously, the nanofiltration properties of PEEK membranes are correlated to the drying of the membranes [21]. Nevertheless, the permeance values reported were relatively low. In order to improve the permeance – without compromising the MWCO – a study on polymer concentration (8–12 wt%) and drying temperatures (20 °C, 40 °C, 80 °C and 120 °C) was performed in order to determine their influence on membrane performance (Table 1).

Table 1: Summary of PEEK membranes PM-B prepared from dopes with different polymer concentrations (8 wt%, 10 wt% and 12 wt%) and dried from water at different temperatures. The viscosity (Pa s) of the membrane dope solution as well as the spindle speed (rpm) used are presented in this table. These membranes were used to test the influence of polymer concentration and drying temperature on permeance and rejection.

Membrane code	Polymer concentration (wt%)	Viscosity (Pa s)	Spindle speed (rpm)	Drying temperature (°C)
PM-B 8 wt% 20 °C	8	7.72±0.04	10	20
PM-B 8 wt% 40 °C				40
PM-B 8 wt% 80 °C				80
PM-B 8 wt% 120 °C				120
PM-B 10 wt% 20 °C	10	25.46±1.86	3	20
PM-B 10 wt% 40 °C				40
PM-B 10 wt% 80 °C				80
PM-B 10 wt% 120 °C				120
PM-B 12 wt% 20 °C	12	58.03±1.58	1	20
PM-B 12 wt% 40 °C				40
PM-B 12 wt% 80 °C				80
PM-B 12 wt% 120 °C				120

As expected, the membranes with lower polymer concentration (8 wt%) presented higher permeance values, in the range of 1.25–2.30 L h^{-1} m^{-2} bar^{-1}, and a MWCO in the range of 795–1295 g mol^{-1} (Fig. 3). Interestingly, the membranes dried at 20 °C and 120 °C were the tightest ones and with lower permeance whereas the ones dried at 40 °C and 80 °C presented a higher MWCO and higher permeance, i.e., there was no trend as a function of the temperature. Both higher polymer concentrations – 10 wt% and 12 wt% – presented lower permeances and lower MWCO (tighter membranes). The permeance of the membranes prepared with 10 wt% of polymer was in the range of 0.42–0.52 L.h^{-1} m^{-2} bar^{-1} and the MWCO was in the range of 395 –495 g mol^{-1}. As for the 12 wt% membranes, the permeance was in the range of 0.18–0.40 L h^{-1} m^{-2} bar^{-1} and the MWCO was in the range of 295–395 g mol^{-1}. It can be seen from the different membranes dried at 120 °C (Fig. 4) that the polymer concentration had a greater influence on membrane performance than drying temperature (Fig. 3). The difference was more noticeable between the 8 wt% and the 10 wt% than between the 10 wt% and the 12 wt%. The difference in performance for different polymer concentrations could be explained by the viscosity of the dope solution, because the 8 wt% polymer dope solution had 3.30 times and 7.51 times lower viscosity than the 10 wt% polymer dope and 12 wt% polymer dope respectively. In contrast, the difference in viscosity between 10 wt% polymer dope solution and 12 wt% polymer dope solution was only 2.28 times. The viscosity of the dope solution (Table 1) could explain the results obtained because higher casting solution viscosities slow down non-solvent in-diffusion and demixing is delayed, resulting in membranes with thicker and denser skin-layers and sublayers with lower porosities. From the SEM images

(Fig. 4 bottom) it was found that membranes PM-B 8 wt% 120 °C had a thinner active layer of approximately 2.0 μm whereas for membranes with higher polymer concentration the active layer had a thickness of approximately 2.9 μm.

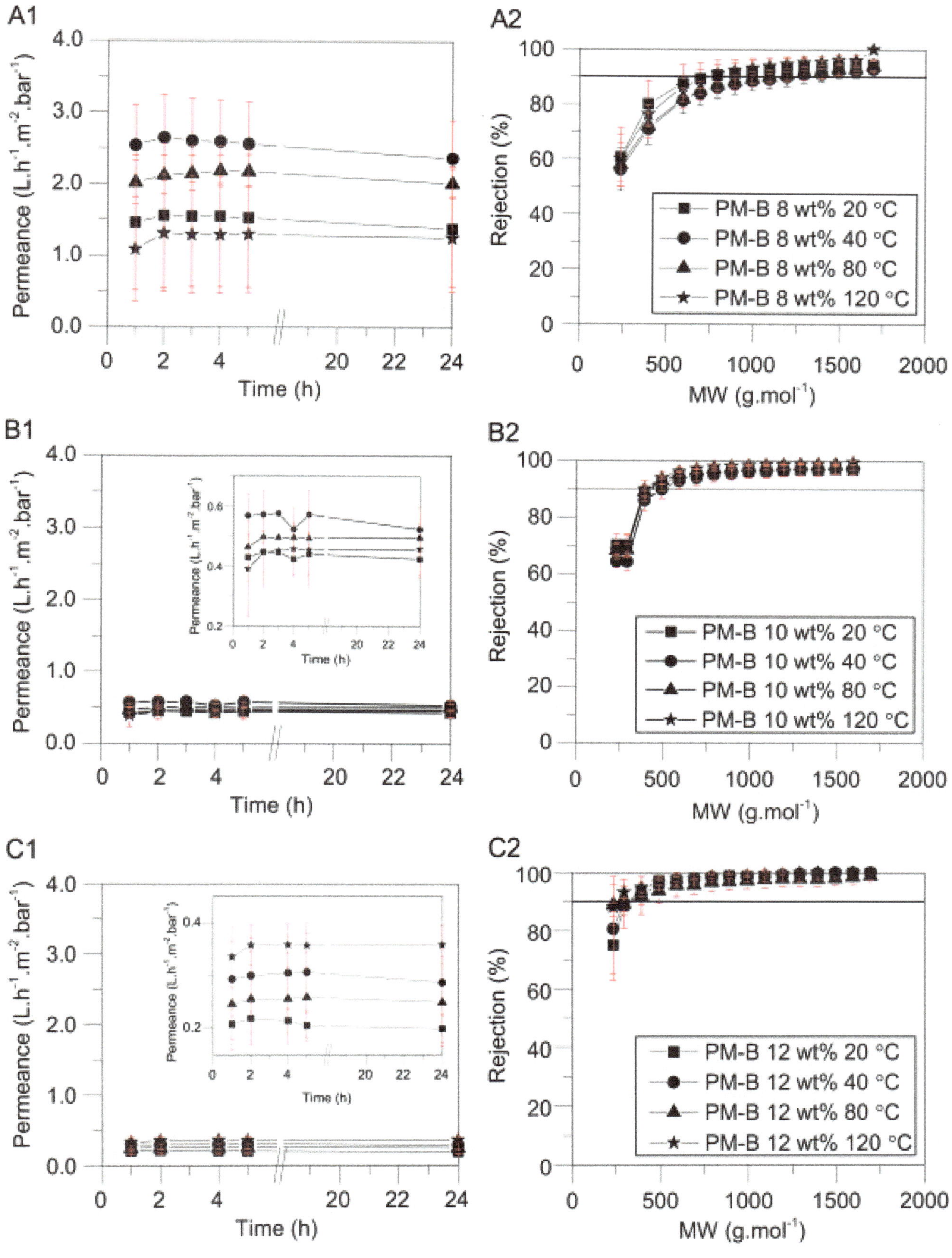

Figure 3: A1, B1 and C1: Permeance values (L h^{-1} m^{-2} bar^{-1}) over a period of 24 h for the different membranes under study. A2, B2 and C2: Rejection values of the different PEEK membranes under study as a function of the molecular weight (M_W, g mol^{-1}) of different polystyrenes after 24 h. All the membranes presented were dried from water at different temperatures (20 °C, 40 °C, 80 °C and 120 °C) prior to their insertion in the cross-flow cells. The membranes were filtered with a solution of THF and PS (1 g L^{-1}). The flow-rate, temperature and pressure were set at 100 L h^{-1}, 30 °C and 30 bar, respectively. The red bars represent the standard deviation of the mean. (For interpretation of the references to color in this figure legend, the reader is referred to the web version of this article.)

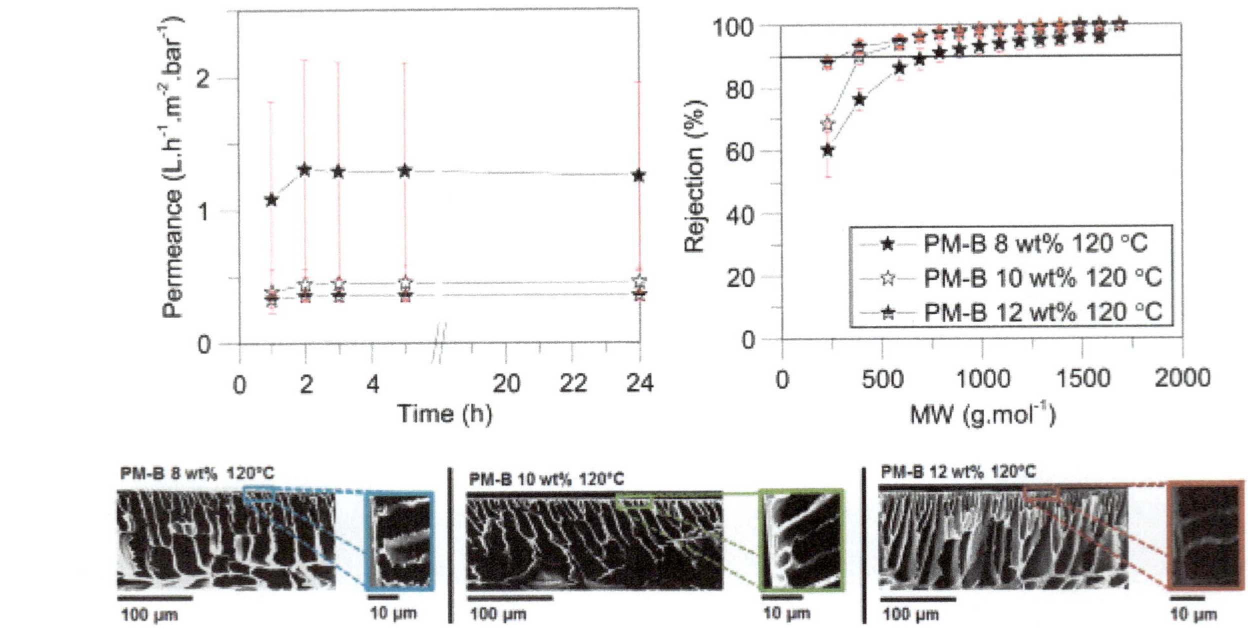

Figure 4: Top left: Permeance values (L h^{-1} m^{-2} bar^{-1}) over a period of 24 h for the different membranes under study. Top right: Rejection values of the different PEEK membranes under study as a function of the molecular weight (M_W, g mol^{-1}) of different polystyrenes after 24 h. All the membranes presented were dried from water at 120 °C prior to their insertion in the cross-flow cells. The membranes were filtered with a solution of THF and PS (1 g L^{-1}). The flow-rate, temperature and pressure were set at 100 L h^{-1}, 30 °C and 30 bar, respectively. The red bars represent the standard deviation of the mean. The membranes dried from water at 120 °C are significantly different ($p \leq 0.05$, *F*-test). Bottom: Cross-section SEM images (magnification 300×) of the different membranes under study: PM-B 8 wt% 120 °C, PM-B 10 wt% 120 °C and PM-B 12 wt% 120 °C. (For interpretation of the references to color in this figure legend, the reader is referred to the web version of this article.)

There is no sharp border line between MWCO of the different membrane separation processes, however in general nanofiltration is considered to cover separations of molecules within the 200–2000 Da range [3]. As such all PEEK membranes reported in this study are nanofiltration membranes. However currently the organic solvent nanofiltration membranes market suffers a lack of "tight" membranes with MWCO at the lowest range of nanofiltration, ~200 Da, that can be used for example for solvent recovery. To the best of our knowledge there is only one OSN membrane on the market claiming MWCO of 150 Da – Duramem150 manufactured by Evonik Industries. That is why we have chosen to investigate further the tightest membrane from the PEEK series in an attempt to manipulate its MWCO and eventually make it tighter. Given the fact that the membranes with a polymer concentration of 12 wt% presented the lowest MWCO, all subsequent studies in this research work were performed using this polymer concentration.

The effect of drying solvent

As already mentioned in Section 2 of the manuscript, according to the literature[14] and [20], the final membrane pore size is greatly influenced by the surface tension of the solvent filling membrane pores prior to drying. To investigate this effect on the PEEK membranes a solvent exchange from water to IPA, MeOH, EtOH, n-hexane, acetone or THF was performed after the phase inversion process in order to change the surface tension and possibly achieve different extents of collapsing in the polymer nodular structure. Water has a surface tension of 72.8 mN m^{-1} while the remaining solvents have similar (and much lower) values of surface tension in the range of 18.4–26.4 mN.m^{-1}(Table 2).

Table 2: Summary of PEEK membranes PM-B 12 wt% prepared from different dopes and with different post-treatments. These membranes were used to test the influence of solvent exchange and drying temperature on permeance and rejection. In addition, properties of the solvents used for the solvent exchange: surface tension (mN m^{-1}), M_W (g mol^{-1}), boiling point (°C), vapour pressure (kPa) and molar volume (cm^3 mol^{-1}). All properties listed were obtained from [25] at 20 °C and 1 bar.

Membrane code	Drying solvent	Drying temperature (°C)	**Solvent properties**			
			Surface tension (mN m^{-1})	Boiling point (°C)	Vapour pressure (kPa)	Molar volume (cm^3 mol^{-1})
PM-B1.1	Water	20	72.8	100	2.33	18.0
PM-B1.2	Water	40				
PM-B1.3	Water	80				
PM-B1.4	Water	120				

PM-B2.1	MeOH	20	22.6	64	16.93	40.6
PM-B2.2	MeOH	40				
PM-B2.3	MeOH	80				
PM-B2.4	MeOH	120				
PM-B3.1	EtOH	20	22.3	78	5.95	58.6
PM-B3.2	EtOH	40				
PM-B3.3	EtOH	80				
PM-B3.4	EtOH	120				
PM-B4.1	IPA	20	21.7	82	4.10	76.9
PM-B4.2	IPA	40				
PM-B4.3	IPA	80				
PM-B4.4	IPA	120				
PM-B5.1	Acetone	20	23.3	56	30.80	73.8
PM-B5.2	Acetone	40				
PM-B5.3	Acetone	80				
PM-B5.4	Acetone	120				
PM-B6.1	THF	20	26.4	66	21.60	81.9
PM-B6.2	THF	40				
PM-B6.3	THF	80				
PM-B6.4	THF	120				
PM-B7.1	n-hexane	20	18.4	69	20.17	131.4
PM-B7.2	n-hexane	40				
PM-B7.3	n-hexane	80				
PM-B7.4	n-hexane	120				

The contact angle water/PEEK was measured to be 60°. We were unable to measure contact angles for the other solvents, since the droplet spread instantaneously, thus these contact angles were assumed as 0°. Therefore, and according to the theory presented by Brown (Eq. (3)) [20], membranes immersed in IPA, MeOH, EtOH should give similar MWCO because of the similarity in surface tension; n-hexane should present higher MWCO (looser membranes) because it has the lowest surface tension and acetone and THF should give tighter membranes (excluding the ones dried from water). According to this method F_c should be higher for water at any given pore radius and therefore, pore collapse in water is expected to occur at a much higher extent. As a result, membranes dried from all the other solvents should be looser than membranes dried from water with the following order (from lower MWCO to higher MWCO membrane): water<THF<acetone<MeOH< EtOH<IPA<n-hexane. Together with the solvent type the effect of drying temperature on the permeance and on the MWCO was also studied. The membranes produced are presented in Table 2.

As mentioned before in Section 3.1.1, the permeance for all membranes dried from water at different temperatures had permeance values in the range of 0.20–0.36 L h^{-1} m^{-2} bar^{-1}. The membranes dried from water at 120 °C had almost double the permeance of membranes dried at 20 °C. This fact could be attributed to residual water that may have been retained in the smallest pores (thus obstructing solvent permeance), whilst above 100 °C (boiling point of water at 1 bar) all residual water may have been completely removed (hence higher permeance). Another interesting result was to determine the effect of temperature on the degree of crystallinity of the membranes dried from water (Fig. 13 Appendix). It can been seen that from PM-B1.1 (dried at 20 °C) to PM-B1.4 (dried at 120 °C) there were no changes in the membrane crystallinity. A membrane dried at 140 °C was also prepared. It showed THF permeance of 0.04 L h^{-1} m^{-2} bar^{-1} but no rejection in the NF range (data not shown), possibly due to defects originating from the partial melting of the backing material, and was not further investigated. Thus all further drying experiments were limited to 120 °C.

As for membranes dried from the alcohols, it can be observed that for MeOH (Fig. 5A1 and A2) the permeance values varied more with the temperature ranging from 1.07 L h^{-1} m^{-2} bar^{-1} (PM-B2.4) to 2.3 L h^{-1} m^{-2} bar^{-1} (PM-B2.3).

From the rejection data (Fig. 5A2) it can be observed that the drying temperature has a greater effect on the MWCO, i.e, the higher the drying temperature the tighter the membrane. For the temperatures of 40 °C and 80 °C the rejection values were in fact quite similar, although the variability of the membranes PM-B2.2 and PM-B2.3 makes it difficult to confirm this result. The loosest membrane, PM-B2.1, has a MWCO beyond the NF range. Membranes PM-B2.2 and PM-B2.3 presented a MWCO of around 1300 g mol^{-1}, but the standard deviation was not narrow enough to validate the result. The tightest membrane, PM-B2.4, had a MWCO around 600 g mol^{-1}.

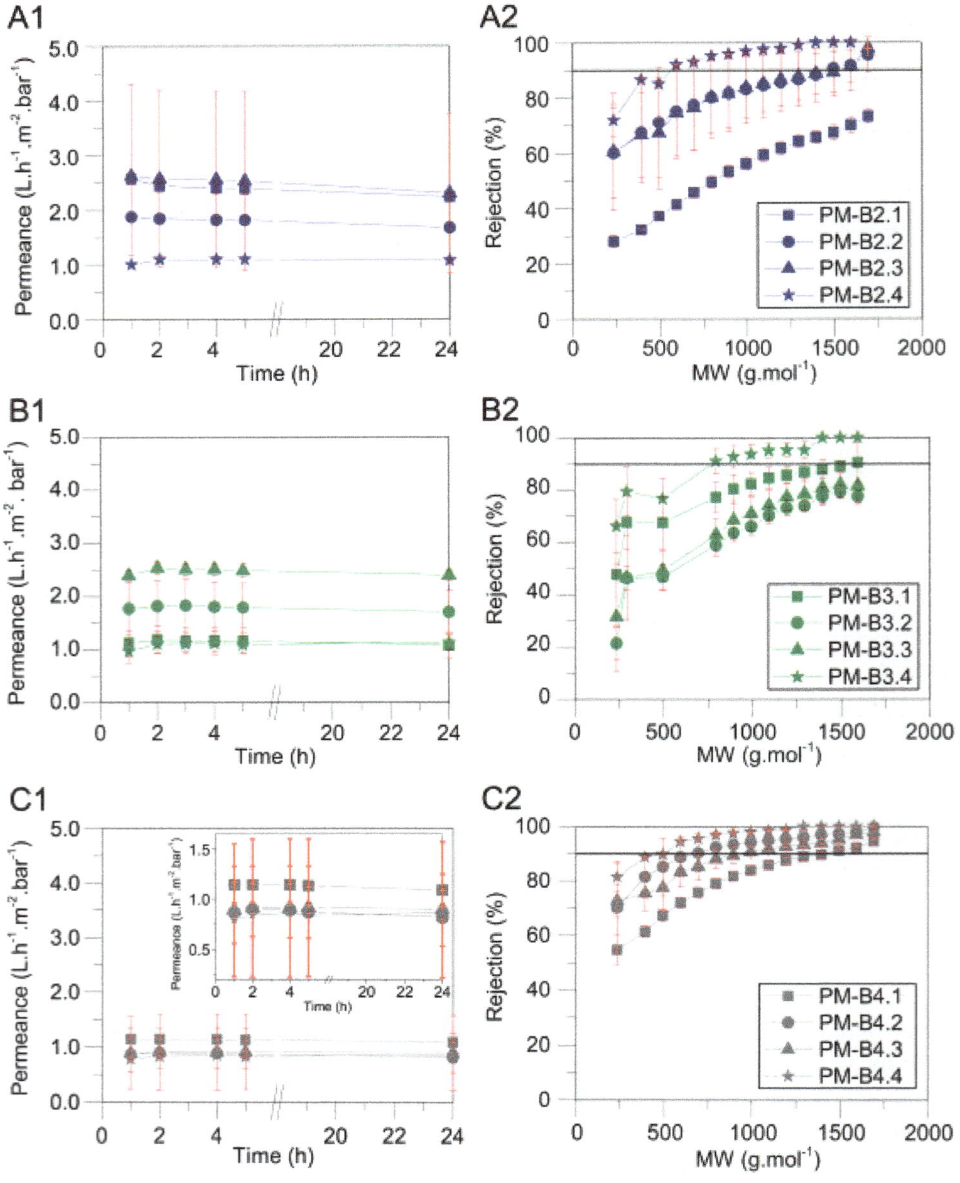

Figure 5: A1, B1 and C1: Permeance values (L h^{-1} m^{-2} bar^{-1}) over a period of 24 h for the different membranes under study. A2, B2 and C2: Rejection values of the different PEEK membranes under study as a function of the molecular weight (M_W, g mol^{-1}) of different polystyrenes after 24 h. Membranes PM-B2.*x*, PM-B3.*x*and PM-B4.*x* (*x*=1, 2, 3 and 4) were dried from MeOH, EtOH and IPA respectively at different temperatures (20 °C, 40 °C, 80 °C and 120 °C) prior to their insertion in the cross-flow cells. The membranes were filtered with a solution of THF and PS (1 g L^{-1}). The flow-rate, temperature and pressure were set at 100 L h^{-1}, 30 °C and 30 bar, respectively. The red bars represent the standard deviation of the mean. (For interpretation of the references to color in this figure legend, the reader is referred to the web version of this article.)

For the membranes dried from EtOH (Fig. 5B1 and B2) the permeance varied from 1.07 L h^{-1} m^{-2} bar^{-1} (PM-B3.1) to 2.1 L h^{-1} m^{-2} bar^{-1} (PM-B3.3). From the rejection data (Fig. 5B2) one can observe that for the temperatures of 40 °C and 80 °C the rejection values were quite similar and both had a MWCO beyond the NF range; membrane PM-B3.1 presented a MWCO of around 1595 g mol^{-1}; and the tightest membrane, PM-B3.4, had a MWCO around 795 g mol^{-1}.

For the membranes dried from IPA (Fig. 5C1 and C2) the permeance was on average 3.5 times higher than the membranes dried from water. In fact, the values of permeance ranged from 0.81 L h^{-1} m^{-2} bar^{-1} (PM-B4.2, dried at 40 °C) to 1.36 L h^{-1} m^{-2} bar^{-1} (PM-B4.4 dried at 20 °C). Analysing the rejection data it can be seen that the higher the drying temperature, the tighter the membrane, with the exception of PM-B4.2 (dried at 40 °C). For temperatures of 40 °C and 80 °C the rejection values were quite similar, although slightly higher for PM-B4.2 (as mentioned above). The membrane with the lowest permeance (PM-B4.4) presented the lowest MWCO and its value was around 500 g mol^{-1}. For membrane PM-B4.1 (membrane with a high permeance) the MWCO was in the upper range of NF with a value around 1400 g mol^{-1}.

In the case of alcohols, we speculate that the boiling points of each of the alcohols are lower than the boiling point of water (Table 2) which allows for more solvent to be removed from the membrane pores at a faster rate; therefore, the drying temperature had more pronounced effect on the properties of the membrane when compared with water.

Fig. 6D and E shows that membranes dried from acetone and THF were affected to a greater extent by the temperature. For both solvents (acetone and THF), the membranes had similar performances at 20 °C to 80 °C, but a substantial difference arose when dried at 120 °C (Fig. 6D2 and E2). In the case of acetone, the membranes dried at 120 °C had a permeance of 2.15 L h^{-1} m^{-2} bar^{-1} which was on average 4.5 times lower than for any other drying temperature considered; the MWCO was 895 g mol^{-1} while for the other drying temperatures the membranes produced were not in the NF range. For the membranes dried from THF over the temperature range of 20–80 °C the standard deviations made it difficult to assess within a confidence interval either permeance and rejection. Nevertheless, for a temperature of 120 °C the membranes presented a permeance of 2.72 L h^{-1} m^{-2} bar^{-1} which was on average 28 times lower than PMB-6.1 and 12 times lower than PMB-6.2 and PMB-6.3. This membrane did not present a MWCO in the NF range, but nevertheless from Fig. 6E2 one can observe that a shift occurred in terms of rejection when comparing PM-B6.4 with the other membranes, presumably due to tightening of the membrane matrix by increasing drying temperature.

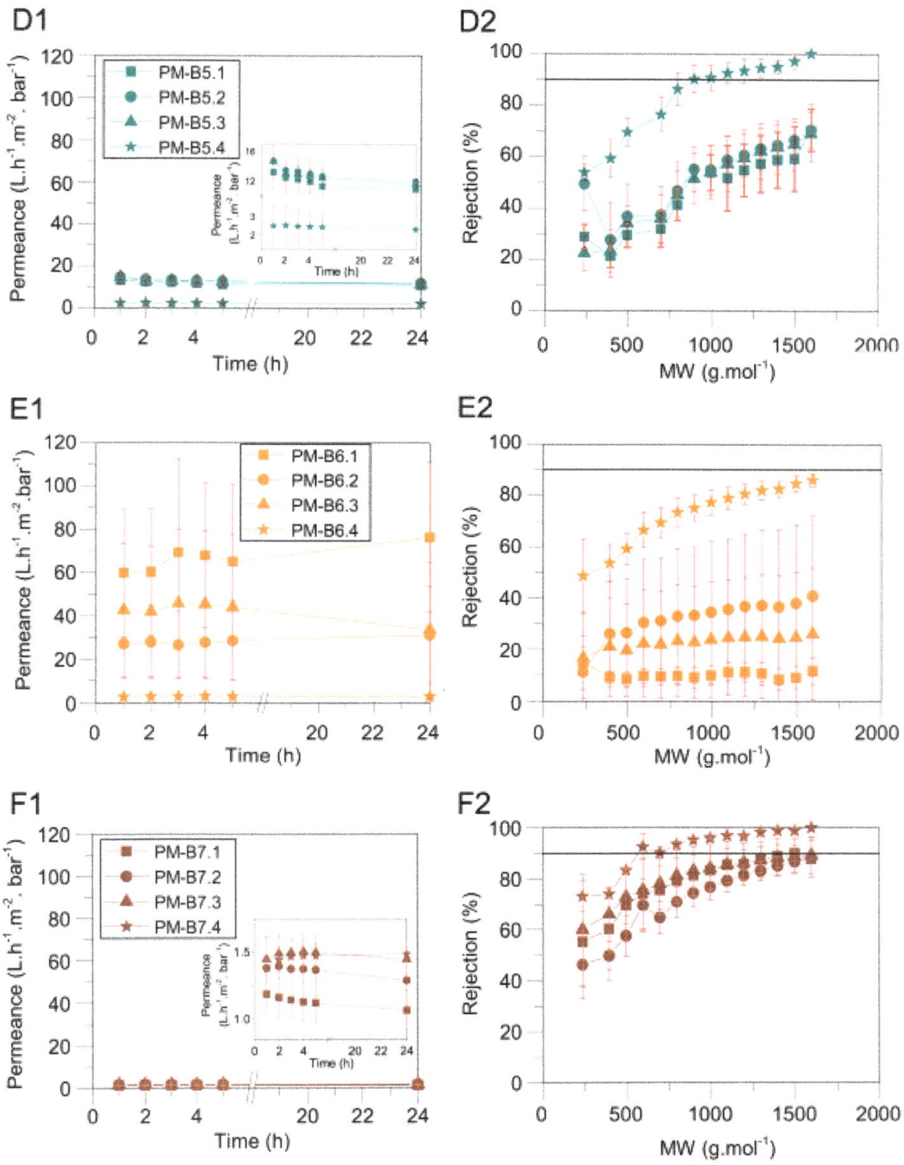

Figure 6: D1, E1 and F1: Permeance values (L h^{-1} m^{-2} bar^{-1}) over a period of 24 h for the different membranes under study. D2, E2 and F2: Rejection values of the different PEEK membranes under study as a function of the molecular weight (M_W, g mol^{-1}) of different polystyrenes after 24 h. Membranes PM-B5.*x*, PM-B6.*x*and PM-B7.*x* (*x*=1, 2, 3 and 4) were dried from acetone, THF and n-hexane respectively at different temperatures (20 °C, 40 °C, 80 °C and 120 °C) prior to their insertion in the cross-flow cells. The membranes were filtered with a solution of THF and PS (1 g L^{-1}). The flow-rate, temperature and pressure were set at 100 L h^{-1}, 30 °C and 30 bar, respectively. The red bars represent the standard deviation of the mean. (For interpretation of the references to color in this figure legend, the reader is referred to the web version of this article.)

For membranes dried from n-hexane the temperature effect was not that pronounced but nevertheless the membranes dried at 120 °C were tighter (MWCO=595 g mol^{-1}) than the ones dried at other temperatures which had similar performances (MWCO around 1400 g mol^{-1}). The permeance ranged from 1.06 L h^{-1} m^{-2} bar^{-1} to 1.49 L h^{-1} m^{-2}.bar^{-1}. It is also important to point out that membranes dried from n-hexane had two solvent exchanges from water to IPA and then to n-hexane. In this particular case water and IPA could still be present in the smaller pores and the drying solvent might not have been pure n-hexane but a mixture of the three (although IPA and water should be present in very small amounts).

In order to investigate whether the solvent was completely removed after 24 h at 20 °C and 120 °C membranes PM-B1.4, PM-B3.4 and PM-B6.4 were further dried using a vacuum oven at 120 °C. From Fig. 7 one can observe that the membranes PM-B1.4 and PM-B3.4 did not change in terms of rejection performance once vacuum drying was applied. Only a slight increase of the permeance was observed, suggesting some residual solvent may still be present in the smallest pores, but this does not significantly affect the separation properties. However, for membrane PM-B6.4 there was as a decrease in permeance of about two times and the membrane was tighter after vacuum oven was applied (MWCO~1395 g mol^{-1}). This shows that for looser membranes there might be rearrangement of the polymeric structure and further nodule collapse from vacuum treatment. Nevertheless, there seems to be a physical limit in the collapse; the membrane did not collapse completely to, for example, the same degree as water dried membrane.

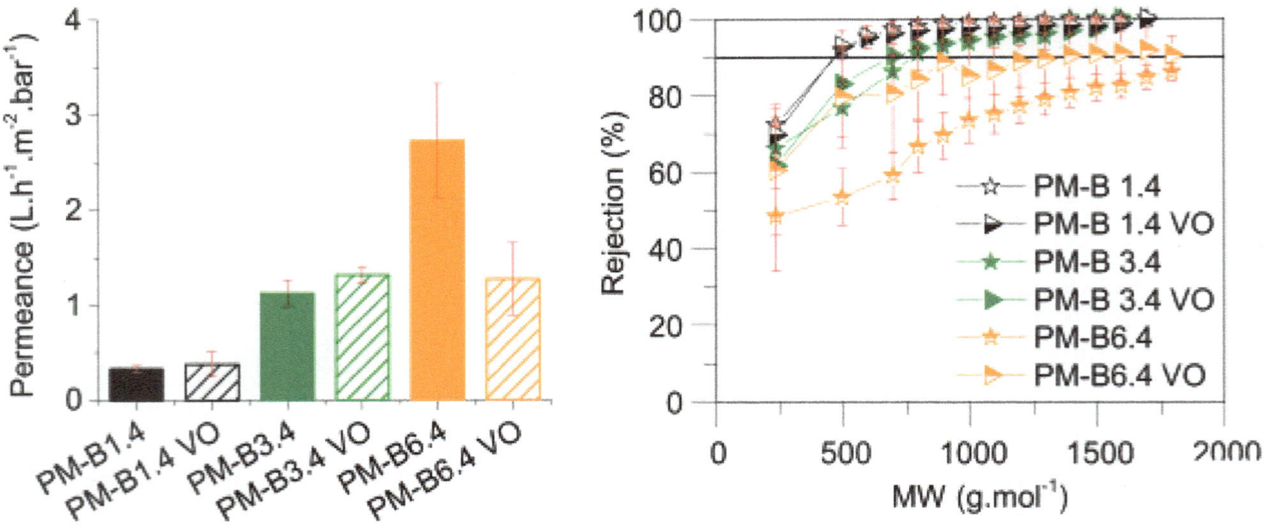

Figure 7: Left: Permeance values (L h^{-1} m^{-2} bar^{-1}) at 24 h for the different membranes under study. Right: Rejection values of the different PEEK membranes under study as a function of the molecular weight (M_W, g mol^{-1}) of different polystyrenes after 24 h. All the membranes presented were at least dried from water, EtOH, acetone and THF at 120 °C and some were further dried in a vacuum oven (code VO) at 120 °C for 24 h prior to their insertion in the cross-flow cells. The membranes were filtered with a solution of THF and PS (1 g L^{-1}). The flow-rate, temperature and pressure were set at 100 L h^{-1}, 30 °C and 30 bar, respectively. The red bars represent the standard deviation of the mean. (For interpretation of the references to color in this figure legend, the reader is referred to the web version of this article.)

The influence of drying time and cool-down rate were also evaluated (Table 3). Experimental data showed that regardless of the drying time, 0.5 h or 24 h, the membrane performance does not change (Fig. 8). This means that an air-drying treatment at 120 °C for 0.5 h was sufficient to provide the same performance when 24 h was applied which constitutes considerable energy savings from a process point of view. In terms of cooling-down rate, there was no difference between instantaneous cool-down (∞) or 40 °C h^{-1} cool-down rate, further indicating that the membrane reaches a "frozen" and stable state once it is heated at 120 °C.

Table 3: Summary of PEEK membranes PM-B prepared from different dopes and with different post-treatments. These membranes were used to test the influence of drying time and cool-down rate on permeance and rejection.

Membrane code	Drying solvent	Drying temperature (°C)	Drying time (h)	Cool-down rate (°C h^{-1})
PM-B1.4A1	Water	120	0.5	40
PM-B1.4A2	Water	120	0.5	∞ (Instantaneous)
PM-B1.4B1	Water	120	24	40
PM-B1.4B2	Water	120	24	∞ (Instantaneous)

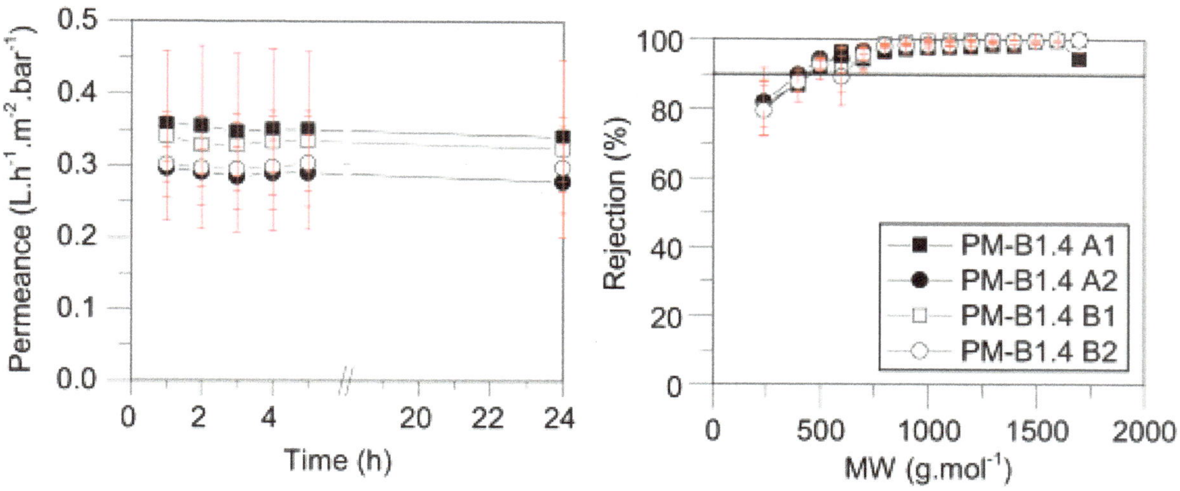

Figure 8: Left: Permeance values (L h^{-1} m^{-2} bar^{-1}) over a period of 24 h for the different membranes under study. Right: Rejection values of the different PEEK membranes under study as a function of the molecular weight (M_w, g mol^{-1}) of different polystyrenes after 24 h. All the membranes presented were dried from water at 120 °C for 0.5 h (A) or for 24 h (B) and cool downed slowly (1) or fast (2) prior to their insertion in the cross-flow cells. The membranes were filtered with a solution of THF and PS (1 g L^{-1}). The flow-rate, temperature and pressure were set at 100 L.h^{-1}, 30 °C and 30 bar, respectively. The red bars represent the standard deviation of the mean. (For interpretation of the references to color in this figure legend, the reader is referred to the web version of this article.)

Apparently the correlation proposed by Brown and adopted by Beerlage [14] and by Gevers et al. [26], can describe only to a limited extent the results obtained. This correlation accounts only for surface tension, but not for other solvent properties, which may be important during membrane drying, such as boiling point, vapour pressure, Hansen solubility parameter, viscosity and molar volume. Brown's correlation is also based on the assumption that a complete solvent exchange has taken place in all of the membrane pores. However this may not be the case if some residual water is retained into the smallest membrane pores, or some of the pores are filled with solvent mixtures with properties (specifically surface tension) different from the pure solvent.

Our findings are in agreement with literature studies. Matsuyama et al. [27] performed and extensive study on the effect of drying on the structure of microporous polyethylene membranes. They observed that polymer film contraction can be attributed to a combination of two physical phenomena: densification of the amorphous regions of the film and collapse of pores due to capillary forces (Fig. 9). They investigated 11 different solvents and concluded that membrane porosity is inversely proportional to the solvent surface tension and the boiling point of the solvent. These authors hypothesised that pore collapse involves rearrangement of the amorphous polymer molecules within the matrix phase. Since such a rearrangement requires time, it is hypothesized that the longer the capillary force is in effect, the greater the time for rearrangement of polymer chains in the matrix phase and the greater the extent of pore collapse. This rather logical hypothesis explains well our experimental results and the difference between the membranes dried from the different solvents. Having a lower boiling point and higher vapour pressure, acetone and THF disappear much faster from the membrane pores, shortening the action time of capillary force applied and the degree of pore collapse. However it somewhat contradicts the general trend for decrease of the membrane MWCO with the increase of the drying temperature observed with all solvents. The latter effect may be attributed to a larger contribution of the second phenomenon – densification of the amorphous regions due to polymer chain relaxation. In any case it is clear that the change from ultrafiltration to nanofiltration in our PEEK membranes is due to a secondary rearrangement of the polymeric chains during the drying–heating–cooling post-manufacturing treatment.

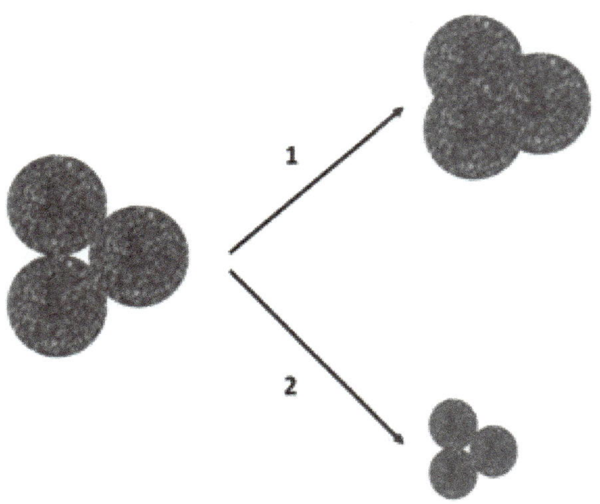

Figure 9: Schematic representation of the two physical phenomena methods proposed by Matsuyama et al. [27]: 1 – collapse of pores due to capillary forces and 2 – densification of the amorphous regions of the film.

Other studies also pointed out similar effects of the solvent boiling point [28] and the drying temperature [29] on the membrane permeance and MWCO. Interestingly a few studies [17], [28] and [29] present a correlation between the polymer-drying solvent affinity (expressed in terms of the Hansen solubility parameter) and the dried membrane properties. For our case study the solubility parameter for PEEK at 20 °C is reported as 9.5 (cal cm^{-3})$^{0.5}$ vs. water – 25.5 (cal cm^{-3})$^{0.5}$, methanol – 14.5 (cal cm^{-3})$^{0.5}$, IPA – 11.5 (cal cm^{-3})$^{0.5}$, EtOH – 13.4 (cal cm^{-3})$^{0.5}$, n-hexane – 6.9 (cal cm^{-3})$^{0.5}$, acetone – 10 (cal cm^{-3})$^{0.5}$ and THF – 9.1 (cal cm^{-3})$^{0.5}$[24] and

[30]. Therefore THF and acetone have the highest affinity to PEEK and should be more difficult to remove from the pores, resulting on them being more open. This seems to be the case since these membranes presented the highest MWCO. However, when considering IPA, which has higher affinity to PEEK than MeOH and should be therefore more difficult to remove from the pores, the opposite is observed: the IPA-dried membrane is tighter than the MeOH one. Again it seems that the contribution of the solvent–polymer affinity factor is not the primary driver and that other factors are dominant during the membrane drying.

It should be also noted that none of the parameters in Eq. (3) is independent of the temperature. In fact both the solvent surface tension and the tensile modulus decrease with the increase of the drying temperature [31] and [32]. The rate of this decrease in both terms of Eq. (3) may actually change the inequality for a given pore radius, and thus alter the degree of pore collapse. This effect may also contribute to the fact that for temperatures between 40 °C and 80 °C the rejection values were quite similar and they did not follow a specific order. It has been shown however that close to the glass-transition temperature PEEK undergoes a sharp decline in the tensile strength (Fig. 15Appendix). Therefore between 120 °C and 140 °C the effect of the surface tension should be more pronounced. In other words, a fair comparison between different drying solvents can be made for a drying temperature of 120 °C, where all solvents were presumably completely evaporated and pore collapse was predominantly due to the surface tension effect. Such comparison is shown in Fig. 10. As expected, membranes dried from water had a higher extent of pore collapse when compared with membranes dried from other solvents. This is in accordance with predictions from Brown›s theory, but does not explain the results obtained for the other solvents.

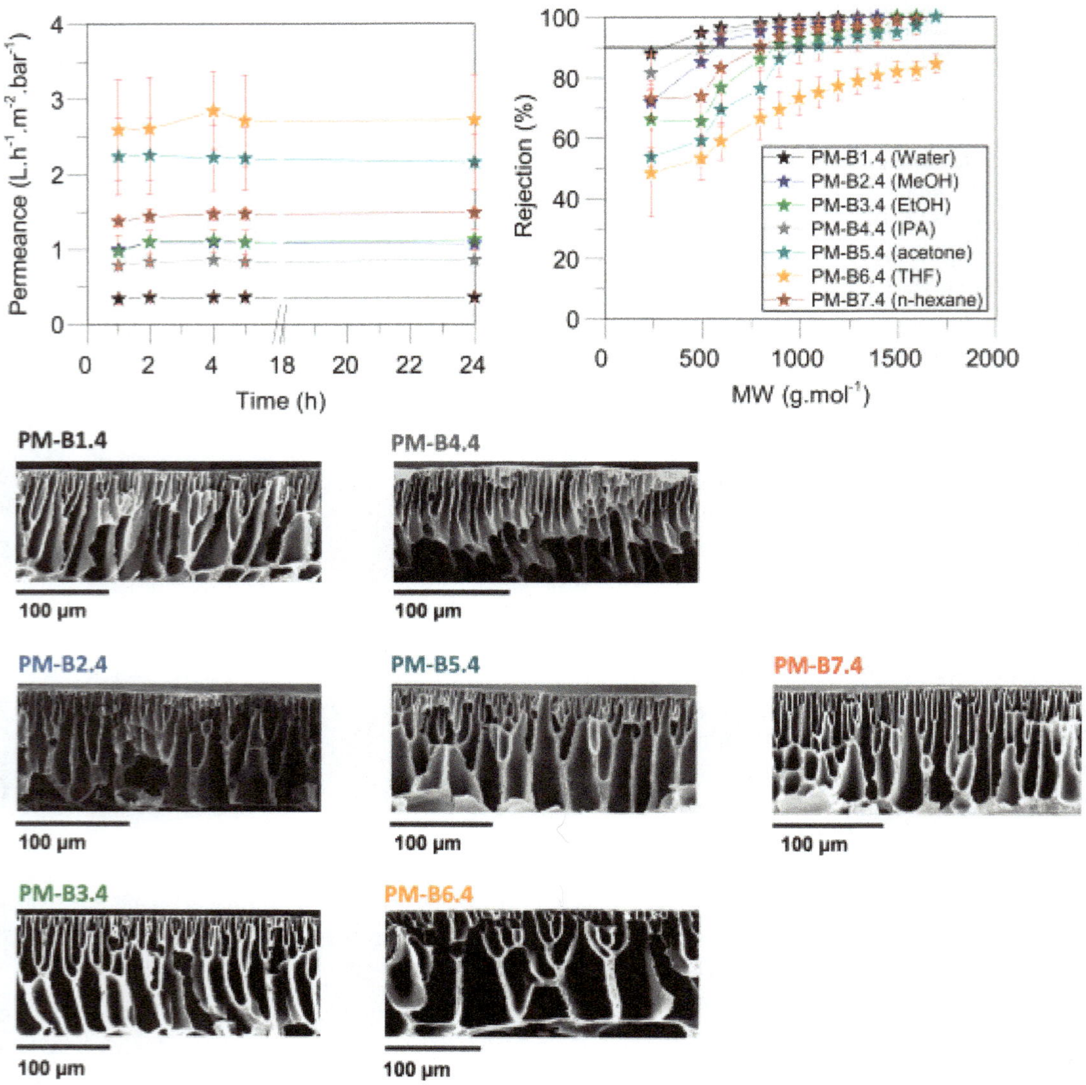

Figure 10: Top left: Permeance values (L h^{-1} m^{-2} bar^{-1}) over a period of 24 h for the different membranes under study. Top right: Rejection values of the different PEEK membranes under study as a function of the molecular weight (M_w, g mol^{-1}) of different polystyrenes after 24 h. Membranes PM-B1.4, PM-B2.4, PM-B3.4, PM-B4.4, PM-B5.4, PM-B6.4 and PM-B7.4 were dried at 120 °C prior to their insertion in the cross-flow cells from water, MeOH, EtOH, IPA, acetone, THF and n-hexane, respectively. The membranes were filtered with a solution of THF and PS (1 g L^{-1}). The flow-rate, temperature and pressure were set at 100 L h^{-1}, 30 °C and 30 bar, respectively. The red bars represent the standard deviation of the mean. The membranes dried from water, MeOH, EtOH, IPA, acetone, THF and n-hexane at 120 °C are significantly different ($p \leq 0.05$, *F*-test). Bottom: Cross-section SEM images (magnification 300×) of the different membranes under study: PM-B1.4, PM-B2.4, PM-B3.4, PM-B4.4, PM-B5.4, PM-B6.4 and PM-B7.4. (For interpretation of the references to color in this figure legend, the reader is referred to the web version of this article.)

MODELLING THE POST-PHASE INVERSION DRYING PROCESS OF PEEK NANOFILTRATION MEMBRANES

As mentioned before, the drying of PEEK membranes is a very complex phenomenon involving interactions between the solvent and the polymeric membrane as well as mass and heat transfer mechanisms combined with capillary forces. To date a satisfying phenomenological theory to describe such phenomenon has not been developed.

In order to understand which are the most important solvent properties that affect the drying process, a genetic algorithm was used in this research work to correlate solvent properties with membrane performance (the solvent permeance and the α-methyl styrene dimer flux were used as performance descriptors). The "weight" of each property was computed in order to assess the relevance of the parameters. The membrane performance at 120 °C was chosen for each solvent, since this was characterised by a more pronounced variation of permeance and MWCO (see Fig. 10). The solvent properties were obtained from literature for 20 °C and no correction of the solvent properties with temperature was implemented. As not all the properties can be easily correlated with temperature, the choice of using all the properties at 20 °C was made to avoid introduction of further sources of error.

The solvent properties were chosen in order to account for heat transfer contribution (vapour pressure), mass transfer contribution (viscosity), capillary forces (surface tension), steric effects (molar volume) and interactions between solvent and polymeric material (Hansen solubility parameter and polarity parameter). Initially, a linear model was used to describe the experimental flux data, with and without constant factor (Eq.(7)). In order to have an overdetermined system, a total of nine solvents was used. In addition to the seven solvents reported in the previous sections, acetonitrile and n-heptane were introduced. Properties and performance results for all these solvents can be found in Table 8 (Appendix). The regression system is overdetermined, as seven model parameters are regressed from performance data of nine different solvents. In order to evaluate the regression performance of the algorithm certain statistical measures have been proposed. The main measures used in literature are the Mean Absolute Error (MAE), the Mean Absolute Percentage Error (MAPE), the Mean Square Error (MSE) and the Root Mean Square Error (RMSE) [33]. In this work we chose the MAPE (defined below, Eq. (8)) as a measure of the error.

$$Y_{model}^{Sj} = b_0 + \sum_{i=1}^{n} b_i X_i^{Sj} \tag{7}$$

$$\text{MAPE}(\%) = \frac{1}{n}\sum_{j=1}^{n}\left|\frac{Y_{exp}^{Sj} - Y_{model}^{Sj}}{Y_{exp}^{Sj}}\right| \times 100 \tag{8}$$

where Y_{model}^{Sj} is the model prediction for performance (solute flux or permeance) in respect to a given solvent j (j =1,...,9), X_i^{Sj} is the solvent property (i =1,...,6), b_0 is the constant factor, b_i is the coefficient associated with each X_i^{Sj} (model coefficient), n is the number of solvents studied and Y_{exp}^{Sj} is the experimental result for performance (solute flux or permeance) in respect to a given solvent j (j =1,...,9). In this work, the values of MAPE correspond to the sum of the contribution from permeance and solute flux, i.e., MAPE*Total*=MAPE*permeance*+ MAPE*rejection*. The coefficients associated with solvent property (b_i) for both solute flux and permeance are presented in Table 4 and the experimental data and model fitting are presented in Fig. 11.

Table 4: Solute flux model coefficients associated with solvent properties.

Solvent property	Coefficient associated with solvent property (bi)				
	Symbol	Solute flux		Permeance	
		Linear model with constant factor	Linear model without constant factor	Linear model with constant factor	Linear model without constant factor
Constant factor	b_0	123	Not applicable	16.8	Not applicable
Vapour pressure(X_1)	b_1	-3.86×10^{-1}	-3.95×10^{-1}	7.47×10^{-3}	3.04×10^{-3}
Surface tension(X_2)	b_2	7.31×10^{-1}	2.08	-6.88×10^{-2}	1.16×10^{-1}
Hansen solubility parameter(X_3)	b_3	−5.00	−12.7	3.55×10^{-1}	-7.14×10^{-1}
Polarity parameter($X4$)	b_4	-4.89×10^{-1}	3.06	-3.16×10^{-1}	1.71×10^{-1}
Molar volume($X5$)	b_5	-4.76×10^{-1}	-1.45×10^{-1}	-5.18×10^{-2}	4.10×10^{-1}
Viscosity($X6$)	b_6	−7.48	−14.4	3.72×10^{-1}	-5.79×10^{-1}

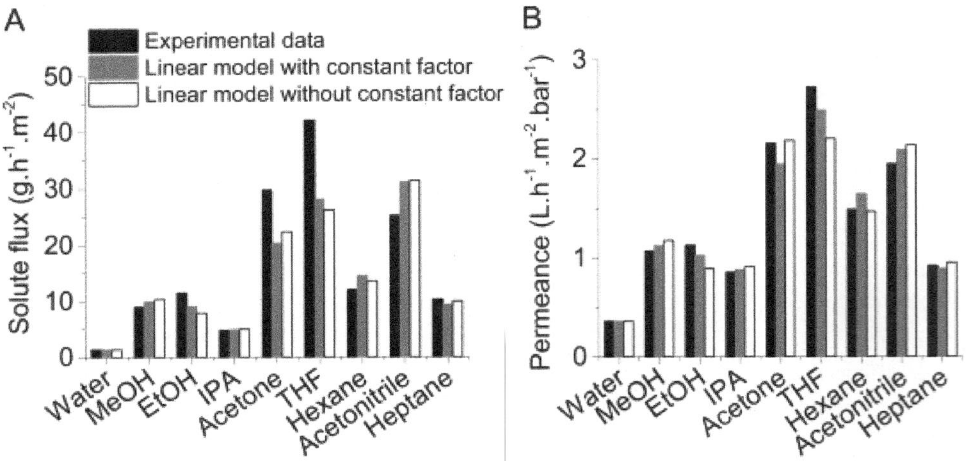

Figure 11: A: Experimental data for solute flux and prediction of solute flux values using linear model with and without constant factor. B: Experimental data for permeance and prediction of permeance values using linear model with and without constant factor.

The MAPE values for linear model with and without constant factor were 9.58% and 9.80% respectively. Based on this result, the linear model with constant factor showed the best fitting and it was chosen for assessing the parameter weight (%). The parameter weight (%) was calculated using the equation below and the results are shown in Table 5.

$$\text{Parameter weight (\%)} = \frac{\left|b_i X_i^{S_j}\right|}{\sum_{i=0}^{n}\left|b_i X_i^{S_j}\right|},\quad i = 0, \ldots, 9;\quad X_0^{S_j} = 1 \tag{9}$$

Table 5: Parameter weight (%) for both solute flux and permeance using the linear model with constant factor. The sign in brackets indicates if a given solvent property had a negative or positive effect on the overall response.

Solvent property	Parameter weight (%)	
	Solute flux	**Permeance**
Vapour pressure	1.91 (−)	0.19 (+)
Surface tension	7.32 (+)	4.54 (−)
Hansen solubility parameter	21.80 (−)	10.14 (+)
Polarity parameter	8.02 (−)	33.89 (−)
Molar volume	13.57 (−)	9.85 (−)
Viscosity	1.96 (−)	0.64 (+)
Constant factor	45.41 (+)	40.70 (+)

The most important parameters are polarity, Hansen solubility parameter and molar volume for both permeance and solute flux. For the solute flux model, all these parameters had a negative effect whereas for the permeance model the Hansen solubilty parameter had a positive effect and both polarity and molar volume had a negative effect. Nevertheless, the constant factor showed higher importance for both models. By carrying on with the three most important parameters, it was possible to develop a linear model which accounts for single effects as well as interaction effects (Eq. (10)).

$$Y_{model}^{S_j} = b_0 + b_3 X_3^{S_j} + b_4 X_4^{S_j} + b_5 X_5^{S_j} + c_1 X_3^{S_j} X_4^{S_j} + c_2 X_3^{S_j} X_5^{S_j} + c_3 X_4^{S_j} X_5^{S_j} \tag{10}$$

c_i denotes coefficients associated with solvent property interactions. The coefficients associated with solvent property (b_i) or with solvent property interactions (c_i) for both solute flux and permeance are presented in Table 6 and the experimental data and model fitting are presented in Fig. 12.

Table 6: Solute flux model coefficients associated with solvent properties and solvent property interactions.

Solvent property/solvent property interaction	Coefficient associated with solvent property/ solvent property interaction (*b*i or *c*i)		
	Symbol	**Solute flux**	**Permeance**
Constant factor	b_0	473.7	30.4
Hansen solubility parameter (***X*3**)	b_3	-3.02×10^{1}	-2.02
Polarity parameter (***X*4**)	b_4	-6.44	-3.99×10^{-1}
Molar volume (***X*5**)	b_5	-1.76	-1.09×10^{-1}
***X*3×*X*4**	c_1	4.05×10^{-1}	2.76×10^{-2}
***X*3×*X*5**	c_2	2.83×10^{-1}	1.28×10^{-2}
***X*4×*X*5**	c_3	-4.02×10^{-2}	-1.42×10^{-3}

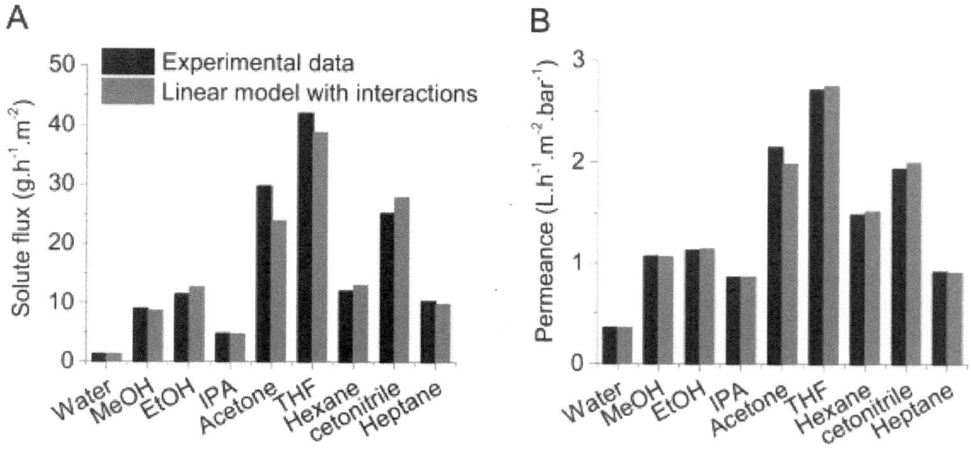

Figure 12: A: Experimental data for solute flux and prediction of solute flux values using linear model with interactions. B: Experimental data for permeance and prediction of permeance values using linear model with interactions.

The regression improved significantly with respect to the model that accounted for single effects only. The MAPE value for linear model with interactions was 1.41% which was much lower than the one obtained for the linear models without property interactions. This result showed that taking into account the three most important parameters and their interactions improves the fitting, suggesting that the properties were not completely independent of each other. The parameter weight/solvent property interaction (%) was calculated using Eq. (9) and the results are shown in Table 7.

Table 7: Parameter weight (%) for both solute flux and permeance using the linear model with interactions. The sign in brackets indicates if a given solvent property/ solvent property interaction had a negative or positive effect on the overall response.

Solvent property/ solvent property interaction	**Parameter weight/solvent property interaction (%)**	
	Solute flux	**Permeance**
Constant factor (b_0)	26.06 (+)	27.80 (+)
Hansen solubility parameter (***X*3**)	19.38 (−)	15.38 (−)
Polarity parameter (***X*4**)	15.62 (−)	15.99 (−)
Molar volume (***X*5**)	7.64 (−)	7.92 (−)
***X*3×*X*4**	12.40 (+)	13.79 (+)
***X*3×*X*5**	12.00 (+)	9.09 (+)
***X*4×*X*5**	6.91 (−)	4.09 (−)

Polarity and Hansen solubility parameter were still the most important parameters but molar volume was surpassed by the interaction between the Hansen solubility parameter and polarity and by the interaction between Hansen solubility parameter and molar volume. The signs of the parameters – which indicates positive or negative effect – were the same for both solute flux and permeance (sovent flux). This means that both solute and solvent flux were affected in the same way by the solvent properties/solvent property interaction. These results showed again how important the interactions between solvent properties were in terms of model fitting.

Overall polymer drying is a very complex and not well understood phenomena. Further extensive investigation is required to elucidate and gain a better control over the PEEK nanofiltration membrane properties. We are confident that the research on this interesting, exceptionally stable material in the area of OSN will continue, and this work presents a first step in this direction.

CONCLUSIONS

The post-fabrication drying process of PEEK membranes was found to be the reason for variation in the separation performance from the ultra to nanofiltration range. Two factors were investigated in an attempt to manipulate membrane MWCO: the concentration of polymer in the dope solution and the solvent filling the pores prior to drying. When varying the polymer dope concentration from 8 wt% to 12 wt% a shift from more open membranes (8 wt%) to tighter membranes (10 wt% and 12 wt%) was observed. A parallel between dope solution viscosity and membrane performance was drawn in order to explain the difference in performance for different polymer concentrations: the higher the viscosity the tighter the membranes produced. The type of solvent filling the membrane pores prior to drying had a pronounced effect on the separation performance. It was possible to vary the MWCO from 295 g mol^{-1} to 1400 g mol^{-1} (in terms of nanofiltration range). Another parameter studied for both factors (polymer concentration and solvent filling prior to drying) was the effect of drying temperature. For membranes dried from water the effect of drying temperature was negligible whereas for membranes dried from other solvents the effect was more pronounced (e.g. acetone and THF). In summary, by increasing the temperature from 20 °C to 120 °C it was possible to further manipulate the MWCO when drying from the same solvent. In order to set some guidance (and understanding) for a phenomenological study of membrane drying a statistical analysis of the presented data was performed in order to assess the relevant solvent properties involved. The Hansen solubility parameter, polarity and their interactions with molar volume were found to be the most important parameters influencing membrane MWCO. Nevertheless, this is just the beginning of a very complex phenomenon that needs to be further pursued in order to be fully understood.

AUTHOR CONTRIBUTIONS

The manuscript was written through contributions of all authors. All authors have given approval to the final version of the manuscript.

ACKNOWLEDGEMENTS

This work was funded by the Novartis Pharma AG, Massachusetts Institute of Technology, USA sub-award Agreement no. 5710002924. The authors would like to acknowledge EPSRC, United Kingdom for the financial support under the Project EP/J014974/1entitled Molecular Builders: Constructing Nanoporous Materials.

APPENDIX

See Fig. 13, Fig. 14 and Fig. 15 and Table 8.

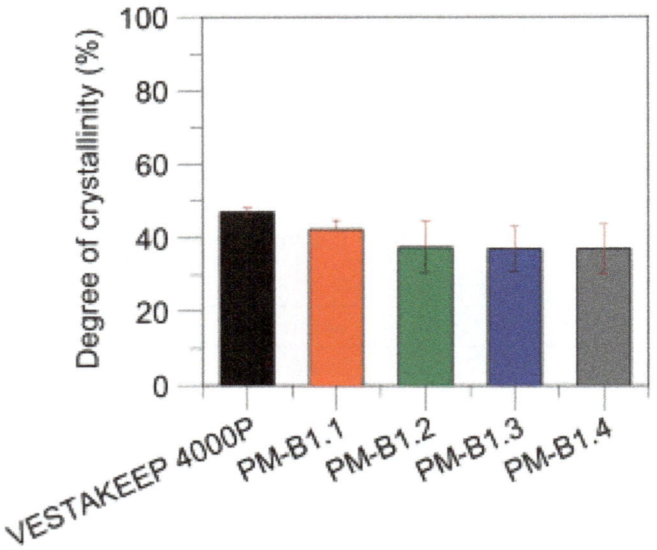

Figure 13: Degree of crystallinity (%) obtained from DSC (2.3.3) for VESTAKEEP 4000P and membranes PM-B dried from water at different temperatures: PM-B1.1, PM-B1.2, PM-B1.3 and PM-B1.4. The error bars represent the standard deviation from two sequential heating cycles.

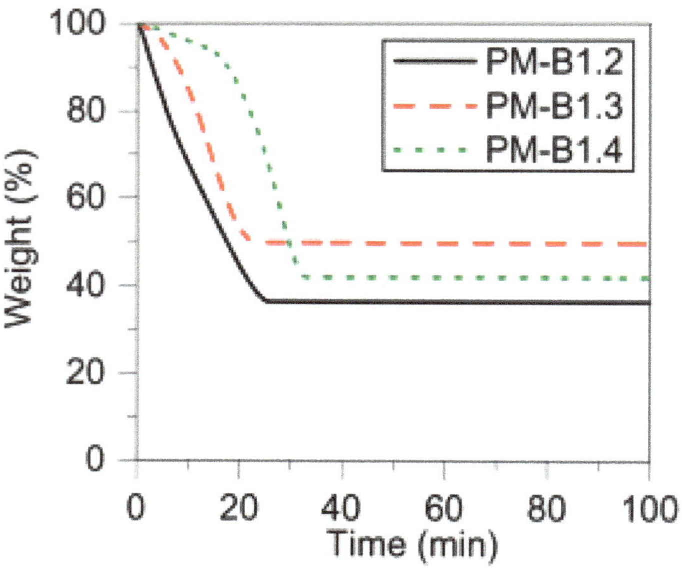

Figure 14: Weight loss (%) as a function of time (min) obtained from TGA analysis for membranes PM-B dried at different temperatures (40 °C, 80 °C and 120 °C).

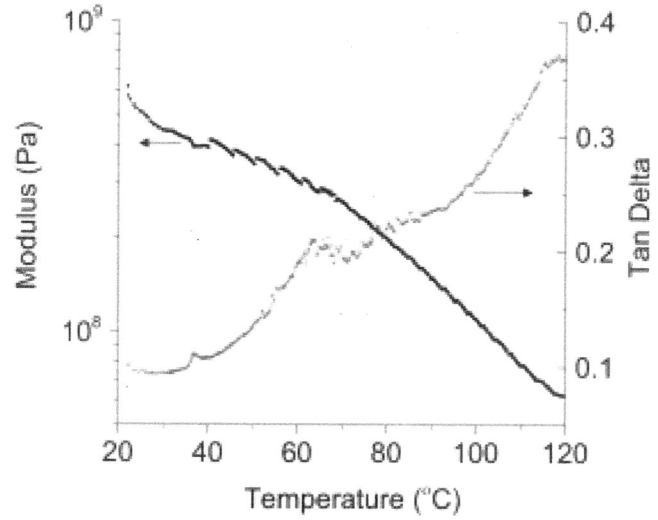

Figure 15: Values of dynamic modulus (stiffness, Pa) and mechanical damping (tan delta) for membrane PM-B 12 wt%. The membrane was inserted while water "wet" at a heating rate of 2 K min^{-1}. The tests were carried out at 1 Hz with a displacement of 0.05 mm.

Table 8: Properties of the solvents used for the solvent exchange: vapour pressure (kPa), surface tension (mN m^{-1}), Hansen solubility parameter (cal cm^{-3})$^{0.5}$, polarity parameter (kcal mol^{-1}), molar volume (cm^3 mol^{-1}) and viscosity (cP). All properties listed were obtained from [34] at 20 °C and 1 bar. In addition, rejection values (%) for the dimer (M_W=236 g mol^{-1}) and permeance values (L h^{-1} m^{-2} bar^{-1}) for PEEK membranes 12 wt% dried at 120 °C from the corresponding solvent, and used for the modelling, are presented.

	Vapour pressure (kPa)	**Surface tension (mN m−1)**	**Hansen solubility parameter (cal cm−3)0.5**	**Polarity parameter (kcal mol−1)**	**Molar volume (cm3 mol−1)**	**Viscosity (cP)**	**Rejection (%)**	**Permeance (L h−1 m−2 bar−1)**
Water	2.33	72.75	25.5	63.1	18	1	88.05	0.36
MeOH	16.933	22.6	14.5	55.4	40.6	0.6	71.9268	1.07
EtOH	5.9466	22.3	13.4	51.9	58.6	1.08	66.13	1.13
IPA	4.1	21.7	11.5	48.4	76.9	2	81.5211	0.86
Acetone	30.8	23.3	10	42.2	73.8	0.33	53.85	2.15
THF	21.6	26.4	9.1	37.4	81.9	0.46	48.44	2.72
Hexane	20.17	18.4	6.9	31	131.4	0.31	73.06	1.49
Acetonitrile	9.6	29.1	11.9	45.6	52.9	0.38	56.57	1.94
Heptane	6.093	19.3	7.5	31.1	147	0.41	62.69	0.92

REFERENCES

1. L. Peeva, J. da Silva Burgal, S. Vartak, A.G. Livingston, Experimental strategies for increasing the catalyst turnover number in a continuous Heck coupling reaction, J. Catal. 306 (0) (2013) 190–201.
2. H. Strathmann, K. Kock, The formation mechanism of phase inversion membranes, Desalination 21 (3) (1977) 241–255.
3. Y.H. See-Toh, M. Silva, A. Livingston, Controlling molecular weight cut-off curves for highly solvent stable organic solvent nanofiltration (OSN) membranes, J. Membr. Sci. 324 (1–2) (2008) 220–232.
4. I. Soroko, A. Livingston, Impact of TiO2 nanoparticles on morphology and performance of crosslinked polyimide organic solvent nanofiltration (OSN) membranes, J. Membr. Sci. 343 (1–2) (2009) 189–198.
5. P. Vandezande, L.E.M. Gevers, I.F.J. Vankelecom, Solvent resistant nanofiltration: separating on a molecular level, Chem. Soc. Rev. 37 (2) (2008) 365–405.
6. M. Bulut, L.E.M. Gevers, J.S. Paul, I.F.J. Vankelecom, P.A. Jacobs, Directed development of high-performance membranes via high-throughput and combinatorial strategies, J. Comb. Chem. 8 (2) (2006) 168–173.
7. Y.H. See-Toh, F.C. Ferreira, A.G. Livingston, The influence of membrane formation parameters on the functional performance of organic solvent nano- filtration membranes, J. Membr. Sci. 299 (1–2) (2007) 236–250.
8. T. Shimoda, H. Hachiya, Process for preparing a polyether ether ketone membrane, 5,997,741, 1999, US Patent 5,997,741, 08/836,034.
9. I.F.J. Vankelecom, K.D. Smet, L.E.M. Gevers, P.A. Jacobs, Nanofiltration membrane materials and preparation, in: A.I. Schäfer, A.G. Fane, T.D. Waite (Eds.), Nanofiltration – Principles and Applications, Elsevier, Oxford, 2006, pp. 33–42.

10. I. Soroko, M. Makowski, F. Spill, A. Livingston, The effect of membrane formation parameters on performance of polyimide membranes for organic solvent nanofiltration (OSN). Part B: Analysis of evaporation step and the role of a co-solvent, J. Membr. Sci. 381 (1–2) (2011) 163–171.
11. T.-H. Young, J.-H. Huang, W.-Y. Chuang, Effect of evaporation temperature on the formation of particulate membranes from crystalline polymers by dry-cast process, Eur. Polym. J. 38 (1) (2002) 63–72.
12. P.J. Rae, E.N. Brown, E.B. Orler, The mechanical properties of poly(ether-etherketone) (PEEK) with emphasis on the large compressive strain response, Polymer 48 (2) (2007) 598–615.
13. H.I. Kim, S.S. Kim, Plasma treatment of polypropylene and polysulfone supports for thin film composite reverse osmosis membrane, J. Membr. Sci. 286 (1–2) (2006) 193–201.
14. M.A.M. Beerlage, Polyimide Ultrafiltration Membranes for Non-aqueous Systems, 1994.
15. P. Manos, Solvent exchange drying of membranes for gas separation, US Patent 4120098 A, 1978.
16. I.-C. Kim, H.-G. Yun, K.-H. Lee, Preparation of asymmetric polyacrylonitrile membrane with small pore size by phase inversion and post-treatment process, J. Membr. Sci. 199 (1–2) (2002) 75–84.
17. P. Manos, Gas separation membrane drying with water replacement liquid, US Patent 4080744 A, US05/698,697, 1978.
18. W. MacDonald, C.-Y. Pan, Method for drying water-wet membranes, US Patent 1974 3,842,515, 1974.
19. W.M. King, M.-W. Tang, C.G. Wensley, Air dried cellulose acetate membranes, US Patent 4855048 A, 1989.
20. G.L. Brown, Formation of films from polymer dispersions, J. Polym. Sci. 22 (102) (1956) 423–434.
21. P. Manos, Membrane drying process, US Patent 4080743 A, 1978.
22. J. da Silva Burgal, L.G. Peeva, S. Kumbharkar, A. Livingston, Organic solvent resistant poly(ether-ether-ketone) nanofiltration membranes, J. Membr. Sci. 479 (0) (2015) 105–116.
23. D.J. Blundell, B.N. Osborn, The morphology of poly(aryl-ether-ether-ketone), Polymer 24 (8) (1983) 953–958.
24. Y.H. See Toh, F.W. Lim, A.G. Livingston, Polymeric membranes for nanofiltration in polar aprotic solvents, J. Membr. Sci. 301 (1–2) (2007) 3–10.
25. I.M. Smallwood, Handbook of Organic Solvent Properties, Elsevier, Amsterdam, 1996.
26. L.E.M. Gevers, S. Aldea, I.F.J. Vankelecom, P.A. Jacobs, Optimisation of a labscale method for preparation of composite membranes with a filled dense top-layer, J. Membr. Sci. 281 (1–2) (2006) 741–746.
27. H. Matsuyama, M.-m Kim, D.R. Lloyd, Effect of extraction and drying on the structure of microporous polyethylene membranes prepared via thermally induced phase separation, J. Membr. Sci. 204 (1–2) (2002) 413–419.
28. X. Jie, Y. Cao, J.-J. Qin, J. Liu, Q. Yuan, Influence of drying method on morphology and properties of asymmetric cellulose hollow fiber membrane, J. Membr. Sci. 246 (2) (2005) 157–165.
29. H. Matsuyama, A. Yamamoto, H. Yano, T. Maki, M. Teramoto, K. Mishima, K. Matsuyama, Effect of organic solvents on membrane formation by phase separation with supercritical CO2, J. Membr. Sci. 204 (1–2) (2002) 81–87.
30. J. Scheirs, Surface-tension and solubility-parameter effects, in: Compositional and Failure Analysis of Polymers: A Practical Approach, WIley, 2000.
31. E.W. Flick, Industrial Solvents Handbook, 5th ed., William Andrew Publishing/ Noyes, New York, 1998.
32. VICTREX Material Properties Guide. http://www.victrex.com/docs/literaturedocs/Victrex_Material%20Properties%20Guide%20%20%203_7_US.pdf.
33. E.S. Gardner, Forecasting: methods and applications (second edition), Makridakis, S., Wheelwright, S.C. and McGee, V.E., New York: Wiley, 1983. In Journal of Forecasting, John Wiley & Sons, Ltd.: 1984; Vol. 3, pp 457–460.
34. G. Wypych, Knovel Solvents – A Properties Database, ChemTec Publishing, 2008.

CHAPTER 13

HIGH-PRECISION MEASUREMENTS OF SEAWATER PB ISOTOPE COMPOSITIONS BY DOUBLE SPIKE THERMAL IONIZATION MASS SPECTROMETRY

Maxence Paul, Luke Bridgestock, Mark Rehkämper, Tina van DeFlierdt, Dominik Weiss

Keywords: Seawater, Lead isotope, Double spike, Thermal ionization mass spectrometry, GEOTRACES, Inter-calibration

INTRODUCTION

Investigations that map the distribution of trace metals and their isotopes in seawater and which study their relationship to and potential impacts on global biogeochemical cycles in the oceans are an overarching goal of modern marine geochemistry and a key objective of the international GEOTRACES program [1]. The global scale impact of human activities on natural Pb budgets was established more than 40 years ago by Chow and co-workers [2], [3], [4] and [5] and massive inputs of anthropogenic Pb into the oceans were soon thereafter identified by Patterson and co-workers [6], [7] and [8], based on the first accurate and precise Pb concentration measurements for seawater. Studies of the anthropogenic impact on marine Pb budgets were further advanced with the first successful Pb isotope analyses of seawater [6], [9], [10], [11], [12] and [13]. Whilst these early measurements were all carried out by TIMS (thermal ionization mass spectrometry), the majority of seawater Pb analyses since the 1990s have been conducted using various types of ICP-MS instrumentation. At present, the most precise Pb isotope analyses of seawater apply either single collector high resolution or multiple collector ICP-MS (HR-ICP-MS and MC-ICP-MS, respectively), most commonly following low-blank pre-concentration of the element by co-precipitation with $Mg(OH)_2$[14] and [15].

These methodological advances are a consequence of the particular importance of seawater Pb isotope data to studies of environmental pollution and marine geochemistry. For example, analyses of corals yielded a seawater Pb isotope record for the North Atlantic over the last 220 years [12], [15] and [16], which revealed significant isotopic variability through time, in accord with known changes in the extent and type of anthropogenic emissions. Similarly, Pb isotope data for the modern ocean are used to infer the distribution of Pb from different anthropogenic and natural sources in surface seawater [9], [13], [17] and [18] and to determine the lateral mixing and advection of subsurface water masses [19], [20], [21], [22] and [23]. Shen and Boyle [12], furthermore, highlighted that temporal changes in the Pb content and isotope composition of different water masses can be applied to trace ventilation times.

In this paper, we present a new method for the accurate and precise determination of seawater Pb concentrations and isotope compositions. The methodology involves pre-concentration of Pb by $Mg(OH)_2$ co-precipitation [24] followed by ion exchange chromatography [25] and [26], and subsequent isotopic analyses by TIMS using a ^{204}Pb–^{207}Pb double spike to achieve precise control of instrumental mass fractionation[27] and [28]. The technique is validated by analyses of both in-house and international seawater reference materials and its utility demonstrated by data obtained for a GEOTRACES seawater depth profile from the South Atlantic Ocean.

A number of previous investigations [28], [29], [30], [31], [32], [33] and [34] have shown that the application of Pb double or triple spikes enables Pb isotope measurements that

Citation: Maxence Paul, Luke Bridgestock, Mark Rehkämper, Tina van DeFlierdt, Dominik Weiss, High-precision measurements of seawater Pb isotope compositions by double spike thermal ionization mass spectrometry, Analytica Chimica Acta, Volume 863, 10 March 2015, Pages 59-69, ISSN 0003-2670, http://dx.doi.org/10.1016/j.aca.2014.12.012.

are significantly more precise and accurate than conventional TIMS or MC-ICP-MS techniques and less susceptible to the analytical artifacts, which can be generated by instrumental mass bias effects. Our study is the first to apply the double spike methodology to seawater Pb isotope analyses. Using these techniques, we are not only able to analyze the commonly measured ratios $^{207}Pb/^{206}Pb$, $^{208}Pb/^{206}Pb$ to high precision but $^{206}Pb/^{204}Pb$, $^{207}Pb/^{204}Pb$ and $^{208}Pb/^{204}Pb$ can also be determined to an uncertainty of better than about ±3‰. Only very scarce seawater data are available for the latter ratios due to the analytical challenges but measurements of non-radiogenic ^{204}Pb are deemed to be important to fully exploit the potential of the Pb isotope system.

EXPERIMENTAL

Samples

A gravimetric solution of the well-characterized NIST standard reference material SRM 981 Pb was employed for calibration of the Pb double spike and to evaluate the precision of the mass spectrometric methods. Four filtered seawater samples that were available in large quantities were utilized as in house reference materials to further assess reproducibility. These encompass a surface water sample (Atlantic – S) and three subsurface water samples, Atlantic – D1, Atlantic – D2 (both from 2000 m depth), and Weddell (400 m depth). These samples were not specifically collected for Pb isotope analyses, and potential contamination during sampling and initial handling hence renders them unsuitable for evaluation of accuracy or an assessment of water mass properties.

To enable a comparison of our results with literature data for assessment of accuracy, we analyzed the international seawater reference samples GSI and GDI. These were collected at water depths of 7 m and 2000 m, respectively, at the Bermuda Atlantic Ocean Time Series station (BATS) during the 2008 GEOTRACES intercalibration cruise on the R/V Knorr (KN193-6). The collection and handling of these samples employed materials and procedures that were specifically designed for the intercalibration of contamination-prone trace metals and isotopes [35].

The seawater for a Pb isotope and concentration depth profile was collected in the South Atlantic Ocean off the African coast at 36.46°S, 13.39°E in October 2010 during the eastern part of the GEOTRACES GA10 transect along 40°S on the R.R.S. Discovery (D357; 18 October–22 November 2010) [1]. This depth profile from station 3 encompasses 12 samples that cover the entire 4500 m of the water column at the sampling location. Station 3 serves as a 'cross over station', which means that it was sampled by three separate GEOTRACES cruises (UK GEOTRACES cruise D357 – this study; UK GEOTRACES cruise JC068 in 2011/2012; MD166 BONUS – GoodHope Cruise in 2008, Boye et al. [36]), thereby enabling a comparison of results for samples from a single location but that were collected at different times and analyzed using different methodologies. The samples from cruise D357 analyzed here were obtained with a titanium rosette deployed from plasma rope and equipped with trace metal clean sample bottles and supplemented by underway surface water sampling (2–3 m depth) using a towed 'fish' (Wyatt et al. [37]). All bottle handling was conducted in a clean container and each sample was transferred into two pre-cleaned 1 L LPDE bottles without filtration or acidification. About one year after collection and two weeks prior to analysis, the samples were acidified to pH 2 with 6 M HCl.

Reagents

Water with a resistivity of >18.2 MΩ cm from a Milli-Q water system was used throughout for reagent preparation and cleaning of labware. Purified concentrated HF (~28 M) and HBr (~8.5 M) were purchased from VWR as Optima (distilled) grade acids whilst concentrated (14.5 M) HNO_3 and 6 M HCl were prepared in-house by subboiling distillation of reagent grade acids in quartz stills. The Pb blanks of these acids were checked regularly and found to be consistently less than 2 pg/mL. The dilute mineral acids that were employed for the separation chemistry were freshly prepared for each batch of samples.

A purified aqueous solution of NH_3 was used to affect precipitation of $Mg(OH)_2$ from seawater and prepared by cold vapor phase equilibration of reagent grade 28% (w/w) aqueous NH_3 with Milli-Q water in a Savillex Teflon elbow. Approximately 2 volumes of 28% aqueous NH_3 were employed for each volume of Milli-Q water, and these were equilibrated for about two weeks to obtain purified aqueous NH_3 (aq-NH_3) at a concentration of about 9 M. During this period, the Teflon elbow was stored double bagged in plastic, to limit contamination and loss of gaseous NH_3. The aq-NH_3 was collected from the still just prior to use and each batch of three samples, plus a full procedural blank, was treated with the same solution. The Pb blank of the aqueous ammonia was stable at less than 0.3 pg/mL during the course of the study.

Preparation and calibration of ^{207}Pb–^{204}Pb double spike

Highly enriched (>99%) metallic ^{204}Pb and ^{207}Pb in wire form were purchased from Isoflex (USA). The wires were separately dissolved in 7.5 M HNO_3 and diluted with water to obtain solutions in 2 M HNO_3. A dilute double spike solution was then prepared with a total Pb concentration of ~10 ng/mL and an optimal molar ratio of $^{207}Pb/^{204}Pb \approx 1$, based on the error propagation modeling of Galer [28] and Rudge [38].

Spike calibration was performed using a gravimetric solution of NIST SRM 981 Pb. In a first step, the concentration of the double spike was determined by two reverse isotope dilution runs that were carried out with a Nu Plasma HR MC-ICP-MS, using external normalization to added Tl for mass bias control [24], [39] and [40]. Initially, a preliminary concentration was obtained, based on an approximate spike isotope compositions, which was calculated from the mixing proportions of the ^{204}Pb and ^{207}Pb solutions, and the respective isotope compositions, as supplied by Isoflex.

The isotope composition of the double spike (Table 1) was then calibrated by analyses of different gravimetrically prepared double spike – NIST SRM 981 mixtures using TIMS[38] and [41]. The instrumental mass fractionation correction [24], [39] and [40], the spike-standard mixing line, and the double spike composition were calculated by over-determining the system of equations (number of double spike – NIST SRM 981 mixtures >2) and solving in the least-square sense, as described by Rudge et al. [38]. Finally, the calibrated double spike isotope composition was applied to the data from the initial reverse isotope dilution runs, to accurately define the double spike concentration.

Table 1: Lead isotope data of this study and reference values for NIST SRM 981 Pb. Also shown is the isotope composition of the Pb double spike used in this study.

Sample	Pb used (ng)[a]	n[b]	$^{206}Pb/^{204}Pb$ ±2sd[c]	$^{207}Pb/^{204}Pb$ ±2sd[c]	$^{208}Pb/^{204}Pb$ ±2sd[c]	$^{207}Pb/^{206}Pb$ ±2sd[c]	$^{208}Pb/^{206}Pb$ ±2sd[c]
This study							
SRM 981 Pb	2	42	16.943 ± 16	15.496 ± 15	36.716 ± 38	0.91460 ± 18	2.16710 ± 61
±2sd relative (ppm)			920	970	1050	200	280
±2se of typical (ppm)[d]			~500	~500	~550	~60	~80
SRM 981 Pb	10	25	16.9419 ± 31	15.4968 ± 38	36.7224 ± 91	0.91470 ± 11	2.16755 ± 20
±2sd relative (ppm)[c]			180	240	250	120	90
±2se of typical run			~90	~100	~100	~40	~60
SRM 981 Reference values[e]							
Galer and Abouchami (TIMS) [57]	~10		16.9405 ± 15	15.4963 ± 16	36.7219 ± 44	0.91475 ± 4	2.16771 ± 10
Thirlwall (TIMS) [31]	~20		16.9409 ± 22	15.4956 ± 26	36.7228 ± 80	0.91469 ± 7	2.16770 ± 21
Todt et al. (TIMS) [58]	80		16.9356 ± 23	15.4891 ± 30	36.7006 ± 113	0.91459 ± 13	2.16701 ± 43
Thirlwall (MC-ICP-MS) [32]			16.9417 ± 29	15.4996 ± 31	36.724 ± 9	0.91488 ± 8	2.16770 ± 24
Baker et al. (MC-ICP-MS)[29]	~50		16.9416 ± 13	15.5000 ± 13	36.7262 ± 31	0.91491 ± 4	2.16781 ± 12
207Pb–204Pb double spike[f]			0.00240	1.03181	0.00820	431.03	3.4267

a Mass of Pb loaded on filament and/or used for a single analysis.

b n = number of analyses.

c 2sd = 2× standard deviation calculated from n separate runs; denotes long-term "external" reproducibility.

d 2se = 2× standard error of the mean; denotes within-run ("internal") precision of the data, based on individual results for measurement cycles.

e All reference values were obtained using either double or triple spike techniques for correction of instrumental mass fractionation.

f Isotope composition of the Pb double spike used in the current study.

Sample preparation for determination of Pb isotope compositions

Lead pre-concentration by co-precipitation with magnesium hydroxide

A number of previous studies have shown that co-precipitation of Pb with $Mg(OH)_2$ is an effective, low-blank method for pre-concentration of Pb from seawater [14] and [42] and this technique is retained in our procedure. Initial tests showed that Pb is efficiently and reproducibly extracted from seawater at yields ≥80% by precipitation with $Mg(OH)_2$ at pH ≈ 10. The yields were, furthermore, not correlated with the exact pH value that was adjusted or the amount of $Mg(OH)_2$ precipitated. The $Mg(OH)_2$ precipitate, however, was also found to scavenge a significant amount of sea salt, but the latter could be removed without noticeable reduction of the Pb yield by rinsing the precipitate with water.

In order to obtain sufficient Pb for precise analyses, even for Pb depleted seawater from the deep ocean, samples of up to about 2 L were processed. The seawater was transferred to 1 L pre-cleaned Teflon separating funnels, to aid the collection of the Mg hydroxide precipitate. Precipitation of $Mg(OH)_2$ was affected by introduction of the purified aq-NH_3 until a final pH value of about 10 was obtained. As each seawater sample had been acidified to a slightly different extent, and because the precipitate forms only gradually, the applied volume of aq-NH_3 was optimized for each sample by slowly adding successive portions of the reagent to the separating funnel. This was achieved by an initial addition of 2–4 mL aq-NH_3, whereupon the samples were stirred vigorously and then left to stand for least 3 h to allow the $Mg(OH)_2$ precipitate to form. If precipitation had not occurred after this time, an additional volume of 0.5 mL of aq-NH_3 was added every 2–3 h until formation of precipitate was observed.

Once about 5 mL of $Mg(OH)_2$ slurry had formed, the precipitate was drained through the bottom of the separating funnel, collected in a 15 mL Teflon centrifuge tube, centrifuged at 1700 rpm for about 10 min and the remaining solution poured off. This step was repeated several times until all precipitate was recovered from the funnel. The solids were then washed with 5 mL H_2O, centrifuged again and finally dissolved directly in the tubes by addition of 5 mL of a 2 M HBr–0.01 M HF acid mixture. To aid dissolution, the centrifuge tubes were shaken whereupon the formation of many small bubbles within the solutions was observed for seawater samples that had not been filtered following collection. Bubble formation is hence presumably related to the presence of residual organic compounds. It is important to remove the bubbles at this stage because they can otherwise lead to the formation of larger bubbles within the resin bed during the subsequent column chemistry, effectively blocking the flow. For removal of the bubbles, the centrifuge tubes were left to stand overnight, sealed and then gently heated on a hotplate for 10 min at 50 °C. After cooling, bubbles should no longer be visible in the solutions, otherwise the heating was repeated.

Lead purification by anion exchange chromatography

The anion exchange purification procedure for Pb (Table 2) is based on previously published methods that apply strongly basic AG1X8 resin and dilute $HBr–HNO_3$ mixtures for elution, with modifications as detailed below.

Table 2: Column chemistry procedure for the separation of Pb from seawater for isotope compositions (IC) and isotope dilution (ID) concentration measurements by TIMS

	Pb IC chemistry 1st stage	Pb IC chemistry 2nd stage	Pb ID chemistry
Resin[a]	100 µL AG1X8, 100–200 mesh	20 µL AG1X8, 100–200 mesh	20 µL AG1X8, 100–200 mesh
Clean resin	3 × 3 mL 0.1 M HNO_3	3 × 1 mL 0.1 M HNO_3	3 × 1 mL 0.1 M HNO_3
Equilibrate resin	2 × 0.1 mL	2 × 0.1 mL	2 × 0.1 mL
	2 M HBr–0.01 M HF	0.2 M HBr–0.5 M HNO_3	0.5 M HBr–0.2 M HNO_3
Load sample	~5 mL in	~1 mL in	2 mL
	2 M HBr–0.01 M HF	0.2 M HBr–0.5 M HNO_3	0.5 M HBr–0.2 M HNO_3
Elute matrix	2 × 0.1 mL	2 × 0.3 mL	2 × 0.3 mL
	2 M HBr–0.01 M HF	0.2 M HBr–0.5 M HNO_3	0.5 M HBr–0.2 M HNO_3
	2 × 0.5 mL		
	0.2 M HBr–0.01 M HNO_3		
	0.2 mL		
	0.03 M HBr–0.5 M HNO_3		
Elute Pb	3 mL	1 mL	1 mL
	0.03 M HBr–0.5 M HNO_3	0.03 M HBr–0.5 M HNO_3	0.03 M HBr–0.5 M HNO_3

a The resin is used in columns prepared from shrink-fit Teflon tubing. The columns have a reservoir volume of ~3 mL and a resin column with an internal diameter of ~3 mm.

Previous work [14], [43] and [44] has shown that silica is trapped by the $Mg(OH)_2$precipitate, which is used for the separation of Pb from seawater. Relatively high Si concentrations of about 10–150 µmol/kg are, furthermore, found in seawater samples from depths >1000 m [45] and [46]. During method development we found that such concentrations suffice to form precipitates of silica on top of the resin column, when HF is not present in sample solutions. Hence, the 1st stage column chemistry of our final method (Table 2) applies a 2 M HBr–0.01 M HF mixture for sample loading, whereby the small concentration of HF present enhances silica solubility to prevent precipitation of silicic acid. The same acid mixture is subsequently also used to completely flush silica from the resin, prior to elution of other matrix elements (Table 2). In 2 M HBr, Pb has a distribution coefficient K_D of about 200 toward the AG1X8 anion exchange resin that was used [25], [47] and [48]. Elution experiments confirmed that, whilst the Pb K_D value is reduced when HF is present, the use of trace HF in the elution scheme (Table 2) does not prevent retention of Pb on the resin bed during sample loading and matrix elution.

Following elution of Pb from the 1st stage columns, the samples were dried and further purified by a second column chemistry step, which is needed to eliminate remaining traces of Mg [49] and improve the ionization of Pb during the TIMS analyses. To this end, the dried samples were redissolved in 0.2 M HBr–0.5 M HNO_3 and processed using established HBr–HNO_3 procedures on small 20 µL resin columns (Table 2), which help to minimize the procedural blank.

TIMS analyses of Pb isotope compositions

The accurate determination of Pb isotope compositions with precise correction of instrumental mass discrimination requires two separate mass spectrometric runs – one of the pure sample and a second analysis of a double spike – sample mixture. To this end, the purified Pb was divided into two aliquots and the double spike was admixed to one, prior to the TIMS measurements. The addition of the double spike after, rather than before, the chemical separation of Pb has a number of advantages. Conveniently, each sample needs to be processed through the (time-consuming) pre-concentration and purification steps only once, rather than twice. Furthermore, the impact of contamination from reagents and handling is minimized and uncertainties arising from different blanks impacting on the spiked and unspiked analyses are reduced.

In detail, our procedure involves splitting the Pb solutions from the 2nd stage column chemistry into a larger and a smaller aliquot. The larger aliquot comprising 2/3 of the sample remained unspiked, whilst Pb double spike was added to the remaining aliquot with 1/3 of the solution. The double spike addition was thereby carried out to obtain a molar ratio of natural (N) to spike-derived Pb (S) that approximates the optimum S/N ≈ 1.226 [38]. This is equivalent to a molar proportion of spike, S/(S + N), of 55.1%, corresponding to $^{206}Pb/^{204}Pb \approx 0.411$ (assuming a natural seawater $^{206}Pb/^{204}Pb = 18.4$). The amount of natural Pb present was thereby estimated from the Pb concentration measurements that were carried out first for all samples, and assuming a Pb recovery of 60% from the chemical separation (see below).

Following aliquoting and spiking, the Pb solutions were evaporated to dryness and redissolved in 1 μL 2 M HNO_3. The samples were loaded on a single, previously outgassed zone refined Re filament (99.999%; H. Cross, USA), dried at 0.4 A, 0.5 μL of silica–gel activator was then added and also dried at 0.4 A. The sample–activator mixture was finally heated at 2 A for ~1 s to evaporate phosphoric acid and fuse the silica gel.

Gerstenberger and Hasse [50] prepared a highly successful activator for Pb isotope analyses by TIMS but this applies silica particles which are no longer commercially available. Hence, a new SiO_2–H_3PO_4 activator mixture was made up for the current study. This mixture was prepared from SiO_2 nanoparticles (20–60 nm; Io-Li-Tec Inc.) and Merck Ultrapur phosphoric acid, whereby a SiO_2/H_3PO_4 weight ratio of 0.43% was applied with respect to the addition of the silica to dilute 0.16 M H_3PO_4. The amount of activator used for analysis was calibrated to optimize ionization yields, based on tests that were carried out with 2 ng NIST 981 Pb. These test revealed that use of 0.5 μL activator solution was preferable to loading with 1 μL of activator – whilst the latter yielded more stable Pb ion beams the former produced ion beam intensities that were typically about 2–3 times higher.

The Pb isotope measurements were carried out on a TRITON TIMS instrument operating in static mode and using Faraday cups fitted with 10^{11} Ω resistors. Prior to each measurement session, a gain calibration was carried out for all amplifiers. For analysis, the filaments were heated to a target temperature of about 1300 °C within 30 min, and isotopic data were then collected in 15 blocks of 15 cycles with an integration time of 8.3 s. Half-mass baseline measurements (30 cycles × 1.05 s integrations following 5 s idle time) and an automatic peak centering routine (followed by 3 s idle time) were carried out prior to each block. Loads of 2 ng NIST 981 Pb provided stable and very reproducible ion beam intensities of 200–300 mV ^{208}Pb. For the seawater samples, the ion beam intensities were more variable and generally lower, whereby 2 ng Pb generally yielded ^{208}Pb ion beams of about 100–200 mV.

The measured isotopic data for the spiked and unspiked sample aliquots were processed using an iterative solver that employs published methods [38] and which was implemented as a macro within a Microsoft Excel spreadsheet. The method uses the exponential law to obtain mass bias corrected Pb isotope data and is employed on a cycle-by-cycle basis to provide an accurate assessment of the final propagated data uncertainties.

Following Doucelance and Manhès [51], the TIMS source was baked for 3 h after the introduction of a new sample turret. This technique was observed to substantially reduce the formation of hydrocarbon species that can form inferring molecular ions during the Pb isotope runs [31]. The baking also significantly improved the vacuum within the ion source (to about 3.5 to 6 × 10^{-8} mbar, without use of a liquid N_2 cold trap) and enhanced Pb ion beam stability, thereby permitting analyses without interblock re-heating. The latter improvement is presumably due to the efficient drying of the silica gel, which is achieved during heating in vacuum condition.

Barium-PO_x ions were mainly observed at temperatures, which are slightly higher compared to those employed for optimal Pb ionization [31]. The formation of such ions, and hence the presence of Ba in samples, is best avoided because they produce minor molecular interferences on ^{204}Pb (from $^{138}Ba^{31}P^{17}O^{18}O^+$, $^{137}Ba^{31}P^{18}O_2^+$) and, even more problematic, hinder the ionization of Pb whereby rapidly decaying Pb ion beams are observed. The presence of $BaPO_x$ ions was routinely monitored during the analyses at mass 201, which corresponds primarily to $^{138}Ba^{31}P^{16}O_2^+$. The measured ion beam intensities at mass 201 were observed to rise during Pb sample runs but only to maximum values of 1–3 mV. At this level, the $BaPO_2$ interference contributes less than 15 ppm to the ^{204}Pb ion beam, such that a correction was deemed unnecessary.

Determination of Pb concentrations

Lead concentrations were determined by isotope dilution (ID), and using a chemical separation method similar to that employed for the isotope ratio measurements but optimized for the processing of smaller seawater samples. In detail, 50 mL of seawater were weighed into clean polyethylene centrifuge tubes. Following addition of the Pb double spike, the mixtures were left to stand for at least 10 days to ensure full spike–sample equilibration. The amount of Pb spike added was based on an initial 'guess' of or any information available for the Pb concentration of the samples, to achieve an optimal S/N ≈ 2.61 (equivalent to $^{206}Pb/^{204}Pb \approx 0.188$) for minimal error propagation during ID data reduction.

Co-precipitation of Pb with $Mg(OH)_2$ was affected by adding 150 μL of purified aq-NH_3. Following separation of the precipitate by centrifugation and dissolution in a dilute HNO_3–HBr mixture, Pb was further purified using a single stage anion exchange chemistry. This employed small resin columns of only 20 μL to minimize the procedural blank (Table 2). The anion exchange chemistry was slightly modified in comparison to the 2nd stage chemistry for Pb isotopes through use of a 0.5 M HBr–0.2 M HNO_3 acid mixture for sample loading and matrix elution, because this was observed to facilitate the dissolution of the $Mg(OH)_2$ precipitate whilst ensuring quantitative retention of Pb on the small resin columns.

The ID TIMS analyses for Pb used the same techniques that were applied for the isotopic measurements but with shorter runs comprising 5 blocks of 15 cycles. During filament loading of some samples for the ID TIMS analyses, the presence of impurities, in the form of salts and possibly also organic compounds, was detected. When present, such impurities were associated with poor Pb ion beam intensities during the TIMS runs and in some extreme cases the analyses were even abandoned. Such contamination may also produce interfering molecular ions that could generate inaccurate ID results. Subsequent test showed that the impurities can be reduced to negligible levels, by allowing the resin bed to drain and dry completely (for about 10 min) after each step of the column chemistry procedure for Pb separation (Table 2) and, in particular, prior to the elution of Pb.

RESULTS AND DISCUSSION

Blanks

The total blank for the Pb isotope measurements was determined by carrying out the complete analytical procedure without an initial seawater sample being present. Over a 24-month period, the mean procedural blank was 28 ± 21 pg (1sd, n = 25; ignoring two abnormally high blanks of ~100 pg encountered during a single two-week period). The individual contributions to this total from (i) filament loading and (ii) reagents were also determined and found to be 1.9 ± 0.5 pg for the former, 5.0 ± 1.4 pg for the mineral acids used, and 2.5 ± 0.2 pg for the aq-NH_3 solutions. The contribution of the anion exchange resin and column chemistry (without reagents) to the total blank could hence be quantified as 17 ± 8 pg, by subtracting the reagent component from separately measured column chemistry blanks. The sum of the individual blank contributions is, therefore, 27 ± 10 pg, in accord with the full procedural blanks values determined during the study. Ignoring the individual contributions (and chemistry yields), a Pb blank of 10–50 pg contributes about 0.05–2.5% to the indigenous Pb budget of samples that were analyzed for isotopic compositions, assuming samples volumes and Pb concentrations of 1–2 L and 10–50 pmol/kg, respectively. The Pb isotope composition of the blank was also measured on several occasions with rather variable results, due to the low ion beam intensities and/or actual differences in blank compositions (e.g., $^{206}Pb/^{204}Pb$ = 18.34 ± 0.62, $^{207}Pb/^{206}Pb$ = 1.196 ± 0.028, $^{208}Pb/^{206}Pb$ = 2.422 ± 0.021; 2se, n = 10). Blank corrections were not applied to the isotopic data because such corrections did not improve the reproducibility of results from replicate sample analyses (see discussion below).

The total procedural blank for the Pb ID concentration measurements was determined using 100 μL seawater samples, as only negligible amounts of Pb (~0.2 to 1 pg) are present in such small volumes. These measurements revealed a relatively stable total Pb blank of 11.7 ± 4.3 pg (1sd, n = 36), which is equivalent to about 1–15% to the indigenous Pb in ID sample aliquots, assuming Pb abundances of 10–50 pmol/kg (for simplicity, this ignores variable chemistry yields and individual blank contributions). Blank corrections were applied to all measured Pb concentrations based on the mean blank value. The uncertainties of the Pb concentration data propagate both the (within-run) precision of the isotopic analyses and the uncertainty of the blank correction.

Chemical separation of Pb

Stable isotope fractionation of Pb during anion exchange chromatography with HCl was recently documented and shown to vary systematically with Pb recovery [29] and [52]. Such fractionation can have a detrimental impact on the accuracy of subsequent Pb isotope measurements and may also occur during Pb purification with other anion exchange elution protocols. As part of this study, it was, therefore, investigated whether incomplete recovery of Pb from the anion exchange chemistry applied here (Table 2) also has a systematic impact on the measured Pb isotope compositions. To this end, multiple 10 ng aliquots of NIST SRM 981 Pb were processed through the complete 2-stage column separation procedure and the purified Pb fractions were then analyzed using standard methods. The data (Fig. 1) illustrate that the column chemistry routinely affords Pb yields of about 50–70%. Such recoveries are in accord with the total Pb yields of about 30–60% that were determined for multiple analyses of five different seawater samples (Table 3 and Table 4). Assuming that the $Mg(OH)_2$ co-precipitation achieves a recovery of ≥80% for seawater Pb (see above), the Pb yield of the column chemistry is constrained to be about 40–75%.

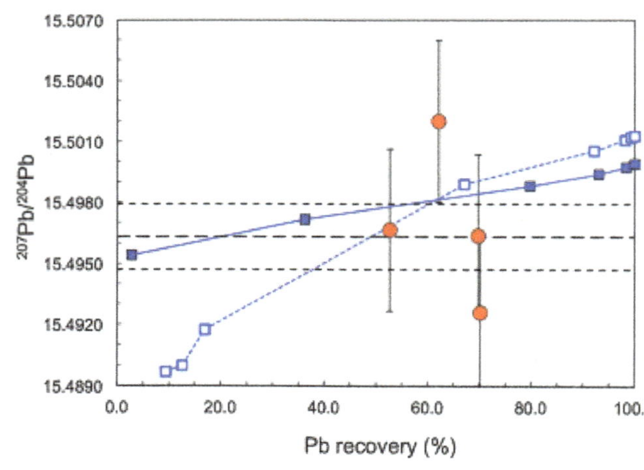

Figure 1: The recovery of Pb and $^{207}Pb/^{204}Pb$ ratios obtained for 10 ng aliquots of NIST SRM 981 Pb that were processed through the 2-stage column chemistry (Table 2) are shown as full red circles with error bars that denote a ±2sd uncertainty of about 240 ppm (Table 3). Shown for comparison are literature data for cumulative recovery and isotope composition of SRM 981 Pb that was collected in successive fractions from anion-exchange resin which was eluted with HCl: full blue squares and solid line–elution with 6 M HCl; open blue squares with hashed line–elution with 2 M and 8 M HCl (Baker [29]. The dashed black lines indicate the $^{207}Pb/^{204}Pb$ reference value determined for SRM 981 Pb (15.4963 ± 16; Table 1). (For interpretation of the references to color in this figure legend, the reader is referred to the web version of this article.)

Table 3: Lead concentration and isotope composition data for the four in-house seawater reference materials and other key information relating to samples.

Sample	*n*-ID[a]	[Pb] ± 1sd pmol/ kg[b]	±1sd (%)	*n*-IC[c]	$^{206}Pb/^{204}Pb$ ±2sd[d]	$^{207}Pb/^{204}Pb$ ±2sd[d]	$^{208}Pb/^{204}Pb$ ±2sd[d]	$^{207}Pb/^{206}Pb$ ±2sd[d]	$^{208}Pb/^{206}Pb$ ±2sd[d]	Sample-size L[e]	To-talPb ng[f]	Meas. Pb (mean) ng[g]	Pbyield (%)[h]
Atlantic – S	6	24.7 ± 1.0	4.0	5	18.429 ± 12	15.639 ± 9	38.307 ± 25	0.84862 ± 98	2.0787 ± 25	~2	~11	3–6 (~4)	30–60
±2sd relative (ppm)[d]					630	560	650	1150	1220				
±2se typi-cal run (ppm)[i]					~300	~300	~300	~40	~50				
Atlantic – D1	11	45.3 ± 0.5	1.1	7	18.499 ± 30	15.625 ± 20	38.287 ± 58	0.84464 ± 50	2.0697 ± 12	1–2	10–20	3–10 (6)	30–50
±2sd relative (ppm)[d]					1610	1280	1520	590	580				
±2se typi-cal run (ppm)[i]					~350	~350	~350	~100	~50				
Atlantic – D2	2	47.4 ± 1.2	0.8	7	18.481 ± 30	15.625 ± 22	38.272 ± 59	0.84540 ± 59	2.0707 ± 10	~2	~20	2–15 (7)	10–70
±2sd relative (ppm)[d]					1730	1500	1650	760	550				
±2se typi-cal run (ppm)[i]					~650	~650	~700	~80	~90				
Wed-dell	8	7.6 ± 1.0	12.3	7	18.397 ± 48	15.628 ± 20	38.293 ± 70	0.84945 ± 147	2.0814 ± 20	~2	~3.8	~2.0	50–60
±2sd relative (ppm)[d]					2600	1300	1830	1730	980				
±2se typi-cal run (ppm)[i]					~600	~600	~600	~60	~90				

a Number of individual sample aliquots analyzed for Pb concentrations.

b Pb concentrations are corrected for a mean total blank of 11.7 ± 4.3 pg. The uncertainties of the concentration data propagate both the (within-run) precision of the isotopic analyses and the uncertainty of the blank correction (see text).

c Number of individual sample aliquots analyzed for Pb isotope compositions.

d 2sd = 2× standard deviation calculated from results of *n* separate runs; denotes long-term "external" reproducibility.

e Volume of individual aliquots used for Pb isotope analyses.

f Original mass of Pb in sample aliquots used for Pb isotope analyses.

g Mass of Pb in individual aliquots used for isotopic analyses, following chemical separation; a mean value is given in parentheses.

h Pb recovery for the chemical separation process.

i 2se = 2× standard error of the mean; denotes within-run ("internal") precision of a run, based on individual results for measurement cycles.

Table 4: Lead concentration (ID) and isotope composition (IC) data for the GEOTRACES intercalibration samples GSI and GDI and reference values from the literature.

Sample	*n*-ID[a]	[Pb] ± 1sd pmol/kg[b]	*n*-IC[c]	$^{206}Pb/^{204}Pb$ ±2sd[d]	$^{207}Pb/^{204}Pb$ ±2sd[d]	$^{208}Pb/^{204}Pb$ ±2sd[d]	$^{207}Pb/^{206}Pb$ ±2sd[d]	$^{208}Pb/^{206}Pb$ ±2sd[d]	Samplesize L[e]	TotalPb ng[f]	Pbyield (%)[g]
GSI											
This study	6	27.9 ± 1.2	3	18.359 ± 26	15.635 ± 7	38.246 ± 30	0.85162 ± 93	2.0833 ± 16	~2	~10	45–60
±2sd relative (ppm)[d]				1430	480	790	1100	750			
±2se typical run (ppm)[i]				~160	~170	~180	~30	~40			
Ref. – MIT			5	18.41 ± 11	*15.67 ± 9*	*38.33 ± 23*	*0.85143 ± 39*	*2.0821 ± 15*	0.25 to −0.5	1.5–3	
±2sd relative (ppm)[d]				5870	5790	6020	460	700			
Ref. – UCSC			2[h]	18.46 ± 0	*15.68 ± 1*	*38.36 ± 7*	*0.84929 ± 50*	*2.0780 ± 40*	~0.5	~3	
Ref. – Consensus		29.5 ± 2.1									
GDI											
This study	5	45.7 ± 2.6	2	18.475 ± 41	15.624 ± 10	38.276 ± 60	0.84568 ± 134	2.0718 ± 14	~2	~17	50–55
±2sd relative (ppm)[d]				2240	650	1570	1590	670			
±2se typical run (ppm)[i]				~190	~200	~210	~30	~60			
Ref. – MIT			4	18.74 ± 17	*15.83 ± 12*	*38.79 ± 31*	*0.84493 ± 214*	2.0696 ± 50	0.25 to −0.5	2.5–5	
±2sd relative (ppm)[d]				9320	7330	8000	2530	2410			
Ref. – UCSC			4	18.54 ± 3	*15.69 ± 2*	*38.43 ± 10*	*0.84640 ± 42*	*2.0729 ± 27*	~0.5	2.5–5	
Ref. – Consensus		42.2 ± 1.3									

The Pb isotope reference data (Ref.) are from the labs of Ed Boyle at the Massachusetts Institute of Technology (MIT) and Russ Flegal at the University of California Santa Cruz (UCSC) [35]. Results in italics were calculated from data given in the literature. The 'Consensus' Pb concentrations are from [53].

a Number of individual sample aliquots analyzed for Pb concentrations.

b Pb concentrations are corrected for a mean total blank of 11.7 ± 4.3 pg. The uncertainties of the concentration data propagate both the (within-run) precision of the isotopic analyses and the uncertainty of the blank correction (see text).

c Number of individual sample aliquots analyzed for Pb isotope compositions.

d 2sd = 2× standard deviation calculated from results of *n* separate runs; denotes long-term "external" reproducibility.

e Volume of individual aliquots used for Pb isotope analyses.

f Original mass of Pb in sample aliquots used for Pb isotope analyses.

g Pb recovery for the chemical separation process.

i 2se = 2× standard error of the mean; denotes within-run ("internal") precision of a run, based on individual results for measurement cycles.

H zDuplicate analyses of Pb separated from a single seawater sample.

In a previous investigation, Reuer et al. [15] reported essentially complete recovery of seawater Pb pre-concentrated by $Mg(OH)_2$ co-precipitation, during purification on AG1X8 resin (as in this study) but using HBr and HCl for elution [49]. Test conducted during the course of the present study showed, however, that the method of Reuer et al.[15] was associated with Pb recoveries similar to those reported here, when the procedure was modified to include additional use of HF to prevent precipitation of silicic acid during sample loading (Table 2). The use of dilute HF thus appears to be associated with either incomplete Pb retention on or elution from the anion exchange resin, regardless of the exact nature of the mineral acids, which are applied in the procedure. Ultimately, the current HBr-HNO_3 method (Table 2) was preferred to the elution method of Reuer et al. [15], because the former applies only dilute mineral acids (which helps to minimize the blank) and routinely produced cleaner Pb separates for subsequent isotopic analyses.

Importantly, the isotope data of Fig. 1 demonstrate that incomplete recovery of Pb from the column separation is not associated with significant shifts in isotopic composition. These results do not, however, rule out small, but barely resolvable, enrichments of heavier Pb isotopes following chemical purification. This follows from the observations of previous studies, which detected only minor Pb isotope fractionations, comparable to or smaller than the reproducibility of our results, as a result of incomplete elution of the element from anion exchange columns (Fig. 1).

Analyses of NIST SRM 981 Pb

Multiple measurements of 2 ng (n = 42) and 10 ng (n = 25) loads of NIST SRM 981 Pb were carried out for a basic validation for our double spike techniques of Pb isotope analysis. Notably, the mean isotopic data acquired for both the 2 ng and 10 ng runs are in excellent agreement with recent reference values that were acquired using various Pb double or triple spikes in conjunction with analyses by either TIMS or MC-ICP-MS (Table 1).

The 2sd between-run ('external') precision of the 10 ng data is ~100 ppm for $^{207}Pb/^{206}Pb$, $^{208}Pb/^{206}Pb$ and ~250 ppm for $^{20x}Pb/^{204}Pb$ (where x = ^{206}Pb, ^{207}Pb, ^{208}Pb hereafter). This is similar to or only slightly worse than the reproducibility achieved by other workers with TIMS or MC-ICP-MS for comparable or somewhat larger Pb samples (10–80 ng; Table 1). Our data for the 2 ng loads, a sample size that is more relevant for seawater analyses (Table 3), is less precise than the 10 ng results, with a reproducibility (±2sd external) of ~250 ppm for $^{207}Pb/^{206}Pb$, $^{208}Pb/^{206}Pb$ and ~1000 ppm for $^{20x}Pb/^{204}Pb$ (Table 3). The significantly larger uncertainty of the $^{20x}Pb/^{204}Pb$ data presumably reflects that the 2 ng Pb analyses were hampered by the measurement of ^{204}Pb at low ion beam intensities (about 5–20 mV) using a Faraday cup and standard 10^{11} Ω amplifiers. The within-run ('internal') precision (quantified by 2se; standard error of the mean) of both the 2 and 10 ng Pb measurements are about a factor of 2–3 better compared to the between-run uncertainties for all Pb isotope ratios (Table 1).

Analyses of in-house seawater reference samples

Multiple Pb concentration and isotope ratio analyses were carried out on four different in-house seawater reference samples (Table 3). These were available in sufficient quantity such that 5–7 repeat isotopic analyses could be carried out to rigorously investigate the reproducibility of the methods.

For the Pb contents, the precision (expressed as ±1sd, as is common for concentration data) varies between less than ±1% to about ±12% (Table 3). Notably, the reproducibility is worse for samples with low Pb contents. As similar samples sizes (of only ~50 mL) were employed for these measurements, the higher uncertainties for the Pb-depleted samples are most likely related to inaccuracies of the Pb blank correction, possibly due to spurious high blanks.

The isotopic results for the in-house seawater reference samples (Table 3) were acquired on individual aliquots of 1–2 L, from which about 2–10 ng of Pb were separated. The data display a ±2sd reproducibility of ~500–1500 ppm for $^{207}Pb/^{206}Pb$, $^{208}Pb/^{206}Pb$ and ~500–2000 ppm for $^{20x}Pb/^{204}Pb$ (Table 3). Clearly, this is inferior to the 2sd precision that was obtained for the comparable 2 ng and 10 ng loads of pure NIST SRM 981 Pb (Table 1). At the same time, the internal (within-run) ±2se precision of the seawater measurements is similar to the NIST SRM 981 results, with uncertainties of ~40–100 ppm for $^{207}Pb/^{206}Pb$, $^{208}Pb/^{206}Pb$ and ~300–700 ppm for $^{20x}Pb/^{204}Pb$ (Table 3). Given this discrepancy, it is most likely that the external 2sd reproducibility of the seawater Pb isotope data is limited by contamination of the samples by blanks of both variable magnitude and isotope composition. This interpretation is supported by two observations. First, the Weddell seawater, which has the lowest Pb concentration and thus the smallest Pb sample sizes, generally features the largest 2sd uncertainties (Table 3). Second, the application of a blank correction did not improve the 2sd reproducibilities for any of the samples, regardless of whether the corrections used mean blank values or only blank data that were linked to a specific batch of samples. In addition, it is also conceivable that the between-run 2sd uncertainties of the samples (Table 3) are, at least in part, limited by minor isotope fractionations that occur during the column chemistry (Fig. 1).

GEOTRACES intercalibration samples GSI and GDI

Multiple Pb isotope and concentration measurements were also carried out for the two GEOTRACES intercomparison samples GSI and GDI (Table 4). For the Pb concentrations, GEOTRACES 'consensus values' were recently published, which are based on analyses of more than 10 independent laboratories [53]. The mean Pb concentrations determined for the samples in this study from analyses of five or six separate aliquots have ±1sd (external) reproducibilities that are similar to those obtained for the in-house seawater reference samples (Table 3), at about ±4 to ±6% (Table 4). The results are, furthermore, identical to the reference values within the quoted uncertainties (Table 4).

The Pb isotope data for GSI and GDI were obtained on 2–3 sample aliquots of 2 L each, with originally about 10–15 ng Pb (Table 4). Following chemical separation, about 5–10 ng Pb were available for isotopic analysis. As for the seawater results of Table 3, the GSI and GDI measurements achieved (within-run) 2se repeatabilities that are similar to those obtained for similar loads of pure SRM 981 Pb, at 30–60 ppm for $^{207}Pb/^{206}Pb$, $^{208}Pb/^{206}Pb$ and 150–200 ppm for $^{20x}Pb/^{204}Pb$ (Table 4). In contrast, the external 2sd uncertainties, of ~650–1500 ppm for $^{207}Pb/^{206}Pb$, $^{208}Pb/^{206}Pb$ and ~500–2200 ppm for $^{20x}Pb/^{204}Pb$, are typically significantly larger (Table 4). Again, this is most likely due to small Pb blank contributions of variable magnitude and isotope compositions and, potentially, minor isotope fractionation during Pb purification. Despite of this, our results for the two samples compare favorably and are generally in agreement with the published reference results.

The reference Pb isotope data for the GEOTRACES samples were obtained at MIT (Massachusetts Institute of

Technology) using MC-ICP-MS and at UCSC (University of California at Santa Cruz) by HR-ICP-MS [35]. These analyses did not apply a Pb double spike and were carried out using significantly smaller samples sizes (of about 0.25–0.5 L) compared to those utilized here.

For GSI, all Pb isotope ratios determined are identical to the MIT results within the 2sd uncertainties. There is also good agreement with the UCSC results for $^{207}Pb/^{206}Pb$,$^{208}Pb/^{206}Pb$ but the UCSC data are higher for $^{20x}Pb/^{204}Pb$ (Table 4). The uncertainties of the UCSC data are only poorly characterized, however, as they are based on only two analyses of a single seawater aliquot. Our new isotopic data for GDI are in excellent agreement with the MIT and USCS results for $^{207}Pb/^{206}Pb$, $^{208}Pb/^{206}Pb$ and in reasonable agreement, just within uncertainty, with the UCSC $^{206}Pb/^{204}Pb$ and $^{208}Pb/^{204}Pb$ ratios (Table 4). Hence, there are small but significant discrepancies for the remaining $^{20x}Pb/^{204}Pb$ data with the reference values, whereby our ratios are slightly lower. There is, however, also only marginal agreement between the MIT and UCSC results for these Pb isotope ratios, which are difficult to determine due to the low natural abundance of ^{204}Pb. This indicates that further analyses are needed to establish more reliable reference data for the $^{20x}Pb/^{204}Pb$ ratios of the GEOTRACES intercalibration samples.

The particular difficulty of ^{204}Pb measurements is also apparent in a comparison of the reproducibilities for the Pb isotope data obtained here and previously [35]. For the $^{20x}Pb/^{204}Pb$ ratios, the multiple MIT analyses reveal a ±2sd uncertainty of ~5000–10,000 ppm, whilst our data for GSI, GDI (Table 4) and the in-house test samples (Table 3) have uncertainties which are at least a factor of 5 better, at ~500–2000 ppm. For $^{207}Pb/^{206}Pb$ and $^{208}Pb/^{206}Pb$ the difference is less apparent, however, as the reproducibility of the MIT results varies between 500 and 2500 ppm, whilst the data of this study (Table 3 and Table 4) typically have an uncertainty of about 500–1500 ppm.

Depth profile for Pb isotopes and concentrations in the eastern South Atlantic Ocean

Twelve seawater samples from a depth profile in the eastern part of the GEOTRACES GA10 section in the South Atlantic Ocean were analyzed (Fig. 2a and Table 5), to (i) demonstrate that our new methodology is suitable for routine acquisition of high precision seawater Pb isotope data and (ii) highlight the important insights, which can be gained from such results.

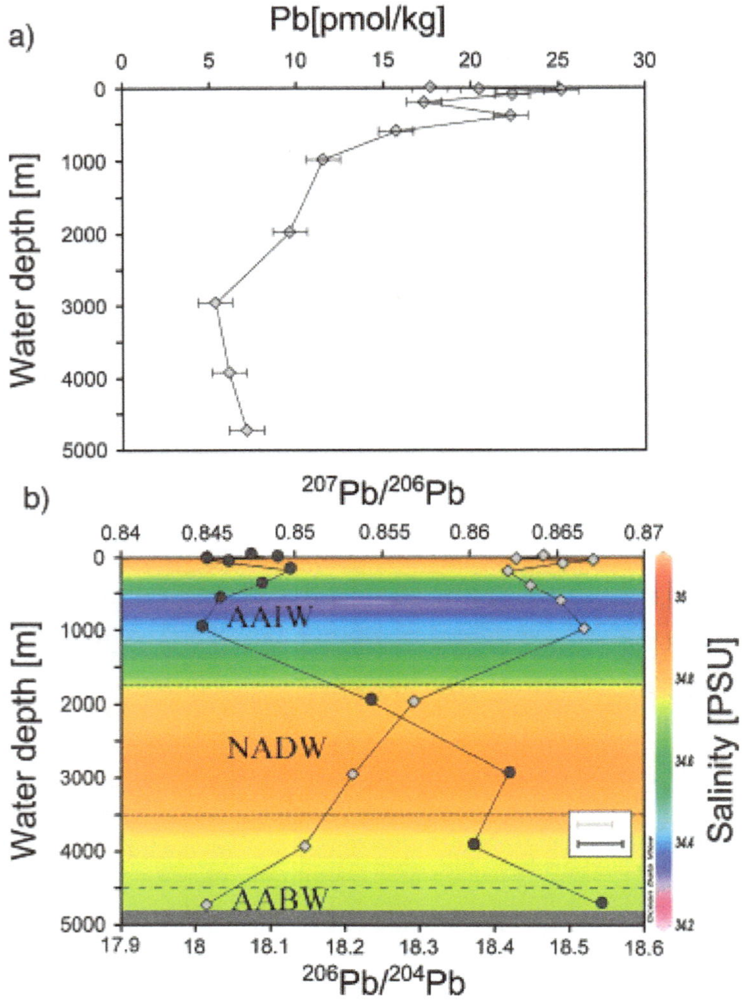

Figure 2: Lead data for the seawater depth profile collected at station 3 (36.46°S, 13.39°E) of the GEOTRACES GA10 cruise. (A) Vertical distribution of Pb concentrations, with uncertainties as reported in Table 4 and (B) vertical distribution of $^{207}Pb/^{206}Pb$ (grey diamonds) and $^{206}Pb/^{204}Pb$ (dark grey dots) ratios, with uncertainties as reported in Table 4. The background colors from Ocean Data View [59] display the salinity of the water column during the time of sample collection [37].

Table 5: Lead isotope and concentration data for a seawater profile from the southeastern Atlantic Ocean (D357 station 3, GEOTRACES transect GA10; latitude: 36.46°S, longitude: 13.39°E; sampled in October 2010).

Sample	Depth (m)	[Pb] pmol/kg[a]	±1sd[b]	$^{206}Pb/^{204}Pb$[c]	$^{207}Pb/^{204}Pb$[c]	$^{208}Pb/^{204}Pb$[c]	$^{207}Pb/^{206}Pb$[c]	$^{208}Pb/^{206}Pb$[c]
Fish 113	2–3	17.7	0.9	18.073	15.620	38.030	0.86428	2.1040
183	25	20.5	0.4	18.111	15.624	38.017	0.86266	2.0991
180	49	25.2	0.5	18.013	15.618	37.922	0.86707	2.1053
177	99	22.4	0.8	18.044	15.614	37.941	0.86533	2.1027
174	198	17.3	0.6	18.125	15.627	38.015	0.86221	2.0975
171	397	22.3	0.6	18.087	15.618	37.970	0.86350	2.0993
170	594	15.7	0.5	18.031	15.600	37.885	0.86518	2.1011
169	989	11.5	0.5	18.007	15.604	37.911	0.86655	2.1052
167	1975	9.6	0.4	18.234	15.623	38.159	0.85677	2.0927
166	2955	5.3	0.4	18.418	15.717	38.452	0.85331	2.0879
164	3931	6.1	0.4	18.370	15.625	38.266	0.85050	2.0830
163	4724	7.1	0.4	18.540	15.664	38.537	0.84492	2.0785

a Pb concentrations are corrected for a mean total blank of 11.7 ± 4.3 pg.

b The uncertainties of the concentration data propagate both the (within-run) precision of the isotopic analyses and the uncertainty of the blank correction (see text).

c For the Pb isotopes, assumed external 2sd reproducibilities of ±1250 ppm for $^{20x}Pb/^{204}Pb$ and of ±800 ppm for $^{207}Pb/^{206}Pb$, $^{208}Pb/^{206}Pb$ are larger than the ±2se internal uncertainties in all cases. Based on this, appropriate 2sd uncertainties are ±0.023 for $^{206}Pb/^{204}Pb$, ±0.020 for $^{207}Pb/^{204}Pb$, ±0.047 to 0.048 for $^{208}Pb/^{204}Pb$, ±0.00068 to 0.00069 for $^{207}Pb/^{206}Pb$ and ±0.0017 for $^{208}Pb/^{206}Pb$.

Lead concentrations are highest in the uppermost part of the water column with a surface (~3 m water depth) concentration of 17.7 pmol/kg, increasing to a relative maximum of 25.2 pmol/kg at 49 m and decreasing again to 22.4 pmol/kg at 397 m depth. From 400 m downwards, the Pb concentrations decrease, until they reach relatively constant and low values of 5–10 pmol/kg below 2000 m (Fig. 2a). A comparison with the dissolved Pb concentrations for a depth profile collected at the exact same location during the second expedition along the South Atlantic GA10 transect (cruise JC068; same station; Schlosser et al., pers. comm.) demonstrates excellent agreement for the entire water column. In contrast, data reported from IPY cruise MD166 (Boye et al. [36]) for the same location show significant deviations from our results, as the dissolved Pb concentrations are elevated by up to a factor of 2 and they define a pattern which is less smooth with depth.

Overall, our Pb concentrations for seawater from the southeast Atlantic Ocean, sampled in 2010, are significantly lower than the values of 29.5 pmol/kg and 42.2 pmol/kg that were reported for 2008 samples from the surface and 2000 m depth at the BATS station in the northwest Atlantic Ocean (Table 4). Lead concentrations in seawater have been strongly affected by anthropogenic Pb emissions over the past century. Such emissions were generally higher in the northern Hemisphere compared to southern regions but peaked in the 1970s. Since then, they have decreased by almost a factor of 10 due to the phasing out of leaded gasoline, first in the US and then in Europe (e.g., Kelly et al. [16]). Consequently, northwest Atlantic surface waters now contain less anthropogenic Pb (and exhibit a different isotopic composition) than deeper North Atlantic waters, which were contaminated in the past [13] and [19]. A comparison of our (sub-) surface Pb concentrations (18–25 pmol/L) with data for eastern South Atlantic surface waters sampled in 1990 (with [Pb] = 11–29 pmol/kg at 21–32°S and 7–13°E; [47]) reveals similar results. This implies that there was no major change in the Pb supply to the southeast Atlantic Ocean between 1990 and 2010.

The $^{207}Pb/^{206}Pb$ isotope ratios for our seawater profile from the South Atlantic exhibit the highest values in the subsurface and at mid depth (0.867 at 49 m and 989 m water depth) and slightly lower values at the surface and at 198 m ($^{207}Pb/^{206}Pb \approx$ 0.862–0.863; Fig. 2b). Below ~1000 m water depth, $^{207}Pb/^{206}Pb$ decreases all the way to the bottom of the profile ($^{207}Pb/^{206}Pb$ = 0.845 at 4723 m water depth). These trends in $^{207}Pb/^{206}Pb$ are mirrored by the $^{206}Pb/^{204}Pb$ ratios, which are lowest at the subsurface ($^{206}Pb/^{204}Pb$ = 18.01; 49 m) and at intermediate depth ($^{206}Pb/^{204}Pb$ = 18.01; 989 m), and increase to $^{206}Pb/^{204}Pb$ = 18.54 at the bottom of the profile (Fig. 2b). The correlation between the seawater $^{207}Pb/^{206}Pb$ and $^{206}Pb/^{204}Pb$ ratios (r^2 = 0.97) observed here is significantly improved compared to the results of previous studies [13], [19], [20],[23] and [54]. We assign this improvement to the accurate and precise determination of the minor isotope ^{204}Pb (and hence $^{20x}Pb/^{204}Pb$ ratios) with our new methodology, and surmise that previous claims of poor correlations may be due to analytical artifacts and should hence be assessed with care.

The major deep and intermediate water masses encountered at station 3 in the eastern South Atlantic are indicated by the salinity contouring in Fig. 2b. In brief, most of the water column is dominated by water masses sourced from high southern latitudes – cold and dense Antarctic bottom water (AABW) at >4500 m depth, and fresh Antarctic intermediate water (AAIW) at ~500–1100 m – whilst saline, nutrient-poor North Atlantic Deep Water (NADW) is present at ~1700–3500 m depth (see Wyatt et al. [37] for further details on the transect hydrography). Since we are presenting the first

Pb isotope depth profile for the eastern South Atlantic, there are no data available for direct comparison. However, our results confirm previously suggested differences [23] in the Pb isotope fingerprints of two main Southern Ocean water masses, AAIW ($^{207}Pb/^{206}Pb$ = 0.865–0.867, $^{206}Pb/^{204}Pb$ = 18.01–18.03; n = 2) and AABW ($^{207}Pb/^{206}Pb$ = 0.845, $^{206}Pb/^{204}Pb$ = 18.54; n = 1). These isotopic differences are a consequence of different water mass formation mechanisms and regions (e.g., Orsi et al. [55]), and hence different proportions of anthropogenic versus natural Pb. prior to ventilation into the deeper ocean. The Pb isotopic composition of NADW in the southeast Atlantic Ocean ($^{207}Pb/^{206}Pb$ = 0.853–0.857, $^{206}Pb/^{204}Pb$ = 18.42–18.23 at ~2000–3000 m water depth; n = 2) appears distinct from NADW in the northwestern North Atlantic (BATS Station at 2000 m; Table 4), and values previously reported for the equatorial western Atlantic Ocean [23]. This points to older ventilation ages in the eastern basins of the Atlantic Ocean compared to the western basins (e.g., England [56]). Further interpretation of the results will, however, have to await the acquisition of Pb isotope and concentration data for additional profiles from the GEOTRACES GA10 transect.

CONCLUSIONS

A new analytical procedure has been developed for Pb isotope analysis of seawater and associated Pb concentration measurements by isotope dilution. This involves Mg hydroxide co-precipitation and anion exchange chromatography for Pb pre-concentration and isotopic analysis by TIMS in conjunction with a ^{207}Pb–^{204}Pb double spike. Analyses of multiple aliquots of six seawater samples demonstrate typical reproducibilities (±2sd) of about 700–1500 ppm and 1000–2000 ppm for $^{207}Pb/^{206}Pb$, $^{208}Pb/^{206}Pb$ and $^{206}Pb/^{204}Pb$, $^{207}Pb/^{204}Pb$, $^{208}Pb/^{204}Pb$, respectively. With this precision, the new data are superior to literature results that were obtained by plasma source mass spectrometry. In particular, they are at least a factor of five more precise for ratios involving the minor ^{204}Pb isotope. Analyses of two GEOTRACES intercalibration samples from the Atlantic Ocean (GDI and GSI) display good agreement with recent results of two other laboratories.

Our methods were applied to samples from a seawater depth profile from the eastern South Atlantic Ocean at ~40°S. For these samples, both Pb concentrations and isotope compositions show a smooth pattern with depth and no obvious outliers. Furthermore, $^{206}Pb/^{204}Pb$ shows a clear correlation with $^{207}Pb/^{206}Pb$, underlining the improvement that was achieved for the challenging Pb isotope analyses, particularly of $^{206}Pb/^{204}Pb$. The depth profile results permit the identification and characterization of several key water masses, including AAIW, AABW and NADW. Considering that our profile is not located far from the formation regions of AAIW and AABW, the data provide the most precise Pb isotope characterization of these water masses published to date, but substantiation of the results by analyses of additional seawater samples from this region is needed.

ACKNOWLEDGEMENTS

We thank all of the MAGIC team for helping to keep spirits high, the mass spec running and the clean labs clean, and are particularly grateful for the vital support provided by Barry Coles and Katharina Kreissig. Ed Boyle volunteered helpful tips during method development, first suggesting the use of trace HF during column chemistry. The scientific party (principal scientist Gideon Henderson), technicians and crew of the R.R.S. Discovery on GEOTRACES cruise D357 are thanked for collection of the South Atlantic seawater profile. The supportive and helpful comments of two anonymous referees and swift editorial handling were also highly appreciated. Financial support for this study was provided by NERC consortium grant NE/H006095/1 to the UK GEOTRACES program and D. Weiss, T. van de Flierdt and M. Rehkämper at Imperial College.

REFERENCES

1. S.W. Group, GEOTRACES – an international study of the global marine biogeochemical cycles of trace elements and their isotopes, Chem. Erde – Geochem. 67 (2007) 85–131.
2. T.J. Chow, M.S. Johnstone, Lead isotopes in gasoline and aerosols of Los Angeles basin, California, Science 147 (1965) 502–503, doi:http://dx.doi.org/10.1126/science.147.3657.502.
3. T.J. Chow, J.L. Earl, Lead isotopes in North American coals, Science 176 (1972) 510–511.
4. T.J. Chow, K.W. Bruland, K. Bertine, A. Soutar, M. Koide, E.D. Goldberg, Lead pollution: records in Southern California coastal sediments, Science 181 (1973) 551–552.
5. T.J. Chow, J.L. Earl, Lead aerosols in the atmosphere: increasing concentrations, Science 169 (1970) 577–580.
6. A.R. Flegal, C.C. Patterson, Vertical concentration profiles of lead in the Central Pacific at 15aN and 20S, Earth Planet, Sci. Lett. 64 (1983) 19–32, doi:http://dx.doi.org/10.1016/0012-821X(83)90049-3.
7. B.K. Schaule, C.C. Patterson, Perturbations of the natural lead depth profile in the sargasso sea by industrial lead, in: C.S. Wong, E. Boyle, K. Bruland, J.D. Burton, E. Goldberg (Eds.), Trace Metals Sea Water, Springer, US, 1983, pp. 487– 503, doi:http://dx.doi.org/10.1007/978-1-4757-6864-0_29.
8. B.K. Schaule, C.C. Patterson, Lead concentrations in the northeast Pacific: evidence for global anthropogenic perturbations, Earth Planet. Sci. Lett. 54 (1981) 97–116, doi:http://dx.doi.org/10.1016/0012-821X(81)90072-8.
9. C.C. Patterson, D. Settle, B. Glover, Analysis of lead in polluted coastal seawater, Mar. Chem. 4 (1976) 305–319.
10. A.R. Flegal, T.F. Duda, S. Niemeyer, High gradients of lead isotopic composition in north-east Pacific upwelling filaments, Nature 339 (1989) 458–460, doi: http://dx.doi.org/10.1038/339458a0.
11. A.R. Flegal, B.K. Schaule, C.C. Patterson, Stable isotopic ratios of lead in surface waters of the Central Pacific, Mar. Chem. 14 (1984) 281–287, doi:http://dx.doi. org/10.1016/0304-4203(84)90048-3.
12. G.T. Shen, E.A. Boyle, Thermocline ventilation of anthropogenic lead in the western North Atlantic, J. Geophys. Res. Oceans 93 (1988) 15715–15732, doi: http://dx.doi.org/10.1029/JC093iC12p15715.
13. A.J. Véron, T.M. Church, C. Patterson, A.R. Flegal,

Use of stable lead isotopes to characterize the sources of anthropogenic lead in North Atlantic surface waters, Geochim. Cosmochim. Acta 58 (1994) 3199–3206, doi:http://dx.doi. org/10.1016/0016-7037(94)90047-7.

14. D. Weiss, E.A. Boyle, V. Chavagnac, M. Herwegh, J. Wu, Determination of lead isotope ratios in seawater by quadrupole inductively coupled plasma mass spectrometry after Mg(OH)2 co-precipitation, Spectrochim. Acta, Part B At. Spectrosc. 55 (2000) 363–374.
15. M.K. Reuer, E.A. Boyle, B.C. Grant, Lead isotope analysis of marine carbonates and seawater by multiple collector ICP-MS, Chem. Geol. 200 (2003) 137–153, doi:http://dx.doi.org/10.1016/S0009-2541(03)00186-4.
16. A.E. Kelly, M.K. Reuer, N.F. Goodkin, E.A. Boyle, Lead concentrations and isotopes in corals and water near Bermuda, 1780–2000, Earth Planet. Sci. Lett. 283 (2009) 93–100.
17. Véron, P. Flament, M.L. Bertho, L. Alleman, R. Flegal, B. Hamelin, Isotopic evidence of pollutant lead sources in Northwestern France, Atmos. Environ. 33 (1999) 3377–3388, doi:http://dx.doi.org/10.1016/S1352-2310(98)00376-8.
18. D. Weiss, E.A. Boyle, J. Wu, V. Chavagnac, A. Michel, M.K. Reuer, Spatial and temporal evolution of lead isotope ratios in the North Atlantic Ocean between 1981 and 1989, J. Geophys. Res. Oceans 108 (2003) 3306, doi:http://dx.doi.org/ 10.1029/2000JC000762.
19. B. Hamelin, J.L. Ferrand, L. Alleman, E. Nicolas, A.J. Véron, Isotopic evidence of pollutant lead transport from North America to the subtropical North Atlantic gyre, Geochim. Cosmochim. Acta 61 (1997) 4423–4428, doi:http://dx.doi.org/ 10.1016/S0016-7037(97)00242-1.
20. A.J. Véron, T.M. Church, I. Rivera-Duarte, A.R. Flegal, Stable lead isotopic ratios trace thermohaline circulation in the subarctic North Atlantic, Deep Sea Res. Part II Top. Stud. Oceanogr. 46 (1999) 919–935.
21. F. von Blanckenburg, H. Igel, Lateral mixing and advection of reactive isotope tracers in ocean basins: observations and mechanisms, Earth Planet. Sci. Lett. 169 (1999) 113–128.
22. M.K. Reuer, D.J. Weiss, Anthropogenic lead dynamics in the terrestrial and marine environment, Philos. Trans. R. Soc. Lond. Ser. Math. Phys. Eng. Sci. 360 (2002) 2889–2904, doi:http://dx.doi.org/10.1098/rsta.2002.1095.
23. L.Y. Alleman, T.M. Church, P. Ganguli, A.J. Véron, B. Hamelin, A.R. Flegal, Role of oceanic circulation on contaminant lead distribution in the South Atlantic, Deep Sea Res. Part II Top. Stud. Oceanogr. 48 (2001) 2855–2876.
24. T. Hirata, Lead isotopic analyses of NIST standard reference materials using multiple collector inductively coupled plasma mass spectrometry coupled with a modified external correction method for mass discrimination effect, Analyst 121 (1996) 1407–1411.
25. F.W.E. Strelow, Distribution coefficients and anion exchange behavior of some elements in hydrobromic-nitric acid mixtures, Anal. Chem. 50 (1978) 1359– 1361, doi:http://dx.doi.org/10.1021/ac50031a041.
26. G.W. Lugmair, S.J.G. Galer, Age and isotopic relationships among the angrites Lewis Cliff 86,010 and Angra dos Reis, Geochim. Cosmochim. Acta 56 (1992) 1673–1694.
27. B. Hamelin, G. Manhes, F. Albarede, C.J. Allègre, Precise lead isotope measurements by the double spike technique: a reconsideration, Geochim. Cosmochim. Acta 49 (1985) 173–182, doi:http://dx.doi.org/10.1016/0016-7037 (85)90202-9.
28. S.J.G. Galer, Optimal double and triple spiking for high precision lead isotopic measurement, Chem. Geol. 157 (1999) 255–274.
29. J. Baker, D. Peate, T. Waight, C. Meyzen, Pb isotopic analysis of standards and samples using a 207Pb–204Pb double spike and thallium to correct for mass bias with a double-focusing MC-ICP-MS, Chem. Geol. 211 (2004) 275–303.
30. N.H. Gale, A solution in closed form for lead isotopic analysis using a double spike, Chem. Geol. 6 (1970) 305–310.
31. M.F. Thirlwall, Inter-laboratory and other errors in Pb isotope analyses investigated using a 207Pb–204Pb double spike, Chem. Geol. 163 (2000) 299– 322.
32. M.F. Thirlwall, Multicollector ICP-MS analysis of Pb isotopes using a 207Pb–204Pb double spike demonstrates up to 400 ppm/amu systematic errors in Tl-normalization, Chem. Geol. 184 (2002) 255–279.
33. J.D. Woodhead, F. Volker, M.T. McCulloch, Routine lead isotope determinations using a lead-207-lead-204 double spike: a long-term assessment of analytical precision and accuracy, Analyst 120 (1995) 35–39.
34. J.D. Woodhead, J.M. Hergt, Application of the 'double spike' technique to Pbisotope geochronology, Chem. Geol. 138 (1997) 311–321.
35. E.A. Boyle, S. John, W. Abouchami, J.F. Adkins, Y. Echegoyen-Sanz, M. Ellwood, et al., GEOTRACES IC1 (BATS) contamination-prone trace element isotopes Cd, Fe, Pb, Zn, Cu, and Mo intercalibration, Limnol, Oceanogr. Methods 10 (2012) 653–665, doi:http:// dx.doi.org/10.4319/lom.2012.10.653.
36. M. Boye, B.D. Wake, P. Lopez Garcia, J. Bown, A.R. Baker, E.P. Achterberg, Distributions of dissolved trace metals (Cd, Cu, Mn, Pb, Ag) in the southeastern Atlantic and the Southern Ocean, Biogeosciences 9 (2012) 3231–3246, doi: http://dx.doi.org/10.5194/bg-9-3231-2012.
37. N.J. Wyatt, A. Milne, A.P. Woodward, T.J. Browning, H.A. Bouman, et al., Biogeochemical cycling of dissolved zinc along the GEOTRACES South Atlantic transect GA10 at 40S, Glob. Biogeochem. Cycles 28 (2014) 44–56, doi:http:// dx.doi.org/10.1002/2013GB004637.
38. J.F. Rudge, B.C. Reynolds, B. Bourdon, The double spike toolbox, Chem. Geol. 265 (2009) 420–431.
39. W.M. White, F. Albarède, P. Télouk, High-precision analysis of Pb isotope ratios by multi-collector ICP-MS, Chem. Geol. 167 (2000) 257–270.
40. M. Rehkämper, A.N. Halliday, Accuracy and long-term reproducibility of lead isotopic measurements by multiple-collector inductively coupled plasma mass

spectrometry using an external method for correction of mass discrimination, Int. J. Mass Spectrom. 181 (1998) 123–133.

41. M.H. Dodson, A theoretical study of the use of internal standards for precise isotopic analysis by the surface ionization technique: part I – general firstorder algebraic solutions, J. Sci. Instrum. 40 (1963) 289–295, doi:http://dx.doi. org/10.1088/0950-7671/40/6/307.

42. J. Wu, E.A. Boyle, Lead in the western North Atlantic Ocean: completed response to leaded gasoline phaseout, Geochim. Cosmochim. Acta 61 (1997) 3279–3283.

43. P.M. Nicolaysen, E. Steinnes, T.E. Sjobakk, Pre-concentration of selected trace elements from seawater by co-precipitation on Mg(OH)2, J. Phys. (IV Fr.) 107 (2003) 945–948.

44. J. Wu, E.A. Boyle, Low blank preconcentration technique for the determination of lead copper, and cadmium in small-volume seawater samples by isotope dilution ICPMS, Anal. Chem. 69 (1997) 2464–2470.

45. G.F. de Souza, B.C. Reynolds, J. Rickli, M. Frank, M.A. Saito, L.J.A. Gerringa, et al., Southern Ocean control of silicon stable isotope distribution in the deep Atlantic Ocean: silicon isotopes in the deep Atlantic, Glob. Biogeochem. Cycles 26 (2012) 1944–9224, doi:http://dx.doi.org/10.1029/2011GB004141.

46. Y. Nozaki, K.K. Turekian, K. Von Damm, 210Pb in GEOSECS water profiles from the North Pacific, Earth Planet. Sci. Lett. 49 (1980) 393–400.

47. W.I. Manton, Separation of Pb from young zircons by single-bead ion exchange, Chem. Geol. Isot. Geosci. Sect. 73 (1988) 147–152.

48. L.I. Guseva, A study of ion-exchange behavior of Pb in dilute HBr solutions, aimed to evaluate the possibility of on-line isolation of element 114 228Ra-212Pb generator, Radiochemistry 49 (2007) 92–96, doi:http://dx.doi. org/10.1134/S106636220701016X.

49. B.S. Kamber, A.H. Gladu, Comparison of Pb purification by anion-exchange resin methods and assessment of long-term reproducibility of Th/U/Pb ratio measurements by quadrupole ICP-MS, Geostand. Geoanal. Res. 33 (2009) 169– 181, doi:http://dx.doi.org/10.1111/j.1751-908X.2009.00911.x.

50. H. Gerstenberger, G. Haase, A highly effective emitter substance for mass spectrometric Pb isotope ratio determinations, Chem. Geol. 136 (1997) 309– 312.

51. R. Doucelance, G. Manhès, Reevaluation of precise lead isotope measurements by thermal ionization mass spectrometry: comparison with determinations by plasma source mass spectrometry, Chem. Geol. 176 (2001) 361–377.

52. J. Blichert-Toft, B. Zanda, D.S. Ebel, F. Albarède, The solar system primordial lead, Earth Planet. Sci. Lett. 300 (2010) 152–163.

53. The GEOTRACES Standards and Intercalibration Committee, Consensus Values for the GEOTRACES 2008, (2013).http://www.geotraces.org/science/intercalibration/322-standards-and-reference-materials.

54. L.Y. Alleman, T.M. Church, A.J. Véron, G. Kim, B. Hamelin, A.R. Flegal, Isotopic evidence of contaminant lead in the South Atlantic troposphere and surface waters, Deep Sea Res. Part II Top. Stud. Oceanogr. 48 (2001) 2811–2827.

55. A.H. Orsi, G.C. Johnson, J.L. Bullister, Circulation, mixing, and production of Antarctic bottom water, Prog. Oceanogr. 43 (1999) 55–109, doi:http://dx.doi. org/10.1016/S0079-6611(99)00004-X.

56. M.H. England, The age of water and ventilation timescales in a global ocean model, J. Phys. Oceanogr. 25 (1995) 2756–2777.

57. S.J.G. Galer, W. Abouchami, Practical application of lead triple spiking for correction of instrumental mass discrimination, Miner. Mag. 62A (1998) 491– 492, doi:http://dx.doi.org/10.1180/minmag.1998.62A.1.260.

58. W. Todt, R.A. Cliff, A. Hanser, A.W. Hofmann, Evaluation of a 202Pb–205Pb double spike for high – precision lead isotope analysis, in: A. Basu, S. Hart (Eds.), Geophysical Monographs Series, American Geophysical Union, Washington, D.C, 1996, pp. 429–437.

59. R. Schlitzer, Ocean Data View, (2002) http://www.awi-bremerhaven.de/GEO/ ODV.

CHAPTER 14

DEGRADING ORGANIC MICROPOLLUTANTS: THE NEXT CHALLENGE IN THE EVOLUTION OF BIOLOGICAL WASTEWATER TREATMENT PROCESSES

Naresh Singhal and Octavio Perez-Garcia

Keywords: Emerging contaminants, organic micropollutants, biocatalysis, metabolic diversity, biodegradation, oxidoreductases

WASTEWATER, A WASTED RESOURCE

The Oxford English Dictionary defines wastewater as superfluous water, or water that has served its purpose. This definition of wastewater is ingrained in our perception and attitude. Yet, this definition would appear to be outdated in light of the emerging view that wastewater represents an unutilized water resource and not a waste product to be discarded. Furthermore, promoting such use of wastewater is neither new nor radical. Virtually all urban societies already make use of water sources laden with wastewater, albeit in varying amounts. For example, consider the Mississippi, the longest river in North America, which serves as both a water source and wastewater receptor for the numerous towns and cities located along its path. As the river meanders downstream, water increasingly enriched by upstream discharges of wastewater serves as a source of potable water for downstream communities. Such observations support the view that wastewater is better defined as a potential water resource that is currently wasted. A third of the world's population lives in water-stressed countries and this proportion is expected to rise to two thirds by 2025 (Service, 2006). Finding alternate water sources is critical to the survival of these water stressed societies. Although wastewater has been overlooked as a potential water resource, it could become an option if appropriate treatment technologies can be developed.

While biological treatment processes are significantly cheaper to build and operate, current designs are not capable of delivering high quality water suitable for a variety of purposes. Despite several advances since the inception of biological wastewater treatment processes, with an initial focus on removing organic carbon through to subsequent extension to include nitrogen and phosphorous removal, these systems continue to only be able to effectively degrade substances present at mg/L concentrations.

Over the past decade the rapid increase in instrument sensitivity has led to wide detection of ng/L levels of biologically active contaminants, referred to as emerging pollutants or organic micropollutants, in the environment. These contaminants include pharmaceuticals, personal care products, steroid hormones, industrial chemicals, and pesticides (Galloway et al., 2010; Oulton et al., 2010; Vandenberg et al., 2013; Richardson and Ternes, 2014; Petrie et al., 2015; Richardson and Kimura, 2016). Discharges from wastewater treatment plants are acknowledged as a major source of these contaminants in receiving environments (Spongberg and Witter, 2008; Bartelt-Hunt et al., 2009; Loganathan et al., 2009). Various studies have highlighted the need to remove the ng/L levels of these contaminants from the treated wastewater discharge due to their negative effect on human and animal health and ecotoxicology effects (Escher et al., 2011)—for example, a recent critical review concluded that doses of bisphenol A up to 4 orders of magnitude lower than the currently prescribed lowest observed adverse effect level of 50 mg/kg/day reliably produced effects in animals (Vandenberg et al., 2013) and another study reported changes in sex hormones

Citation: Singhal N and Perez-Garcia O (2016) Degrading Organic Micropollutants: The Next Challenge in the Evolution of Biological Wastewater Treatment Processes. Front. Environ. Sci. 4:36. doi: 10.3389/fenvs.2016.00036.

associated with exposure to bisphenol A in men (Galloway et al., 2010). The cost of treating impacted environments can be large. London's drinking water is reported to be tainted with synthetic estrogens from excreted contraceptive pill residues in the Thames that have resulted from the water having gone through the equivalent of six people before reaching London. The cost of cleaning up Britain's contaminated waterways is estimated to exceed £30bn (McKie, 2012). Development of appropriate technologies is needed to convert wastewater into a water resource, potentially even for meeting potable needs. This paper aims to stimulate a discussion on ways to enhance the performance of biological treatment systems to achieve this vision.

CURRENT WASTEWATER TREATMENT APPROACHES

The biological treatment approach, specifically the activated sludge process, is the most commonly used wastewater treatment technology. While it was originally designed to only remove organic carbon, it was subsequently extended to remove nitrogen and phosphorous. Transitioning between aerobic, anoxic, and anaerobic conditions is key to achieving biological nutrient removal. Doing so triggers the utilization of different electron acceptors and donors, therefore promoting the transformation of C, N, and P compounds. This strategy has continued to be exploited with the development of new system configurations, such as the Anammox (Lackner et al., 2014), SHARON (Hellinga et al., 1998; van Dongen et al., 2001), and Nareda (van der Roest et al., 2011) processes. Different redox environments influence organic micropollutant removal in different ways. Various studies have reported that organic micropollutant removal occurs to varying extent in biological treatment systems via a combination of biosorption and biodegradation. Maximum achievable removal for micropollutant such estrogens, nonylphenolics and metals was recorded at the highest sludge retention times (SRT) and hydraulic retention time (HRT) studied (Petrie et al., 2014). This micropollutant removal is related to the concomitant reduction in food: microorganism ratio. Nevertheless, studies have concluded that micropollutant degradation is insensitive to different SRTs (Falås et al., 2016). Other aspects influencing micropollutant removal in biological treatment processes are heterotrophic activity (Majewsky et al., 2010), pH (Gulde et al., 2014), and suspended/attached growth configuration (Falås et al., 2013). Although the heterogeneity of micropollutants in wastewaters makes removal difficult to predict since their chemistry is so diverse. For instance micropollutants can be broadly classified as easily, moderate and poorly degradable. The reality is that the biological processes are not designed to remove these pollutants, resulting in their incomplete removal and detection in final effluents and receiving surface waters (Joss et al., 2006; Racz and Goel, 2010). No specific strategies appear to have been successfully developed to enhance micropollutant removal by biological systems. Instead, organic micropollutant removal is typically achieved using physicochemical processes. These processes include adsorption in to organic matrices, passive effluent treatment in wetlands and aquifers, advanced oxidation processes (e.g., ozonation, UV treatment, photocatalysis, and Fenton oxidation), membrane filtration (nanofiltration and reverse osmosis membranes), and membrane biological reactor (Bolong et al., 2009; Rossner et al., 2009; Oulton et al., 2010; Luo et al., 2014). The shortcomings of these treatment systems are high investment and maintenance costs, generation of toxic residuals, and complex treatment procedures (Grassi et al., 2012). Operational difficulties of these physicochemical processes also need to be considered. For instance, wetlands and aquifers require large surface areas and large HRTs, which significantly complicates its implementation in urban areas. Similarly, advanced oxidation processes require chemicals not easily available (e.g., O_3, F+2F2+, H_2O_2) and the non-selective nature of these reactions can result in formation of daughter products that are more toxic than the parent micropollutant (Rosal et al., 2009). Other concerns may also need consideration: energy and chemical inputs of ozonation processes are substantial (Cañizares et al., 2009); the Fenton process requires rigorous pH control (Chong et al., 2012); specially configured reactors are needed to maximize light exposure by UV and photocatalysis based water treatment, or water turbidity could significantly influence treatment efficiency (Chong et al., 2012); and membrane filtration and membrane biological reactors treatments routinely suffer from membrane fouling and biofilm formation (Guo et al., 2012).

ENHANCING BIOLOGICAL WASTEWATER TREATMENT

Exploiting Microbial Diversity

Complex microbial consortia composed by bacteria-bacteria, bacteria-archea and bacteria-fungi can be developed to enhance biological degradation of micropollutants. Oxidase enzymes have great potential as biocatalysts for micropollutant and organic waste breakdown. Two of such class of enzymes are the oxygenase Cytochromes P450 (CYPs or Cyt P450), a highly efficient group of monooxygenases responsible for the destruction of drugs and toxins in organisms, and the laccases (EC 1.10.3.2), a class of copper-containing oxidase enzymes used by microorganisms to break down lignin) (Riva, 2006; Kumar, 2010). Both enzymes have been shown to efficiently degrade a vast array of organic micropollutants in pure enzyme assays (Harms et al., 2011; Lah et al., 2011). Despite the demonstrated effectiveness of these enzymes, their use in wastewater treatment is just starting to be investigated and has not been implemented in pilot or full scale (Lah et al., 2011). Current biological systems are reliant on using prokaryotic bacteria, which can rapidly oxidize organic carbon but generally do not express Cyt P450. Additionally, no applications of bacterial laccases have yet been realized due to their limited characterization (Werck-Reichhart and Feyereisen, 2000). In comparison, many fungi produce Cyt P450 (Lah et al., 2008; Kelly and Kelly, 2013) and only fungal laccases are used currently in biotechnological applications (Sharma et al., 2007; Harms et al., 2011). A system integrating suspended phase bacteria with activated fungi to transform conventional and emerging categories of pollutants can potentially take advantage of both organisms. Fungi can be activated by inducing the production of high levels of oxidase enzymes and at the same time stimulating enzyme activity (e.g., by providing H_2O_2 as cofactor for bacterial Cyt P450). By controlling the carbon source and electron acceptor regimens, it is possible to induct biocatalyst expression and activation

(Price et al., 2013; Chubukov et al., 2014). As whole-cell catalyzed reactions are reportedly 10- to 100-fold slower than reactions catalyzed by free enzymes (Sotirova et al., 2008), cell membranes can be permeabilized to enhance extracellular laccase secretion and promote cellular influx of pollutants to membrane bound Cyt P450. In addition, gel encapsulated of fungi or bacterial cells can be implemented for improved performance. Alternatively, iterating between different redox conditions can promote the growth of different microbial populations to improve micropollutant biodegradation (Falås et al., 2016).

Exploiting Metabolic Diversity

New bioprocesses to treat wastewater can be developed by exploiting the diverse metabolic capabilities of microbes. Traditionally, only metabolic capabilities in relation to aerobic heterotrophy, aerobic nitrification and anoxic denitrification, and phosphate accumulation have been used in biological wastewater treatment systems. Exploiting alternative metabolic capabilities such as heterotrophic sulfate-reduction (Zhang and Wang, 2014), autotrophic sulfur-oxidizing denitrification (Hao et al., 2014; Pokorna and Zabranska, 2015), anaerobic methane-oxidation denitrification (Raghoebarsing et al., 2006), partial nitrite reduction to nitrous oxide for energy generation (Scherson et al., 2013) and electron shuttle redox biotransformation (Van der Zee and Cervantes, 2009) could improve micropollutant removal from wastewaters. These metabolic capabilities can also be exploited to remove inorganic micropollutants (e.g., heavy metals and radioactive elements) in which redox transformations can alter the solubility and precipitate the contaminants (Groudev et al., 1999; Gadd, 2010).

Exploiting Biocatalyst Diversity

Many enzymes are "promiscuous" biocatalysts capable of transforming a variety of substrates that share structural similarity with their primary substrate. These promiscuous or generalized enzymes: (i) are frequently not essential, (ii) maintain low metabolic fluxes, and (iii) require less regulation of enzyme activity to control metabolic flux in dynamic environments than do specialized enzymes (Nam et al., 2012). This non-specificity makes them promising biocatalysts for organic micropollutant degradation. Bacteria are constantly developing new catabolic pathways in order to either access sources of carbon, energy and nutrients or simply to detoxify new compounds. However, unraveling these degradation processes in nature is made difficult by the large number of chemicals, their occurrence at mostly low concentrations, and the number of unknown chemicals resulting from bacterial transformation and biodegradation. Genomics has revealed that many microorganisms have far greater potential to produce specialized enzymes and metabolites than was thought from classic bioactivity screens; however, realizing their degradation potential has been hampered by the fact that many specialized metabolite biosynthetic gene clusters are not expressed in laboratory cultures (Keller et al., 2005; Miller et al., 2010; Mora-Pale et al., 2014; Rutledge and Challis, 2015). Theoretically bacteria can utilize the full space of catabolic biochemical reaction types and initiate several pathways to degrade micropollutants (Kolvenbach et al., 2014). For example, multiple biodegradation mechanisms have been discovered for bisphenol A (BPA), an industrial chemical found in a variety of plastics and epoxies and a putative endocrine disrupting compound. Similarly, enzymatic biodegradation pathways may be discoverable for other compounds using biocatalyst and daughter products screening techniques such as metagenomics (Fernández-Arrojo et al., 2010), metaproteomics, and metabolomics (Villas-Bôas and Bruheim, 2007; Helbling et al., 2010, 2012).

IMPLEMENTATION

Current designs focus on removing conventional pollutants (organic and inorganic substances present) that are present at much larger concentrations than micropollutants to be metabolized as a source of carbon and energy. The low concentrations of organic micropollutants suggests that these substances are more likely to be cometabolized. Hence, system design will need to focus on ways of stimulating the production of enzymes that could degrade the micropollutants. Current wastewater treatment system designs contain the basic ingredients required to formulate new processes for efficiently degrading micropollutants using the strategies discussed in the sections above. Biological treatment systems contain a wide variety of active microorganisms, and employ different redox environments are used to stimulate a wider range of chemical transformations. Rapid degradation of micropollutants could be achieved by stimulating the production of enzymes (e.g., oxidoreductases and laccases) by using hard-to-degrade substrates or by employing other environmental stresses. Liquids from anaerobic digestate could serve as carbon source that promotes such activity. The performance of such systems could be further enhanced by employing operational strategies that avert a lowering of enzyme production resulting from adaptation of microbial consortium to the imposed stress. Two plant configurations could potentially be developed using these approaches: (i) main stream micropollutant removal (i.e., modification of existing bioprocesses) or (ii) add-on reactors for micropollutant removal (i.e., dedicated tanks to remove micropollutants in the effluent of treatment plant).

Conclusions

Numerous organic micropollutants are present in wastewater. To convert wastewater into high quality water will require their removal. Current wastewater treatment systems are not designed to degrade organic micropollutants. Physicochemical treatment systems are effective but expensive while biological approaches cheap but ineffective and inconsistent. The next stage in the evolution of biological wastewater treatment processes could be based on discovering and employing novel metabolic traits of unconventional microbes (e.g., sulfur or iron oxidizing bacteria), or inducing the synthesis of enzymes capable of degrading micropollutants. Implementation of these new bioprocesses will involve the use of different waste streams used as substrate and new process operation strategies such as dynamic substrates control.

CONFLICT OF INTEREST STATEMENT

The authors declare that the research was conducted in the absence of any commercial or financial relationships that could be construed as a potential conflict of interest.

REFERENCES

1. Bartelt-Hunt, S. L., Snow, D. D., Damon, T., Shockley, J., and Hoagland, K. (2009). The occurrence of illicit and therapeutic pharmaceuticals in wastewater effluent and surface waters in Nebraska. *Environ. Pollut.* 157, 786–791. doi: 10.1016/j.envpol.2008.11.025
2. Bolong, N., Ismail, A. F., Salim, M. R., and Matsuura, T. (2009). A review of the effects of emerging contaminants in wastewater and options for their removal. *Desalination* 239, 229–246. doi: 10.1016/j.desal.2008.03.020
3. Cañizares, P., Paz, R., Sáez, C., and Rodrigo, M. A. (2009). Costs of the electrochemical oxidation of wastewaters: a comparison with ozonation and Fenton oxidation processes. *J. Environ. Manag.* 90, 410–420. doi: 10.1016/j.jenvman.2007.10.010
4. Chong, M. N., Sharma, A. K., Burn, S., and Saint, C. P. (2012). Feasibility study on the application of advanced oxidation technologies for decentralised wastewater treatment. *J. Cleaner Prod.* 35, 230–238. doi: 10.1016/j.jclepro.2012.06.003
5. Chubukov, V., Gerosa, L., Kochanowski, K., and Sauer, U. (2014). Coordination of microbial metabolism. *Nat. Rev. Microbiol.*12, 327–340. doi: 10.1038/nrmicro3238
6. Escher, B. I., Baumgartner, R., Koller, M., Treyer, K., Lienert, J., and McArdell, C. S. (2011). Environmental toxicology and risk assessment of pharmaceuticals from hospital wastewater. *Water Res.* 45, 75–92. doi: 10.1016/j.watres.2010.08.019
7. Falås, P., Longrée, P., la Cour Jansen, J., Siegrist, H., Hollender, J., and Joss, A. (2013). Micropollutant removal by attached and suspended growth in a hybrid biofilm-activated sludge process. *Water Res.* 47, 4498–4506. doi: 10.1016/j.watres.2013.05.010
8. Falås, P., Wick, A., Castronovo, S., Habermacher, J., Ternes, T. A., and Joss, A. (2016). Tracing the limits of organic micropollutant removal in biological wastewater treatment. *Water Res.* 95, 240–249. doi: 10.1016/j.watres.2016.03.009
9. Fernández-Arrojo, L., Guazzaroni, M. E., López-Cortés, N., Beloqui, A., and Ferrer, M. (2010). Metagenomic era for biocatalyst identification. *Curr. Opin. Biotechnol.* 21, 725–733. doi: 10.1016/j.copbio.2010.09.006
10. Gadd, G. M. (2010). Metals, minerals and microbes: Geomicrobiology and bioremediation. *Microbiology* 156, 609–643. doi: 10.1099/mic.0.037143-0
11. Galloway, T., Cipelli, R., Guralnick, J., Ferrucci, L., Bandinelli, S., Corsi, A. M., et al. (2010). Daily bisphenol a excretion and associations with sex hormone concentrations: results from the inchianti adult population study. *Environ. Health Perspect.* 118, 1603–1608. doi: 10.1289/ehp.1002367
12. Grassi, M., Belgiorno, V., and Lofrano, G. (2012). "Removal of emerging contaminants from water and wastewater by adsorption process," in *Emerging Compounds Removal from Wastewater*, ed Lofrano G. (New York, NY: Springer), p15–p37.
13. Groudev, S. N., Bratcova, S. G., and Komnitsas, K. (1999). Treatment of waters polluted with radioactive elements and heavy metals by means of a laboratory passive system. *Minerals Eng.* 12, 261–270. doi: 10.1016/S0892-6875(99)00004-7
14. Gulde, R., Helbling, D. E., Scheidegger, A., and Fenner, K. (2014). pH-dependent biotransformation of ionizable organic micropollutants in activated sludge. *Environ. Sci. Technol.* 48, 13760–13768. doi: 10.1021/es5037139
15. Guo, W., Ngo, H.-H., and Li, J. (2012). A mini-review on membrane fouling. *Bioresour. Technol.* 122, 27–34. doi: 10.1016/j.biortech.2012.04.089
16. Hao, T.-W., Xiang, P., Mackey, H. R., Chi, K., Lu, H., Chui, H., et al. (2014). A review of biological sulfate conversions in wastewater treatment. *Water Res.* 65, 1–21. doi: 10.1016/j.watres.2014.06.043
17. Harms, H., Schlosser, D., and Wick, L. Y. (2011). Untapped potential: exploiting fungi in bioremediation of hazardous chemicals.*Nat. Rev. Microbiol.* 9, 177–192. doi: 10.1038/nrmicro2519
18. Helbling, D. E., Hollender, J., Kohler, H.-E., Singer, H., and Fenner, K. (2010). High-throughput identification of microbial transformation products of organic micropollutants. *Environ. Sci. Technol.* 44, 6621–6627. doi: 10.1021/es100970m
19. Helbling, D. E., Johnson, D. R., Honti, M., and Fenner, K. (2012). Micropollutant biotransformation kinetics associate with WWTP process parameters and microbial community characteristics. *Environ. Sci. Technol.* 46, 10579–10588. doi: 10.1021/es3019012
20. Hellinga, C., Schellen, A. A. J. C., Mulder, J. W., van Loosdrecht, M. C. M., and Heijnen, J. J. (1998). The sharon process: an innovative method for nitrogen removal from ammonium-rich waste water. *Water Sci. Technol.* 37, 135–142. doi: 10.1016/S0273-1223(98)00281-9
21. Joss, A., Zabczynski, S., Göbel, A., Hoffmann, B., Löffler, D., McArdell, C. S., et al. (2006). Biological degradation of pharmaceuticals in municipal wastewater treatment: proposing a classification scheme. *Water Res.* 40, 1686–1696. doi: 10.1016/j.watres.2006.02.014
22. Keller, N. P., Turner, G., and Bennett, J. W. (2005). Fungal secondary metabolism - From biochemistry to genomics. *Nat. Rev. Microbiol.* 3, 937–947. doi: 10.1038/nrmicro1286
23. Kelly, S. L., and Kelly, D. E. (2013). Microbial cytochromes P450: biodiversity and biotechnology. Where do cytochromes P450 come from, what do they do and what can they do for us? *Philos. Trans. R. Soc. B* 368, 20120476. doi: 10.1098/rstb.2012.0476
24. Kolvenbach, B. A., Helbling, D. E., Kohler, H.-P. E., and Corvini, P. F.-X. (2014). Emerging chemicals and the evolution of biodegradation capacities and pathways in bacteria. *Curr.Opin. Biotechnol.* 27, 8–14. doi: 10.1016/j.copbio.2013.08.017
25. Kumar, S. (2010). Engineering cytochrome P450 biocatalysts for biotechnology, medicine and bioremediation. *Expert Opin. Drug Metab. Toxicol.* 6, 115–131. doi: 10.1517/17425250903431040
26. Lackner, S., Gilbert, E. M., Vlaeminck, S. E., Joss,

A., Horn, H., and van Loosdrecht, M. C. M. (2014). Full-scale partial nitritation/anammox experiences – an application survey. *Water Res.* 55, 292–303. doi: 10.1016/j.watres.2014.02.032

27. Lah, L., Kraševec, N., Trontelj, P., and Komel, R. (2008). High diversity and complex evolution of fungal cytochrome P450 reductase: cytochrome P450 systems. *Fungal Genet. Biol.* 45, 446–458. doi: 10.1016/j.fgb.2007.10.004

28. Lah, L., Podobnik, B., Novak, M., Korošec, B., Berne, S., Vogelsang, M., et al. (2011). The versatility of the fungal cytochrome P450 monooxygenase system is instrumental in xenobiotic detoxification. *Mol. Microbiol.* 81, 1374–1389. doi: 10.1111/j.1365-2958.2011.07772.x

29. Loganathan, B., Phillips, M., Mowery, H., and Jones-Lepp, T. L. (2009). Contamination profiles and mass loadings of macrolide antibiotics and illicit drugs from a small urban wastewater treatment plant. *Chemosphere* 75, 70–77. doi: 10.1016/j.chemosphere.2008.11.047

30. Luo, Y., Guo, W., Ngo, H. H., Nghiem, L. D., Hai, F. I., Zhang, J., et al. (2014). A review on the occurrence of micropollutants in the aquatic environment and their fate and removal during wastewater treatment. *Sci. Total Environ.* 473–474, 619–641. doi: 10.1016/j.scitotenv.2013.12.065

31. Majewsky, M., Gallé, T., Zwank, L., and Fischer, K. (2010). Influence of microbial activity on polar xenobiotic degradation in activated sludge systems. *Water Sci. Technol.* 62, 701–707. doi: 10.2166/wst.2010.925

32. McKie, R. (2012). Drug and water firms attack EU plan to reduce ethinylestradiol pollution. *BMJ* 344:4029. doi: 10.1136/bmj.e4029

33. Miller, L. D., Mosher, J. J., Venkateswaran, A., Yang, Z. K., Palumbo, A. V., Phelps, T. J., et al. (2010). Establishment and metabolic analysis of a model microbial community for understanding trophic and electron accepting interactions of subsurface anaerobic environments. *BMC Microbiol.* 10:149. doi: 10.1186/1471-2180-10-149

34. Mora-Pale, M., Sanchez-Rodriguez, S. P., Linhardt, R. J., Dordick, J. S., and Koffas, M. A. G. (2014). Biochemical strategies for enhancing the in vivo production of natural products with pharmaceutical potential. *Curr. Opin. Biotechnol.* 25, 86–94. doi: 10.1016/j.copbio.2013.09.009

35. Nam, H., Lewis, N. E., Lerman, J. A., Lee, D., Chang, R. L., Kim, D., et al. (2012). Network context and selection in the evolution to enzyme specificity. *Science* 337, 1101–1104. doi: 10.1126/science.1216861

36. Oulton, R. L., Kohn, T., and Cwiertny, D. M. (2010). Pharmaceuticals and personal care products in effluent matrices: a survey of transformation and removal during wastewater treatment and implications for wastewater management. *J. Environ. Monit.*12, 1956–1978. doi: 10.1039/c0em00068j

37. Petrie, B., Barden, R., and Kasprzyk-Hordern, B. (2015). A review on emerging contaminants in wastewaters and the environment: current knowledge, understudied areas and recommendations for future monitoring. *Water Res.* 72, 3–27. doi: 10.1016/j.watres.2014.08.053

38. Petrie, B., McAdam, E. J., Lester, J. N., and Cartmell, E. (2014). Assessing potential modifications to the activated sludge process to improve simultaneous removal of a diverse range of micropollutants. *Water Res.* 62, 180–192. doi: 10.1016/j.watres.2014.05.036

39. Pokorna, D., and Zabranska, J. (2015). Sulfur-oxidizing bacteria in environmental technology. *Biotechnol. Adv.* 33, 1246–1259. doi: 10.1016/j.biotechadv.2015.02.007

40. Price, M. N., Deutschbauer, A. M., Skerker, J. M., Wetmore, K. M., Ruths, T., Mar, J. S., et al. (2013). Indirect and suboptimal control of gene expression is widespread in bacteria. *Mol. Sys. Biol.* 9, 660. doi: 10.1038/msb.2013.16

41. Racz, L., and Goel, R. (2010). Fate and removal of estrogens in municipal wastewater. *J. Environ. Monit.* 12, 58–70. doi: 10.1039/B917298J

42. Raghoebarsing, A. A., Pol, A., van de Pas-Schoonen, K. T., Smolders, A. J. P., Ettwig, K. F., Rijpstra, W. I. C., et al. (2006). A microbial consortium couples anaerobic methane oxidation to denitrification. *Nature* 440, 918–921. doi: 10.1038/nature04617

43. Richardson, S. D., and Kimura, S. Y. (2016). Water analysis: emerging contaminants and current issues. *Anal. Chem.* 88, 546–582. doi: 10.1021/acs.analchem.5b04493

44. Richardson, S. D., and Ternes, T. A. (2014). Water analysis: emerging contaminants and current issues. *Anal. Chem.* 86, 2813–2848. doi: 10.1021/ac500508t

45. Riva, S. (2006). Laccases: blue enzymes for green chemistry. *Trends Biotechnol.* 24, 219–226. doi: 10.1016/j.tibtech.2006.03.006

46. Rosal, R., Gonzalo, M. S., Boltes, K., Letón, P., Vaquero, J. J., and García-Calvo, E. (2009). Identification of intermediates and assessment of ecotoxicity in the oxidation products generated during the ozonation of clofibric acid. *J. Hazard. Mat.* 172, 1061–1068. doi: 10.1016/j.jhazmat.2009.07.110

47. Rossner, A., Snyder, S. A., and Knappe, D. R. U. (2009). Removal of emerging contaminants of concern by alternative adsorbents. *Water Res.* 43, 3787–3796. doi: 10.1016/j.watres.2009.06.009

48. Rutledge, P. J., and Challis, G. L. (2015). Discovery of microbial natural products by activation of silent biosynthetic gene clusters. *Nat. Rev. Microbiol.* 13, 509–523. doi: 10.1038/nrmicro3496

49. Scherson, Y. D., Wells, G. F., Woo, S.-G., Lee, J., Park, J., Cantwell, B. J., et al. (2013). Nitrogen removal with energy recovery through N_2O decomposition. *Energy Environ. Sci.* 6, 241–248. doi: 10.1039/C2EE22487A

50. Service, R. F. (2006). Desalination freshens up. *Science* 313, 1088–1090. doi: 10.1126/science.313.5790.1088

51. Sharma, P., Goel, R., and Capalash, N. (2007). Bacterial laccases. *World J. Microbiol. Biotechnol.* 23, 823–832. doi: 10.1007/s11274-006-9305-3

52. Sotirova, A. V., Spasova, D. I., Galabova, D. N., Karpenko, E., and Shulga, A. (2008). Rhamnolipid-biosurfactant permeabilizing effects on gram-positive and gram-negative bacterial strains. *Curr. Microbiol.* 56, 639–644. doi: 10.1007/s00284-008-9139-3

53. Spongberg, A. L., and Witter, J. D. (2008). Pharmaceutical compounds in the wastewater process stream in Northwest Ohio. *Sci. Total Environ.* 397, 148–157. doi: 10.1016/j.scitotenv.2008.02.042

54. Vandenberg, L. N., Ehrlich, S., Belcher, S. M., Ben-Jonathan, N., Dolinoy, D. C., Hugo, E. R., et al. (2013). Low dose effects of bisphenol A: an integrated review of *in vitro*, laboratory animal, and epidemiology studies. *Endocr. Disrupt.* 1, 25078. doi: 10.4161/endo.26490

55. van der Roest, H. F., de Bruin, L. M. M., Gademan, G., and Coelho, F. (2011). Towards sustainable waste water treatment with Dutch Nereda® technology. *Water Pract. Technol.* 6, 1–2. doi: 10.2166/wpt.2011.059

56. Van der Zee, F. P., and Cervantes, F. J. (2009). Impact and application of electron shuttles on the redox (bio) transformation of contaminants: a review. *Biotechnol. Adv.* 27, 256–277. doi: 10.1016/j.biotechadv.2009.01.004

57. van Dongen, U., Jetten, M. S., and van Loosdrecht, M. C. (2001). The SHARON-Anammox process for treatment of ammonium rich wastewater. *Water Sci. Technol.* 44, 153–160.

58. Villas-Bôas, S. G., and Bruheim, P. (2007). The potential of metabolomics tools in bioremediation studies. *OMICS A J. Integr. Biol.* 11, 305–313. doi: 10.1089/omi.2007.0005

59. Werck-Reichhart, D., and Feyereisen, R. (2000). Cytochromes P450: a success story. *Genome Biol.* 1:reviews3003. doi: 10.1186/gb-2000-1-6-reviews3003

60. Zhang, M., and Wang, H. (2014). Organic wastes as carbon sources to promote sulfate reducing bacterial activity for biological remediation of acid mine drainage. *Minerals Eng.* 69, 81–90. doi: 10.1016/j.mineng.2014.07.010

CHAPTER 15

RECLAIMED WATER AS A RESERVOIR OF ANTIBIOTIC RESISTANCE GENES: DISTRIBUTION SYSTEM AND IRRIGATION IMPLICATIONS

Nicole Fahrenfeld, Yanjun Ma, Maureen O'Brien and Amy Pruden

Keywords: Antibiotic resistance genes, water reuse, reclaimed water distribution systems, irrigation

INTRODUCTION

Water reuse is an increasingly common sustainable water management practice motivated by climate change, urbanization, energy efficiency, and environmental protection (US Environmental Protection Agency [USEPA], 2012). Reclaimed or recycled wastewater is treated by municipalities for a variety of purposes, including non-potable urban reuse (Grant et al., 2012; USEPA, 2012). In the United States Environmental Protection Agency (USEPA) guidelines on water reuse, the presence of antibiotics as trace organic contaminants in wastewater is noted and a need for more information is acknowledged to reduce the proliferation of antibiotic resistance and protect public health (USEPA, 2012).

Antibiotic resistance proliferation is currently outpacing the development of novel antibiotics, calling for effective strategies to mitigate the spread of antibiotic resistance (Carlet et al., 2012). Bacterial resistance to antibiotics is partially conferred through antibiotic resistance genes (ARGs), which code for specific antimicrobial functions such as efflux pumps (Webber and Piddock, 2003). ARG contamination has been quantified in a variety of environmentally relevant matrices, including wastewater treatment plant (WWTP) effluent, which is known to contribute to ARG loadings in surface waters (Pruden et al., 2013; Storteboom et al., 2010; LaPara et al., 2011). Some states require chlorine or UV disinfection for reused water (USEPA, 2012) and certain disinfectants (free chlorine, O_3, and UV) are capable of reacting with nucleic acids during treatment and therefore may potentially reduce ARGs, as recently reviewed by Dodd (2012). However, McKinney and Pruden (2012) recently demonstrated in a controlled lab study that typical UV doses applied at WWTPs are capable of reducing antibiotic resistant strains of bacteria, but not ARGs. Others have noted little reduction in ARGs following UV effluent treatment in full-scale WWTPs (Auerbach et al., 2007; Kim et al., 2010).

While WWTPs are now well-established as a reservoir of ARGs (Auerbach et al., 2007; Kim et al., 2010; Czekalski et al., 2012), and some have considered effect of irrigation with reclaimed water (McLain and Williams, 2012; Negreanu et al., 2012), there is a void of studies focused on the potential for re-growth in treated wastewater distribution systems ("purple" pipes). In one study examining soil irrigated with treated wastewater, no differences in the microbiome or ARG levels were observed compared to soil irrigated with fresh water (Negreanu et al., 2012). In fact, in Llobregat (NE Spain), reclaimed water emitted to a river had lower concentrations of indicator organisms than the stream water (Rubiano et al., 2012). In contrast, re-growth of indicator organisms has been observed between the point of entry (POE) and point of use (POU) in reclaimed water systems (Ryu et al., 2005), raising the question of whether ARGs can also increase during distribution. In a study examining drinking water distribution systems, antibiotic resistant bacteria were shown to decrease between POE and POU, but ARGs were observed to increase (Xi et al., 2009).

Citation: Fahrenfeld N, Ma Y, O'Brien M and Pruden A (2013) Reclaimed water as a reservoir of antibiotic resistance genes: distribution system and irrigation implications. Front. Microbiol. 4:130. doi:10.3389/fmicb.2013.00130.

In this study, ARG occurrence patterns were evaluated in the POU water in three arid western U.S. recycled water distribution systems using quantitative polymerase chain reaction (qPCR). Depending on access, POE, POU biofilm, and soil irrigated with recycled water were also examined. Samples were also screened by qPCR for the potential presence of known waterborne pathogenic bacteria and indicators, *Legionella pneumophila*, *Escherichia coli*, and *Pseudomonas aeruginosa*. To simulate the effect of reused water for irrigation, a series of batch laboratory soil microcosm studies were performed to compare irrigation with secondary, chlorinated, and dechlorinated effluent from a representative conventional WWTP.

MATERIALS AND METHODS

Water Reuse Systems

Samples were collected from three non-potable reclaimed wastewater distribution systems in the western U.S., together served by four WWTPs (Table 1). Water samples (-POE or -POU) were collected in sterile centrifuge tubes. Biofilm (-F) was collected with a sterile swab, and packed in a sterile centrifuge tube. A soil sample was collected from a field irrigated with reclaimed wastewater (-S), in a sterile centrifuge tube. All samples were shipped overnight on ice and stored frozen until extraction. Water samples were freeze-dried prior to DNA extraction (FreeZone Plus, Labconco, Kansas City, MO, USA).

Table 1: Summary of reclaimed WWTP tertiary treatment characteristics.

	Aa[1]	Ab[1]	B	C
Capacity (Mgd)	6	4	1.1	0.6
Filtration	Media	Dual media	Carbon filter	Sand + activated carbon
Disinfection	Chlorine	UV	Chlorine	Chlorine

All WWTPs studied use conventional primary and secondary treatment processes.
[1]*Aa and Ab are separate WWTPs that emit water to a commingled distribution system.*

Batch Microcosms

Aerobic, batch microcosms were prepared to investigate two water reuse scenarios on historically manured soil: (1) a single irrigation event ("batch irrigation") and (2) repeated irrigation events ("periodic irrigation"). Soil was collected (upper 7.5–10 cm) in winter from historically manured corn fields near Virginia Tech campus using a soil probe. Soil was air dried and sieved (2 mm), and an aliquot was preserved for DNA extraction. For each study, microcosms were prepared in 250 mL flasks in triplicate with 50 g of soil and incubated at room temperature on a shaking table to maintain aerobic conditions. Slurry samples were collected weekly without sacrifice. Secondary, chlorinated, and dechlorinated WWTP effluents were collected before each irrigation treatment event from a representative 4.5 Mgd Domestic WWTP.

Batch irrigation soil was initially treated with 80 mL of freshly collected WWTP effluent fractions. Slurry samples (~0.4 g wet) were collected and an equal volume of deionized (DI) water was added to each flask to maintain soil moisture. WWTP effluent fractions (2–4 L) were filtered through 0.22 μm membrane and total DNA was extracted from the filter, as described below. The periodic irrigation soil was initially treated with 100 mL of freshly collected WWTP effluent fractions. Slurry samples (10 mL) were collected, centrifuged at 3,300 × *g* for 5 min, and 0.4 g of the pellet was used for DNA extraction. Fresh WWTP effluent fractions were added to the flasks to replace the volume removed during each sampling. WWTP effluent fractions (60 mL) were freeze-dried, as above, and DNA was extracted from the residuals.

Molecular Techniques

DNA was extracted from freeze-dried water/slurry, 0.4 g soil, or swabs using a FastDNA® SPIN Kit for Soil (MP Biomedicals, Solon, OH, USA) and diluted 1:50 or 1:100 for the water reuse field study and 1:30 for the irrigation studies prior to downstream analysis. qPCR was performed to quantify 16S rRNA (Suzuki et al., 2000), *sul*1 (Aminov et al., 2001), *sul*2 (Aminov et al., 2001), *tet*(A) (Aminov et al., 2002), *tet*(O) (Aminov et al., 2001), *erm*F (Chen et al., 2007), *van*A (Dutka-Malen et al., 1995), *mec*A (McKinney and Pruden, 2012), *L. pneumophila*-specific *mip* (Nazarian et al., 2008), *E. coli*-specific *gad*AB (Chen et al., 2006), and *P. aeruginosa*-specific *ecfX*/*gyr*B (Anuj et al., 2009) genes for the water reuse field study. Reaction matrix and PCR protocols were as previously described (Ma et al., 2011; Wang et al., 2012). For the irrigation study, 16S rRNA, *sul*1,*sul*2, *tet*(O), and *tet*(W) genes were monitored. All standard curves of qPCR were constructed from serial dilutions of cloned genes ranging from 10^8 to 10^2 gene copies/μL. Samples were analyzed in triplicate with a standard curve and negative control included in each run. Limits of quantification with respect to sample volume varied depending on the volume processed and the dilution of DNA extract, ranging from -1.4 to 0.6 $\log_{10}$ gene copies/mL, 3.1–5.1 $\log_{10}$ gene copies per swab, and 0.5–2.5 $\log_{10}$ gene copies/g of soil. Additionally, cloning and sequencing of qPCR product was performed for assays that had not been validated previously (*van*A) to demonstrate specificity of PCR product (GenBank accession number KC792557–KC792573).

Statistics

Cluster analysis was performed on transformed (square root) 16S rRNA gene normalized ARG profiles from the reclaimed wastewater systems and significance testing was carried out using the SimProf test in PrimerE (Plymouth, UK). To compare between wastewater treatment effluent fractions and treatments in the irrigation study, data was Box–Cox transformed. Transformed data were compared using ANOVA, and significant differences ($p < 0.05$) were determined using Tukey's honest significance test, as implemented in R (http://www.r-project.org/). Multiple comparisons for the distribution system and irrigation studies were performed on Box–Cox transformed data using least square means comparison with a Satterthwaite estimation of degrees of freedom in SAS, again using Tukey adjustment for multiple comparisons. Linear modeling for the irrigation study data was performed in Microsoft Excel.

RESULTS

ARG Occurrence Patterns in Recycled Water

Antibiotic resistance genes were detected in all water reuse samples (Figures 1 and 2). The most frequently detected ARGs were *van*A, *erm*F, and *sul*2 with frequencies of detection of 100, 73, and 65% (n = 23), respectively. *tet*(O) had the lowest frequency of detection (17%, n = 23) other than*mec*A, which was not detected in any of the samples. For System A, significant differences in ARG concentrations were observed between POE and POU for *sul*1 and *tet*(A), $p < 0.001$ for both. Concentrations of *van*A, *erm*F, and *sul*2, 16S rRNA, and *tet*(O) were not significantly different between POE and POU (p = 0.33–0.99).

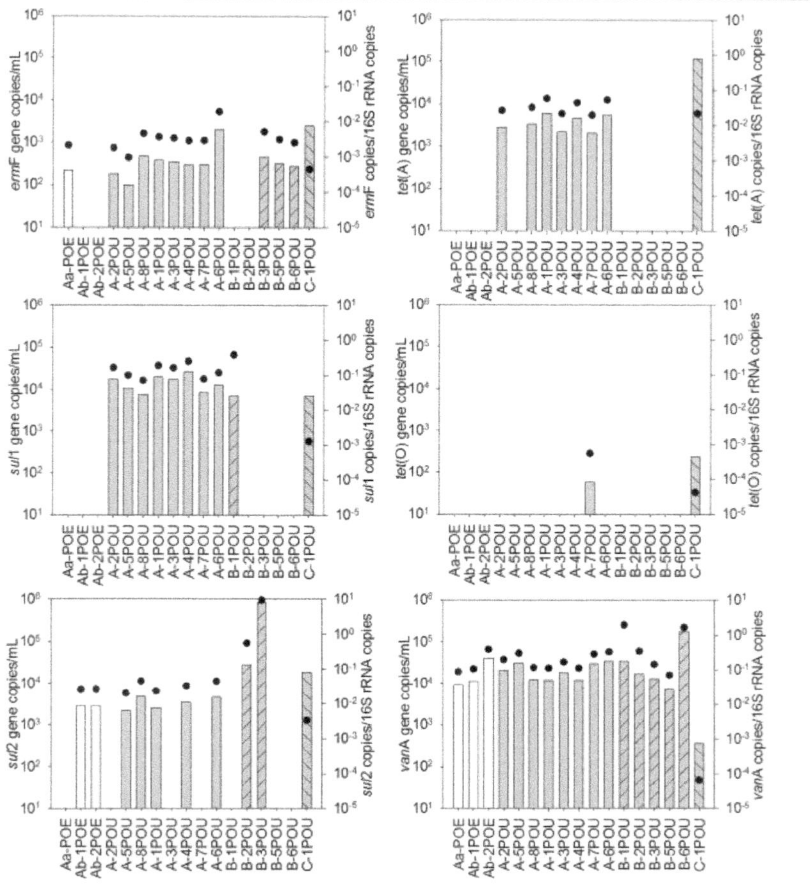

Figure 1: Absolute (bars) and 16S rRNA gene-normalized (symbols) levels of ARGs in water samples collected at point of entry (POE) for WWTPs Aa and Ab and at point of use (POU) for systems A, B, and C, numbers differentiate between sample locations.

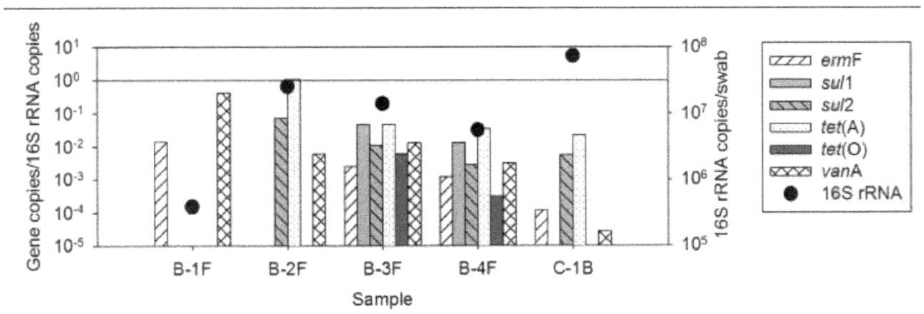

Figure 2: Quantification of ARGs in biofilm samples available from recycled water distribution systems B and C. Bars represent absolute ARG copies per swab, symbols represent 16S rRNA gene copies per swab.

In the irrigated soil sample (B-7S), 9.5, 7.3, 7.2, and 5.6 $\log_{10}$ gene copies/g of soil were quantified for 16S rRNA, *sul*1, *sul*2, and *van*A, respectively; *erm*F, *tet*(A), and *tet*(O) were below detection. Normalizing ARGs to 16S rRNA gene copy numbers indicated that *sul*1 and *sul*2 were one to two orders of magnitude lower and *van*A three to five orders of magnitude lower in the soil sample than observed in water and biofilm samples.

The occurrence of ARGs varied among the five biofilm samples examined, ranging from two to six classes of ARGs (Figure 2). However, individual ARG levels in the biofilm did not correlate with corresponding bulk water ARGs among the available paired samples (Figure 3). For example, the most abundant ARG, *tet*(A), in B-2F was below detection in B-2POU.

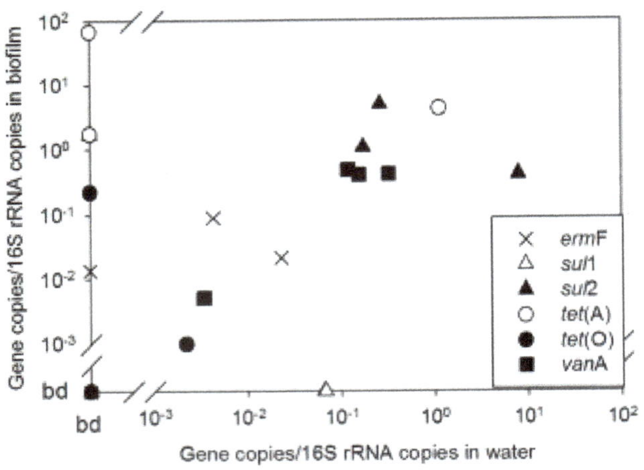

Figure 3. Biofilm versus bulk water ARG copies normalized to 16S rRNA gene copies in available paired biofilm and water samples. ARGs were below detection in several samples, indicated on the *x* and *y* axes as bd.

Cluster analysis performed on 16S rRNA gene normalized ARG profiles considered both the kinds of ARGs detected and their frequencies and resulted in several significantly different sample clusters (Figure 4). Generally, three distinct clusters were formed primarily by POE, WWTP-A POU, and WWTP-C POU samples, while the WWTP-B POU samples were interspersed among the three clusters, and a fourth cluster consisting only of WWTP-B samples. The three WWTP-A POE samples formed a significantly different cluster from POU samples in its corresponding distribution system (50% similarity). POU samples from System A (n = 7) formed a cluster with 78–96% similarity, except A-5POU which was significantly different (71%) and clustered with some samples from System B (n = 5). No patterns were observed in the biofilm and water clustering patterns for paired samples (n = 4). Biofilms did not cluster with one another nor did biofilm samples cluster with paired water sample with similarities ranging from 16% for F2 and F3 (significantly different) to 100% for F1 (no difference). The soil sample (B-7S) clustered with other B system biofilm and water samples, which were only 20% similar to the remainder of the samples selected.

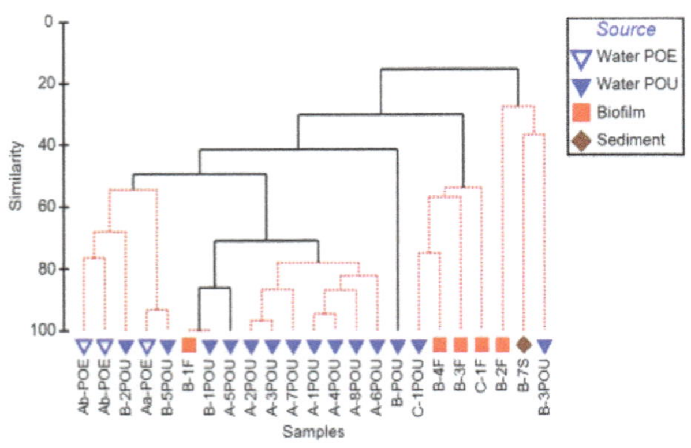

Figure 4: Cluster analysis of 16S rRNA gene normalized ARG copy numbers, considerate of class and relative abundance of each ARG measured in the distribution systems (Aa, Ab, B, C) and across environmental matrices (POE, point of entry water; POU, point of use water; F, biofilm; S, soil). Solid branches indicate significantly different clusters ($p < 0.05$).

Waterborne Pathogen and Fecal Indicator Screening

Quantitative PCR screening for *E. coli* and *L. pneumophila* through *gad*AB and *mip* resulted in positive detections for 48% of samples: 35 and 17% for *gad*AB and *mip*, respectively (Table 2).*ecrf*X/*gyr*B, corresponding to *P. aeruginosa*, were below detection in all samples.

Table 2: Log copies of *E. coli* (*gad*AB) and *L. pneumophila* (*mip*) specific genes per milliliter (POE and POU's), swab (F's), or gram soil (S).

Sample	*gadAB*	*mip*
Ab-POE	3.7	bd*
A-1POU	4.2	bd
A-2POU	5.1	1.8
A-3POU	4.4	bd
A-8POU	5.3	bd
B-1POU	bd	1.4
C-1POU	6.6	bd
B-1F	bd	4.6
B-2F	6.0	bd
C-1F	6.9	bd
B-7S	bd	5.6

**bd, below detection.*

ARGs during Simulated Land Application

WWTP Effluent Fraction ARG Loads

Levels of ARGs in secondary, chlorinated, and dechlorinated WWTP effluent are compared in Figure 5. *sul*1 was the most frequently detected ARG (100%), followed by *sul*2 (71%), *tet*(W) (71%), and *tet*(O) (61%; n = 21). Comparing the ARGs and 16S rRNA gene levels across effluent fractions, secondary effluent was significantly higher than chlorinated and dechlorinated (both $p < 0.0001$) effluents, which were not significantly different from each other ($p = 0.45$). Comparing by gene across effluent fractions, secondary effluent and chlorinated effluent were significantly different in terms of 16S rRNA and *sul*2 gene copies ($p < 0.0001$ for both). Interestingly, when ARGs were normalized to 16S rRNA gene copies, significant differences were not observed among the secondary, chlorinated, or dechlorinated effluents (p = 0.29–0.91). This suggests no preferential destruction of specific gene types by chlorination.

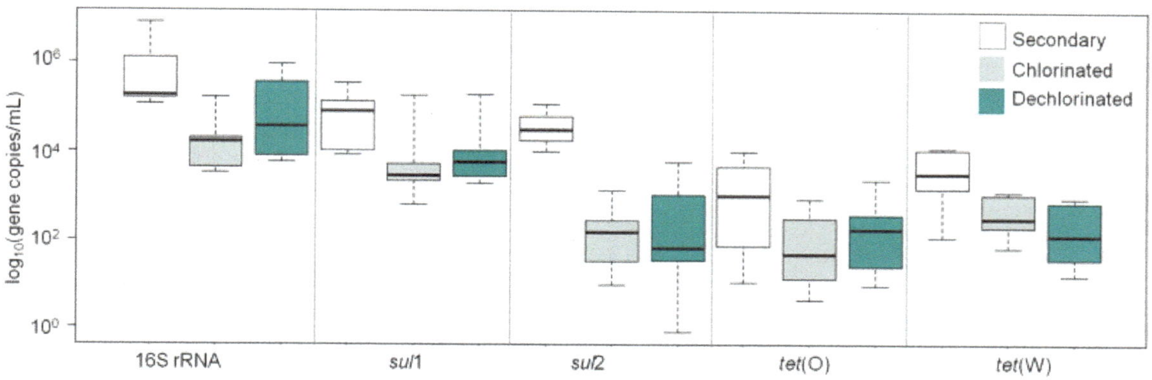

Figure 5: Censored boxplot of 16S rRNA genes, *sul*1, *sul*2, *tet*(O), and *tet*(W) gene copies per milliliter in secondary (no fill), chlorinated (gray), and dechlorinated (blue) domestic WWTP effluents applied in the batch (2–4 L 0. 22 **μm** filtered) and periodic (60 mL freeze-dried) irrigation study ($n = 6$ *tet*genes, 7 *sul* and 16S rRNA genes).

Irrigation Studies

Results of the batch irrigation study (Figure 6) indicated no difference between soil irrigated with secondary, chlorinated, or dechlorinated effluent, or DI water in terms of ARG concentration with time for *sul*1, *sul*2, *tet*(O), or *tet*(W) ($p > 0.405$ for all, except for *sul*2 deionized $p = 0.05$ and secondary effluent $p = 0.006$).

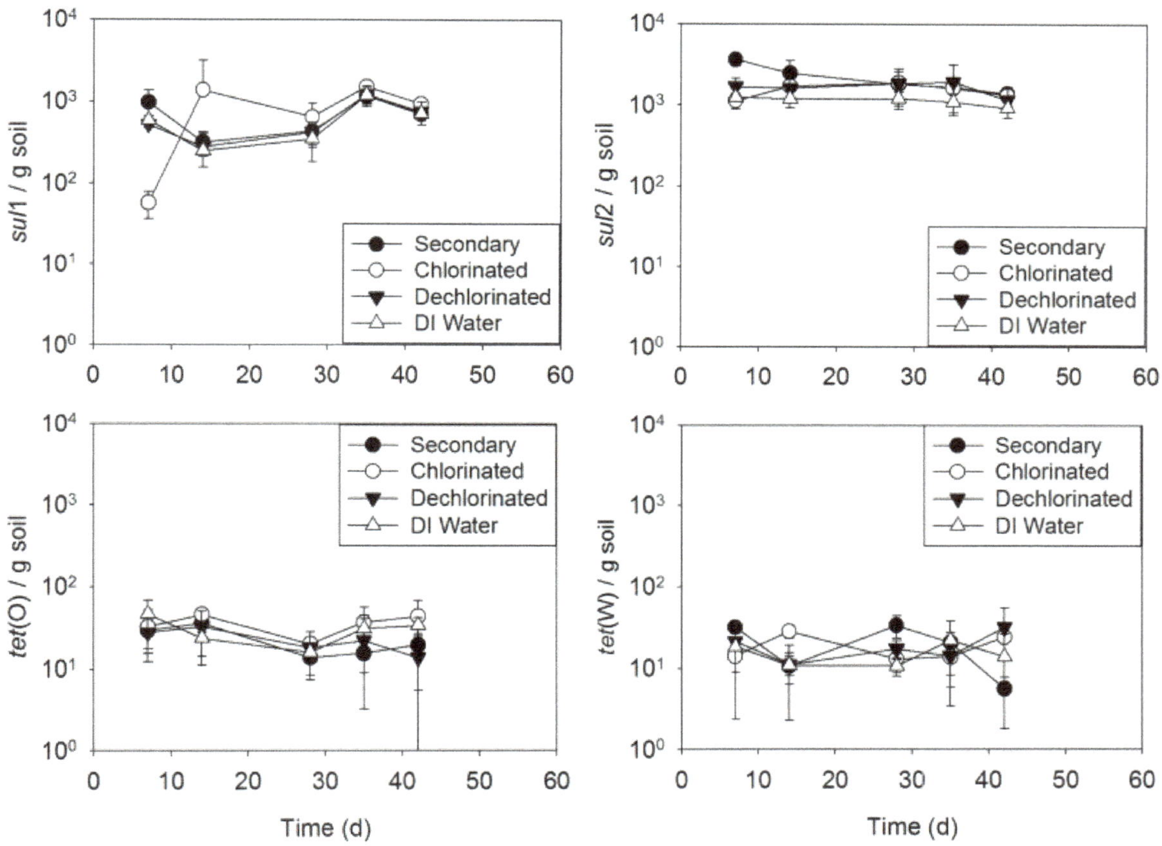

Figure 6: *sul*1, *sul*2, *tet*(O), and *tet*(W) genes in soils subject to one time (batch) irrigation with secondary, chlorinated, dechlorinated domestic WWTP effluents and deionized (DI) water, per gram of soil (slurry). Error bars represent standard deviation of triplicate microcosms.

Soil periodically irrigated resulted in a significant increase in *sul*2 when receiving secondary effluent, compared to irrigation with the other water types (p ranging from 0.032 to $p < 0.0001$; Figure 7). Additionally, soil irrigated with secondary effluent had significantly higher *sul*1 copies than that irrigated with chlorinated effluent, dechlorinated effluent, or DI water (all $p < 0.0001$). No significant difference across time was observed among the irrigation water types for *tet*(O) (p = 0.13–1.0) or *tet*(W) (p = 0.74–1.0). Linear modeling of gene copies versus time revealed an increasing trend of *sul*2 copies with time in soil irrigated with secondary effluent ($R^2 = 0.92$).

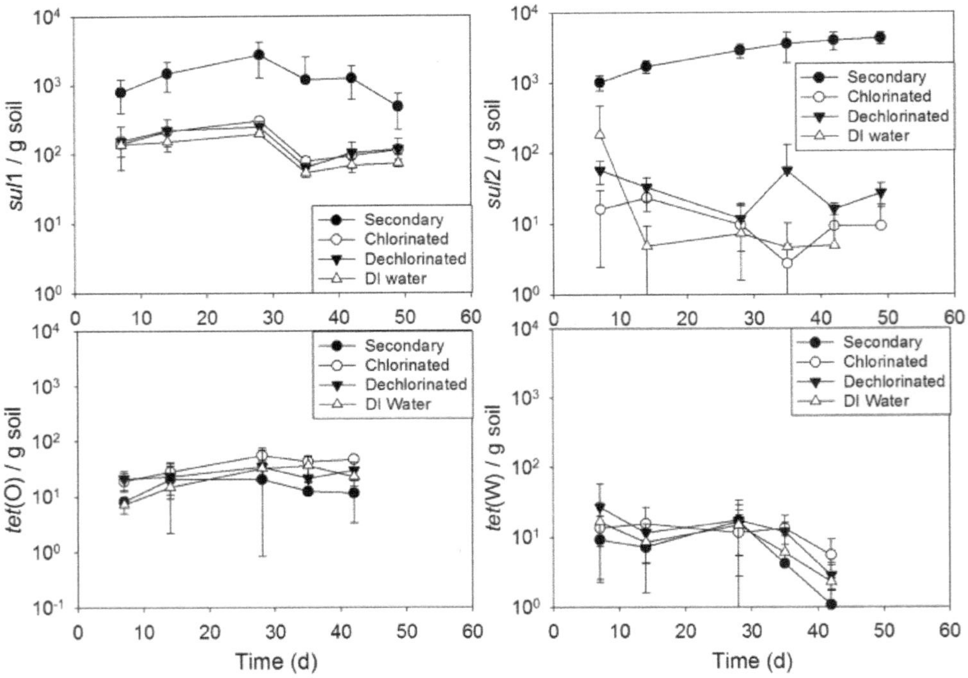

Figure 7: *sul*1, *sul*2, *tet*(O) and *tet*(W) genes in soils periodically irrigated with secondary, chlorinated, dechlorinated domestic WWTP effluents and deionized (DI) water, normalized to soil mass (wet weight). Error bars represent standard deviation of triplicate microcosms.

DISCUSSION

This study explored the occurrence of several ARGs in three reclaimed water distribution systems in the U.S. Given Limited access to such systems, only one sampling event was possible. Nonetheless, the results provide important baseline information to support future research, including insight into the kinds of ARGs, bacteria, and applications that may be of concern. To the knowledge of the authors, this is the first study specifically investigating the potential for ARGs to persist or amplify within the reclaimed water distribution pipes.

ARG Occurrence in Recycled Water Distribution Systems

Several ARGs were detected at the POU in this study, many of which were below detection at the POE. This highlights the need to consider microbiological processes occurring in reclaimed water distribution systems that may be contributing to ARG amplification and suggests that focus on the water quality at the POU may be the most appropriate for assessing risk. Re-growth is a well-known phenomenon even in drinking water distribution systems (Xi et al., 2009), and has also recently been documented in a recycled water system (Ryu et al., 2005).

Several studies have documented levels of ARGs in WWTP effluents, providing a reference for comparison. Generally, *erm*F, *sul*1, *sul*2, *tet*(A), and *tet*(O) levels measured in the POE, POU, and WWTP effluent applied in the irrigation study were below or at the lower end of ranges reported by others in WWTP effluents (Kim et al., 2010; Czekalski et al., 2012; Negreanu et al., 2012). However, the irrigated soil in this study carried higher levels of *sul*1 (7.3 $\log_{10}$ copies/g) and *sul*2 (7.2 $\log_{10}$copies/g) than reported by others in a field study of irrigation with recycled water (5.1–6.7 and 3.5–4.7 $\log_{10}$ copies/g, respectively; Negreanu et al., 2012).

The occurrence of *van*A is of particular interest given that it confers resistance to vancomycin, a last-resort life-saving antibiotic. Vancomycin is commonly prescribed to treat methicillin-resistant *Staphylococcus aureus* (MRSA) infections, but has been losing effectiveness due to increased resistance among staphylococci (Stevens, 2006). *Van* A was detected in every sample in this study using primers targeting a 732-bp product. Interestingly, using *van* A primers designed for longer target products (1030 bp; Clark et al., 1993), *van* A was detectable at several POU sites, but not in any POE samples (data not shown). Detection differences between the two PCR primer sets may indicate that *van* A was partially damaged during disinfection, preventing amplification of longer PCR products. Long-amplicon PCR has recently been demonstrated to provide enhanced detection of DNA damage events (McKinney and Pruden, 2012). Few studies have reported detection of *van*A in environmental samples, providing little reference for comparison. However, *van*A has been reported in drinking water biofilms and wastewater (Schwartz et al., 2003), but was below detection in wastewater reclaimed for groundwater recharge environments (Böckelmann et al., 2009). To the knowledge of the authors, this is the first report of the presence of *van*A in distributed recycled water.

Molecular Detection of Waterborne Pathogens and Indicators

Opportunistic pathogens residing in water systems, such as *L. pneumophila* and *P. aeruginosa*, are now the primary source of waterborne disease outbreak in developed countries (Brunkard et al., 2011). However, there is a need for epidemiological studies to better quantify the precise contributions of various water systems to human disease (Pruden et al., 2013). Of interest to the present study was whether such organisms may be present in recycled water, which could be of special concern for bacterial pathogens because they are capable of developing antibiotic resistance.*P. aeruginosa* is an example of an opportunistic pathogen that colonizes taps and is prone to multi-antibiotic resistant forms (Trautmann et al., 2005). *E. coli* was also of interest as a fecal indicator.

The *mip* gene, specific to *L. pneumophila*, was detectable at levels comparable to those recently observed in chloraminated drinking water distribution systems (Wang et al., 2012). Thus, there could be concern for aerosolization of *L. pneumophila* during spray irrigation. Future research is suggested to more closely examine this potential transmission pathway. *gad*AB, specific to *E. coli*, detected at POU in this study, was comparable to levels previously observed in manure runoff (4.6–4.9 $\log_{10}$ copies/mL, assuming one gene copy per cell) and higher than previously observed in WWTP effluent (2.74 $\log_{10}$ copies/mL) (Grant et al., 2001). This combined with the observation of*gad*AB below detection at POE, but detectable at POU, is further evidence of bacterial re-growth within reclaimed water distribution systems.

Effect of Disinfection on ARGs

Chlorination is commonly applied to WWTP effluents during warm seasons, in which case the effluent must be dechlorinated prior to discharge. Chlorination had a significant impact on 16S rRNA and *sul*2 gene copies, which is consistent with the expectation that chlorination would have a moderate reactivity with nucleic acids (Dodd, 2012). Although occasionally ARGs detected in dechlorinated effluent were not detected in chlorinated effluent, the levels of detected genes were not significantly different. The levels of ARGs in secondary, chlorinated, and dechlorinated WWTP effluent fractions were comparable to those detected at the POU in the field study (Figure 5).

ARG Fate during WWTP Effluent Land Application

In the lab study, periodic irrigation with secondary effluent increased the prevalence of*sul*1 and*sul*2 in historically manured soil compared to soil irrigated with chlorinated, dechlorinated, or DI water. This could be due to direct inputs of extracellular ARGs, intracellular ARGs, or horizontal gene transfer to native soil bacteria (Dodd, 2012). Interestingly, *tet*(O) or *tet*(W) levels in the soil slurry were not affected by irrigation. This highlights that different ARGs have different environmental fates, as has been observed recently with respect to *sul*1 and *tet*(W) in a watershed-scale study (Storteboom et al., 2010; Pruden et al., 2012). Differences in ARG fate likely relate to host properties and their overall propensity for horizontal gene transfer. Because 16S rRNA gene copy levels were relatively consistent with time across the soil irrigation treatments (data not shown), changes in total bacterial population sizes were not likely a factor in the observed differences.

A recent field study carried out in Israel suggested that irrigation does not significantly affect the soil microbiome, and increases were not observed in *sul*1, *sul*2, *tet* (O), *erm* F, or *erm* B in soil subjected to long-term (6–12 years) irrigation with secondary effluent compared to freshwater irrigation (Negreanu et al., 2012). Given the difference in controls, irrigation frequency, soil type, climate, and wastewater chemistry, direct comparison between the studies is difficult. It is suspected that the difference in soil moisture and incubation time may likely be an important difference. Negreanu et al. (2012) irrigated with 4 L of water/m^2 soil/day which experienced a combination of infiltration, run off, and evaporation. In this study, soil-water slurries were used with irrigation waters that incubated with soil rather than infiltrating, allowing for greater contact time. Therefore, the microcosm results presented here may indicate that irrigating at high rates with secondary effluent may still result in amplified soil ARGs. The present study is consistent with the observation of increased ARG copies directly under irrigation drippers compared to soils 50 m from drippers (Negreanu et al., 2012). Further, soil type has been noted to be a critical factor in determining the level of impact of land application of biosolids containing ARGs (Munir and Xagoraraki, 2011), and may also affect the fate of ARGs applied by irrigation.

Understanding land use scenarios that affect soil ARGs is of utmost importance given that the resistome of multi-drug resistant soil bacteria was recently shown to match that from diverse human pathogens (Forsberg et al., 2012). Heavy irrigation with secondary effluent is shown here to be capable of increasing soil ARGs. In practice, given that ARG prevalence can increase between POE and POU within distribution systems, increased ARG levels in irrigation waters and therefore soils are expected. Given that spray irrigation of recreational fields with treated wastewater is common practice, there is considerable potential for human contact with aerosols and soil.

CONCLUSION

This study brings to light the occurrence of ARGs at the POU in recycled water irrigation systems, including *van*A, which is of significant concern to human health. Differences between POE and POU ARG occurrences underscore the need to take into consideration re-growth that occurs in the distribution system when estimating overall exposure and risk. Based on the lab microcosm study, amplified levels of ARGs in soil irrigated with recycled water is possible. Molecular data in this study also indicates the potential presence of waterborne bacterial pathogens, such as *L. pneumophila*. As ARGs are emerging contaminants, risk assessment is in its infancy and no guidance yet exists on safe levels (Pruden, 2011). In addition to direct contact with water and aerosols during recreational activity, a greater concern may be overall contribution to the global pool of antibiotic resistance and ultimately reducing the effectiveness of available antibiotics for treating human disease.

CONFLICT OF INTEREST STATEMENT

The authors declare that the research was conducted in the absence of any commercial or financial relationships that could be construed as a potential conflict of interest.

ACKNOWLEDGMENTS

This work was funded by the Virginia Tech Institute for Critical Technology and Applied Science Award TSTS 11-26 and NSF Research Experience for Undergraduates site award #1062860. Any opinions, findings, and conclusions or recommendations expressed in this paper are those of the author(s) and do not necessarily reflect the views of the sponsors. Gratitude is also extended to Chad McKinney and Mark Mazzochette for laboratory assistance and Andrew Hoegh for assistance with statistics.

REFERENCES

1. Aminov, R., Garrigues-Jeanjean, N., and Mackie, R. (2001). Molecular ecology of tetracycline resistance: development and validation of primers for detection of tetracycline resistance genes encoding ribosomal protection proteins. *Appl. Environ. Microbiol.* 67, 22–32.
2. Aminov, R. I., Chee-Sanford, J. C., Garrigues, N., Teferedegne, B., Krapac, I. J., White, B. A., et al. (2002). Development, validation, and application of PCR primers for detection of tetracycline efflux genes of Gram-negative bacteria. *Appl. Environ. Microbiol.* 68, 1786–1793.
3. Anuj, S. N., Whiley, D. M., Kidd, T. J., Bell, S. C., Wainwright, C. E., Nissen, M. D., et al. (2009). Identification of *Pseudomonas aeruginosa* by a duplex real-time polymerase chain reaction assay targeting the *ecfX* and the *gyrB* genes. *Diagn. Microbiol. Infect. Dis.* 63, 127–131.
4. Auerbach, E. A., Seyfried, E. E., and McMahon, K. D. (2007). Tetracycline resistance genes in activated sludge wastewater treatment plants. *Water Res.* 41, 1143–1151.
5. Böckelmann, U., Dörries, H.-H., Ayuso-Gabella, M. N., Salgot de Marçay, M., Tandoi, V., Levantesi, C., et al. (2009). Quantitative PCR monitoring of antibiotic resistance genes and bacterial pathogens in three European artificial groundwater recharge systems. *Appl. Environ. Microbiol.* 75, 154–163.
6. Brunkard, J. M., Ailes, E., Roberts, V. A., Hill, V., Hilborn, E. D., Craun, G. F., et al. (2011). Surveillance for waterborne disease outbreaks associated with drinking water – United States. *MMWR Surveill. Summ.* 60, 38–75.
7. Carlet, J., Jarlier, V., Harbarth, S., Voss, A., Goossens, H., and Pittet, D. (2012). Ready for a world without antibiotics? The pensières antibiotic resistance call to action. *Antimicrob. Resist. Infect. Control* 1, 11.
8. Chen, J., Yu, Z., Michel, F. C., Wittum, T., and Morrison, M. (2007). Development and application of real-time PCR assays for quantification of *erm* genes conferring resistance to macrolides–lincosamides–streptogramin B in livestock manure and manure management systems. *Appl. Environ. Microbiol.* 73, 4407–4416.
9. Chen, Y.-C., Higgins, M. J., Maas, N. A., and Murthy, S. N. (2006). DNA extraction and *Escherichia coli* quantification of anaerobically digested biosolids using the competitive touchdown PCR method. *Water Res.* 40, 3037–3044.
10. Clark, N. C., Cooksey, R. C., Hill, B. C., Swenson, J. M., and Tenover, F. C. (1993). Characterization of glycopeptide-resistant enterococci from U.S. hospitals. *Antimicrob. Agents Chemother.* 37, 2311–2317.
11. Czekalski, N., Berthold, T., Caucci, S., Egli, A., and Buergmann, H. (2012). Increased levels of multiresistant bacteria and resistance genes after wastewater treatment and their dissemination into Lake Geneva, Switzerland. *Front. Microbiol.* 3:106. doi: 10.3389/fmicb.2012.00106
12. Dodd, M. C. (2012). Potential impacts of disinfection processes on elimination and deactivation of antibiotic resistance genes during water and wastewater treatment. *J. Environ. Monit.* 14, 1754–1771.
13. Dutka-Malen, S., Evers, S., and Courvalin, P. (1995). Detection of glycopeptide resistance genotypes and identification to the species level of clinically relevant enterococci by PCR. *J. Clin. Microbiol.* 33, 24–27.
14. Forsberg, K. J., Reyes, A., Wang, B., Selleck, E. M., Sommer, M. O. A., and Dantas, G. (2012). The shared antibiotic resistome of soil bacteria and human pathogens. *Science* 337, 1107–1111.
15. Grant, M. A., Weagant, S. D., and Feng, P. (2001). Glutamate decarboxylase genes as a prescreening marker for detection of pathogenic *Escherichia coli* groups. *Appl. Environ. Microbiol.* 67, 3110–3114.
16. Grant, S. B., Saphores, J.-D., Feldman, D. L., Hamilton, A. J., Fletcher, T. D., Cook, P. L. M., et al. (2012). Taking the "waste" out of "wastewater" for human water security and ecosystem sustainability. *Science* 337, 681–686.
17. Kim, S., Park, H., and Chandran, K. (2010). Propensity of activated sludge to amplify or attenuate tetracycline resistance genes and tetracycline resistant bacteria: a mathematical modeling approach. *Chemosphere* 78, 1071–1077.
18. LaPara, T. M., Burch, T. R., McNamara, P. J., Tan, D. T., Yan, M., and Eichmiller, J. J. (2011). Tertiary-treated municipal wastewater is a significant point source of antibiotic resistance genes into Duluth–Superior harbor. *Environ. Sci. Technol.* 45, 9543–9549.
19. Ma, Y., Wilson, C. A., Novak, J. T., Riffat, R., Aynur, S., Murthy S., et al. (2011). Effect of various sludge digestion conditions on sulfonamide, macrolide, and tetracycline resistance genes and class I Integrons. *Environ. Sci. Technol.* 45, 7855–7861.
20. McKinney, C. W., and Pruden, A. (2012). Ultraviolet disinfection of antibiotic resistant bacteria and their antibiotic resistance genes in water and wastewater. *Environ. Sci. Technol.* 46, 13393–13400.
21. McLain, J. T., and Williams, C. (2012). Assessing environmental impacts of treated wastewater through monitoring of fecal indicator bacteria and salinity in irrigated soils. *Environ. Monit. Assess.* 184, 1559–1572.
22. Munir, M., and Xagoraraki, I. (2011). Levels of antibiotic resistance genes in manure, biosolids, and fertilized soil. *J. Environ. Qual.* 40, 248–255.
23. Nazarian, E. J., Bopp, D. J., Saylors, A., Limberger, R. J., and Musser, K. A. (2008). Design and implementation of a protocol for the detection of *Legionella* in clinical and environmental samples. *Diagn. Microbiol. Infect. Dis.* 62, 125–132.
24. Negreanu, Y., Pasternak, Z., Jurkevitch, E., and Cytryn, E. (2012). Impact of treated wastewater irrigation on antibiotic resistance in agricultural soils. *Environ. Sci. Technol.* 46, 4800–4808.
25. Pruden, A. (2011). "Unique challenges in the health risk assessment of antibiotic resistance genes in the

environment," in *Society for Environmental Toxicology and Chemistry (SETAC) 32nd Annual Meeting*, November 13–16, Boston, MA.

26. Pruden, A., Arabi, M., and Storteboom, H. N. (2012). Correlation between upstream human activities and riverine antibiotic resistance genes. *Environ. Sci. Technol.* 46, 11541–11549.
27. Pruden, A., Edwards, M., and Falkinham, J. O. III. (2013). "State of the Science of opportunistic pathogens in premise plumbing: methodology, microbial ecology, and epidemiology," in *Water Research Foundation Project 4379 Final Report*. Denver, CO: Water Research Foundation.
28. Pruden, A., Pei, R., Storteboom, H., and Carlson, K. H. (2006). Antibiotic resistance genes as emerging contaminants: studies in Northern Colorado. *Environ. Sci. Technol.* 40, 7445–7450.
29. Rubiano, M.-E., Agulló-Barceló, M., Casas-Mangas, R., Jofre, J., and Lucena, F. (2012). Assessing the effects of tertiary treated wastewater reuse on a Mediterranean river (Llobregat, NE Spain), part III: pathogens and indicators. *Environ. Sci. Pollut. Res.*19, 1026–1032.
30. Ryu, H., Alum, A., and Abbaszadegan, M. (2005). Microbial characterization and population changes in nonpotable reclaimed water distribution systems. *Environ. Sci. Technol.* 39, 8600–8605.
31. Schwartz, T., Kohnen, W., Jansen, B., and Obst, U. (2003). Detection of antibiotic-resistant bacteria and their resistance genes in wastewater, surface water, and drinking water biofilms. *FEMS Microbiol. Ecol.* 43, 325–335.
32. Stevens, D. L. (2006). The role of vancomycin in the treatment paradigm. *Clin. Infect. Dis.* 42, S51–S57.
33. Storteboom, H., Arabi, M., Davis, J. G., Crimi, B., and Pruden, A. (2010). Tracking antibiotic resistance genes in the South Platte River basin using molecular signatures of urban, agricultural, and pristine sources. *Environ. Sci. Technol.* 44, 7397–7404.
34. Suzuki, M. T., Taylor, L. T., and DeLong, E. F. (2000). Quantitative analysis of small-subunit rRNA genes in mixed microbial populations via 5′-nuclease assays. *Appl. Environ. Microbiol.* 66, 4605–4614.
35. Trautmann, M., Lepper, P. M., and Haller, M. (2005). Ecology of *Pseudomonas aeruginosa* in the intensive care unit and the evolving role of water outlets as a reservoir of the organism. *Am. J. Infect. Control* 33, S41–S49.
36. US Environmental Protection Agency (USEPA). (2012). *Guidelines for Water Reuse*. Washington, DC: USEPA.
37. Wang, H., Edwards, M., Falkinham, J. O., and Pruden, A. (2012). Molecular survey of the occurrence of *Legionella* spp., *Mycobacterium* spp., *Pseudomonas aeruginosa*, and *Amoeba* hosts in two chloraminated drinking water distribution systems.*Appl. Environ. Microbiol.* 78, 6285–6294.
38. Webber, M. A., and Piddock, L. J. V. (2003). The importance of efflux pumps in bacterial antibiotic resistance. *J. Antimicrob. Chemother.* 51, 9–11.
39. Xi, C., Zhang, Y., Marrs, C. F., Ye, W., Simon, C., Foxman, B., et al. (2009). Prevalence of antibiotic resistance in drinking water treatment and distribution systems. *Appl. Environ. Microbiol.* 75, 5714–5718.

CHAPTER 16

EFFICIENT PURIFICATION AND CONCENTRATION OF VIRUSES FROM A LARGE BODY OF HIGH TURBIDITY SEAWATER

Guowei Sun, Jinzhou Xiao, Hongming Wang, Chaowen Gong , Yingjie Pan, Shuling Yan, Yongjie Wang

Keywords: Marine viruses, Concentration, Tangential flow filtration, Ultrafiltration, High turbidity seawater

METHOD DETAILS

Sample preparation

More than 300 L of subsurface water (2 m depth), characterized by highly suspended matter contents (approximately 3.4 g L^{-1} (wet weight)), was sampled from Yangshan Deep-Water Port, South East Shanghai, China. Among them, 500 mL was fixed *in situ* by adding 0.02 μm filtered formalin (37–40% (w/v) formaldehyde solution) (Sangon, Shanghai, China) to 2% (v/v) final concentration for enumeration of viruses in the original water sample. The collected water samples were kept on ice and delivered to the laboratory as quickly as possible after finishing sampling.

Enumeration of viruses in original seawater sample

In order to assess the recovery efficiency of the method established in this study for purification and concentration of marine viruses from high turbidity seawater, viral particles in the original seawater samples (Table 1 and Fig. 1A) were initially determined by epifluorescence microscopy after SYBR Green I staining according to the procedure described in [1] and [2]. All enumeration of viral particles in this study was performed by using the identical protocols.

Citation: Guowei Sun, Jinzhou Xiao, Hongming Wang, Chaowen Gong, Yingjie Pan, Shuling Yan, Yongjie Wang, Efficient purification and concentration of viruses from a large body of high turbidity seawater, MethodsX, Volume 1, 2014, Pages 197-206, ISSN 2215-0161, http://dx.doi.org/10.1016/j.mex.2014.09.001.

Table 1: Viral abundance and recovery efficiency after concentration of original high turbidity seawater samples.

Method	Viral abundance						Recovery efficiency[a](% ± SD)	Concentration factor (10^4)
	Original sample (10^6 VLPs mL^{-1} ± SD) [10^{12} VLPs ± SD]	0.3 and 0.2 μm filtrate (10^6 VLPs mL^{-1} ± SD) [10^{11} VLPs ± SD]	TFF concentrate (10^8 VLPs mL^{-1} ± SD) [10^{11} VLPs ± SD]	Membrane-rinsing concentrate (10^7 VLPs mL^{-1} ± SD) [10^{10} VLPs ± SD]	Sediment (10^8 VLPs g^{-1} ± SD) [10^{11} VLPs ± SD]	Plus-70 reconcentrate (10^{10} VLPs mL^{-1} ± SD) [10^{11} VLPs ± SD]		
Standard TFF	(6.95 ± 0.46) [1.04 ± 0.07]	(4.08 ± 0.34) [6.12 ± 0.50]	(3.82 ± 0.21) [3.82 ± 0.21]			(7.81 ± 0.78) [3.12 ± 0.31]	30.04 ± 3.19	3.75
This study		(3.19 ± 0.12) [4.79 ± 0.18]	(3.11 ± 0.28) [3.11 ± 0.28]	(5.98 ± 0.54) [5.98 ± 0.54]	(7.51 ± 0.39) [3.83 ± 0.20]	(16.30 ± 1.05) [6.98 ± 0.45]	67.10 ± 5.49	3.50

The mean values and standard deviations were calculated based on triplicate counts.

A ZThe final viral abundance compared to the viral abundance in original samples.

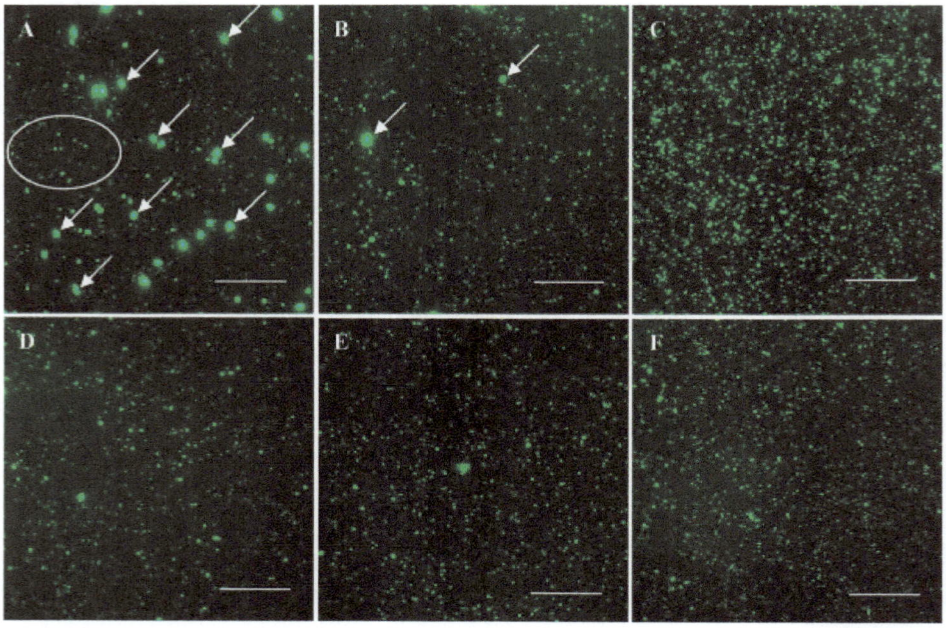

Figure1: Epifluorescence-microscope image of each step's samples filtered onto a Whatman 0.02 μm Anodisc filter and stained with SYBR Green I. (A) Original seawater; (B) sample "A" filtered by 0.3 μm and 0.2 μm filters; (C) sample "B" concentrated by 50 kDa TFF ultrafilter and diluted 100 times; (D) eluant samples of membrane rinsing process diluted 10 times; (E) viral concentrate from settled matter diluted 100 times; (F) samples "C + D + E" reconcentrated by 30 kDa Centricon Plus-70 ultrafilter (Millipore) and diluted 15,000 times. The arrows indicate prokaryotes and the ellipse indicates >30 KDa virus-like particles. Scale bar = 20 μm. Note: a small number of prokaryotes appear in (B), and their possible origins are from the instruments and environment. They were removed after TFF by using the 0.22 μm cut-off filter unit.

Concentration of viruses from the high turbidity seawater

The procedures for purification and concentration of virus particles from water samples are outlined in Graphical Abstract. The major procedures are as followed:

1. *Sedimentation by natural gravity.* 150 L of seawater samples were maintained in the dark for 12 h at 4 °C. After sedimentation, the samples were divided into two parts: water and settled matter fractions. The water fraction (approximately 150 L) was subsequently subjected to viral concentration, and the settled matter (approximately 509.8 g) was stored at −80 °C before viruses were isolated and concentrated.
2. *Removal of all particles or cells larger than 0.2 μm.* The water fraction (approximately 150 L) was successively filtered through 0.3 and 0.2 μm pore-sized filters in a stainless steel filter holder, a high performance and throughput filtration system (Millipore, MA, USA) equipped with the reusable cartridge filter with a large surface area, under a low entry pressure (<0.2 bar) driven by a peristaltic pump (Millipore, MA, USA).

Afterwards, the "viral fraction" seawater (the filtrate, approximately 150 L) was obtained. The number of viral particles in the filtrate (Table 1 and Fig. 1B) was determined by epifluorescence microscopy after SYBR Green I staining.

3. *TFF concentration.* The "viral fraction" seawater (approximately 150 L) was subsequently concentrated by using a large-scale TFF system with 50 kDa cut-off tangential flow filter (Millipore, MA, USA) (see Graphical Abstract). The intake pressure driven by the peristaltic pump was below 10 p.s.i. (approximately 0.7 bar) to protect viruses from being destroyed, resulting in loss of virus yield [3]. When the volume of "viral fraction" seawater was less than 1 L, the viral concentrate was transferred into a sterilized container (*i.e.* viral concentrate (I), see Graphical Abstract). The number of viral particles in 'viral concentrate (I)' (Table 1 and Fig. 1C) was determined by epifluorescence microscopy after SYBR Green I staining.
4. *Membrane rinsing.* 20 L of TFF permeate (virus-free seawater) (see Graphical Abstract) was used to rinse the tangential flow filter until the water volume was less than 1 L. This part of the eluant, containing viruses trapped on the filter membrane and connection hoses during the first TFF, was also incorporated into the viral concentrate (I) (we call this step "membrane rinsing"). Precautions and operating techniques in TFF are described in detail in [3]. The number of viral particles in the eluant samples (Table 1 and Fig. 1D) was determined by epifluorescence microscopy after SYBR Green I staining.
5. *Isolation of viruses from settled matter.* Approximately 4 g of frozen settled matter was transferred to a 50 mL sterile centrifuge tube containing 5 mL virus-free seawater (sterile 50 kDa TFF permeate). Then 0.7 mL of tetrasodium pyrophosphate solution (10 mM final concentration) was added, and the slurry was shaken for 1 min with a ShakeMaster (Eppendorf, Hamburg, Germany) at maximum speed. After incubation for 15 min in the dark, the slurry was sonicated for 3 min (3× 1 min) using a 60 W sonication probe (Xinzhi, Ningbo, China). To prevent overheating, sonication treatment was performed at intervals of 1 min with 30 s interruptions in an ice bath. The sonicated slurry was centrifuged for 5 min at 700 × *g* and 4 °C, and the supernatant was filtered through a 0.22 μm sterile filter. DNase I (0.5 U mL^{-1} final concentration, Fermentas, Vilnius, Lithuania) was added to the filtrate, which was incubated in the dark for 30 min at room temperature [4]. Subsequently, DNase was inactivated according to the manufacturer's instructions. The number of viral particles in the filtrate (Table 1 and Fig. 1E) was determined by epifluorescence microscopy after SYBR Green I staining.
6. *Viral reconcentration and deionization.* Previous studies indicated that $\geq 10^9$ mL^{-1} of viral particles was required for exploring the diversity of marine viruses based on metagenomic methods [5], [6] and [7]. In addition, it is generally known that the extra salt ions largely affect the downstream viral enumeration and molecular biological analysis, for example, viral genomic DNA extraction, PCR reaction. Accordingly, the Centricon Plus-70 centrifugal filter device (Millipore, MA, USA) was used to reconcentrate the viral concentrate obtained from the water and settled matter samples (see Graphical Abstract, viral concentrate (I) was obtained from the water fraction using TFF concentration, and viral concentrate (II) was isolated from the settled matter fraction). The Centricon Plus-70 centrifugal filter device was firstly rinsed with 0.02 μm filtered and autoclaved MilliQ water to remove the humectant according to the manufacturer›s instructions. Approximately 70 mL viral concentrate was added to each sample filter cup. Samples were centrifuged for 15 min at 3500 × *g* and 4 °C using a benchtop swinging bucket rotor (Xiangyi, Hunan, China) followed by three runs of rinsing, using 60 mL of 0.02 μm filtered and autoclaved MilliQ water for each run and a sample filter cup for deionization. The salinity of the filtrate was measured by using a conductivity meter (HACH, NYC, USA) to verify whether salt ions were removed. Finally, the viral concentrates were recovered by centrifuging at 900 × *g* and 4 °C for 2 min. The number of viral particles in the concentrate (Table 1 and Fig. 1F) was determined by epifluorescence microscopy after SYBR Green I staining.

Verification of viral particle purity and structural integrity

1. *PCR amplification analysis and DNase I treatment.* Prior to viral nucleic acid extraction, it was necessary to check whether viral concentrates were contaminated with microbial, eukaryotic cells and/or extracellular nucleic acids. The 16S rRNA and 18S rRNA gene fragments were amplified using universal primer pairs, (i) 27F(5′-AGAGTTTGATCCTGGCTCAG-3′) and 1492R (5′-GGTTACCTTGTTACGACTT-3′) (specific for bacteria [8]), (ii) 340F (5′-CCCT AYGGGGYGCASCAG-3′) and 1000R (5′-GGCCATGCACYWCYTCTC-3′) (specific for archaea [9]) and (iii) 18SF (5′-CCGCAGCTAGGAATAATGGAATAGGAC-3′) and 18SR (5′-GTTAGCATGCCAGAGTCTCGTTCGT-3′) (specific for eukaryotes [10]). PCR amplification was performed in a total volume of 25 μL reactant containing 12.5 μL 2× PCR MasterMix (TIANGEN, Beijing, China), 0.4 μM of each primer, 9.5 μL ddH_2O, and 1 μL of final viral concentrate as template. PCR programs were listed as follows: (i) For amplifying bacterial 16S rRNA gene fragment – initial denaturation at 95 °C for 4 min, followed by 35 cycles of denaturation at 94 °C for 45 s, annealing at 55 °C for 45 s, extension at 72 °C for 1 min, and a final extension at 72 °C for 10 min; (ii) For amplifying archaeal 16S rRNA gene fragment – initial denaturation at 98 °C for 2 min, followed by 35 cycles of denaturation at 95 °C for 30 s, annealing at 57 °C for 30 s, extension at 72 °C for 90 s, and a final extension at 72 °C for 7 min; (iii) For amplifying eukaryotic 18S rRNA gene fragment – initial denaturation at 94 °C for 5 min, followed by 35 cycles of denaturation at 94 °C for 1 min, annealing at 55 °C for 1 min, extension at 72 °C for 1 min, and a final extension at 72 °C for 10 min. Subsequently, 5 μL of each PCR product was electrophoresed at 120 V for 30 min in 1.5% (w/v) agarose gel and 1× Tris-acetate-EDTA (TAE) buffer. PCR products were

visualized and photographed by using a GEL imaging system (Bio-Rad Laboraties, CA, USA).

The results indicated the existence of eukaryotic, microbial cells, and/or extracellular DNA contaminations (Fig. 2). To remove any potential cell (eukaryotic and microbial) contaminations introduced into the viral concentrate during the experiments as well as extracellular DNA from the lysed cells, the final viral concentrate was treated using a 0.22 µm cut-off filter and DNase I (4 U mL^{-1} final concentration). As a consequence, no obvious amplicons were observed for both 16S and 18S rRNA genes (Fig. 2).

2. *Examination of viral morphotypes.* Viral morphotypes were observed by using transmission electron microscopy. The negative staining of viral particles referred to the methods described in [11], [12] and [13] with modifications. A drop (10 µL) of the viral concentrate was placed on a sheet of parafilm. A copper grid was floated on the drop for 15 min. The grid was then removed, and its edge was blotted with a piece of clean filter paper. Subsequently, the grid was stained with 2% phosphotungstic acid in 60 mM Sørensen phosphate buffer (pH 6.5) for 2 min. Excess phosphotungstic acid was removed as described above followed by air-drying for a few minutes. The grids were examined under a Philips TECNAI 12 transmission electron microscope at an acceleration voltage of 100 kV.

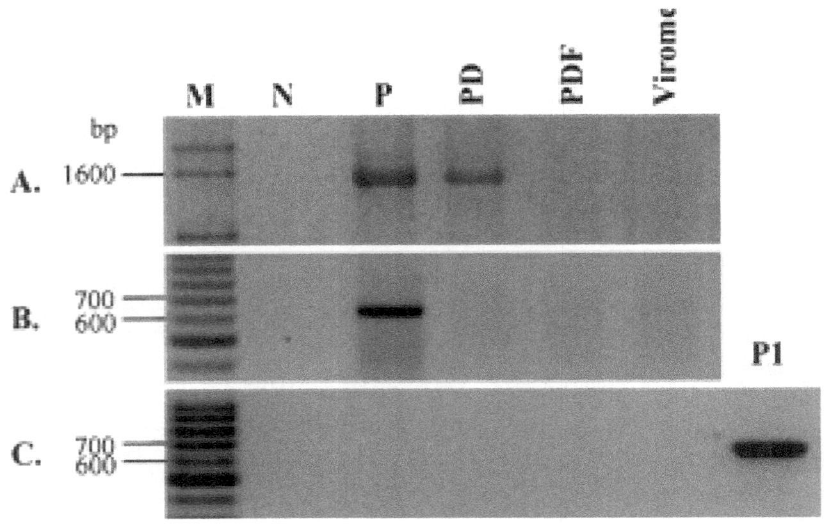

Figure 2: Agarose gel electrophoresis images of PCR-amplified bacterial 16S rRNA gene (A), eukaryotic 18S rRNA gene (B), and archaeal 16S rRNA gene (C) fragments. Abbreviations are as follows: M, DNA marker; N, negative control; P, final viral concentrate reconcentrated by using Centricon Plus-70 centrifugal filter device; PD, sample "P" was treated by DNase I; PDF, sample "PD" was filtered by 0.22 µm cut-off filter; P1, positive control.

Most of the observed viral particles had a distinct head-and-tail morphology (Fig. 3), which are typical features of bacterial DNA viruses. The results of transmission electron microscopy revealed structural integrity of the purified and concentrated viral particles.

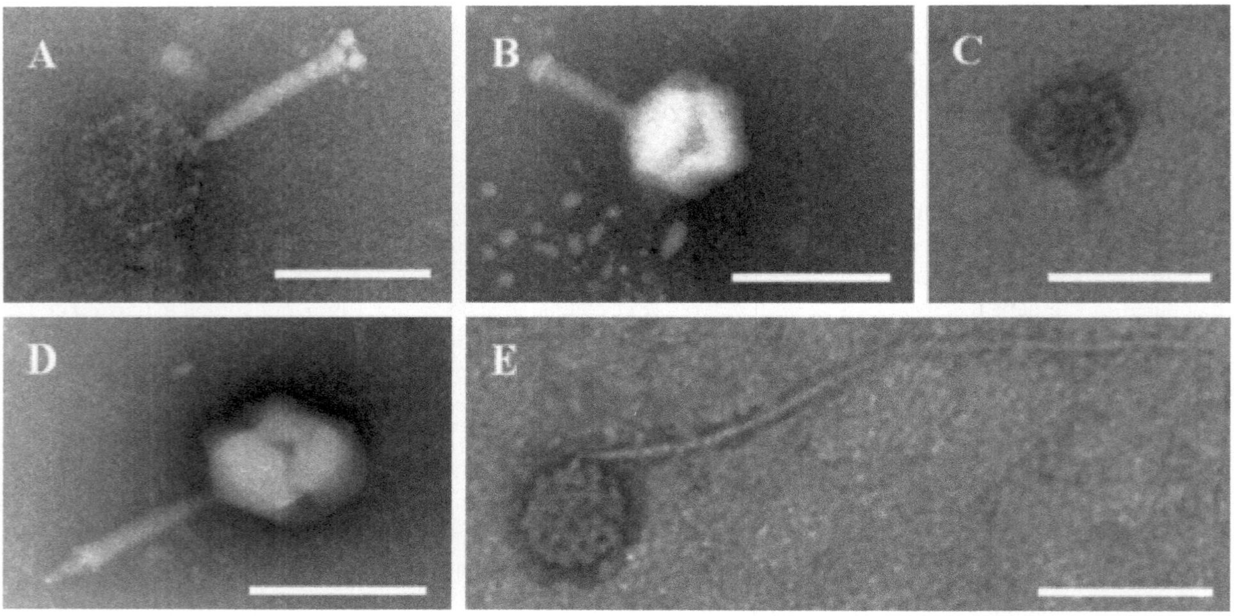

Figure 3: Transmission electron micrographs of phosphotungstic acid stained viruses isolated from the high turbidity seawater. (A, B and D) Myoviruses; (C) Podoviruses; (E) Siphoviruses. Scale bar = 100 nm.

Verification of virome purity

1. *Viral genomic DNA extraction.* The viral genomic DNA was extracted from the final viral concentrate by using a DNeasy Blood & Tissue Kit (QIAGEN, Dusseldorf, Germany) according to 'Nonnucleated' procedure of 'Protocol: Purification of Total DNA from Animal Blood or Cells (Spin-Column Protocol)' with modifications. In brief, 20 μl proteinase K was firstly added to 200 μl of the final viral concentrate, and the following procedures were the same as described in the kit.
2. *PCR amplification.* The 16S and 18S rRNA genes were amplified using the obtained viral genomic DNA as the template by using above identical protocols. On agarose gel there were no obvious amplicons detected for both 16S (Fig. 2A and C) and 18S rRNA genes (Fig. 2B), which indicated that the extracted viral DNA was basically free DNA contamination of cellular organisms, and subsequently confirmed the purity of the final viral concentrate.

Analysis of viral recovery efficiency

The viral recovery efficiency was calculated as the percentage of the number of viruses in final viral concentrate divided by the number of viruses in the original sample, which was determined with triplicate counts.

Efficiency of the optimized method

Approximately 1 L viral concentrate (viral concentrate (I), see Graphical Abstract) was obtained from the water fraction after TFF concentration in less than 20 h. The number of concentrated viral particles was 3.11×10^8 VLPs mL^{-1} (Table 1 and Fig. 1C).

For the membrane rinsing process, approximately 1 L eluant was obtained after washing with 20 L TFF permeate, and the number of viral particles was 5.98×10^7 VLPs mL^{-1}(Table 1 and Fig. 1D). This part of the eluant, containing viruses trapped on the filter membrane during the first TFF, was incorporated into the viral concentrate (I).

A total of 509.8 g settled matter was separated from 150 L seawater, subjected to viral isolation (see Graphical Abstract), and contained 7.51×10^8 VLPs g^{-1} (Table 1 and Fig. 1E). The total number of viral particles isolated from the settled matter (3.83×10^{11} VLPs, see Table 1) was slightly higher than that recovered from water fraction (3.11×10^{11} VLPs, see Table 1). Obviously, in order to isolate the most viral particles from high turbidity seawater, the flocculation of viral particles to suspended matter needs to be considered seriously and cannot be ignored.

To reduce the volume of viral concentrate and obtain a high density of viral particles, the whole viral concentrates were reconcentrated by using a Centricon Plus-70 centrifugal filter device. In the end, approximately 4.28 mL viral concentrate (16.30×10^{10} VLPs mL^{-1}) (Table 1 and Fig. 1F) was obtained from 2.5 L viral concentrates (1 L from TFF concentration of the water fraction, 1 L from membrane rinsing, and 500 mL from the settled matter). After washing, the salt ions were removed from the viral concentrates. Compared to the original seawater samples, the viral recovery efficiency was 67.10%, and the concentration factor were $>3.5 \times 10^4$ (150 L divided by 4.28 mL) regarding water volume and 2.3×10^4 (16.30×10^{10} VLPs mL^{-1} divided by 6.95×10^6 VLPs mL^{-1}) in consideration of viral density.

Comparison of the optimized and standard methods

For the standard TFF concentration method, 150 L of water samples were successively prefiltered by gravity through eight layers of gauze and 100, 50, 25, and 10 μm pore-sized nylon fiber filters [14], and the filtrates were stored in sterile containers at 4 °C for subsequent virus concentration. The settled matter was discarded, and the rinsing process for the TFF filter membrane was not applied. Other steps involved in the concentration and reconcentration of viruses were the same as those for "water fraction" (see Graphical Abstract).

As a result, compared to the standard TFF method with a viral recovery efficiency of 30.04% (Table 1), the recovery efficiency of the optimized concentration method (67.10%) (Table 1) increased by over two times. This suggests that the method developed in this study is much more efficient in terms of purification and concentration of viruses from large body of high turbidity seawater.

SUMMARY

In this study, the optimized method for viral concentration from high turbidity seawater is characterized by high viral recovery efficiency (67.10%), high concentration factor ($>3.50 \times 10^4$), high viral particle densities ($>10^{11}$ VLPs mL^{-1}), high throughput (150 L seawater per 36 h), good structural integrity of viral particles, and the lack of other contaminants, for example particles and cells larger than 0.22 μm, extracellular nucleic acids, organic materials, and inorganic ions. The isolated viruses can be used directly for subsequent analysis, for example transmission electron microscopy observation and metagenomic sequencing. Accordingly, this optimized viral concentration method paves the way for routine investigation of viruses in a large body of highly turbid seawater and other water bodies.

ADDITIONAL INFORMATION

Background

Viruses represent the most abundant and dynamic biological components in the global ecosystems, including the marine environments [15] and [16]. Since viruses lack a universally conserved gene, for example the ribosomal RNA genes in cellular organisms (such as 16S rRNA gene of prokaryotes and 18S rRNA gene of eukaryotes), and since most viral hosts are nonculturable [17], traditional molecular technologies, such as denaturing gradient gel electrophoresis and clone library, are not suitable to the study of viral diversity, which make it even harder to understand marine viral diversity. To surmount these difficulties, recently, the viral particles have been isolated from seawater followed by diversity analysis with pulsed-field gel electrophoresis (PFGE) [18],[19] and [20], random amplified polymorphic DNA (RAPD) technique [21] or metagenomic approaches [6], [22] and [23]. Viral metagenomics requires high-density (viral particle densities of $\geq 10^9$ mL^{-1}) and large-scale concentrations of seawater to obtain enough genetic material for sequencing [5] and [6]. An efficient and reliable method for isolation and concentration of marine viruses from seawater samples is thus particularly important for downstream molecular biological analysis.

Currently, there are dozens of approaches to viral isolation and concentration from water samples, which include adsorption-elution [24], [25] and [26], chemical flocculation[14] and [27], ultracentrifugation [19] and [28], and ultrafiltration (such as tangential flow filtration (TFF)) [7]. However, the first three methods have certain inherent defects, including (i) selective adsorption of viruses to a solid matrix with focus on the detection of specific viruses [7] and [29], (ii) some elution buffers interfering with downstream viral enumeration and molecular biological analysis [30], and (iii) limited throughput and low viral concentration efficiency [14], [31] and [32]. By contrast, TFF appears to be the most efficient technology for concentrating viruses from large volumes of water samples, especially in the marine environment.

However, for high turbidity seawater samples (*i.e.* seawater samples with highly suspended matter contents), such as coastal and estuary subsurface seawater [22],[33] and [34], which have been studied frequently during recent years, direct collection of viruses by using TFF encounters difficulties or even causes damage to the expensive TFF system. The general solution to this problem in previous studies [7] and [14] has involved prefiltering raw natural seawater samples before TFF with a series of pore-size nylon fiber filters or wound polypropylene sediment filters to remove all settled matter and particles by gravity or by a pump-driven system. However, this procedure has drawbacks including time wasting (filter clogging causes low flow rate), increasing the chance of contamination, and low viral concentration efficiency because of lysis and settled matter adsorption of viruses. For this reason, we focus on TFF and centrifugal ultrafiltration technology and develop an efficient viral purification and concentration method for high turbidity seawater samples. The obtained viral concentrate is ideal for subsequent analysis by epifluorescence microscopy, transmission electron microscopy, and metagenomics.

ACKNOWLEDGMENTS

This work was supported by the National Natural Science Foundation of China(41376135 and 81341082), the Program for Professor of Special Appointment (Eastern Scholar) 20101222 from Shanghai Municipal Education Commission, Doctoral Fund of Ministry of Education of China (20133104110006), and Innovation Program of Shanghai Municipal Education Commission, China (14ZZ144). *MethodsX* thanks the (anonymous) reviewers of this article for taking the time to provide valuable feedback.

REFERENCES

1. R.T. Noble, J.A. Fuhrman, Use of SYBR Green I for rapid epifluorescence counts of marine viruses and bacteria, Aquat. Microb. Ecol. 14 (2) (1998) 113–118.
2. Patel, R.T. Noble, J.A. Steele, M.S. Schwalbach, I. Hewson, J.A. Fuhrman, Virus and prokaryote enumeration from planktonic aquatic environments by epifluorescence microscopy with SYBR Green I, Nat. Protoc. 2 (2) (2007) 269–276.
3. R.V. Thurber, M. Haynes, M. Breitbart, L. Wegley, F. Rohwer, Laboratory procedures to generate viral metagenomes, Nat. Protoc. 4 (4) (2009) 470–483.
4. C.A. Suttle, Enumeration and isolation of viruses, in: Handbook of Methods in Aquatic Microbial Ecology, 1993, 121–134.
5. S.R. Bench, T.E. Hanson, K.E. Williamson, D. Ghosh, M. Radosovich, K. Wang, K.E. Wommack, Metagenomic characterization of Chesapeake Bay virioplankton, Appl. Environ. Microbiol. 73 (23) (2007) 7629–7641.
6. M. Breitbart, P. Salamon, B. Andresen, J.M. Mahaffy, A.M. Segall, D. Mead, F. Azam, F. Rohwer, Genomic analysis of uncultured marine viral communities, Proc. Natl. Acad. Sci. U.S.A. 99 (22) (2002) 14250–14255.
7. K.E. Wommack, T. Sime-Ngando, D.M. Winget, S. Jamindar, R.R. Helton, Filtration-based methods for the collection of viral concentrates from large water samples, in: Manual of Aquatic Viral Ecology (MAVE), 2010, 110–117.
8. W.G. Weisburg, S.M. Barns, D.A. Pelletier, D.J. Lane, 16S ribosomal DNA amplification for phylogenetic study, J. Bacteriol. 173 (2) (1991) 697–703.
9. S. Gantner, A.F. Andersson, L. Alonso-Sa´ez, S. Bertilsson, Novel primers for 16S rRNA-based archaeal community analyses in environmental samples, J. Microbiol. Methods 84 (1) (2011) 12–18.
10. M. Breitbart, F. Rohwer, Method for discovering novel DNA viruses in blood using viral particle selection and shotgun sequencing, Biotechniques 39 (5) (2005) 729–736.
11. H. Montanie´, H.J. Hartmann, C. Crottereau, C. Trichet, Virus-like particle analysis in a seston-rich coastal pond using transmission electron microscopy, Aquat. Microb. Ecol. 28 (2) (2002) 105–115.
12. H.W. Ackermann, H.M. Krisch, A.M. Comeau, Morphology and genome sequence of phage f 1402: a dwarf myovirus of the predatory bacterium Bdellovibrio bacteriovorus, Bacteriophage 1 (3) (2011) 138–142.
13. B. Zechmann, G. Zellnig, Rapid diagnosis of plant virus diseases by transmission electron microscopy, J. Virol. Methods 162 (1–2) (2009) 163–169.
14. J. Colombet, A. Robin, L. Lavie, Y. Bettarel, H.M. Cauchie, T. Sime-Ngando, Virioplankton 'pegylation': use of PEG (polyethylene glycol) to concentrate and purify viruses in pelagic ecosystems, J. Microbiol. Methods 71 (3) (2007) 212–219.
15. J.A. Fuhrman, Marine viruses and their biogeochemical and ecological effects, Nature 399 (6736) (1999) 541–548.
16. L.M. Proctor, J.A. Fuhrman, Viral mortality of marine bacteria and cyanobacteria, Nature 343 (1990) 60–62.
17. F. Rohwer, R.V. Thurber, Viruses manipulate the marine environment, Nature 459 (7244) (2009) 207–212.
18. R.A. Sandaa, A. Larsen, Seasonal variations in virus–host populations in Norwegian coastal waters: focusing on the cyanophage community infecting marine Synechococcus spp, Appl. Environ. Microbiol. 72 (2) (2006) 4610–4618.

19. G.F. Steward, J.L. Montiel, F. Azam, Genome size distributions indicate variability and similarities among marine viral assemblages from diverse environments, Limnol. Oceanogr. 45 (8) (2000) 1697–1706.

20. K.E. Wommack, J. Ravel, R.T. Hill, J.S. Chun, R.R. Colwell, Population dynamics of Chesapeake bay virioplankton: totalcommunity analysis by pulsed-field gel electrophoresis, Appl. Environ. Microbiol. 65 (1) (1999) 231–240.

21. C. Winter, M.G. Weinbauer, Randomly amplified polymorphic DNA reveals tight links between viruses and microbes in the bathypelagic zone of the Northwestern Mediterranean Sea, Appl. Environ. Microbiol. 76 (20) (2010) 6724–6732.

22. A.I. Culley, A.S. Lang, C.A. Suttle, Metagenomic analysis of coastal RNA virus communities, Science 312 (5781) (2006) 1795–1798.

23. G.F. Steward, C.M. Preston, Analysis of a viral metagenomic library from 200m depth in Monterey Bay, California constructed by direct shotgun cloning, Virol. J. 8 (2011) 287.

24. S.R. Farrah, D.R. Preston, G.A. Toranzos, M. Girard, G.A. Erdos, V. Vasuhdivan, Use of modified diatomaceous earth for removal and recovery of viruses in water, Appl. Environ. Microbiol. 57 (9) (1991) 2502–2506.

25. S. Kamata, S. Suzuki, Concentration of marine birnavirus from seawater with a glass fiber filter precoated with bovine serum albumin, Mar. Biotechnol. 5 (2) (2003) 157–162.

26. H. Katayama, A. Shimasaki, S. Ohgaki, Development of a virus concentration method and its application to detection of enterovirus and norwalk virus from coastal seawater, Appl. Environ. Microbiol. 68 (3) (2002) 1033–1039.

27. S.G. John, C.B. Mendez, L. Deng, B. Poulos, A.K. Kauffman, S. Kern, J. Brum, M.F. Polz, E.A. Boyle, M.B. Sullivan, A simple and efficient method for concentration of ocean viruses by chemical flocculation, Environ. Microbiol. Rep. 3 (2) (2011) 195–202.

28. S.M. Short, C.M. Short, Diversity of algal viruses in various North American freshwater environments, Aquat. Microb. Ecol. 51 (1) (2008) 13–21.

29. S.L. Percival, R.M. Chalmers, M. Embrey, P.R. Hunter, J. Sellwood, P. Wyn-Jones, Microbiology of Waterborne Diseases, Elsevier, Amsterdam, the Netherlands, 2004.

30. K.E. Williamson, K.E. Wommack, M. Radosevich, Sampling natural viral communities from soil for culture-independent analyses, Appl. Environ. Microbiol. 69 (11) (2003) 6628–6633.

31. J.A. Fuhrman, X. Liang, R.T. Noble, Rapid detection of enteroviruses in small volumes of natural waters by real-time quantitative reverse transcriptase PCR, Appl. Environ. Microbiol. 71 (8) (2005) 4523–4530.

32. T.M. Fumian, J.P. Leite, A.A. Castello, A. Gaggero, M.S. Caillou, M.P. Miagostovich, Detection of rotavirus A in sewage samples using multiplex qPCR and an evaluation of the ultracentrifugation and adsorption-elution methods for virus concentration, J. Virol. Methods 170 (1–2) (2010) 42–46.

33. Y. Bettarel, T. Bouvier, M. Agis, C. Bouvier, T.V. Chu, M. Combe, X. Mari, M.N. Nghiem, T.T. Nguyen, T.T. Pham, O. Pringault, E. Rochelle-Newall, J.P. Torre´ton, H.Q. Tran, Viral distribution and life strategies in the Bach Dang Estuary, Vietnam, Microb. Ecol. 62 (1) (2011) 143–154.

34. J.C. Auguet, H. Montanie´, P. Lebaron, Structure of virioplankton in the Charente Estuary (France): transmission electron microscopy versus pulsed field gel electrophoresis, Microb. Ecol. 51 (2) (2006) 197–208.

CHAPTER 17

HIGH SALINITY TOLERANCE OF THE RED SEA CORAL FUNGIA GRANULOSA UNDER DESALINATION CONCENTRATE DISCHARGE CONDITIONS: AN IN SITU PHOTOPHYSIOLOGY EXPERIMENT

Riaan van der Merwe, Till Röthig, Christian R. Voolstra, Michael A. Ochsenkühn, Sabine Lattemann and Gary L. Amy

Keywords: Desalination, salinity tolerance, Fungia granulosa, Symbiodinium, coral reef, Red Sea, marine monitoring, environmental impact assessment

INTRODUCTION

A growing demand of freshwater in semi-arid and arid regions (e.g., Arabian Peninsula) leads to the construction of an increasing number of seawater desalination plants, especially the low energy consuming seawater reverse osmosis (SWRO) desalination plants (Fritzmann et al., 2007). As a result, more hypersaline concentrate discharge (brine) reaches the marine environments (Lattemann and Höpner, 2008). Environmentally safe disposal of this brine is one of the key factors determining the environmental impacts of a desalination plant. The highest salinity that marine organisms can cope with in a desalination discharge area is defined as a salinity tolerance threshold. It depends on the species and the exposure time to elevated salinity levels (Voutchkov, 2009).

Euryhaline marine organisms can commonly tolerate changes in salinity (in contrast to stenohaline species) and series of small increments are generally better tolerated than direct exposure to high salinities (Voutchkov, 2009). Effects of concentrate discharges depend on exposure intensities, frequencies, the environment the brine is released into, and the brine temperature (Roberts et al., 2010). Accordingly, the effects of discharged brine can range from no significant impacts on microbial abundance or plankton communities, to widespread alterations in community structures of seagrass, invertebrates, soft-sediment infauna, and corals (Roberts et al., 2010 and references therein; van der Merwe et al., 2014).

Hermatypic corals are the key stone species of coral reefs, which are among the most diverse and productive ecosystems on this planet (Moberg and Folke, 1999). Coral health and survival fundamentally depend on the interaction between coral host and photosynthetic algae (zooxanthellae, genus *Symbiodinium*) that can be found in the endodermal tissues of reef-building corals. Unfortunately, research on the effects of brine discharge on corals is scarce. Mabrook (1994)reported corals disappearing from coastal areas in the Red Sea (Egypt) as a result of desalination plants discharge, but no reproducible data are presented. Corals are generally considered stenohaline osmoconformers and very sensitive to the effects of desalination plant discharge (Ferrier-Pages et al., 1999; Manzello and Lirman, 2003; Elimelech and Phillip, 2011).

Besides a desalination discharge context, more data on coral salinity tolerance are available; especially for decreased salinities. Generally, changes in salinity may affect metabolism

Citation: Van der Merwe R, Röthig T, Voolstra CR, Ochsenkühn MA, Lattemann S and Amy GL (2014) High salinity tolerance of the Red Sea coral Fungia granulosa under desalination concentrate discharge conditions: an in situ photophysiology experiment. Front. Mar. Sci. 1:58. doi: 10.3389/fmars.2014.00058.

and/or photophysiology of the coral animal and/or the corals' algal symbionts due to salinity stress (Muscatine, 1967; Chartrand et al., 2009). Moberg et al. (1997) suggested that photosynthetic rates are lowered in proportion to the reductions in salinity (10–20 PSU), whereas respiration rates were either slightly decreased or unaltered for two hermatypic corals upon hyposaline treatments.Hoegh-Guldberg and Smith (1989) concluded moderately reduced salinity (30 PSU) exposure for 4–10 days at several temperatures does not induce bleaching in *Stylophora pistillata* and*Seriatopora hystrix*. At lower salinities (i.e., at 23 PSU) the authors reported death within 48 h. Only a limited number of studies have included the impacts of hypersalinity on corals (and the possible effect it might have on dinoflagellate symbiont functionality within the host). Generally, the effects of hyper- and hyposaline treatments are similar (Muthiga and Szmant, 1987; Lirman and Manzello, 2009). Severity of effects observed commonly coincides with salinity concentration, exposure time, coral species, and the speed of salinity change.

In this study, our objective was to determine long-term effects (4 weeks) of a strong salinity increase on the coral *Fungia granulosa*. To do this, we transplanted corals along a 25 m transect from a SWRO facility discharge structure, and determined salinity, temperature, oxygen, and light levels regularly. At the same time, we measured the (photo) physiological state of the algal symbiont via PAM fluorometry and checked for signs of visual bleaching. PAM fluorometry quantifies the photosynthetic efficiency with which light energy is converted into chemical energy at photosystem II (PSII) level. Evaluating the chlorophyll fluorescence can indicate an organism's photosynthetic efficiency under changing or stressful conditions, e.g., varying salinity regimes (Chartrand et al., 2009), and therefore, serves as a stress indicator. Additionally, we assessed the salinity tolerance of cultured *Symbiodinium* to understand the contribution of the algal symbiont to salinity tolerance of the coral holobiont.

MATERIALS AND METHODS

Experiment Overview

The existing SWRO facility (i.e., the submerged discharge location) at the King Abdullah University of Science and Technology (KAUST) was selected for this case study. The plant is located on the KAUST campus and is designed to provide all potable water needs. Under current operational conditions, the raw water intake is about 2825 $m^3 h^{-1}$ with a recovery rate of 39%, resulting in an average brine flow of 1723.25 $m^3 h^{-1}$ (41,358 $m^3 d^{-1}$) that is discharged to the Red Sea. The submerged outfall (discharge structure) is located at a water depth of 18 m, approximately 2.8 km from the pump station (22° 17.780N, 39° 04.444E). The concentrate is pumped through a 1.2 m diameter pipeline to the offshore structure where the concentrate is pushed up in a concrete riser and discharged horizontally through four discharge screens (1800 × 1000 mm) approximately 6 m above the seafloor.

Experimental Setup

We conducted an *in situ* salinity stress experiment on the coral *F. granulosa* collected from Fsar reef (22° 13.945N, 39° 01.783E, approximately 9 km from the study site) over 29 days (15.01.14–13.2.14). At Fsar reef salinity, light conditions, and effective quantum yields for *F. granulosa* were measured for reference purposes. Corals were handled with latex gloves and 18 specimens were collected from 16 to 19 m (similar depth to experimental study area) into separate zip lock bags. Corals were transported to the study site in shaded opac plastic boxes filled with ambient sea water. All specimens were placed on the roof of the discharge structure (Figure 1), labeled with nylon fishing line and under-water paper tags, and left for acclimation for 20 h. Three specimens were then randomly selected and placed at each of the 6 stations at the discharge screen (station 1) and along a 25 m transect (stations 2–6) (Figures 2A,B).

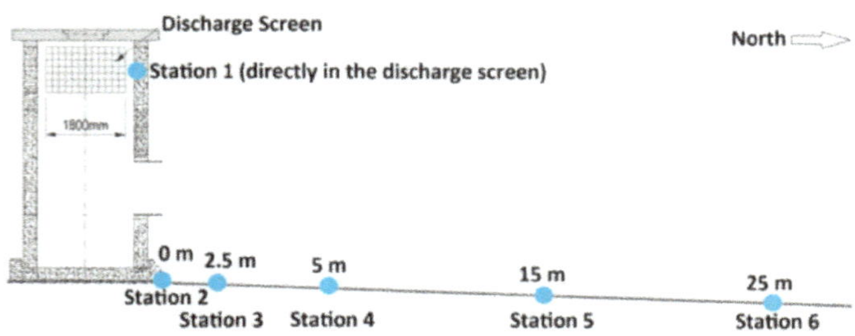

Figure 1: Discharge structure and stations assayed in this experiment.

Figure 2: (A) *Fungia granulosa* attached to the discharge screen. **(B)** Tagged *Fungia granulosa* specimen; blue temp logger visible on right hand side (station 6).

Data Collection

We collected data on temperature, dissolved oxygen (DO), and salinity. For each sampling time point (T0–T6) we sampled between 11:00 and 12:00 h. We measured the effective quantum yield ($\Delta F/Fm'$) and visually assessed all specimens for signs of bleaching. Temperature was logged continuously in 10 min intervals with HOBO Pendant® temperature data loggers at each station over the entire experiment. For DO and salinity measurements, water samples were collected during each dive from each station. Water samples for salinity measurements were collected using 50 mL Falcon Conical Centrifuge Tubes and 1 L low-density polyethylene (LDPE) cubitainer were used for DO samples. Salinity and DO were measured for all stations immediately after each dive. DO measurements were conducted using a WTW (Multi) 3500i Multi-Parameter Water Quality Meter with a CellOx® 325 DO electrode and salinity with a WTW Cond 3310 m with TetraCon®325. Salinity and DO were analyzed for significant differences via One-Way ANOVA using Statistica 10 (StatSoft Inc. 2011, version 10). A diving PAM fluorometer (DIVING-PAM, Walz, Germany) was used to measure photosynthetically active radiation (PAR) at each station and to evaluate the effective quantum yields ($\Delta F/Fm'$) of the symbiotic algae of each coral specimen. The effective quantum yield ($\Delta F/Fm'$) of photochemical energy conversion in PSII for each measurement was calculated based on F and Fm' according to Genty et al. (1989):

$$\Phi\,\mathrm{PSII} = (Fm' - F)\,/Fm' = \Delta F/Fm'.$$

F and Fm' measurements for each specimen were conducted between 11:00 and 12:00 h to ensure comparable daytime conditions (in regard to physiology and prevailing light regime). All $\Delta F/Fm'$ measurements were taken in triplicate for each coral and specimens were only collected after the last sampling event (29 days).

Symbiodinium Culturing and Salt Gradient Stress Experiment

Symbiodinium microadriaticum CCMP2467 (clade A1 https://ncma.bigelow.org) was cultured in Guillard and Ryther F/2 media suspension without silicium in an incubator at a temperature of 26°C and with a light intensity set at 4 PAR (as measured by diving-PAM) (Guillard and Ryther, 1962). The F/2 media was newly prepared from sea water (obtained from 100 m depth in the Red Sea) and completed with 0.5 ml $NaNO_3$, NaH_2PO_4, vitamins and trace metals following Guillard and Ryther (1962). The cells were kept in the exponential growth phase with a cell density between 10^5 and 10^6 cells mL^{-1} for 2 weeks prior to experiments. For the comparability of experiments, the cells were then directly transferred into a freshly prepared salt adjusted F/2 media with a cell density of ~1×10^5 cells mL^{-1}. A salt gradient experiment was conducted in concentrations ranging from 25 to 55 PSU (in increments of 5 PSU). The F/2 media were prepared by adding appropriate amounts of double-distilled water (ddH_2O) for a diluted salinity range of 25–35 PSU. In order to obtain elevated salinity levels on the order of 45–55 PSU, the media was spiked with NaCl (Sigma). Cells were then transferred into sterilized plastic culture flasks in equal volumes (400 mL) of adjusted F/2 media to reach a cell density of ~1×10^5 cells mL^{-1}. Cells were sampled over a 7 day period at 7 different time points (0, 1, 2, 3, 5, and 7 d) at a temperature of 28°C and under a 9 PAR light intensity. The 0 d sample was used as a control and only withdrawn from cells at the 40 PSU (ambient) condition. Sample volumes for fluorescence activated cell-sorting (FACS) analysis were 1 mL, which were directly harvested by centrifugation (5430 R centrifuge, Eppendorf) at 10,000× g for 10 min at 4°C. For cell fixation, 700 μL media were withdrawn, adding 100 μL of 40% formaldehyde in dd$H_2$0 to each of the samples in order to reach a final concentration of 10%. Cells were thoroughly resuspended by vortexing and kept at 4°C until further analysis. Following cell fixation, *Symbiodinium* cells were again harvested by centrifugation, supernatant discarded, and washed once with 500 μL buffer solution phosphate buffered saline (PBS). PBS was removed completely and 500 μL SYBR Green DNA (2× conc., Life Technologies) staining in PBS was added. Cells were resuspended by vortexing and stained for 1 h at room temperature at 400 rpm in a Thermo Mixer® (Eppendorf). After staining, cells were again pelleted by centrifugation, washed once with PBS and finally resuspended in 1 mL PBS. For FACS measurements 200 μL of each sample were transferred into a 96 well flat bottom well plate and measured. FACS measurements were conducted on a BD LSR Fortessa™ cell analyzer (BD Bioscience, US) using the 405 nm violet laser and QDot655 filters for chlorophyll fluorescence. SYBR Green fluorescence was excited via the 488 nm blue laser and emission detected via Alexa Fluor® 488 filters for total nucleotide detection. FACS data was analyzed by FlowJo 7.5 flow cytometry analysis software.

RESULTS

Ecological Conditions

Water temperature along the transect showed a range between 24.26 and 28.46°C (Table 1). Water temperatures recorded at the discharge screen (26.3 ± 0.78°C) were on average higher than at the other stations (25.93–26.08°C). The average temperatures were also slightly decreasing with increasing distance from the discharge structure. DO varied between 5.75 and 6.37 mg L^{-1}. Average DO levels were 6.07 ± 0.17 mg L^{-1}. We found no significant differences in DO between the stations ($P_{ANOVA} \geq 0.05$). In contrast to DO, salinity data showed significant differences (Figure 3). At the discharge screen (station 1) salinity differed significantly from the other stations ($P_{ANOVA} \leq 0.05$). We found no significant difference between all other stations ($P_{ANOVA} \geq 0.05$) with an average of 41.3 ± 0.7 PSU. However, salinity decreased slightly with increasing distance from the discharge (Figure 3). Control measurements at Fsar reef (site of coral collection) showed a salinity of 39 PSU.

Table 1: Temperature data (HOBO Pendant® Temperature Data Loggers) from all transect stations.

	Station 1 (discharge)	Station 2 (0 m)	Station 3 (2.5 m)	Station 4 (5 m)	Station 5 (15 m)	Station 6 (25 m)
AVG	26.30	26.08	26.02	25.99	25.84	25.83
±*SD*	±0.78	±0.63	±0.62	±0.64	±0.73	±0.95
Min	24.74	24.64	24.55	24.45	24.35	24.26
Max	28.46	27.76	27.67	27.57	27.57	27.57

Logging interval 10 min; temperature [°C] mean ± SD; minimal and maximal measured temperature.

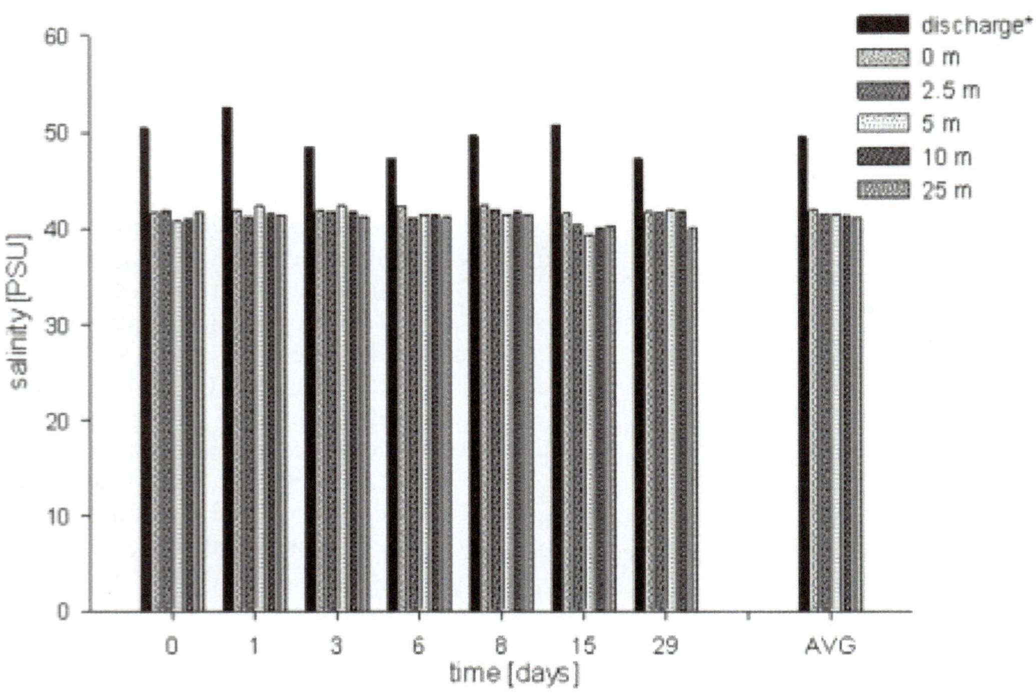

Figure 3: Salinity concentrations measured for each sampling event at each station (6 time points, 6 stations). Average salinities: 49.4 ± 2.0 PSU (discharge screen), 41.8 ± 0.3 PSU (0 m), 41.3 ± 0.6 PSU (2.5 m), 41.3 ± 1.0 PSU (5 m), 41.2 ± 0.6 PSU (15 m) and 41.0 ± 0.7 PSU (25 m), respectively. AVG shows the average salinity at each station.

Coral Salinity Tolerance

The effective quantum yield ($\Delta F/Fm'$) was measured for all 18 coral colonies in triplicates for all time points (i.e., T0–T6) (Figure 4). Control measurements from corals in their natural reef environment showed average $\Delta F/Fm'$ of 0.700 ± 0.010 (at 30–50 PAR; triplicate measurements on three specimens). $\Delta F/Fm'$ at all stations were constant and in the same range as the controls. Average reads were 0.705 ± 0.009 (station 1), 0.694 ± 0.008 (station 2), 0.687 ± 0.007 (station 3), 0.687 ± 0.019 (station 4), 0.693 ± 0.017 (station 5), and 0.690 ± 0.02 (station 6). We found a drop in $\Delta F/Fm'$ at T1 for stations 4–6 which corresponds with elevated light conditions on this day compared to the other stations and sampling events (65–82 PAR; Figures 4D–F). PAR levels ranged between 4 and 82, with average reads of 14 ± 4 (station 1), 17 ± 5 (station 2), 33 ± 13 (station 3), 45 ± 27 (station 4), 34 ± 27 (station 5), and 28 ± 21 (station 6). By trend, PAR levels at station 1 and 2 and to a smaller extent at station 3 were lower and more stable compared to the other stations (Figure 4). This pattern is caused by a shading effect of the discharge structure which kept light levels in its direct surrounding lower and more stable. No PAR was measured at time point T5. During the study we did not observe bleaching characteristics on the measured coral specimens.

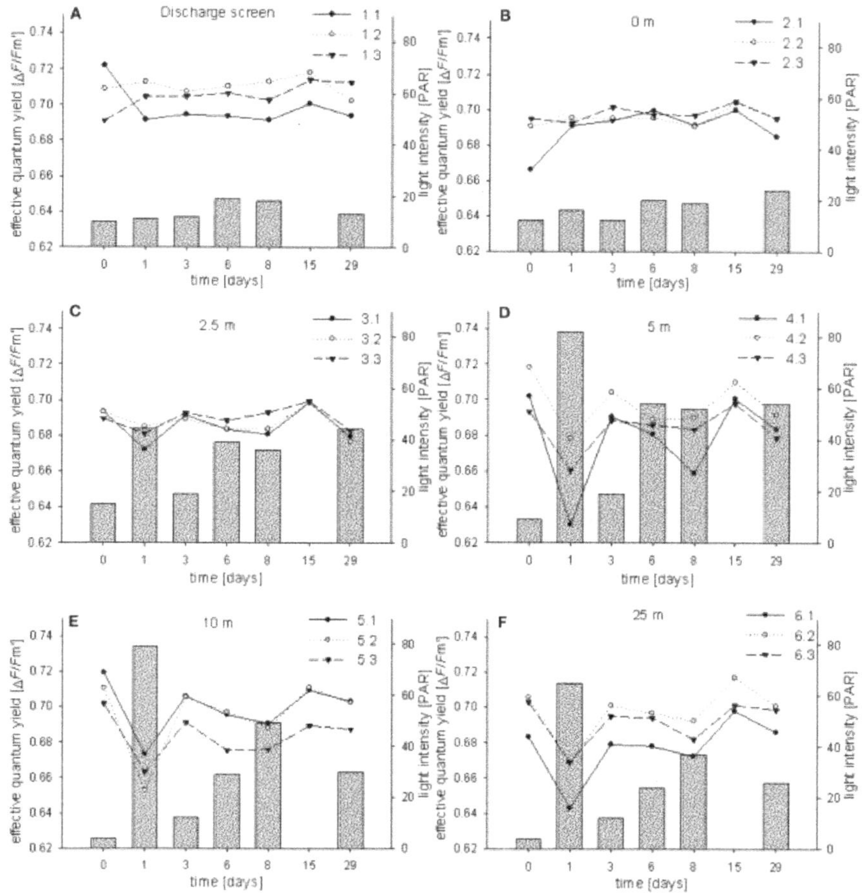

Figure 4: Effective PSII quantum yields ($\Delta F/Fm'$) of individual *Fungia granulosa*specimens during the 29-day transplantation experiment at each station (A–F). Bars show photosynthetic active radiation (PAR).

Coral Symbiont Salinity Tolerance

To confirm our *in vivo* observations, i.e., the absence of measurable detrimental effects of high salinity on the (photo) physiology of *F. granulosa*, we investigated the response of cultured coral symbionts, i.e., *Symbiodinium microadriaticum*, to a range of salinity levels. Chlorophyll a levels and growth rate of *S. microadriaticum* were investigated at salinities between 25 and 55 PSU and samples were taken at 7 time points (0, 1, 2, 3, 5, and 7 d). FACS cell counts and chlorophyll a level measurements showed that *Symbiodinium* cells reached the exponential growth phase after 4 days of incubation at salinities between 30 and 50 PSU (Figure 5). The quickest cell proliferation was observed at 35 PSU. For concentration levels of 25 and 55 PSU, respectively, the cell growth rates were inhibited. Measuring chlorophyll levels by FACS showed no differences (data not shown).

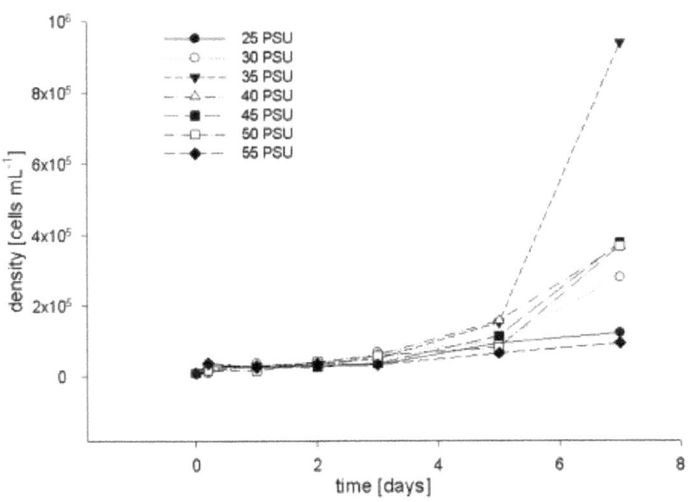

Figure 5: Coral symbiont cell growth determined by FACS after incubation in salt adjusted F/2 media at salinity levels ranging from 25–55 PSU (incubated at 28°C, under a light intensity of 9 PAR).

DISCUSSION

In this study we (1) characterized physicochemical conditions along a 25 m transect at a SWRO facility discharge structure in the Red Sea. We measured (2) photosynthetic characteristics of *F. granulosa* as a response to a sudden and strong increase and continued elevated levels in salinity. We also checked for (3) indications of bleaching for the coral specimens exposed to the highest salinity levels directly at the discharge screen. Last, we (4) exposed *Symbiodinium* cultures to low and high levels of salinity to test salinity tolerance of the coral symbiont.

With regard to the physicochemical conditions, we measured a decrease in salinity and temperature with increasing distance from the discharge structure. Of note, we already measured distinct lower values at stations 2 and 3 (seafloor, 0 and 2.5 m) compared to station 1 (discharge screen). This indicates a quick natural mixing of the brine in the study area to the point of showing little to no discernible salinity abnormalities (within short distances). These findings are in line with the literature (Roberts et al., 2010). Light levels were dependent on water clarity and measured light levels were in a similar range to what we measured at the collection site (Fsar reef, 30–50 PAR). Stations 1–3 were shaded by the discharge structure and showed lower and more stable PAR levels. These light patterns were also reflected in our $\Delta F/Fm'$ measurements, i.e., the photosynthetic characteristics of *F. granulosa* in response to changes in salinity (Figure 6). Measurements especially at station 1 and 2 (but also station 3) were more stable; in contrast stations 4–6 displayed stronger variations with a drop after 1 day (T1). This drop corresponds to noticeable clearer water and accordingly higher PAR values for the unshaded stations 4–6. All results are on the order of the "baseline" yields measured under natural conditions considering variations in light intensity and suggest no discernable effect on PSII level of the dinoflagellate symbionts of the corals. Average $\Delta F/Fm'$ measures recorded at station 1 were higher than at the other stations (0.705 compared to 0.687–0.694). This again corresponds with lower average PAR values (14 compared to 17–45), but also underlines that the significantly higher salinity at station 1 does not negatively impact the photosynthetic efficiency of the coral's symbionts. However, effective quantum yield measurements, reflecting (chronic) photoinhibition, may have limitations. Measurements from bleached corals may result in apparent healthy PAM yields, rather depending on the physiological state of associated *Symbiodinium* and less on the symbiont density or total number of cells (Fitt et al., 2001). Bleaching may exhibit a loss of zooxanthellae and/or pigments and can indicate a breakdown in the essential symbiotic relationship between coral host and algal symbiont (Brown, 1997). Taking this into account, we visually inspected the corals during each sampling dive. We could not detect any apparent colony changes or loss in coloration (i.e., pigment loss) and thus exclude bleaching as a consequence of increased salinity during the duration of the experiment.

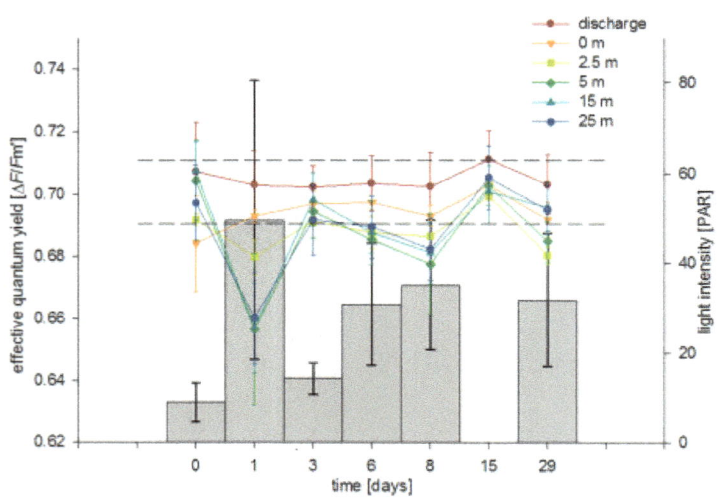

Figure 6: Average effective PSII quantum yields ($\Delta F/Fm'$) from corals at each station over a 29-day exposure period. Bars show average PAR values for each sampling event. Error bars represent standard deviation, dashed lines indicate $\Delta F/Fm'$ range from control coral colonies measured in their natural habitat.

Our results demonstrate a high salinity tolerance of the Red Sea coral *F. granulosa*. In contrast, other studies observed substantial loss of pigmentation and/or symbionts at considerably smaller salinity changes. For example, Ferrier-Pages et al. (1999) used a Red Sea clone of *Stylophora pistillata* that has been maintained in aquaria for several months. At moderately increased salinity levels (+2 PSU), coral colonies showed significant effects on photosynthesis, respiration, and protein content. Interestingly, the increased salinity of 40 PSU corresponds with the salinity in *S. pistillata's* natural habitat. In contrast, Lirman and Manzello (2009) found notable tolerance to salinity in *Siderastrea radians*, collected from Biscayne Bay, Florida where the authors measured highly variable levels of salinity. The authors also reported on *Porites furcata* (collected from the same bay) to be highly salinity tolerant (Manzello and Lirman, 2003). In a study from Chartrand et al. (2009), *S. radians* was collected from several different sites with different levels of natural salinity variations. The authors found a correlative trend of local salinity regimes and hyposaline stress tolerance levels. Colonies originating from less stable environments showed higher photosynthetic efficiency in hyposaline treatments than corals from more stable surroundings. These observations are in line with a recently conducted study, where Barshis et al. (2013) found "front-loading" of genes to confer higher temperature tolerance levels in corals exposed to temperature-variable environments in comparison to cooler, stable environments.

So far, literature mostly assumed corals to be stenohaline osmoconformers (Hoegh-Guldberg and Smith, 1989; Ferrier-Pages et al., 1999; Kerswell and Jones, 2003). In contrast, Chartrand et al. (2009) stated that a threshold response is indicative of the coral maintaining and successfully regulating its internal osmotic balance, which would contradict corals to be stenohaline osmoconformers. This is supported by a broad range of salinity tolerance levels found in different coral species depending on their original environment, which also indicates differently effective osmotic regulation. Furthermore, Mayfield and Gates (2007) discuss potential mechanisms involved in corals maintaining their osmotic balance. Accordingly, osmoregulatory processes presumably play a role in the performance of *F. granulosa* in this experiment. Osmoregulatory processes might be reflected by an initial reaction to the sudden salinity increase. This period presumably happened in our experiment between T0 and T1 and could be addressed by short-term studies. Since ambient salinity levels in the Red Sea are higher than in most other oceans (Douabul and Haddad, 1970) salinity tolerance in Red Sea corals might generally be higher than for corals in other oceans.

Photophysiological resilience of *F. granulosa* toward high salinity is supported by our*Symbiodinium* culture study. Inhibited cell growth at extreme changes in salinity (i.e., 25 and 55 PSU) might be caused by limits in the cellular salinity regulation capabilities of *Symbiodinium*cells, namely Na^+ pumps (Na^+-ATPase) (Goiran et al., 1997). Similar to plants, high NaCl exposure levels could lead to degradation of chlorophyll a as observed in sunflower leaves (Santos, 2004). Continuously lowered levels of active chlorophyll might result in a reduced energy uptake via photosynthesis and therefore, also directly influence cell growth. However, salinity levels up to 50 PSU (at par to highest measured *in situ* salinity levels) did not seem to inhibit cell growth in coral symbiont cultures. This demonstrates a wide plasticity (30–50 PSU) of *Symbiodinium* in regard to salinity changes, which has been found for hyposaline treatments before (Chartrand et al., 2009). This resilience furthermore indicates that the algal symbionts may generally not be determining the acclimation potential of the coral holobiont toward salinity changes.

In conclusion, we found a quick mixing of discharged brine with surrounding waters based on salinity and temperature measurements. The photophysiology of the coral *F. granulosa* exposed to the discharge environment along a 25 m transect was not influenced by rapid and prolonged changes in salinity (but varied according to changes in light conditions). Our data characterize *F. granulosa* coral holobionts to be remarkably resilient toward increased salinity levels, which are potentially brought about by acclimation to increased salinity levels in the Red Sea environment. Additionally, we showed that cell cultures of *Symbiodinium* only displayed inhibited cell growth at very high and low salinity levels. Based on our data we suggest *F. granulosa* from the Red Sea to possess a great acclimation potential to salinity changes, also in regard to future ocean scenarios.

CONFLICT OF INTEREST STATEMENT

The authors declare that the research was conducted in the absence of any commercial or financial relationships that could be construed as a potential conflict of interest.

ACKNOWLEDGMENTS

We would like to thank CMOR for assistance and support in field operations. Research in this study was supported by King Abdullah University of Science and Technology (KAUST).

REFERENCES

1. Barshis, D. J., Ladner, J. T., Oliver, T. A., Seneca, F. O., Traylor-Knowles, N., and Palumbi, S. R. (2013). Genomic basis for coral resilience to climate change. *Proc. Natl. Acad. Sci. U.S.A.* 110, 1387–1392. doi: 10.1073/pnas.1210224110
2. Brown, B. E. (1997). Coral bleaching: causes and consequences. *Coral Reefs* 16, S129–S138. doi: 10.1007/s003380050249
3. Chartrand, K., Durako, M., and Blum, J. (2009). Effect of hyposalinity on the photophysiology of *Siderastrea radians*. *Mar. Biol.* 156, 1691–1702. doi: 10.1007/s00227-009-1204-3
4. Douabul, A., and Haddad, A. M. (1970). *The Red Sea and Yemen's Red Sea Environments*. Hassell and Assoc., AMSAT and UNOPS.
5. Elimelech, M., and Phillip, W. A. (2011). The future of seawater desalination: energy, technology, and the environment. *Science*333, 712–717. doi: 10.1126/science.1200488
6. Ferrier-Pages, C., Gattuso, J.-P., and Jaubert, J. (1999). Effect of small variations in salinity on the rates of photosynthesis and respiration of the zooxanthellate coral *Stylophora pistillata*. *Mar. Ecol. Prog. Ser.* 181, 309–314. doi: 10.3354/meps181309
7. Fitt, W., Brown, B., Warner, M., and Dunne, R. (2001). Coral bleaching: interpretation of thermal tolerance limits and thermal thresholds in tropical corals. *Coral Reefs* 20, 51–65. doi: 10.1007/s003380100146
8. Fritzmann, C., Löwenberg, J., Wintgens, T., and Melin, T. (2007). State-of-the-art of reverse osmosis desalination. *Desalination*216, 1–76. doi: 10.1016/j.desal.2006.12.009
9. Genty, B., Briantais, J.-M., and Baker, N. R. (1989). The relationship between the quantum yield of photosynthetic electron transport and quenching of chlorophyll fluorescence. *Biochim. et Biophys. Acta Gen. Subj.* 990, 87–92. doi: 10.1016/S0304-4165(89)80016-9
10. Goiran, C., Allemand, D., and Galgani, I. (1997). Transient Na^+ stress in symbiotic dinoflagellates after isolation from coral-host cells and subsequent immersion in seawater. *Mar. Biol.* 129, 581–589. doi: 10.1007/s002270050199
11. Guillard, R. R. L., and Ryther, J. H. (1962). Studies of marine planktonic diatoms: I. *Cyclotella nana* Hustedt, and *Detonula confervacea* (cleve) Gran. *Can. J. Microbiol.* 8, 229–239. doi: 10.1139/m62-029

12. Hoegh-Guldberg, O., and Smith, G. J. (1989). The effect of sudden changes in temperature, light and salinity on the population density and export of zooxanthellae from the reef corals *Stylophora pistillata* Esper and *Seriatopora hystrix* Dana. *J. Exp. Mar. Biol. Ecol.* 129, 279–303. doi: 10.1016/0022-0981(89)90109-3

13. Kerswell, A. P., and Jones, R. J. (2003). Effects of hypo-osmosis on the coral *Stylophora pistillata*: nature and cause of 'low-salinity bleaching'. *Mar. Ecol. Prog. Ser.* 253, 145–154. doi: 10.3354/meps253145

14. Lattemann, S., and Höpner, T. (2008). Environmental impact and impact assessment of seawater desalination. *Desalination*220, 1–15. doi: 10.1016/j.desal.2007.03.009

15. Lirman, D., and Manzello, D. (2009). Patterns of resistance and resilience of the stress-tolerant coral *Siderastrea radians*(Pallas) to sub-optimal salinity and sediment burial. *J. Exp. Mar. Biol. Ecol.* 369, 72–77. doi: 10.1016/j.jembe.2008.10.024

16. Mabrook, B. (1994). Environmental impact of waste brine disposal of desalination plants, Red Sea, Egypt. *Desalination* 97, 453–465. doi: 10.1016/0011-9164(94)00108-1

17. Manzello, D., and Lirman, D. (2003). The photosynthetic resilience of *Porites furcata* to salinity disturbance. *Coral Reefs* 22, 537–540. doi: 10.1007/s00338-003-0327-0

18. Mayfield, A. B., and Gates, R. D. (2007). Osmoregulation in anthozoan–dinoflagellate symbiosis. *Comp. Biochem. Physiol. A: Mol. Integr. Physiol.* 147, 1–10. doi: 10.1016/j.cbpa.2006.12.042

19. Moberg, F., and Folke, C. (1999). Ecological goods and services of coral reef ecosystems. *Ecol. Econ.* 29, 215–233. doi: 10.1016/S0921-8009(99)00009-9

20. Moberg, F., Nyström, M., Kautsky, N., Tedengren, M., and Jarayabhand, P. (1997). Effects of reduced salinity on the rates of photosynthesis and respiration in the hermatypic corals *Porites lutea* and *Pocillopora damicornis*. *Mar. Ecol. Prog. Ser.* 157, 53–59. doi: 10.3354/meps157053

21. Muscatine, L. (1967). Glycerol excretion by symbiotic algae from corals and tridacna and its control by the host. *Science* 156, 516–519. doi: 10.1126/science.156.3774.516

22. Muthiga, N. A., and Szmant, A. M. (1987). The effects of salinity stress on the rates of aerobic respiration and photosynthesis in the hermatypic coral *Siderastrea siderea*. *Biol. Bull.* 173, 539–551. doi: 10.2307/1541699

23. Roberts, D. A., Johnston, E. L., and Knott, N. A. (2010). Impacts of desalination plant discharges on the marine environment: a critical review of published studies. *Water Res.* 44, 5117–5128. doi: 10.1016/j.watres.2010.04.036

24. Santos, C. V. (2004). Regulation of chlorophyll biosynthesis and degradation by salt stress in sunflower leaves. *Sci. Hortic.* 103, 93–99. doi: 10.1016/j.scienta.2004.04.009

25. van der Merwe, R., Hammes, F., Lattemann, S., and Amy, G. (2014). Flow cytometric assessment of microbial abundance in the near-field area of seawater reverse osmosis concentrate discharge. *Desalination* 343, 208–216. doi: 10.1016/j.desal.2014.01.017

26. Voutchkov, N. (2009). Salinity tolerance evaluation methodology for desalination plant discharge. *Desalination Water Treat.* 1, 68–74. doi: 10.5004/dwt.2009.126

CHAPTER 18

DRIVERS AND ECONOMIC ASPECTS FOR THE IMPLEMENTATION OF ADVANCED WASTEWATER TREATMENT AND WATER REUSE IN A PVC PLANT

David Prieto, Nathalie Swinnen, Laura Blanco, Daphne Hermosilla ,
Peter Cauwenberg, Ángeles Blanco, Carlos Negro

Keywords: Water reuse, Polyvinyl alcohol, PVC effluent, Membrane bioreactor, Reverse osmosis, Economic evaluation

INTRODUCTION

Water stress is nowadays a major risk for industry and it is expected to aggravate within the next decades due to population growth coupled with industrialization and urbanization. If an integral sustainable water management is not assured, the increasing demand for water will have serious consequences on the environment. As the resource is becoming scarce, tensions among urban, industrial and agricultural stakeholders will intensify and, in periods of severe droughts, industry may lose the right to use water with serious effects on the competitiveness of companies in water stressed regions such as Spain [1]. To contain this risk, the use of alternative water sources becomes an indispensable element in industrial water management.

Under this scenario, industries need to become more independent of the supply of fresh water for their production processes. For the last four years the EU FP7 project E4Water has promoted methods for sustainable water use in the chemical sector by demonstrating the technical and economic feasibility of advanced wastewater treatment and reuse in different plants.

Polyvinyl chloride (PVC) is the third most important polymer, slightly behind polyethylene and polypropylene. It is used in most industrial sectors (e.g. packaging, automotive, building, agriculture, medical care) and main applications include pipes, flooring, window and door frames, as well as electric cables. The production in Europe amounts to around 5 million tons [2] and the demand is increasing. Within the three different processes used in the manufacture of PVC, suspension, emulsion and bulk polymerization, the suspension process is the most applied one for large-scale productions (>80% of total production).

PVC is produced in batch, by polymerization of vinyl chloride monomer (VCM) accompanied by catalyst at a certain temperature and pressure in aqueous medium. Poly(vinyl alcohol) (PVA) is used as dispersing agent during the polymerization step. VCM is produced by thermal cracking of ethylene dichloride (EDC). The chlorine used in the manufacture of EDC is derived electrolytically from NaCl by the chlor-alkali process. Finally, PVC particles are separated by centrifugation [3].

The average water consumption required to produce the polymer is 3 m^3/t PVC according to the BREF [3]. VCM contaminated water (e.g. water used for the cleaning of

---

Citation: David Prieto, Nathalie Swinnen, Laura Blanco, Daphne Hermosilla, Peter Cauwenberg, Ángeles Blanco, Carlos Negro, Drivers and economic aspects for the implementation of advanced wastewater treatment and water reuse in a PVC plant, Water Resources and Industry, Volume 14, June 2016, Pages 26-30, ISSN 2212-3717, http://dx.doi.org/10.1016/j.wri.2016.03.004.

reactors containing VCM, transfer lines and suspension or latex stock tanks), pass through a water stripper to remove the VCM which is recycled while the water is sent to the waste water treatment plant (WWTP) as well as the final effluent containing residual PVC particles and PVA. This effluent is characterized by being slightly alkaline (ammonia), having low chemical oxygen demand (COD), high PVA/COD ratio, and containing both aluminum and solids made of fine PVC particles, as shown in Section 2.1. The WWTPs of PVC plants are usually based on a two-step process comprising flocculation and removal of suspended solids by sedimentation or flotation. In some plants, the treated water is reused for rinsing purposes, although the water demand for rinsing is small compared to the water required for the polymerization of the PVC [3]. Thus water consumption can be reduced, but more than 100 m^3/h are still discharged after the physico-chemical treatment in a plant producing 35 t of PVC/h. Moving towards a more sustainable water use in the PVC industry and reducing its dependence on external water sources requires a further closure of the water circuit, which itself implies the need to implement additional wastewater treatment processes integrated in the current WWTP to allow the reuse of the final effluent. The water generated through the upgraded treatment plant shall be reused as process water in the polymerization process step that takes 30% of the plant water consumption. However, a very high quality demineralized water is required for this reuse option: polyvinyl alcohol (PVA)<1 mg L^{-1}; NH_4^+-N<2 mg L^{-1}; biochemical oxygen demand (BOD_5)~0; electrical conductivity (EC)<10 µS cm^{-1}; total organic carbon (TOC)<10 mg L^{-1} and aluminum (Al)<0.1 mg L^{-1}. Hence, the new treatment must remove the residual PVA and other COD, ammonia, salinity and aluminum.

PVA has been found to interfere in membrane based separation processes such as microfiltration (MF) or ultrafiltration (UF) producing fouling and foaming [4]. Moreover PVA is difficult to remove by microorganisms due to its polymeric structure. Because the biodegradability of these effluents with BOD_5/COD values below 0.01 is very poor, this compound is frequently removed chemically instead of biologically [5]. These chemical processes, however, entail high costs, result in large amounts of solid waste and produce final effluents of poor quality [6]. In previous studies, Blanco et al. have demonstrated that the residual PVA can be successfully removed by aerobic biological treatment if the adaptation and treatment conditions are adequate [7]. It has also been demonstrated that under anoxic-aerobic conditions nitrogen could be efficiently removed up to 80%. This is important since ammonium diffuses through reverse osmosis (RO) membranes deteriorating the permeate quality below the final quality requirements and thus necessitating additional polishing which might question the economic feasibility of the full treatment. Finally, own laboratory experiments have shown that the quality of the effluent from the combination of membrane bioreactor (MBR) with RO technology meets the requirements for water reuse in the PVC polymerization steps.

MBR is a well-established technology for treating industrial effluents. It couples the activated sludge process at a higher sludge concentration with membrane separation (MF or UF) and produces a permeate free of particles and almost disinfected, which can feed the RO directly. Moreover, MBRs have a small footprint, produce less sludge, and achieve superior effluent quality compared to conventional activated sludge, due to the longer sludge retention time, that allows the direct reuse of the treated water for a range of applications [8].

After the technical validation of the proposed solution to treat the PVC plant effluent at pilot scale, the next step for the implementation of this sustainable water alternative at industrial scale is the evaluation of the economic aspects of the alternative studied.

The present study had the objectives (a) to estimate the costs of the industrial water treatment plant based on the aerobic MBR/RO treatments validated at pilot scale and (b) to analyse the potential drivers for the implementation of this solution at industrial scale in different PVC plants.

MATERIALS AND METHODS

PVC plant

A PVC plant operated by INOVYN Spain in Martorell with a capacity of 290 kt/y has been selected for this study for three main reasons:

1) The plant is located in the Llobregat basin near Barcelona. This is a region where water scarcity is a pressing issue. Both fresh water abstraction and wastewater disposal are regulated by legal permits that impose very stringent limitations. Currently, the authorities are increasing the restrictions on water abstraction from existing wells, what sometimes forces the industry to operate with drinking water, which increases the costs and creates tensions between the stakeholders. Furthermore this water source is limited in cases of severe droughts.
2) The plant produces PVC in suspension with a high share of recycled water and therefore with a low fresh water consumption of 2 m^3/t of PVC. Compared to other plants this is achieved by a water recovery system that reuses the water extracted from centrifugal decanters (first PVC drying step before fluidized bed drying) for equipment rinsing purposes.
3) Due to increasing demand of PVC products, an expansion of the production capacity of the plant would be required leading to a higher water demand. Due to the above limitations an alternative water source is the effluent of the plant after an advanced treatment. In this way the production capacity could be expanded without increasing the water demand.

Within INOVYN Spain site, the main uses of water are: 60% for the mercury electrolysis, where chlorine is obtained; 10% for the monomer plant, where VCM is synthetized, and 30% for the polymerization process.

For the PVC process unit studied, the current WWTP includes a physico-chemical process, where PVC colloids are removed from the water. Afterwards, the clarified water is disposed of in the sea through a marine outfall.

The effluent has a pH of 7.9 and the quality is (average values in mg L^{-1}): 38 of NH_3, 0.97 of Al, 211 of Na^+, 0.08 of VCM, 285 of Cl^-, 0.28 of Cl_2, 117 of suspended solids, and 331 of COD.

As mentioned above, the treatment objective for the pilot plant is to produce water from the current effluent meeting the requirements for the polymerization process, i.e. conductivity below 10 μS/cm. To achieve this quality, a RO treatment is required with a pre-treatment by MBR to eliminate PVA, COD and ammonia as previously validated at laboratory scale [7].

Pilot Plant

The flow diagram of the pilot plant is given in Fig. 1. The MBR pilot was equipped with a ZeeWeed 500D system from Zenon (GE, Oroszlány, Hungary) with an outside/in UF hollow fiber membrane (PVDF with a nominal pore size of 0.04 μm). The plant consists of two biological reactors (Bioreactor 1 with the possibility of working in both aerobic or anoxic conditions and Bioreactor 2, fully aerobic)and the filtration unit,with a total effective volume of 20 m³. Three membrane modules with an effective filtration area of 94.8 m² (31.6 m² each) were used for permeate production. In addition, a cleaning in place (CIP) system and a heat exchanger for temperature control were installed. For the pilot phase an aero cooler was used. Even though the feed flow comes from the physico-chemical treatment, a safety filter (Y-strainer of 5 mm) was installed for the pilot trial. To start up, activated sludge was obtained from a mixture of 25% municipal biological treatment sludge and 75% sludge from the internal WWTP in the VCM process unit.

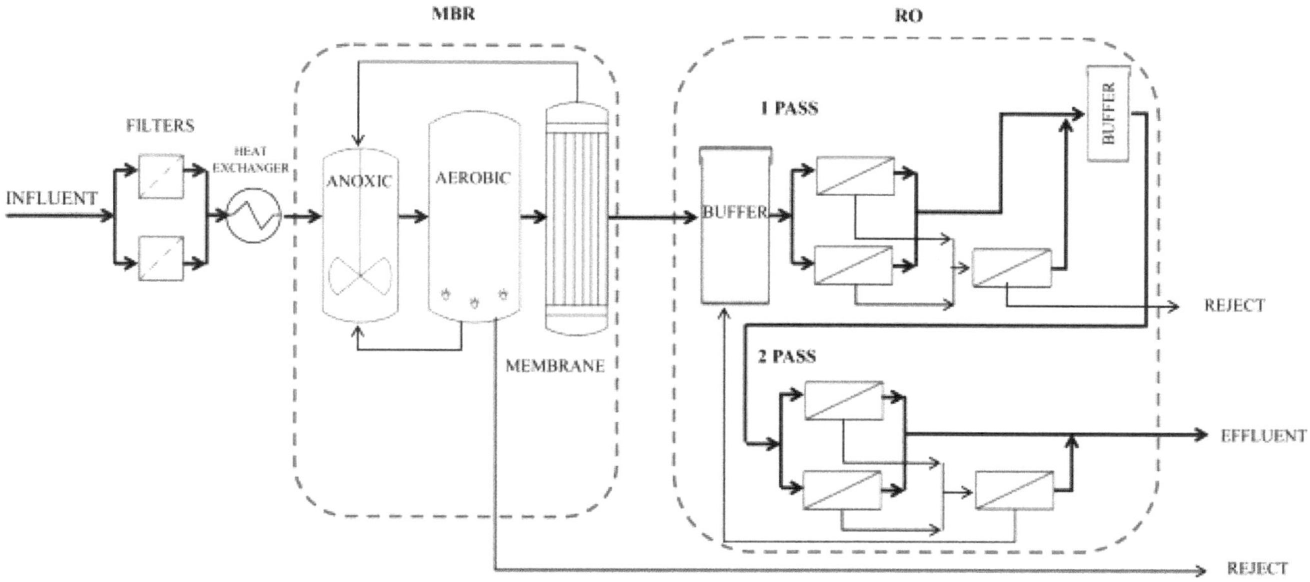

Figure 1: Pilot plant.

Downstream of the MBR, a storage tank was installed to collect the MBR permeate for the backpulse cycles. In this backwashing, the MBR permeate is pumped back through the membranes to remove reversible fouling and avoid a fast decrease in permeability. The cyclic mode of operation consisted of alternated filtration and backpulsing periods. CIP was performed once per day (30 min) during the trial. Cleaning consisted of successive cycles of backpulsing, chemical reagent addition (citric acid and sodium hypochlorite), relaxation with aeration and final backpulsing for chemicals removal. The main purpose was to keep high permeability values and reduce harsh chemical cleanings or recovery cleanings which means longer MBR stops and fast membrane aging. After acid CIP cleanings, effluent was directly used for feeding the RO treatment; when sodium hypochlorite was used, stream was discharged until free chloride was under detection limit, thus membranes would not be damaged.

The MBR permeate was treated in a two-pass RO system supplied by Nalco. The FILMTEC™ LC HR-4040 spiral wound membranes were supplied by Dow Chemical. The first pass treatment consisted of three pressure vessels with five RO elements each one, with a total area of 130.99 m². Concentrated streams from the first two pressure vessels feed the third pressure vessel. The permeate was used for backflushing and for feeding the second pass treatment with three pressure vessels of three RO elements each one, with a total area of 78.59 m². Concentrated streams from the first two vessels fed the third pressure vessel. Part of the concentrated stream was recirculated to feed both, first and second pass treatments, in order to increase the overall recovery (Fig. 1). The permeate was analysed before it was reused in the polymerization trials (pH, conductivity, TOC, COD, PVA, Al).

Different phases were tested in the RO pilot: recoveries in first and second pass were set in the range of 60–75% and 75–85%, respectively, concentrate stream from the second pass was totally recirculated to both first and second pass treatments as well as part of the first pass reject. Permeate flow and flux were set in the range of 1.6–1.7 m³ h⁻¹ and 12–21 LMH, respectively (Fig. 1). Final reject stream that was not recirculated was directly discharged and mixed with the current effluent of the plant. As flow and Trans Membrane Pressure (TMP) kept very stable during the piloting, no chemical cleanings were done; however, biocide shocks (PC-11, from Nalco) were occasionally performed especially when MBR stopped for maintenance/problems to keep RO membranes in good condition and to avoid membrane biofouling.

All chemicals and nutrients used were supplied by MERCK KGaA (Darmstadt, Germany) and PANREAC S.A. (Barcelona, Spain). Sample preservation and all analyses were performed according to the standard methods for the examination of water and wastewaters [9], [10] and [11].

Feasibility study

An economic analysis of the new industrial wastewater treatment plant was performed using net present value (NPV) and discounted payback period (DPP) calculations to account for the time value of money. For that, after finishing the pilot studies, a computational simulator using Excel software was built to make a full scale-up extrapolation in terms of raw materials and utilities consumption, i.e., to determine variable costs (Fig. 2 and Table 2). The main values considered for this analysis are: consumer price index of 2% (European Central Bank objective), depreciation time of 15 years (suggested by AEMA MBR constructor) using linear depreciation method for the DPP estimation, energy cost between 70 and 80 €/MWh (Inovyn Spain average value for 2015), fresh demineralized water production cost between 1 and 2 €/m^3, 35–55 €/h for person-months and a possible technical production of 290 kt/year.

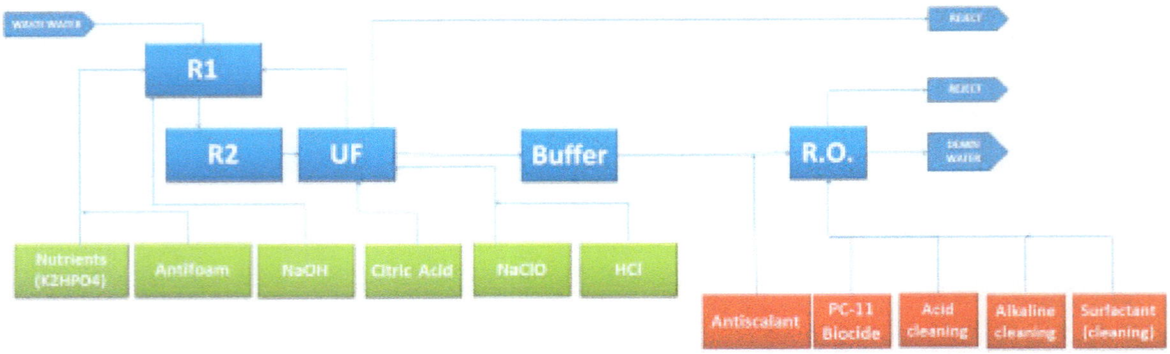

Figure 2: MBR+RO simulator to estimate chemical consumption in the industrial plant.

RESULTS AND DISCUSSION

Pilot Trials

In the MBR, for the design of a treatment with denitrification and nitrification processes, using a single biological tank with separated anoxic plus aerobic zones, the hydraulic retention time should be at least 6 h and solids retention time in the range of 15–20 days[12]. Mixed liquor suspended solids (MLSS) content in the pilot was in the range of 8–10 mg L^{-1}, the optimal flux was in the range of 22–26 LMH (during the piloting, flux was increased from 10 to 26 LMH) and recirculation rate was 4–6. The pilot plant was operated for 15 months. Samples were taken for lab analysis 3–4 times per week. Once it was optimized the quality of the final effluent, after the MBR and two pass RO, was high enough to enable the use of water in the PVC polymerization step, main average values are summarize in Table 1[11] This reclaimed water was tested in real polymerization trials and it was confirmed that it can be successfully reused in the polymerization plant without affecting the process or the final PVC quality.

Table 1: Effluent quality in the pilot plant.

PVA	0.73±1.4 mgL^{-1}
BOD_5	2.5±1.2 mgL^{-1}
Conductivity	3.8±1.5 µScm^{-1}
TOC	0.94±0.64 mgL^{-1}
NH_4^+	1.2±1.7 mgL^{-1}

An acclimatization period of about 1 month was needed in the MBR for achieving an almost total PVA removal, starting with a non-adapted activated sludge culture. Once the MBR was running stably, removal of PVA and BOD_5 was close to 100%, and removal of COD and TOC was in the range of 85–95% and 85–90%, respectively. CIP cleanings will be necessary for daily/weekly basic/acid cleaning and pH control by continuous caustic soda addition and phosphorus dosage for assuring a good biological treatment and foaming control will be also required, with occasional antifoam addition to avoid solid losses as well as operational problems in pipes and equipment. At stable conditions and with adequate pH and dissolved oxygen control, ammonia was completely nitrified to nitrate, matching the quality required for the MBR effluent. Details of the pilot trials are given by Blanco el al. [11].

The operation of the hollow fiber UF was very stable during the piloting. Most of the membrane fouling was reversible as permeability was very well restored after recovery cleanings. Daily CIP seemed to avoid irreversible fouling on the membrane surface.

Recirculation ratio (recirculation/MBR effluent) had a significant influence on the permeability due to the high quantity of solids accumulated in the membranes during the period when recirculation ratio was 3. This means that higher values will have to be used in real installations to avoid irreversible fouling in the membranes and reduce the need for frequent chemical cleanings.

As mentioned before, two pass RO treatment was needed for obtaining permeate with conductivity lower than 10 µS cm^{-1}. Moreover, RO permeate quality was within specifications independently of the quality of the MBR effluent. Due to the high quality of the MBR permeate and the use of high salt rejection polyamide membranes, RO showed a very good performance during the piloting, with no significant transmembrane pressure drop values (ranging from 6.5 to 7.9 bar when recovery was increased) [11].

Table 2 shows the average chemical consumptions of the industrial plant (~100 m^3/h) estimated based on the pilot plant data.

Table 2: Specific consumptions taken from the pilot data and estimated in the simulation of the industrial plant.

	Nutri-ents	Anti-foam	NaOH	Citric acid	Na-ClO	HCl	Antis-calant	Bio-cide	Acid clean-ing	Alka-line clean-ing	Surfactant
Spec. Cons. (mg/L)	3.1	8.3	167	3.9	8.8	0.8	5	25	0.4	0.6	0.3

Regarding the RO, two pass plus two stages are needed to achieve the targeted water quality for polymerization. Antiscalant and biocide additions were required as shown inTable 2. The recovery rates were kept in the range of 60–75% to reach the foreseen membrane lifetime and minimize cleaning requirements. Concentrates will be discharged because they are below the current limits.

Economic feasibility study

The economic feasibility was evaluated on the basis of CAPEX, OPEX and DPP.

CAPEX

Based on the offers received from different suppliers, the capital expenditure for a turnkey project for the requested specifications amounts to 1.5 M€, plus 1 M€ of erection and installation costs in terms of connection to the site, utilities supply and site preparation. Within the turnkey scope, civil works, bioreactors, membranes and all pumping and instrumentation equipment are included.

OPEX

Operational costs include electrical energy, personnel costs (one lab technician and one maintenance technician), exploitation costs (based on the chemical consumption shown in Table 2) and plant consumables reposition. According to the performed simulations, operational costs are about 0.455 M€/year, where fixed costs represent 34% of the total amount (0.153 M€/year) and variable costs 66% (0.292 M€/year). In terms of specific operational expenditure, cost is 0.69 €/m³, with 0.24 €/m³ for fixed cost and 0.45 €/m³ for variable costs.

Discounted payback period

DPP calculation has been done taking into account following constructive and financial estimations: erection period time of 1 year, depreciation time of 15 years and inflation of 2% starting from 2015.

Price of demineralized water has been estimated between 1 €/m³ and 2 €/m³ for the year 2015. Table 3 shows that, according to stated estimations and based on different offers, DPP is 5 years, i.e., the year where NPV becomes positive, with an initial investment of 2.5 M€.

Table 3: NPV (k€) estimation based on INOVYN Spain data reference.

	2015	2016	2017	2018	2019	
INVESTMENT	2550	0	0	0	0	
COSTS	0	463	472	481	491	
DEPRECIATION	0	173	177	180	184	
SAVINGS	0	1266	1291	1317	1344	
NPV	**−2550**	**−1920**	**−1278**	**−622**	**46**	

With the aim of reducing investment costs, i.e., CAPEX, a model based simulation was performed for different treated flowrates and DPP was estimated (Table 4). The data gathered in Table 3, other estimated parameters such as fresh demineralized water cost, electrical energy cost, personnel cost and maintenance cost were kept constant.

Table 4: Project profitability based on modular CAPEX investment.

Treated flowrate (m³/h)	DPP (years)
5	Not profitable
50	15
75	7
100	5

Afterwards, another simulation was performed to make a DPP value comparison depending on fresh demineralized water production cost (Table 5).

Table 5: Project profitability based on fresh demineralized water production price.

Fresh demineralized water price (€/m³)	DPP (years)
0.5	Not profitable
1	23
1.5	7
2	4
2.5	4

Nevertheless, this technology cannot be considered only as a project where a quick payback is sought, but also as a strategic one. Table 6 shows the time to recover the investment in case of production reduction due to water use restrictions resulting from water scarcity.

Table 6: Payback time in months compared with production reduction forced by a drought.

Production restriction (%)	Investment recovery (months)
36	1
25	2
15	3
10	4
5	8
3	13

If a drought period comes, water scarcity inflicts large damages at the business in terms of product and customers loss. Hence, looking to Table 6, it can be seen that a reduction in the production of 36% forced by a drought may give the investment back in 1 month. Even for a production reduction of 3%, the payback time would be one year approximately.

CONCLUSIONS

Based on a previous pilot plant study that confirms that water reclamation from PVC effluents is possible through coupling a MBR and a double pass RO process, the economic aspects for the implementation of the advanced treatment and water reuse in a PVC plant has been assessed.

Regarding the economic figures and based on INOVYN Spain CAPEX standards, the feasibility of the treatment train is not clear from a sole savings point of view, as the NPV is 5 years. Besides, the investment required is rather high, which does not match with free cash flow philosophy of the companies. Subcontracting option was not considered in this study, but it can be taken as an option if subcontractor offers recycled water at a price lower than fresh demineralized water production cost.

If DPP is lower than depreciation time of the active, the alternative considered here can be interesting from an economic point of view. Hence, for sites where fresh demineralized water production cost is higher than 1.5 €/m^3, and with a treatment flow rate greater than 50 m^3/h, this recovery system is potentially profitable.

However this alternative cannot be seen only by considering the payback time because it is also a strategy for the PVC production plant to become more independent from external water supply. It has been calculated that in the scenario of a drought that causes a 3% production restriction, the investment return would only be 1 year.

ACKNOWLEDGEMENTS

The research leading to these results has received funding from the 7th Framework Program of the European Union within the E4Water Project under the Grant Agreement no.608490.

REFERENCES

1. R. Ordóñez, D. Hermosilla, I. San Pío, A. Blanco, Replacement of fresh water use by final effluent recovery in a highly optimized 100% recovered paper mill, Water Sci. Technol. 62 (2010) 1694–1703.
2. Plastics Europe, Plastics – the Facts 201. ⟨http://www.plasticseurope.org/Docu ment/plastics-the-facts-2013.aspx⟩, 2013 (accessed 02.03.16).
3. European Commission, Reference Document on Best Available Techniques in the Production of Polymers, European Commission, Brussels, 2007.
4. U. Balyana, B. Sarkara, Modeling of permeate flux decline during ultrafiltration of polyvinyl alcohol in a batch cell, Desalination Water Treat. 52 (40–42) (2014) 7495–7506.
5. H. Yu, G. Gu, L. Song, Degradation of polyvinyl alcohol in sequencing batch reactors, Environ. Technol. 17 (1996) 1261–1267.
6. Q. Yang, W. Zhang, H. Zhang, Y. Li, C. Li, Wastewater treatment by alkali bacteria and dynamics of microbial communities in two bioreactors, Bioresour. Technol. 102 (2011) 3790–3798.
7. L. Blanco, D. Hermosilla, A. Blanco, N. Swinnen,, D. Prieto, C. Negro, Assessment of the performance of membrane bioreactors applied to the treatment of industrial effluents containing poly(vinyl alcohol), Ind. Eng. Chem. Res. 54 (20) (2015) 5442–5449.
8. R. Ordóñez, D. Hermosilla, N. Merayo, A. Gascó, C. Negro, A. Blanco, Application of multi-barrier membrane filtration technologies to reclaim municipal wastewater for industrial use, Sep. Purif. Rev. 43 (2014) 263–310.
9. APHA, AWWA, WPCF, Standard methods for the examination of water and wastewater, Washington DC, 2005.
10. J.H. Finley, Spectrophotometric determination of polyvinyl alcohol in paper coatings, Anal. Chem. 33 (1961) 1925–1927.
11. L. Blanco, D. Hermosilla, A. Blanco, D. Prieto, N. Swinnen, C. Negro, MBRþRO combination for PVC production effluent reclamation in the resin polymerization step – a case study. Submitted to Ind. Eng. Chem. Res. 2016.
12. S. Judd, The MBR Book: Principles and Applications of Membra

CHAPTER 19 SOLAR UV-TREATMENT OF WATER SAMPLES FOR STRIPPING-VOLTAMMETRIC DETERMINATION OF TRACE HEAVY METALS

Gelaneh Woldemichael, Taffa Tulu, Gerd-Uwe Flechsig

Keywords: Voltammetry, Sample handling, Polarography, Environmental analysis

INTRODUCTION

Investigations of environmental pollution in Africa

Heavy metals usually exist only in trace amounts in the environment; however, they may accumulate unnoticed up to worrying levels. The contamination effects of such heavy metals on human health have been well studied in the past. One major source for heavy metal contaminations are mining activities [1].

Such industries have been flourishing in Africa since the colonial ages. Increasing efforts in environmental studies have revealed major problems [2]. The Niger River has been investigated recently, and extreme pollution by heavy metals such as zinc, copper, chromium, nickel and cadmium was confirmed [3, 4]. Also in India, heavy metal concentration in rivers are sometimes extreme with levels up to the mg/L range for cobalt, copper, nickel, zinc, and lead [5].

In spite of relatively low levels of industrial development in many African countries such as Ethiopia, there have been for decades increasing immission loads of toxic heavy metals in the various environment compartments including soil, air and water. Over the last decade, increased activities in both industry and urbanization have led to rising heavy metals inputs all over Ethiopia. In particular, the levels of arsenic, chromium, mercury, cadmium, zinc, and lead are of great concern. Awash River is one of the economically most important rivers in Ethiopia. Awash River originates from the Central Highland of Ethiopia, west of Addis Ababa near Ginchi Town. It flows in southwest direction passing the southern side of Addis Ababa City where it joins Akaki River. The Akaki River originates from Addis Ababa with many other creeks, which are rich in industrial and municipal wastes. Mojo River, another carrier of immense industrial wastes particularly from tannery, joins Awash River near Koka Dam. Awash River continues its course eastward in the Great East African Rift Valley thousands of kilometers across Afar depressions and ends in the Afar desert.

Awash is probably the most economically utilized river in Ethiopia. It provides drinking water to more than five million people living in urban areas like Adama, Wonji, and the surrounding rural populations. Some of the pastoralist populations like Afar and Kereyu are totally dependent on the

Citation: Gelaneh Woldemichael, Taffa Tulu, Gerd-Uwe Flechsig, Solar UV-treatment of water samples for stripping-voltammetric determination of trace heavy metals, Heliyon, Volume 2, Issue 3, March 2016, Article e00091, ISSN 2405-8440, http://dx.doi.org/10.1016/j.heliyon.2016.e00091.

Awash River for all of their activities. The Awash River irrigates many large-scale farmlands and plantations like Wonji, Nura Era, Merti Jeju, Metehara and a large number of small-scale farms and plantations for local population supporting millions of lives. Awash is the first river on which modern hydroelectric power was build (Koka 1, 2 and 3). Fresh water fishing from the river basin and Koka artificial lake is one of the largest fish suppliers to Addis Ababa, Adama, Bishoftu, Mojo and surrounding towns.

Environmental heavy metal analysis in developing and emerging countries

Trace metal analysis requires techniques that can detect ppb and ppt levels, while coping with sample matrix components like salts and dissolved organic matter. State of the art are therefore most sensitive methods like inductively coupled plasma optical spectrometry (ICP-OES), inductively coupled plasma mass spectrometry (ICP-MS) and graphite furnace atomic absorption spectrometry (GF-AAS) coupled with extraction and enrichment steps. These instrumental techniques are mostly not available at universities and in government laboratories in developing and emerging countries. If atomic absorption spectrometers are available, these are mostly equipped with flame atomization (F-AAS) [3], which requires enrichment of lower ppb traces by water evaporation [5, 6]. This step is usually coupled with open or pressurized nitric acidic digestion procedures [4, 7, 8], which makes the method prone to contamination and increased blank values. This way, trace concentrations in the two-digit ppb range can be determined. However, procedures like these are prone to errors, tedious, time- and energy consuming. If western laboratories are involved, samples are often transported over long distances [9]. The methods and procedures available so far do not allow nationwide environmental monitoring of heavy metal pollution in developing countries.

Voltammetric water analysis assisted by UV digestion

Despite low cost and great performance in trace level metal determination, little has been published about application of stripping voltammetry in developing and emerging countries. However, even this affordable technique requires sample pretreatment by UV oxidation [10, 11, 12]. This technique allows for complete degradation of dissolved organic matter (DOM) that otherwise would disturb voltammetric measurements by complexing metal ions and interfering surfactant properties. At the same time, UV digestion brings only minimal contamination (blank values) by using very small amounts of hydrogen peroxide and nitric acid. A review revealed that usually powerful UV lamps with up to 1000 watts have been used [10]; however, the method was successfully tested even with a lab-made UV irradiation device comprising three 20 watts low-pressure Hg lamps for wine sample pretreatment to allow F-AAS determination of iron and manganese [13].

Solar UV water treatment

The term SODIS (solar water disinfection) designates a procedure that allows disinfection of drinking water by simplest and widely available means [14, 15, 16]. UV-A transparent plastic bottles (mostly PET) are filled with filtrated water from lakes and rivers and left in the sunlight for at least 6 h [17]. During this time, UV-A will cause dissolved oxygen to form active species such as superoxide, ozone and hydroxyl radicals that will inactivate more than 99% of infectious microbes. Addition of H_2O_2 and lowering the pH accelerate the process [18]. Literature on this method has been reviewed recently [19].

During the past couple of years we have introduced a method of water-pretreatment that bases on solar UV photooxidation of complexing dissolved organic matter. This way, the most sensitive techniques of stripping voltammetry have been successfully applied to determine Zn, Cd, Pb, Cu by anodic stripping-voltammetry (ASV) [20], as well as uranium[21], Co, and Ni [22] by adsorptive stripping-voltammetry (AdSV) in river water samples of the Warnow River (Germany). At that location, a maximum solar UV-A intensity of 3.5 mW/cm^2 could be observed. Still, the effect upon complexing dissolved organic matter was remarkable, in particular for adsorptive stripping-voltammetric determination of Ni, Co, and U.

Here we present a pilot field study that aimed at testing the feasibility of the new sample pretreatment method for heavy metal monitoring in the waters of an Ethiopian River. The two objectives were:

a) Optimization of the time required to complete the digestion process of interfering dissolved organic matter (DOM) in the river water samples at locations with relatively high solar UV-A intensity close to the equator followed by systematic evaluation of the results by comparing with pre-treatment by classic UV digestion based on intense hard artificial UV-C.

b) Determination of 7 heavy metals (cadmium, cobalt, copper, lead, nickel, uranium and zinc) in different locations on Awash River, Ethiopia by applying the newly developed sample pre-treatment method.

EXPERIMENTAL

Sample collection and preparation

Water samples from Awash River were collected on March 17, 2012 in Ethiopia, Oromia Regional State, East Shewa Zone at locations given in Table 1 and on the map (Fig. 1). No specific permissions were required for our activities at these public locations, as they are freely accessible. This field study did not involve endangered or protected species. As a sub-equatorial region at an elevated altitude, significantly higher solar UV-A intensity (averaged 5.4 mW/cm^2 at the time of sampling) than in Rostock (Germany) could be used and tested in this study. The altitude at the sampling sites ranges from 1542 to 1595 meters above sea level. After filtration and acidification to pH 2 with 1 mL HNO_3 (65%, TraceSelect grade) per liter sample and addition of 100 μM H_2O_2, the filtered samples were solar UV-irradiated in thin and clean UV-A-transparent 20 × 15 cm^{2}polyethylene bags for 12 or 24 h as described below. The usual irradiation sessions were scheduled between 10 am and 4 pm due to the maximum sun light intensity. An aluminum reflector was used to increase the irradiation efficiency. All samples were then packed and shipped to the University of Rostock (Germany) for voltammetric determination of trace metals applying analytical equipment and procedures as described below [20, 21, 22]. Before and after this 2-day period of transport, the samples were stored at 5 °C. For the reference measurements with the 24 h SoUV samples, an additional classical UV digestion in quartz glass tubes was performed at Rostock University by means of a 30 watts, 254 nm low-pressure mercury lamp.

Table 1: Sampling sites, Awash River, Ethiopia.

Sampling site	GPS coordinates	Altitude (m)
Koka Dam	8°24′24.02″N, 39°01′11.66″E	1595
Sire Robi	8°28′45.82″N, 39°10′52.41″E	1553
Wonji Kuruftu	8°28′24.37″N, 39°12′47.98″E	1553
Wonji Drain	8°23′39.38″N, 39°19′16.11″E	1545
Awash Melkasa	8°24′04.91″N, 39°20′03.28″E	1542

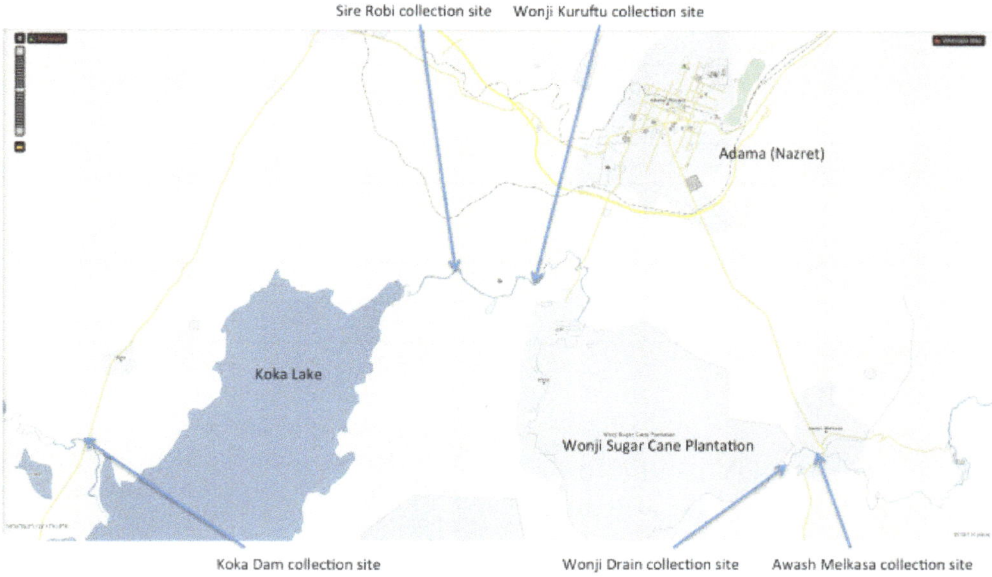

Figure 1: Map of the 5 sampling sites at the Awash River, Ethiopia, reproduced from Wikimapia with permission by license CC-BY-SA.

Materials

Chemicals were delivered by Merck and Fluka (Germany). Nitric acid (65%), acetic acid, hydrochloric acid (30%), potassium chloride, ammonium hydroxide (25%), and sodium hydroxide were all of TraceSelect analytical grade. Metal stock solutions were purchased as 1000 mg/L AAS standard solutions. All other chemicals were of analytical or reagent grade. Ultrapure water was prepared by means of an SG Water system (2005, SG Water GmbH, Germany, now Evoqua LLC; 18 MΩ*cm, TOC < 2 ppb). Amidosulfonic acid [23] and hydroxylammonium sulfate [24] at a level of 100 mg/L were added, if needed to remove any remaining nitrite after the irradiation step, as this interfering ion could be formed during UV irradiation in the presence of nitrate [25]. The following reagent and buffer stock solutions were prepared: Acetate-KCl buffer (1 M KCl containing 0.5 M equimolar sodium acetate/acetic acid), ammonia buffer (1 M NH_4Cl, 2 M NH_3), 0.1 M dimethylglyoxime (DMG), and 5 mM chloranilic acid.

Equipment

Solar UV-A intensity at sampling sites was measured by means of a UV light meter LT Lutron YK-35UV, equipped with a filter covering the range from 290 to 390 nm. Voltammetry was performed using a Metrohm 663 VA Stand equipped with Hg multi-mode-electrode, glassy carbon rod counter electrode and Ag/AgCl (3 M KCl) reference electrode. The 663 VA Stand was controlled by a µAutolab I (Ecochemie B.V., Utrecht, The Netherlands) via an IME interface (Metrohm AG, Herisau, Switzerland). The Autolab was controlled by a PC with GPES 4.9 software. To use the automatic hanging mercury drop electrode mode in this configuration, the rotary switch at the 663 VA Stand was turned to position "SMDE" [sic], as advised by Metrohm.

Voltammetric measurements

Measurements were conducted by inverse (stripping) differential-pulse voltammetry (DPV) at a hanging mercury drop electrode (HMDE) according to the German Industry Standard DIN 38406 parts 16 and 17. The voltammetric parameters are given in Table 2. All samples were purged with nitrogen for 10 min in order to remove oxygen. A 0.05 M acetate buffer with 0.1 M KCl at pH 4.6 was used for the determination of Zn, Cd, Pb, and Cu according to DIN 38406–16. All four metals were accumulated cathodically as amalgams in the HMDE, and then stripped-off in one anodic DPV scan. For the AdSV determination of nickel and cobalt at pH 9.5 according to DIN 38406–16, we added 1 mL ammonia buffer and 2 mL DMG stock solution to 20 mL of the sample.

Both metal ions were accumulated as DMG-complexes at the surface of the HMDE, and then stripped-off in one cathodic DPV scan. Determination of uranium by AdSV with 120 µM chloranilic acid as ligand was conducted at pH 2, adjusted by adding nitric acid [26]. The uranium(VI) analyte formed a complex with chloranilic acid that was accumulated by adsorption at the surface of the HMDE, followed by a cathodic DPV stripping scan (DIN 38406–17).

Table 2: Differential-pulse stripping-voltammetric parameters for ASV of Zn(II), Cd(II), Pb(II), Cu(II), and AdSV of Ni(II), Co(II) and U(VI) in Awash River water samples.

Voltammetric parameters	Analytes		
	Zn(II), Cd(II), Pb(II), Cu(II)	Ni(II), Co(II)	U(VI)
Method	ASV	AdSV	AdSV
Deposition potential (V)	-1.15	-0.7	0.1
Deposition time (s)	120	120	30
Equilibration time (s)	10	5	5
Modulation amplitude(V)	0.0202	0.0202	0.05
Modulation time (s)	0.05	0.05	0.05
Interval time (s)	0.5	0.10	0.10
Start potential (V)	-1.15	-0.8	0.04
End potential (V)	0.05	-1.2	-0.21
Voltage step(V)	0.049	0.049	0.006
Scan rate (V/s)	0.01	0.013	0.012

RESULTS AND DISCUSSION

Optimization of time of solar UV irradiation and evaluation of solar UV digestion efficacy

One of the main objectives of this study was to investigate the dose of solar UV (SoUV) irradiation necessary to obtain the same analytical results as with the reference method based on hard artificial UV light. Accordingly, original river water samples from Sire Robi collection site have been divided into three aliquots each in order to test the following three different irradiation procedures: a) 12 h SoUV, b) 24 h SoUV, and c) 24 h SoUV + 6 h artificial UV irradiation.

The irradiation procedures of the sample aliquots were carried out based on previously developed methods [20, 21, 22]. The heavy metal concentrations obtained after 12 h SoUV, 24 h SoUV and 24 h SoUV + 6 h artificial UV-irradiated samples are presented inTable 3. These values suggest that 12 h of solar UV irradiation at the above conditions already could degrade large portions of the matrices. However, the decomposition was incomplete and this was confirmed by the results obtained after 24 h SoUV irradiation. As a result of this doubled SoUV dose, the amount of free metal ions available for voltammetric determination increased by 17% for Zn(II), 66.7% for Cd(II), 37.9% for Pb(II) and 37.7% for Cu(II). The regarding results for AdSV of Ni(II), Co(II) and U(VI) revealed an increase by 26.4%, 9.8% and 18.2%, respectively.

Table 3: Effect of UV irradiation procedure upon the determined heavy metal concentrations in Awash River water samples collected from Sire Robi.

UV irradiation	Concentration (µg/L)						
	Zn(II)	Cd(II)	Pb(II)	Cu(II)	Ni(II)	Co(II)	U(IV)
12 h SoUV	115.89	0.12	3.06	26.61	4.66	0.51	0.18
24 h SoUV	135.53	0.20	4.22	36.65	5.89	0.56	0.22
24 h SoUV + 6 h UV	137.85	0.20	4.24	37.01	5.97	0.55	0.22
% increase due to additional SoUV irradiation	16.9	66.7	37.9	37.7	26.4	9.8	22.2
% increase due to additional UV irradiation	1.02	0	0.47	0.98	1.36	0	0

In order to evaluate the efficacy of a total 24 h solar UV irradiation, we performed reference measurements with the third aliquot of the water sample from Sire Robi collection site after an additional step of 6 h artificial UV irradiation with a 30 W low-pressure Hg lamp of 254 nm (i.e. equivalent to classic 60 min UV digestion with 180 W lamp). Such additional treatment of the samples revealed no further significant increase of the detected concentrations of the metal ions under consideration (Table 3). This confirms that 24 h of solar-UV irradiation at average intensity of 5.4 mW/cm^2 or more will lead to sufficient depletion of interfering dissolved organic matter (DOM) in river water. The results of this experiment showed that further hard UV irradiation has no significant effect on the amount of free metal ions available for voltammetric determination of all 7 tested metals in river water or similar other natural water samples.

Analytical performance of the used equipment and method

Prior to the determination of the 7 trace metals in all the river water samples, we examined the method detection limit (MDL) for each metal. This was a necessary step in order to evaluate the range with signal to noise ratio (S/N) greater than 5, i.e. to make sure that the observed concentrations for each trace metal was significantly above the noise level, as the latter would be affected by the particular matrix, protocol, and analyte. The investigated MDL values are presented in Table 4 in comparison with instrumental detection limits (IDL) advised by the manufacturer of the instrument (Metrohm-Autolab). These results demonstrate that the MDL values of our voltammetric protocols for all the investigated metal ions were well below the concentrations found in the river water samples, with one exception: The cobalt(II) level in two samples was found to be below our MDL (Table 6). This allowed us to apply our protocols for the monitoring of 7 trace heavy metals in Awash River even at the ng/L level.

Table 4: Instrumental detection limits (IDL) of the manufacturer and method detection limits (MDL) determined according to the procedure given in experimental section.

Analyte	IDL (μg/L)	MDL(μg/L)	Spiked concentration (μg/L)
Cadmium	0.050	0.073	0.100
Cobalt	0.050	0.089	0.100
Copper	0.050	0.260	0.100
Lead	0.050	0.240	0.500
Nickel	0.050	0.058	0.100
Uranium	0.025	0.120	0.100
Zinc	0.050	0.274	1.000

Table 5: Comparison of percentage variation of the concentration of heavy metals of two sampling sites upstream and downstream of Koka power station.

Sampling Site	Metal ion concentrations (μg/L)						
	Zn(II)	Cd(II)	Pb(II)	Cu(II)	Ni(II)	Co(II)	U(VI)
Koka Dam (upstream)	126.62	0.21	3.37	25.45	4.67	0.57	0.19
Sire Robi (downstream)	137.85	0.20	4.24	37.01	5.97	0.55	0.22
Concentration Variation (downstream) %	8.86	-4.84	25.85	45.42	28.36	-3.52	3.00

Table 6: Variation of the concentration of heavy metals in Awash River.

Element	Background concentrations (μg/L)	Background-corrected sample concentrations (μg/L)				
		Awash Melka-sa	Wonji Drain	Wonji Ku-ruftu	Sire Robi	Koka Dam
Zn	28.09 ± 0.86	**441** ± 11	26.12 ± 2.62	76.70 ± 7.11	137.9 ± 1.60	126.62 ± 1.13
Cd	ND[a]	**2.87** ± 0.05	0.06 ± 0.02	0.11 ± 0.03	0.20 ± 0.01	0.21 ± 0.03
Pb	1.59 ± 0.05	**10.93** ± 0.50	0.52 ± 0.05	1.14 ± 0.21	4.24 ± 0.09	3.37 ± 0.17
Cu	6.25 ± 0.72	4.33 ± 0.06	11.70 ± 0.80	25.71 ± 2.27	**37.0** ± 3.7	25.45 ± 0.56
Ni	2.21 ± 0.01	0.75 ± 0.09	3.35 ± 0.51	**20.44** ± 0.29	5.97 ± 0.94	4.67 ± 0.45
Co	ND[a]	ND[a]	0.21 ± 0.06	ND[a]	0.55 ± 0.03	**0.57** ± 0.04
U	0.84 ± 0.01	**1.41** ± 0.17	0.56 ± 0.47	1.35 ± 0.21	0.22 ± 0.13	0.19 ± 0.25

Maximum values highlighted.

a Not detected, i.e. below the method detection limit.

Toxicological review of Awash River in respect to cadmium, cobalt, copper, lead, nickel, uranium and zinc

The analytical results of the samples collected from different sites of Awash River were employed to evaluate the degree of contamination of the river with respect to the target heavy metals based on international (WHO) water quality standards. We also evaluated the status of health hazards (risks) associated with heavy metal pollution of Awash River. In general the determination of these heavy metals helped us to analyze toxicological impacts related to this major heavy metals contamination.

The analytical results of the 7 heavy metals under consideration of the two closest sampling sites upstream and downstream of the Koka reservoir and the Power Station are summarized in Table 5. All analytical results were corrected for the observed background concentrations (blank values) caused by all possible contaminations (following the sampling) such as reagents, plastic bottles (sample containers), polyethylene bags for irradiation, the filter paper, the electrochemical cell, and the electrolyte buffer. Critical evaluation of the variation of the concentrations of the samples collected from Koka reservoir and Sire Robi is demonstrated in Table 5. This investigation reveals that the concentrations of Cu(II), Pb(II), Ni(II), U(VI) and Zn(II) increased downstream by 45.4%, 25.9%, 28.4%, 3.0% and 8.9%, respectively. On the contrary, the concentrations of Cd(II) and Co(II) decreased downstream by 3.5% and 4.9% respectively. Koka Dam samples were collected from the upstream of the Dam just 200 meters before the river joins Koka artificial lake. The time of sampling corresponded to the dry season at that location implying relatively low flow rate, and water volume, and hence, a high DOM content. The other collection site (Sire Robi) is located downstream beyond the outlet of the dam (about 3 km from Koka hydroelectric power station) just at the Adama drinking water treatment plant. The rationale behind the decrease/or increase in the concentrations of metal ions downstream can be explained by considering various factors. For instance, the physical and chemical processes taking place in the water; the velocity (flow rate) of the water; turbulences; water volume; and landscape. Aquatic fauna and flora are some of the factors that affect migration of the metal contaminants and the matrices or the fate and transport of the complexing DOM, and hence, the metal ions. There can be significant number of flora in the reservoir feeding on Cd(II) and Co(II). This needs further investigation. In this particular case, the velocity and turbulence of the water decreases quickly as soon as the river enters the lake.

The observed small variations for Cd(II), Co(II), and U(VI) levels were below 5%, and can probably be attributed to the usual variations of manual sample collection. On the contrary, the concentrations of Cu(II), Pb(II), Ni(II) and Zn(II) significantly increased at the downstream collection site. These metals are all found in alloys like brass and bronze as used in components (bearings) of power plant machineries. One of the most probable causes, therefore, could be the interaction of the river water with such metal components, lubricants and other materials such as paints at the hydroelectric power station.

Table 6 presents a comprehensive view of all investigated analytes and sampling sites. The agrochemicals used by the local farmers could be another source of contamination. Artificial phosphate fertilizers for instance, are known to contain relatively high uranium levels [27]. The nearby Tannery (Ethiopian Tannery) also discharges its waste into the lake. This deadliest waste can contribute to the unexpected sharp increase in the concentration of some metals. In addition to these, there are extensive horticultural fields, which use agrochemicals intensively. The leaching of these chemicals into the lake could lead to the increase of the concentration of the metals by polluting the river water. However, the reason why the concentration of the metal pollutants increased downstream requires further investigation to identify the point and non-point sources and to recommend possible solutions.

Further investigation also revealed that some metal ions increased in concentration much more than expected while others decreased downstream. For instance Zn(II) increased from 126.62 to 440.85 μg/L as we move from Koka Dam (upstream Koka Lake) to Awash Melkasa. On the contrary, the Cu(II) concentration decreased from 25.45 to 6.25 μg/L in the same direction. In general, the determination of the concentration of these heavy metals helped us to analyze the degree of pollution and hence their toxicological impact on humans and animals based on the WHO safe drinking water guideline quality standards (Table 7). Deep investigation of the concentration data of the metal ions in the river would lead to the conclusion that it is moderately risky for drinking with respect to lead, while it is safe to drink this water with respect to cadmium, copper, nickel, cobalt, uranium and zinc. In conclusion, local people living around Awash River and the Adama City dwellers who drink water from Awash River and Koka Lake are at moderate risk of lead intake and the regarding health hazards. We suggest further studies to identify the root causes of the contamination and to recommend remedies.

Table 7: Heavy metals' pollution status in Awash River according to World health organization (WHO) Guidelines for Drinking-Water Quality, Geneva.

Element	Level in drinking water (μg/L)	Maximum observed concentration (μg/L)	WHO guideline value (μg/L)	Toxicological evaluation
Cadmium	1.0	2.87	3.0	endangered
Cobalt	Not specified	0.57	Not specified	safe
Copper	5–30,000	37.0	2000.0	safe
Nickel	<20.0	20.44	70.0	safe
Lead	5.0	10.93	10.0	endangered

Uranium	<1.0	1.41	15.0	safe[a]
Zinc	10.0–50.0	441	3000.0	safe

a Guideline value for tap water in Germany is now 10 µg/L, whereas maximum 2 µg/L has been recommended for infants (both tap and mineral water) [28].

CONCLUSIONS

This study has demonstrated that solar UV irradiation assisted by low pH and addition of hydrogen peroxide can serve as a means for water sample digestion. A solar irradiation time of 24 h in equatorial regions is sufficient to allow accurate voltammetric metal determination. This was evaluated by reference measurements including an additional classic UV digestion step with hard and intense artificial UV light (254 nm, 30 W, 6 h). Affordable UV-transparent materials such as thin polyethylene bags are suitable sample containers during irradiation. The presented protocol can be recommended for ASV determination of trace elements such as cadmium, copper, lead and zinc, as well as AdSV of nickel, cobalt, and uranium in river water and similar natural water samples at remote places, where high-power electric UV lamps or commercial UV digestion units cannot be used or made available. The effect of UV irradiation treatment for nickel, cobalt, and uranium was even larger compared to anodic stripping voltammetry of the other metals in the same river water samples, as reported earlier. The approach reported here should be very useful for mobile electrochemical heavy metal analysis in regions where frequent sunny weather conditions are available as proved by the investigation of Awash River water samples from selected locations in Ethiopia. The method is applicable in real sample analysis of heavy metals contamination. This preliminary field study suggests that Awash River water is moderately risky for drinking with respect to lead and cadmium, while it is in the range of safe drinking water with respect to copper, nickel, cobalt, uranium, and zinc. Future developments of our approach will aim at replacement of the HMDE with bismuth electrodes, as well as shortening of the solar irradiation time and development of solar-powered automatic monitoring stations.

DECLARATIONS

Funding statement

This work was supported by Ministry of Education of Ethiopia, Adama Science and Technology University and German Academic Exchange Service, DAAD (Engineering Education Capacity Building Program, EECBP), as well as German Research Foundation (DFG Heisenberg Fellowship, FL 384/7-2).

Competing interest statement

The authors declare no conflict of interest.

Additional information

No additional information is available for this paper.

REFERENCES

1. M.C. Navarro, C. Pérez-Sirvent, M.J. Martínez-Sánchez, J. Vidal, P.J. Tovar, J. Bech, Abandoned mine sites as a source of contamination by heavy metals: A case study in a semi-arid zone, J. Geochem. Explor. 96 (2008) 183–193.
2. J. Nyamangara, C. Bangira, T. Taruvinga, C. Masona, A. Nyemba, D. Ndlovu, Effects of sewage and industrial effluent on the concentration of Zn, Cu, Pb and Cd in water and sediments along Waterfalls stream and lower Mukuvisi River in Harare, Zimbabwe, Phys. Chem. Earth 33 (2008) 708–713.
3. O.S. Olatunji, O. Osibanjo, Determination of selected heavy metals in inland freshwater of lower River Niger drainage in North Central Nigeria, Afr. J. Environ. Sci. Technol. 6 (2012) 403–408.
4. M.I. Yahaya, Z.M. Agbendeh, Physiochemical characterization and level of potentially toxic metals in surface water around yauri abattoir, north western Nigeria, Afr. J. Pure. Appl. Chem. 9 (2015) 33–38.
5. Begum, M. Ramaiah, Harikrishna, I. Khan, K. Veena, Heavy Metal Pollution and Chemical Profile of Cauvery River Water, E-J. Chem. 6 (2009) 47–52.
6. D. Dibakar, P.P. Sumita, B.K. Gopal, Assessment of heavy metals and physico chemical parameters of water samples of the vicinity of the municipality dumping sites of Karimganj district, Assam, India, Int. J. Environ. Sci. 2 (2012) 1408–1416.
7. H. Lokeshwari, G.T. Chandrappa, Impact of heavy metal contamination of Bellandur Lake on soil and cultivated vegetation, Curr. Sci. 91 (2006) 622–627.
8. S.M.I. Sajidu, W.R.L. Masamba, E.M.T. Henry, S.M. Kuyeli, Water quality assessment in streams and wastewater treatment plantsof Blantyre, Malawi, Phys. Chem. Earth 32 (2007) 1391–1398.
9. E. Oyoo-Okoth, W. Admiraal, O. Osano, V. Ngure, M.H.S. Kraak, E.S. Omutange, Monitoring exposure to heavy metals among children in Lake Victoria, Kenya: Environmental and fish matrix, Ecotoxicol. Environ. Saf. 73 (2010) 1797–1803.
10. J. Golimowski, K. Golimowska, UV-photooxidation as pretreatment step in inorganic analysis of environmental samples, Anal. Chim. Acta 325 (1996) 111–133.
11. F.A.J. Armstrong, P.M. Williams, J.D.H. Strickland, Photo-oxidation of organic matter in sea water by ultraviolet radiation, analytical and other applications, Nature 211 (1966) 481–483.
12. E.P. Achterberg, C.M.G. van den Berg, In-line ultraviolet-digestion of natural water samples for trace metal determination using an automated voltammetric system, Anal. Chim. Acta 291 (1994) 213–232.
13. W.N.L. dos Santos, G.C. Brandao, L.A. Portugal, J.M.

David, S.L.C. Ferreira, A photo-oxidation procedure using UV radiation/H2O2 for decomposition of wine samples: Determination of iron and manganese content by flame atomic absorption spectrometry, Spectrochim. Acta B 64 (2009) 601–604.

14. Acra, Y. Karahagopian, Z. Raffoul, R. Dajani, Disinfection of oral rehydration solutions by sunlight, Lancet 316 (1980) 1257–1258.

15. Minero, E. Pelizzetti, S. Malato, J. Blanco, Large solar plant photocatalytic water decontamination: Effect of operational parameters, Sol. Energy 56 (1996) 421–428.

16. Robert, S. Malato, Solar photocatalysis: a clean process for water detoxification, Sci. Total Environ. 291 (2002) 85–97.

17. M. Berney, H.U. Weilenmann, T. Egli, Flow-cytometric study of vital cellular functions in Escherichia coli during solar disinfection (SODIS), Microbiol. 152 (2006) 1710–1729.

18. M.B. Fisher, C.R. Keenan, K.L. Nelson, B.M. Voelker, Speeding up solar disinfection (SODIS): effects of hydrogen peroxide, temperature, pH, and copper plus ascorbate on the photo inactivation of E.coli, J. Water Health 6 (2008) 35–51.

19. K.G. McGuigan, R.M. Conroy, H.M. Mosler, M. du Preez, E. Ubomba-Jaswa, P. Fernandez-Ibanez, Solar water disinfection (SODIS): A review from bench-top to roof-top, J. Hazard Mater. 235–236 (2012) 29–46.

20. G. Woldemichael, T. Tulu, G.U. Flechsig, Solar UV Photooxidation as Pretreatment for Stripping Voltammetric Trace Metal Analysis in River Water, Int. J. Electrochem. (2011) 7 481370.

21. G. Woldemichael, T. Tulu, G.U. Flechsig, Solar UV-assisted sample preparation of river water for ultra-trace determination of uranium by adsorptive stripping voltammetry, Microchim. Acta 179 (2012) 99–104.

22. G. Woldemichael, G.U. Flechsig, Solar UV-Assisted Pretreatment of River Water Samples for the Voltammetric Monitoring of Nickel and Cobalt Ultratraces, Adv. Chem. 2014 (2014) 7 938419

23. M. Kolb, P. Rach, J. Schäfer, A. Wild, Investigations of oxidative UV photolysis, I. Sample preparation for the voltammetric determination of Zn. Cd. Pb. Cu. Ni, and Co in Waters, Fresenius J. Anal. Chem. 342 (1992) 341–349.

24. T.M. Florence, Development of Physiko-Chemical Speciation Procedures to Investigate the Toxicity of Copper, Lead, Cadmium and Zink towards Aquatic Biota, Anal. Chim. Acta 141 (1982) 73–94.

25. J. Agger, Entwicklung und Anwendung eines vollautomatischen voltammetrischen Analysensystems zur Bestimmung von Schwermetallspuren in Lebensmitteln und biologischen Materialien, PhD Dissertation, Chapter 12.1, University of Hamburg, German, 1998.

26. S. Sander, G. Henze, Adsorption voltammetric techniques for the determination of uranium(VI) with 2.5-dichloro-3,6-dihydroxy-1,4- benzoquinone as complex forming reagent, Fres. J. Anal. Chem. 349 (1994) 654–658.

27. S.K. Sahu, P.Y. Ajmal, R.C. Bhangare, M. Tiwari, G.G. Pandit, Natural radioactivity assessment of a phosphate fertilizer plant area, J. Radiat. Res. Appl. Sci. 7 (2014) 123–128. [28] J. Alexander, D. Benford, A. Cockburn, et al., Scientific Opinion of the Panel on Contaminants in the Food Chain on a request from German Federal Institute for Risk Assessment (BfR) on uranium in foodstuff, in particular mineral water, The EFSA Journal 1018 (2009) 1–59.

INDEX

A

Adsorptive stripping-voltammetry (AdSV) 204
Analytical hierarchical process (AHP) 64
Antibiotic resistance genes (ARGs) 171

B

Ballast water treatment systems (BWTS) 111
Bermuda Atlantic Ocean Time Series station (BATS) 152
Biochemical oxygen demand (BOD5) 198
Bisphenol A (BPA) 167
Bovine serum albumin (BSA) 122
Built Own and Operate (BOO) 69

C

Carbon allotropes such as carbon nanotubes (CNTs) 74
Cellulose triacetate (CTA) 133
Centricon Plus-70 182, 183, 184, 185
cetyltrimethyl ammonium bromide (CTAB) 33
Cleaning in place (CIP) 105, 107
Conceptual by introducing carbon nanotubes (CNTs) 29
Continuing population growth 73
Cracking of ethylene dichloride (EDC) 197

D

Differential-pulse voltammetry (DPV) 205
Differential scanning calorimetry (DSC) 135
Dimethylglyoxime (DMG) 205
Disability-Adjusted Life Year (DALY) 21
Disinfection byproducts (DBP) 89
Dissolved organic carbon (DOC) 81, 89
Dissolved organic matter (DOM) 204, 206
Dissolved oxygen (DO) 191

E

Ecotoxicological 102, 104, 108
Electrical conductivity (EC) 63
Electrochemical double-layer capacitors (EDLCs) 36
Electrochemical double-layer (EDL) 38

F

Flow cytometry (FCM) 112
Food and Drug Administration (FDA) 113, 115, 119
Fungia granulosa 189, 190

G

Graphene oxide frameworks (GOFs) 29
Graphene oxide (GOx) 55, 56
Graphite furnace atomic absorption spectrometry (GF-AAS) 204

H

Hanging mercury drop electrode (HMDE) 205
Heterotrophic plate counts (HPC) 92
High performance size exclusion chromatography (HPSEC) 90
Hydraulic retention time (HRT) 166

I

Inductively coupled plasma optical emission spectrometry (ICP-OES) 58
Inductively coupled plasma optical spectrometry (ICP-OES) 204
Instrumental detection limits (IDL) 207
International Maritime Organization (IMO) 111
Isotope compositions (IC) 154
Isotope dilution (ID) 154, 155

K

King Abdullah University of Science and Technology (KAUST) 190, 195

L

Liquefied natural gas (LNG) 106
Lithium amine borane (LiAB) 31
L low-density polyethylene (LDPE) 191
Low pressure (LP) 111

M

Maleic acid (MA) 133
Mean Absolute Error (MAE) 145
Mean Absolute Percentage Error (MAPE) 145
Melamine resin (MR) 38
Membrane bioreactors (MBRs) 121
Membrane filtration (MF) 90
microfiltration (MF) 198
Micropollutants 165, 166, 167, 168, 169
Ministry for Development of Northern Kenya and other Arid Lands (MDONK) 12
Most probable number (MPN) 113
Multi dimensional scaling (MDS) 64
Multi-dimension scaling (MDS) 64
Multi-effect evaporation thermal vapor compression (METVC) 1

N

Nano-enabled membrane 74, 75, 76
Nano-enabled membranes 73, 74, 75, 76
Nano Filtration (NF) 63
Nevertheless 73, 75
N-Methyl-2-pyrrolidone (N-NMP) 133
No biological effect was detected (NOEC) 107
Nonculturable 185
Nonnucleated 185
Non-purgeable organic carbon (NPOC) 93

O

Organic matter (OM) 89
Organic Solvent Nanofiltration (OSN) 133
Oxygen evolution reaction (OER) 35
Oxygen reduction reaction (ORR) 35

P

Partial Least Squares (PLS) 14, 17, 18, 19
Payments for Ecosystem Services (PES) 17
Phosphotungstic acid 184
Photomultiplier tubes (PMTs) 91
Photosynthetically active radiation (PAR) 191
Point of entry (POE) 171, 173
Point of use (POU) 171, 173
Polyethylene glycol (PEG) 34, 133
Polyvinyl chloride (PVC) 197
Polyvinylidene fluoride (PVDF) 80, 122
Public-Private Partnerships (PPP) 69
Pulsed-field gel electrophoresis (PFGE) 185

R

Random amplified polymorphic DNA (RAPD) 185
Reduced graphene oxide (RGO) 25
Reverse-Osmosis (RO) 63
Root Mean Square Error (RMSE) 145

S

Scanning electron microscopy (SEM) 57
Seawater reverse osmosis (SWRO) 73, 189
Sludge retention times (SRT) 166
Sodium absorption ratio (SAR) 63
Soluble microbial products (SMP) 121, 123
Species Sensitivity Distribution (SSD) 104
Specific ultraviolet absorbance (SUVA) 80
Supporting Information (SI) 56

T

Tangential flow filtration (TFF) 186
Tetrahydrofuran (THF) 133, 134, 135
Total coliforms (TC) 90
Total dissolved solids (TDS) 63
Total organic carbon (TOC) 80, 90
Transition metal oxide (TMO) 32
Trans-membrane pressure (TMP) 79, 123
Tris-acetate-EDTA (TAE) 183
Turnover frequency (TOF) 30

U

United States Environmental Protection Agency (USEPA) 171

V

Variable-Importance-in-the-Projection (VIP) 14
Vinyl chloride monomer (VCM) 197

W

waste water treatment plant (WWTP) 198
Whole effluent toxicity (WET) 102

X

X-ray photoelectron spectroscopy (XPS) 57

Z

Zeta potential (ZP) 80